Lecture Notes in Civil Engineering 603

Lecture Notes in Civil Engineering (LNCE) publishes the latest developments in Civil Engineering—quickly, informally and in top quality. Though original research reported in proceedings and post-proceedings represents the core of LNCE, edited volumes of exceptionally high quality and interest may also be considered for publication. Volumes published in LNCE embrace all aspects and subfields of, as well as new challenges in, Civil Engineering. Topics in the series include:

- Construction and Structural Mechanics
- Building Materials
- Concrete, Steel and Timber Structures
- Geotechnical Engineering
- Earthquake Engineering
- Coastal Engineering
- Ocean and Offshore Engineering; Ships and Floating Structures
- Hydraulics, Hydrology and Water Resources Engineering
- Environmental Engineering and Sustainability
- Structural Health and Monitoring
- Surveying and Geographical Information Systems
- Indoor Environments
- Transportation and Traffic
- Risk Analysis
- Safety and Security

To submit a proposal or request further information, please contact the appropriate Springer Editor:

- Pierpaolo Riva at pierpaolo.riva@springer.com (Europe and Americas);
- Swati Meherishi at swati.meherishi@springer.com (Asia—except China, Australia, and New Zealand);
- Wayne Hu at wayne.hu@springer.com (China).

All books in the series now indexed by Scopus and EI Compendex database!

Andrii Bieliatynskyi · Dmytro Komyshev ·
Wen Zhao

Editors

Proceedings of Conference on Sustainable Traffic and Transportation Engineering in 2023

 Springer

Editors
Andrii Bieliatynskyi
North Minzu University
Yinchuan, Ningxia, China

Dmytro Komyshev
North Minzu University
Yinchuan, China

Wen Zhao
Northeast University School of Resources
Northeastern University
Shenyang, China

ISSN 2366-2557 ISSN 2366-2565 (electronic)
Lecture Notes in Civil Engineering
ISBN 978-981-97-5813-5 ISBN 978-981-97-5814-2 (eBook)
https://doi.org/10.1007/978-981-97-5814-2

This work was supported by Guangzhou KEO Info Technology Co., Ltd.

This Springer imprint is published by the registered company Springer Nature Singapore Pte Ltd.
The registered company address is: 152 Beach Road, #21-01/04 Gateway East, Singapore 189721, Singapore

If disposing of this product, please recycle the paper.

Preface

This proceedings gathers the original, peer-reviewed, select research papers from a wide range of sources, including the Conference on Sustainable Traffic and Transportation Engineering in 2023 (CSTTE 2023), held in Yinchuan, China, from November 22–24, 2023, as well as many universities, research institutes, and business units.

CSTTE 2023 aimed to bring together the latest research results, innovative technologies, and practical experiences in the fields of sustainable traffic and transportation engineering from around the world. We hope to promote the modernization of traffic engineering, transportation engineering, and related fields and establish a sustainable transportation system that is compatible with the needs of economic and social development and the capacity of resources and environment.

During the conference, in addition to the main venue, there were also many sessions, providing a broad communication platform for experts and scholars. The agenda of the conference included the opening ceremony, keynote speeches, oral presentations, poster presentations, group discussions, academic exchanges, and the closing ceremony. The experts and scholars talked about the frontier issues of sustainable traffic and transportation engineering, shared the latest research achievements, and discussed the future development direction in relevant domains.

One of the most important outcomes of CSTTE 2023 is its proceedings. Focusing on the discussion of modern cutting-edge construction techniques and theoretical research in transportation engineering, it gathers the latest research, innovations, and applications in the fields of sustainable traffic and transportation engineering. The topics it covers include new technologies for the construction of roads, bridges, tunnels, airports, ports, railways, embankments, reservoirs, sluices, etc., and various new materials and their new theories and technologies for the construction of transportation facilities.

The publication of this proceedings is not only an acknowledgment and recognition of the academic research results, but also an important contribution to the development of the global transportation engineering field. By sharing and exchanging the latest research results and experiences, it can promote the cooperation and development of the field of transportation engineering on a global scale and enhance the overall level and competitiveness of the industry.

Here, we would like to express our heartfelt thanks to the experts, scholars, staff, and all contributors who participated in the conference. It is your hard work and selfless dedication that made this conference smoothly held and a complete success.

The Committee of CSTTE 2023

Committee Member

Academic Committee Chairman

Xiangsheng Chen   Shenzhen University & Chinese Academy of Engineering, China

Academic Committee Vice Chairman

Bieliatynskyi Andrii	North Minzu University, China
Dmytro Komyshev	North Minzu University, China
Guoliang Bai	Xi'an University of Architecture and Technology, China
Qijun Yu	Hefei University of Technology, China

Academic Committee Member

Weizhong Chen	Institute of Rock and Soil Mechanics, Chinese Academy of Sciences, China
Shaojun Fu	Shaanxi Key Laboratory of Safety and Durability of Concrete Structures, China
Wenyu He	Hefei University of Technology, China
Congying Li	Xi'an University of Architecture and Technology, China
Naifei Liu	Xi'an University of Architecture and Technology, China
Aimin Sha	Chang'an University, China
Zhanping Song	Xi'an University of Architecture and Technology, China
Changqing Wang	Northwestern Polytechnical University, China
Jingfeng Wang	Hefei University of Technology, China
Zhiliang Wang	Hefei University of Technology, China
Nailiang Xiang	Hefei University of Technology, China
Jianyang Xue	Xi'an University of Architecture and Technology, China
Wen Zhao	Northeastern University, China
Chengwen Zhong	Northwestern Polytechnical University, China

| Jianjun Zhou | State Key Laboratory of Shield and Tunneling Technology, China |
| Shengxi Zhou | Northwestern Polytechnical University, China |

Organizational Committee Chairman

Lixin Zhang	North Minzu University, China
Jingfeng Wang	Hefei University of Technology, China
Jianyang Xue	Xi'an University of Architecture and Technology, China

Organizational Committee Vice Chairman

Hefang Jing	North Minzu University, China
Changjian Wang	Hefei University of Technology, China
Zhanping Song	Xi'an University of Architecture and Technology, China
Huayan Yao	North Minzu University, China
Naifei Liu	Xi'an University of Architecture and Technology, China
Zhanwu Ma	North Minzu University, China

Contents

Modeling and Simulation in Civil Engineering

Research on Engineering Material Preparation and Mechanical Properties

Experimental Study on the Effect of Redispersible Latex Powder on Pavement Performance of Road Base

Bo Wang[1], Huimin Hu[2(⊠)], and Jianglong Yao[2]

[1] Anhui Transportation Holding Group Co. Ltd., Anhui 230088, China
[2] School of Civil and Hydraulic Engineering, Hefei University of Technology, Anhui 230009, China
2738465338@qq.com

Abstract. Through unconfined compressive strength test, flexural tensile strength test, freeze-thaw test, dry shrinkage test and SEM, the improvement effect and mechanism of redispersible latex powder on the mechanical properties of recycled cement stabilized macadam were studied. The results show that the flexural and tensile strength of recycled cement stabilized macadam mixture increased and the compressive strength decreased slightly after adding latex powder. With the addition of latex powder, the BDR value of cement stabilized macadam increased, and the frost resistance of the mixture improved. Adding latex powder can reduce the shrinkage strain and shrinkage coefficient of cement stabilized macadam, and significantly improve its crack resistance. SEM images show that the latex powder polymer forms a network-like connection structure in the mixture, which can absorb the shrinkage deformation of mineral particles and hydration products in the material, and significantly reduce the shrinkage strain of the specimen.

Keywords: Road engineering · Recycled aggregate · Redispersible latex powder · Cement stabilized macadam · Crack resistance

1 Introduction

In road maintenance or reconstruction, a large number of waste cement stabilized macadam base materials will be produced by milling or digging. Reasonable utilization of waste cement stabilized macadam base material can not only alleviate the difficulty of stone shortage in engineering construction, but also benefit the ecological civilization construction and the realization of double carbon goal in China. Lu et al. [1] found that adding a certain amount of fly ash into the recycled cement mixture can improve the strength and frost resistance of the recycled cement mixture, and adding a small amount of emulsified asphalt into the recycled old cement stabilized macadam base material can improve the frost resistance and flexibility of the recycled cement mixture. Wang et al. [2, 3] added emulsified asphalt to recycled waste cement stabilized crushed stone, and found that it can improve the dry shrinkage performance of recycled cement stabilized

A. Bieliatynskyi et al. (Eds.): CSTTE 2023, LNCE 603, pp. 3–14, 2024.
https://doi.org/10.1007/978-981-97-5814-2_1

crushed stone mixture, enhance the flexibility of the mixture, and suppress the generation of cracks in the base layer. The research results of Lu et al. [4, 5] showed that the strength and crack resistance of recycled cement stabilized macadam mixture can be improved by adding fiber materials.

Redispersible latex powder is a kind of high molecular polymer powder, which can be redispersed when exposed to water, and forms organic mucosa on the surface of solid materials such as crushed stone to enhance the integrity of the structure. At the same time, the redispersible latex powder has the advantages of dissolving in water and being easy to mix evenly. In China, there are related studies on adding latex powder to cement mortar [6–8]. Adding latex powder to cement mortar can improve the bonding strength and flexural performance of cement mortar. At present, there is no research on its application in recycled cement stabilized macadam base. This paper studied the influence of latex powder on the road performance of recycled cement stabilized macadam, especially the effect of improving its crack resistance.

2   Properties of Recycled Materials for Waste Base Course

2.1   Morphology of Recycled Aggregate

Recycled aggregate is produced by milling cement stabilized macadam base during highway overhaul, as shown in Fig. 1. Most of the surface of recycled aggregate is covered with cement mortar, and the cross sections of coarse aggregate become more (full of edges and corners) and the surface is rough. The cement mortar particles in the recycled materials of milling base course are small and the strength is low. The particle size of recycled aggregate used in the test ranges from 0 to 26.5 mm, and 78% of the aggregate particles are larger than 4.75 mm.

(a) Not screened   (b) Screened

Fig. 1. Recycled aggregate

2.2   Technical Index of Recycled Aggregate

According to "Test Methods of Aggregate for Highway Engineering" (JTG E42–2005), the technical indexes of recycled aggregate and natural aggregate are shown in Table 1.

The water absorption and crushing value of recycled aggregate are obviously higher than those of natural aggregate.

Table 1. Technical indexes of aggregate

Aggregate type	Apparent relative density	Bibulous rate/%	Water content /%	Mud content/%	Needle-like particle content/%	Crushing value of stone/%
Natural aggregate	2.752	1.405	0.06	0.64	18.1	12.74
Recycled aggregate	2.706	3.099	0.11	0.89	19.1	22.70

3 Experimental Design

3.1 Raw Materials

According to the gradation of recycled materials, natural aggregate is still needed in the test, which can be divided into 1 # (20 ~ 30 mm), 2 # (10 ~ 20 mm), 3 # (5 ~ 10 mm) and 4 # (0 ~ 5 mm) according to the particle size. P.O42.5 cement was used, and its technical indexes are shown in Table 2.

Table 2. Technical indexes of cement

Standard consistency water consumption/%	Initial setting time/min	Final setting time/min	Compressive strength/MPa		Flexural strength/MPa	
			7d	28d	7d	28d
28	204	284	38.7	53.0	6.6	8.3

The 8020 redispersible latex powder produced by Anhui Wanwei Company was used in this study. The technical parameters of latex powder are shown in Table 3. Redispersible latex powder is a kind of high molecular polymer powder, and its appearance is milky white powder.

3.2 Mix Proportion

According to "Technical Guidelines for Construction of Highway Roadbases" (JTG/T F20–2015), C-B-1 was selected as the gradation type, which can be used for expressways and first-class highways. The designed gradation of recycled cement stabilized mixture is shown in Table 4, and the content of recycled aggregate is 44%. The compaction test shows that the optimal moisture content of recycled cement stabilized crushed stone is 5.6%, and the maximum dry density is 2.260 g/cm^3.

Table 3. Technical indexes of latex powder

Project	Technical index of 8020 latex powder	Specification requirements
Solid content	99 ± 1%	≥ 98.0
PH value	6–8	5.0 ~ 9.0
Bulk density/g·L^{-1}	500 ± 50	300 ~ 600
Average particle size D50/μm	60–80	≤ 100
Minimum film forming Temperature/°C	0	M ± 2
Vitrification temperature/°C	5	N ± 2

Table 4. Grading of cement stabilized macadam mixture

Mesh size/mm	26.5	19	16	13.2	9.5	4.75	2.36	1.18	0.6	0.3	0.15	0.075
Synthetic gradation of natural aggregate mixture/%	100	83.1	76.3	70.9	61.6	41.6	23.9	16.7	10.6	7.0	3.8	2.0
Synthetic gradation of recycled aggregate mixture/%	100	83.8	76.6	66.4	57.6	38.5	22.0	15.5	9.9	6.9	3.8	2.1
Lower limit of C-B-1 grading passing rate/%	100	82.0	73.0	65.0	53.0	35.0	22.0	13.0	8.0	5.0	3.0	2.0
Upper limit of C-B-1 grading passing rate/%	100	86.0	79.0	72.0	62.0	45.0	31.0	22.0	15.0	10.0	7.0	5.0

3.3 Performance Test

Strength test

The size of the compressive strength specimen is Φ100 mm × 100 mm cylindrical specimen. The bending tensile strength specimen is 100 mm × 100 mm × 400 mm mid beam specimen. The bending tensile strength test piece was formed by static pressure method, and it was conducted on a universal testing machine after 28 days of curing.

Freeze-Thaw performance test

The test method referred to section T0858–2009 in JTG E51–2009. We made same cylinder specimens as compressive strength test with 6 specimens in each group. After curing for 7 days, three specimens were tested for direct compressive strength, and the other three specimens were tested for compressive strength after freeze-thaw cycle. Freeze-thaw temperature setting: -18°C cryogenic box for 16 h, 20 °C water tank for 8 h, freeze-thaw cycle once.

Dry shrinkage performance test

The test was based on section T0854–2009 of the JTG E51–2009. Same beam specimens as flexural-tensile test (3 specimens in each group) were made. To assess the dry shrinkage properties of cement stabilized macadam material at the initial stage of construction, the test began after being cured for 1 day in the standard curing room. A dial indicator was inserted into a specially designed mold, as shown in Fig. 2. The specimen's weight was measured daily to determine the water content loss, while both water content loss and deformation shrinkage were continuously recorded each day.

Fig. 2. Specimen of dry shrinkage test

4 Result Discussions

4.1 Effect of Latex Powder on Compressive Strength of Mixture

The unconfined compressive strength of cement stabilized macadam for 7 days is shown in Fig. 3(a). The compressive strength of cement stabilized macadam with recycled materials is slightly higher than that of natural aggregate. From the physical point of view, the reason is that the angularity of recycled aggregate increases and the surface is rougher, which leads to the increase of internal friction. In terms of microstructure, many micro-cracks on the surface will inhale new cement particles, which makes the hydration of the contact area more sufficient and the interface structure more compact.

After adding latex powder, the compressive strength of cement stabilized macadam decreased slightly. The reason is that after the emulsion is cured, the elastic modulus of the film formed is small, which can not play a rigid supporting role when the whole test block is compressed, so that its compressive strength is reduced.

4.2   Effect of Latex Powder on Flexural Tensile Strength of Mixture

Natural aggregate, recycled aggregate, recycled aggregate + latex powder (8%) were mixed with cement stabilized macadam mixture. After 28 days of curing in standard curing room, the flexural and tensile strength of different mixture specimens was measured. The test results are shown in Fig. 3(b).

The flexural-tensile strength of recycled aggregate cement stabilized macadam is 5.9% higher than that of natural aggregate cement stabilized macadam, and the flexural-tensile strength of recycled aggregate + latex powder mixture is 9.8% higher than that of natural aggregate mixture. The reasons for adding latex powder to improve the flexural strength of recycled cement stabilized macadam are as follows: The latex powder emulsion plays a filling role and improves the internal structure of hardened cement paste. In addition, the film with high adhesive force formed by emulsion dehydration connects the hydrated products of cement more effectively, and the two interweave each other to form a firmer and more flexible three-dimensional network connection structure, which improves the flexural strength of the sample.

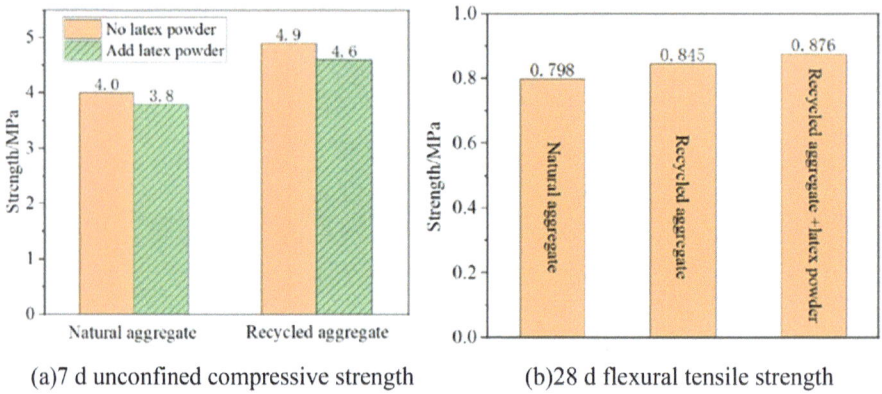

(a)7 d unconfined compressive strength   (b)28 d flexural tensile strength

Fig. 3. Strength test results

4.3   Effect of Latex Powder on Frost Resistance of Mixture

When cement stabilized macadam base is applied in a seasonal frozen area, its bearing capacity decreases after several freezing and thawing seasons. As a result, it may not be sufficient to support the vehicle load transmitted by the pavement, leading to cracking and other forms of damage. In order to assess the frost resistance of cement stabilized macadam, a freeze-thaw cycle test was conducted to evaluate the impact of latex powder. The test was conducted in accordance with the method specified in section T0858–2009 of the JTG E51–2009. The compressive strength loss (BDR) of the mixture was calculated using formula (1) from this section, and the results are presented in Table 5.

$$BDR = R_{DC}/R_C \times 100 \tag{1}$$

In formula (1): *BDR* is the compressive strength loss of the specimen after freeze-thaw cycle, %; R_{DC} is the compressive strength of the specimen after freeze-thaw cycle, MPa; R_C is the compressive strength of the contrast specimen, MPa.

Table 5. Compressive strength loss of specimens after freeze-thaw cycles

Project	7 d unconfined compressive strength/Mpa		BDR/%
	Average intensity before freezing and thawing	Average intensity after freezing and thawing	
Natural aggregate	5.29	4.74	89.60
Recycled aggregate	4.91	4.46	90.84
Recycled aggregate + latex powder	4.59	4.25	92.59

The BDR value of cement stabilized crushed stone increased with the addition of latex powder, and its frost resistance improved. The cement stabilized macadam recycled base material has a few voids inside. When the base material passes through the freeze-thaw cycle, the water in the voids inside the base material will crush the void wall of the mixture because of the frost heaving action, thus reducing the strength of the mixture. Adding latex powder will fill some voids in the mixture, and the void wall has certain elasticity, which will have less freeze-thaw damage. The effect of frost resistance improvement will be more obvious after multiple freeze-thaw cycles.

4.4 Effect of Latex Powder on Dry Shrinkage Performance of Mixture

Four groups (natural aggregate, natural aggregate + latex powder, recycled aggregate, recycled aggregate + latex powder) of mid beam specimens (3 specimens in each group) were made, to compare the effects of latex powder on dry shrinkage performance of natural aggregate and recycled aggregate mixture. According to T0854–2009 in JTG E51–2009, the total dry shrinkage coefficient of the specimen can be calculated by formula (2) ~ ((6). The changes of total water loss rate, total dry shrinkage strain and total dry shrinkage coefficient with time are shown in Fig. 4, Fig. 5 and Fig. 6 respectively.

$$\omega_i = (m_i - m_{i+1})/m_p \tag{2}$$

$$\delta_i = \left(\sum\nolimits_{j=1}^{4} X_{i,j} - \sum\nolimits_{j=1}^{4} X_{i+1,j} \right)/2 \tag{3}$$

$$\varepsilon_i = \delta_i/l \tag{4}$$

$$\alpha_{di} = \varepsilon_i/\omega_i \tag{5}$$

$$\alpha_d = \sum \varepsilon_i / \sum \omega_i \tag{6}$$

In formula (2) ~ (6): ω_i is the i-th water loss rate, %; δ_i is the i-th dry shrinkage of observation, mm; ε_i is the i-th dry shrinkage strain, %; α_{di} is the i-th shrinkage factor, %; m_i is the weighing mass of the i-th standard specimen, g; $X_{i,j}$ is the reading of the j-th dial indicator in the i-th test, mm; l is the length of standard specimen, mm; m_p is the constant weight of standard specimen after drying, g.

The dry shrinkage strain and dry shrinkage coefficient of cement stabilized macadam mixture without latex powder are bigger than those of the mixture with latex powder. After adding latex powder, the dry shrinkage coefficient of natural aggregate mixture decreased by 53.8% in 7 days and 55.8% in 14 days. The dry shrinkage coefficient of recycled aggregate mixture decreased by 57.4% in 7 days and 28.5% in 14 days. Adding latex powder will significantly reduce the dry shrinkage coefficient of cement stabilized macadam, and improve the dry shrinkage performance of the material.

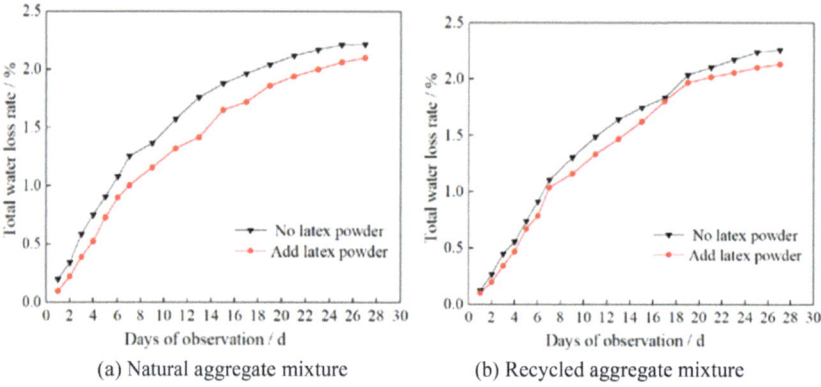

Fig. 4. Total water loss rate

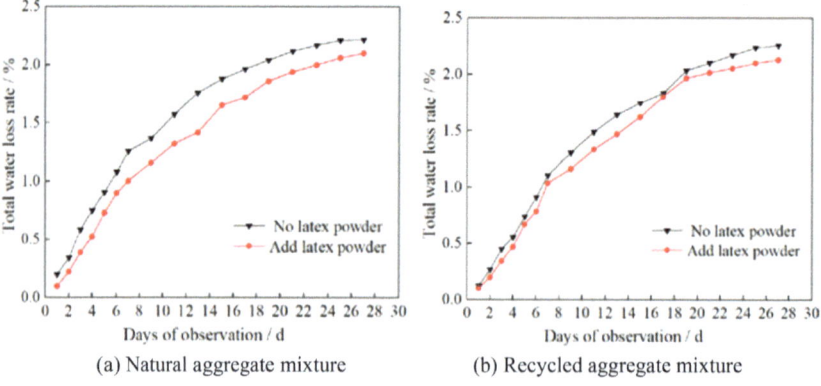

Fig. 5. Total drying shrinkage strain

In order to analyze the mechanism of improving dry shrinkage performance of cement mixture with latex powder, the cement mortar part of recycled mixture specimen and

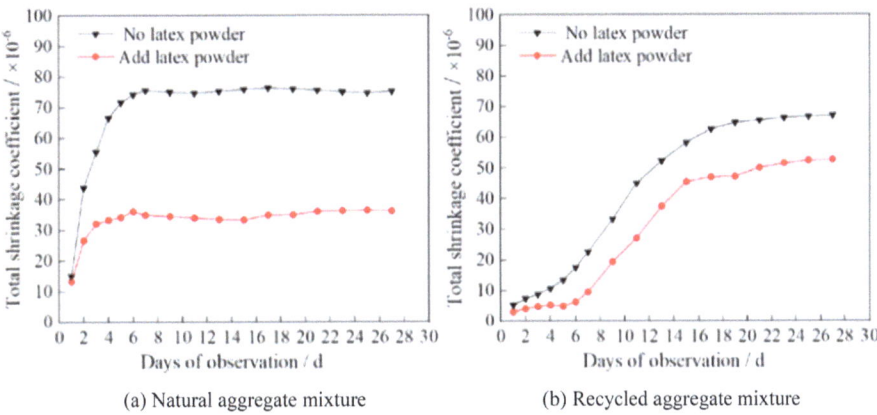

(a) Natural aggregate mixture   (b) Recycled aggregate mixture

Fig. 6. Total dry shrinkage coefficient

recycled mixture with latex powder was taken respectively, and the two samples were observed by electron microscope. SEM test pictures are shown in Fig. 7 and Fig. 8.

The mixture samples without latex powder have more macropores and channels under microscopic conditions. In the sample of latex powder mixture (Fig. 7b), it can be observed that the polymer film formed after the reaction of latex powder covers the hydration products of cement, and the mortar formed by latex powder and cement binds the mineral particles, which reduces the voids between the particles.

(a) No latex powder   (b) Add latex powder

Fig. 7. SEM pictures of mucilage part of mixture (1000 times)

In the 3000 times SEM picture of the latex powder mixture sample, the network-like connection structure can be observed, which is the network-like connection structure formed by cement hydrate, latex powder polymer, etc. (Fig. 8b). This network structure has certain flexibility due to the addition of latex powder. The flexible latex powder cement skeleton connection structure (elastic skeleton) improves the flexural strength of cement macadam mixture. At the same time, the deformable joint structure can absorb the shrinkage deformation of the material, thus reducing the shrinkage deformation of the specimen and the dry shrinkage strain.

The volume shrinkage of cement stabilized macadam materials will be caused by capillary action, adsorption, intermolecular force, interlayer water between mineral crystals or gels and carbonization shrinkage due to the decrease of water in cement stabilized macadam mixture. Based on the analysis of dry shrinkage test results and micro-electron microscope test results of cement-stabilized macadam with latex powder, there are two main reasons why latex powder can improve the dry shrinkage performance of cement-stabilized macadam mixture: (1) The latex powder polymer formed after adding latex powder fills some pore channels, and the latex film formed by latex particles coagulates on the surface of particles, which also hinders the migration of water and makes it difficult to evaporate and lose water. (2) In the latex powder mixture, the latex powder polymer particles form a network-like connection structure. This network-like connection structure (elastic skeleton) with deformation ability can absorb the shrinkage deformation caused by water loss in the material and reduce the volume shrinkage of the specimen. Therefore, the test data show that the dry shrinkage strain of the specimen is obviously reduced. Comparing the dry shrinkage test water loss rate and dry shrinkage strain data, the network structure of latex powder polymer particles is the main factor.

(a) No latex powder   (b) Add latex powder

Fig. 8. SEM pictures of mucilage part of mixture (3000 times)

5   Conclusions

(1) The compressive strength and flexural strength of the cement stabilized macadam with recycled aggregate are higher compared to those with natural aggregate. After the addition of latex powder, the flexural and tensile strength of the cement stabilized macadam mixture increased, while the compressive strength decreased slightly. The result of freeze-thaw test shows that the BDR value of cement stabilized macadam with latex powder increased, and the frost resistance of mixture increased.

(2) Adding latex powder can reduce the shrinkage strain and shrinkage coefficient of cement stabilized macadam, and significantly improve its crack resistance. After adding latex powder, the dry shrinkage coefficient of natural aggregate mixture decreased by 53.8% in 7 days and 55.8% in 14 days, and the drying shrinkage coefficient of recycled aggregate mixture decreased by 57.4% in 7 days and 28.5% in 14 days.

(3) There are two main reasons why latex powder can improve the dry shrinkage performance of cement mixture: First, the formed latex powder polymer fills part of pore channels, and the latex film formed by latex particles coagulates on the surface of particles, which also hinders the migration of water and makes it difficult for water to evaporate and lose. The second is the network connection structure formed by latex powder in the mixture. This network connection structure (elastic skeleton) with deformation ability can absorb the shrinkage deformation caused by water loss in the material and reduce the volume shrinkage of the specimen, so the test data show that the dry shrinkage strain of the specimen is obviously reduced. Comparing the dry shrinkage test water loss rate and dry shrinkage strain data, the network connection structure is the main factor.

Acknowledgement. Project funded by Anhui Transportation Holding Group Co. LTD. Project Number: JKKJ-2022–03.

References

1. Lu, P., Ji, R.: Research on the performance of cement emulsified asphalt recycled old cement stabilized crushed stone base material. Highway. Mot. Transp. **5**, 163–165 (2013)
2. Wang, L., Li, C., Wei, X., Liu, Z.: Experimental study on dry shrinkage of emulsive regeneration old cement stabilized macadam. J. Jiangsu Vocat. Tech. Coll. Archit. **18**(4), 9–12 (2018)
3. Jiang, L., Li, C., Li, H., Wei, X., Liu, Z.: Experimental study on the mechanical properties of the emulsive regenerated discarded cement stabilized macadam. J. Guangdong Vocat. Tech. Coll. **18**(1), 10–13 (2019)
4. Lu, P., Fan, X.: Experimental study on the effect of polyester fiber on the strength of cement cold recycled stabilized crushed stone. Zhongwai Highway **33**(4), 321–324 (2013)
5. Liu, J., Huang, L., Li, X., Shao, H.: Effect of polypropylene fiber on properties of cold recycled mixture of water stabilized base course in alpine region. J. Univ. Shanghai Sci. Technol. **43**(5), 460–467 (2021)
6. Wang, P., Zhao, G., Zhang, G.: Mechanism of redispersible polymer powder in cement mortar. J. Silic. **46**(2), 256–262 (2018)
7. Hou, Y., Zhang, Y., Huang, T.: Influence of redispersible latex powder on the performance of dry-mixed mortar. J. Beijing Univ. Architect. **37**(4), 1–8 (2021)
8. Li, Y., Luo, Z., Jia, Y., Yin, H., Yang, J.: Research of the modification of dispersible powder on cement - based repair mortar Concrete, **12**, 120–122, 126 (2015)

Study on the Durability of Silica Fume Concrete in High Sulfate Environment of Plateau

Zhimin Chen[1], Zheng Zhang[1(✉)], Mingyang Yi[1], Qianlong Yuan[2], Dianqiang Wang[2], and Junhui Liu[2]

[1] School of Civil Engineering, Lanzhou Jiaotong University, Lanzhou 730070, China
2328786454@qq.com
[2] China Construction Second Engineering Bureau Co., LTD., Beijing, China

Abstract. To investigate the effect of sulfate corrosion on the compressive strength of silica fume concrete under freeze-thaw conditions, different concretes with silica fume contents of 0%, 5%, 10%, and 15% were exposed to dry-wet erosion with 5% sulfate, freeze-thaw cycles with clear water, and dry-wet erosion with 5% sulfate followed by freeze-thaw cycles for a period of 75 days. The changes in compressive strength under different conditions and at different time intervals were analyzed. The results indicate that the extent of strength damage follows the order: dry-wet+freeze-thaw > freeze-thaw > dry-wet. The combined effect of dry-wet+freeze-thaw accelerates the deterioration of concrete strength. As the silica fume content increases, the rate of concrete compressive strength loss gradually stabilizes. For the 0% and 5% silica fume concretes, the loss of strength under the combined effect of dry-wet+freeze-thaw is more pronounced for 60 days. However, as the silica fume content increases beyond 5%, the rate of strength loss decreases by more than 10%. Therefore, the increase in silica fume content leads to a decrease in the rate of concrete strength loss.

Keywords: silica fume · Wet and dry · Freezing and thawing · Sulfate erosion · compressive strength

1 Introduction

The project is located in the northeastern part of the Qinghai-Tibet Plateau. The average daily temperature is below 0 °C, and the lowest temperature reaches −18 °C. The annual precipitation is 600 mm. There are significant daily temperature differences and high annual evaporation rates, which lead to sodium sulfate corrosion, sulfate corrosion and freeze-thaw damage under the action of various reasons [1, 2]. These factors are the main reasons for the deterioration and shortened service life of concrete structures in the high-altitude saline areas of northwest China [3]. In addition, freeze-thaw cycles cause hydrostatic pressure and seepage pressure from liquid water to enter the concrete, causing expansion pressure and accelerating concrete degradation [4]. The addition of fly

Q. Yuan, D. Wang and J. Liu—Contributed equally to this work

© The Author(s) 2024
A. Bieliatynskyi et al. (Eds.): CSTTE 2023, LNCE 603, pp. 15–29, 2024.
https://doi.org/10.1007/978-981-97-5814-2_2

ash improves the irregularity and average pore size of concrete pores, thereby improving the freeze-thaw resistance of concrete. But it also reduces the compressive strength of concrete [5]. On the other hand, the addition of slag reduces the porosity of cement store and increases the number of gel pores, thereby improving the frost resistance of concrete. The addition of silica powder fills the pores, reduces the pore size of concrete, and improves its frost resistance [6–8]. When combined with silica powder, it has the advantages of abundant output, low transportation costs, and high quality.

This paper studies the effect of silica fume addition on sodium sulfate attack on concrete exposed to freeze-thaw cycles (KSD). Various tests were conducted on concrete samples with different silica fume contents to explore the effects of different sulfate dry-to-wet attack and freeze-thaw cycle time ratios [9]. This study uses gray system theory and establishes a GM (1,1) model to predict the service life of concrete structures under the action of salt freezing. Three key indicators: mass loss rate, relative dynamic elastic modulus, and compressive strength are used as indicators [10]. These findings provide valuable insights into the potential application of silica fume concrete on highways in the Northwest and similar environmental areas.

2   Raw Materials and Test Method

2.1   Material Selection

Table 1. Chemical composition of silica fume%

Denomination	SiO_2	Fe_2O_3	MgO	CaO	SO_3
Silica fume	96.4	0.7	1.3	0.6	0.9

At present, scholars have conducted extensive research on the sulfate corrosion resistance of slag powder cement concrete. The results show that the corrosion resistance coefficient of anti-corrosion concrete using anti-sulfate corrosion agents is higher than that of ordinary concrete. These findings provide valuable data and guidance for assessing the durability and predicting the service life of concrete structures exposed to high concentrations of sulfate [10–12].

In order to study the durability and service life of silica fume concrete, the experimental materials were designed by comparing the characteristics of concrete resistance to sulfate attack and frost resistance. The test raw materials were chosen for this purpose.

Cement: ordinary P·O 42.5grade cement.

Aggregate: polished gravel, 5–15 mm continuous gradation.

Sand: river sand, fineness modulus is 2.84, medium sand.

Water reducer: polycarboxylates highperformance waterreducing admixture

Airentraining agent: LYYQ airentraining agent.

Silica fume: the main chemical composition is shown in Table 1 and the performance index in Table 2.

Table 2. Silica fume performance table

Numbering	Waterbinder ratio	Percentage of sand	Concrete raw material consumption/(kg/m^3)							
			Cement (kg)	Silica Fume (kg)	Sand (kg)	Calculus (kg)	Water (kg)	Water reducing admixture (kg)	Air entraining agent (kg)	Slump constant (mm)
S0	0.35	0.40	480	0	650	975	168	1.44	3.6	202
S5			457	23	650	975	168	2.00	3.6	208
S10			436	44	650	975	168	2.67	3.6	213
S15			417	63	650	975	168	3.34	3.6	200

2.2 Mixture Ratio

In this study, a total of 120 cube specimens measuring 100 mm × 100 mm × 100 mm and 36 prism specimens measuring 100 mm × 100 mm × 400 mm were used, each group consisted of 3 specimens. Reduce the uncertainty of large data gaps by using multiple specimens to obtain more scientific data. The concrete specimens were prepared in accordance with the' Standard for Test Methods of Long-Term Performance and Durability of Ordinary Concrete'(GB/T 50082-2009). After 28 days of standard curing, the specimens were subjected to an alternate sulfate sodium erosion-freeze-thaw cycle test. The test procedure is detailed in Table 3.

Table 3. Mix proportion and working performance of concrete

Denomination	Water content w (%)	Density ρ(g/cm^3)	Specific surface (m^2/kg)	Porosity (%)	Ignition loss (%)
Silica fume	0.4	2.1	18200	≥91	2.3

It was observed during mixing at a water-cement ratio of 0.35 that the addition of water-reducing admixture resulted in a decrease in concrete slump. This effect was more pronounced when the dose was 10% and above, causing some separation and bleeding when the dose reached 15%.

2.3 Sulfate Sodium Erosion-Freeze-Thaw Cycle Alternating Test

To accelerate the deterioration rate of concrete and facilitate the observation of test silica fume, the method of dry-wet cycle is selected for sulfate attack. Once the specimen reaches the desired curing age, a freeze-thaw cycle test is conducted following the sulfate dry-wet cycle test. In the dry-wet test, the specimen is soaked in a 5%sodium sulfate solution for 16 h. The drying process involves a temperature of 80 °C for 6 h, followed by a cooling time of 2 h, completing a cycle of 24 h. The freeze-thaw cycle test includes freezing the specimen for 4 h at a temperature of 18 ± 2 °C and then melting it for 2 h at a temperature of 5 ± 2 °C.This cycle is repeated for a total of 80 days, with 1 sulfate attack and freeze-thaw cycle alternate test(referred to as KSD cycle)performed every 16 days. A total of 5 alternate cycles are conducted, with the lateral fundamental frequency and quality of the concrete specimens determined every 10 days. The compressive strength is measured at intervals of 16 days. The KSD1 test consists of a dry-wet cycle of 4 days followed by a freeze-thaw cycle of 12 days. The KSD2 test involves a dry-wet cycle of 8 days and a freeze-thaw cycle of 8 days. Lastly, the KSD3 test comprises a dry-wet cycle of 12 days followed by a freeze-thaw cycle of 4 days.

3 Test Results and Analysis

3.1 Mass Loss Rate

The mass loss rate of silica fume concrete under three kinds of sulfate sodium erosion freeze-thaw cycle alternating tests is shown in Fig. 1.

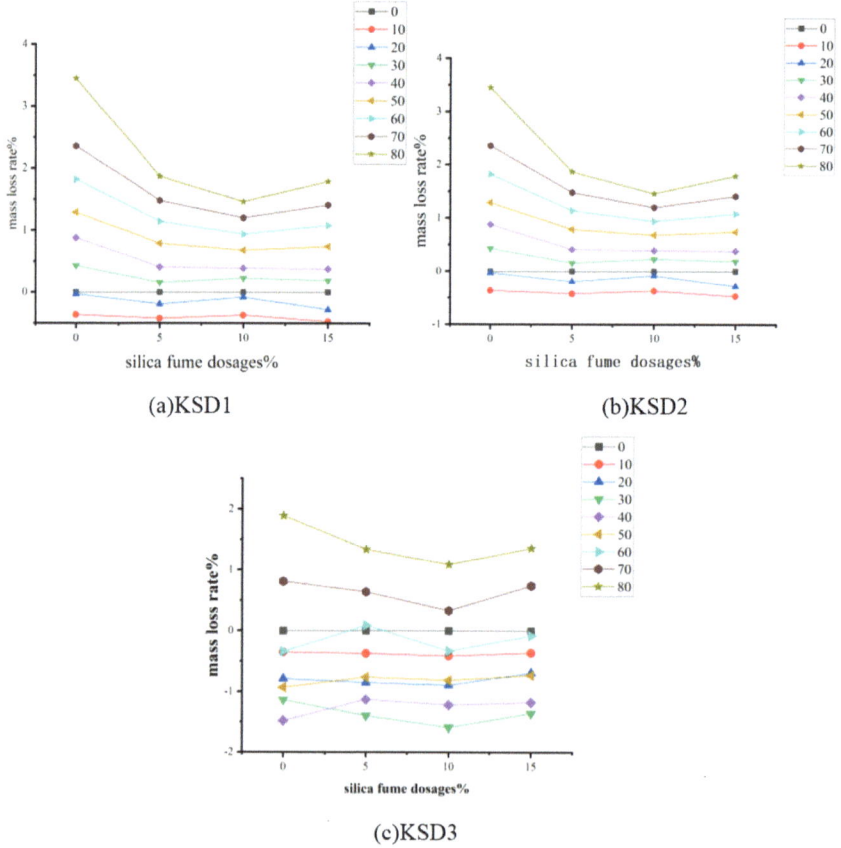

Fig. 1. Change of mass loss rate of silica fume concrete under KSD test

As can be seen from Fig. 1, the trend of silica fume content in silica fume concrete under the KSD test changes from high to low, and then gradually stabilizes. In the

KSD1, KSD2 and KSD3 tests, the most significant changes occurred after about 80 freeze-thaw cycles. When the silicon powder content was 5%, the mass loss rates were 1.87%, 1.44% and 1.34% respectively, and then the decline rate gradually Slow down. When the silicon powder content reaches 10%, the mass loss rates are 1.46%, 1.24% and 1.09% respectively, reaching the lowest point during the experiment. Subsequently, as the silicon powder content increased, the mass loss rate increased slightly and stabilized at 1.79%, 1.47% and 1.36% respectively.

Sodium sulfate solution (5%) was used in the three KSD tests to evaluate the mass loss rate of concrete mixed with silica fume at different proportions. The results showed that both percentages had significantly lower mass loss rates compared to concrete without silica fume over 80 days. Compared with concrete without silica fume, the mass loss rate reduction rates of KSD1, KSD2 and KSD3 are 57.7%, 47.5% and 42.3% respectively. Chemical reactions between sulfates and concrete during the early stages of the trial led to the production of slate and gypsum, improving the quality of the concrete. But late in the trial, frost heave cracking and concrete surface cracks resulted, ultimately reducing the quality of the concrete by slowing mortar shedding.

3.2 Relative Dynamic Elastic Modulus

The results of relative dynamic elastic modulus of silica fume concrete under three kinds of sulfate sodium erosion-freeze-thaw cycle alternating tests are shown in Fig. 2.

It can be seen from Fig. 2 that as the duration of sulfate dry and wet erosion and freeze-thaw cycles increases, the relative dynamic elastic modulus of concrete decreases slowly. The relative dynamic elastic modulus of silica fume concrete increases first and then decreases. The increase was most significant after 80 freeze-thaw cycles. Subsequently, as the silicon powder content continued to increase, the relative dynamic elastic modulus began to decrease, reaching 84%, 85.9% and 90.3% respectively. The figure clearly shows that when the silicon fume content is 10%, the optimal relative dynamic elastic modulus is achieved. In the early stages of the test, the concrete was not affected by sulfate intrusion, and the chemical reactions of concrete deterioration and sulfate production led to the formation of slabs and gypsum, promoting an increase in the relative dynamic modulus of elasticity. However, during the later stages of the test, the concrete experienced swelling and osmotic pressures due to alternating wet and dry conditions, as well as freeze-thaw cycles. These factors begin to deteriorate the concrete, accelerating the decrease in its relative dynamic modulus of elasticity.

3.3 Compressive Strength

The compressive strength of silica fume concrete under three kinds of sulfate sodium erosion-freeze-thaw cycle alternate tests is shown in Fig. 3.

It can be seen from Fig. 3 that the compressive strength of concrete increases with the increase of silica fume content, but decreases with the extension of test time. The most obvious changes were observed at 80 freeze-thaw cycles. Compared with 10% and 15% silica fume concrete, the compressive strength of silica fume-free concrete in KSD1, KSD2, and KSD3 increased by 17.7%, 15%, and 10.4% respectively. However, compared with 10% and 15% silica fume concrete, the increases are only 0.7%, 1%

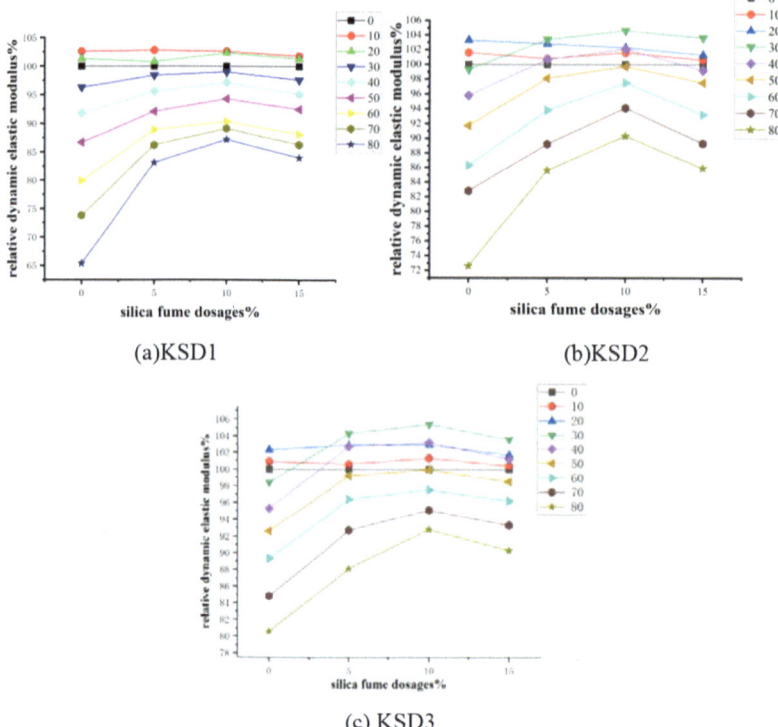

Fig. 2. Variation of relative dynamic elastic modulus of silica fume concrete under KSD test

and 1.3%. This shows that the compressive strength benefit provided by the silica fume content reaches saturation after 10%. When the silica fume content reaches 10%, the compressive strength benefit is less obvious. The water holding capacity and cohesion of concrete are weakened, and the frost resistance and sulfate corrosion resistance of concrete are reduced.

4   Prediction of Concrete Life Based on Grey Theory

4.1   The Establishment and Error Test of GM(1,1)model

As a new and commonly used prediction model, grey theory is widely applied in civil engineering projects due to its simplicity, few samples, and ability to predict complex factors [13–15]. It can be used for various purposes such as high-rise building settlement, slope stability analysis, precast concrete (PC) cost prediction, and geotechnical engineering. The prediction model of grey theory can be classified into five types: GM(1,1)model, GM(0,N)model, GM(1,N)model, grey linear regression model, and grey Markov model application [16]. In this study, the freeze-thaw + sulfate erosion alternating test only considers one dynamic time factor, thus the GM(1,1)model is employed, which refers to the first-order and one-variable grey model.

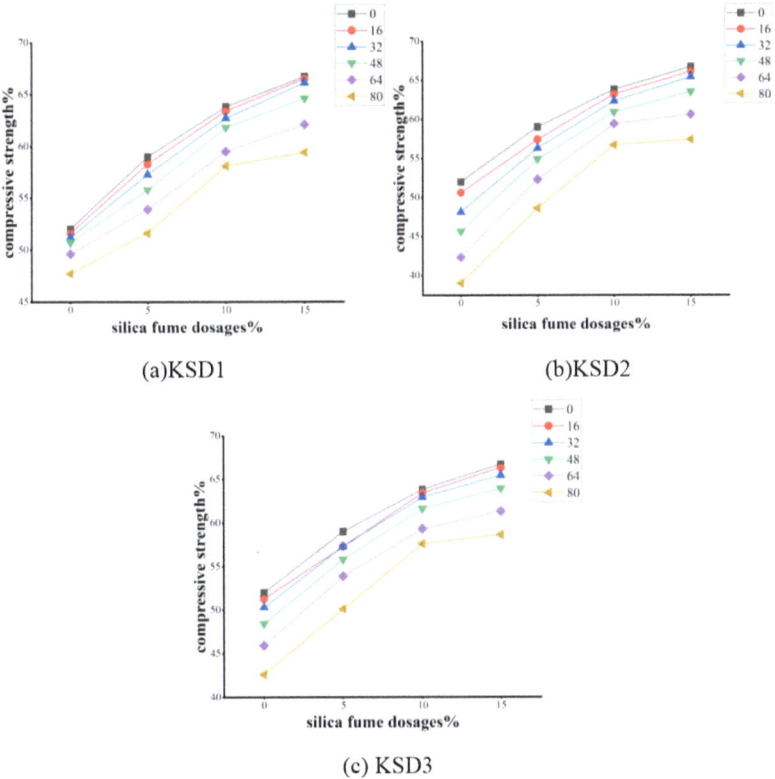

Fig. 3. Change of compressive strength of silica fume concrete under KSD test

The modeling process is as follows:

$$X^{(0)} = \{x^{(0)}(1), x^{(0)}(2), x^{(0)}(3), \ldots, x^{(0)}(n)\} \tag{1}$$

The time series X(0)of the initial data of the study is accumulated to X(1):

$$X^{(1)} = \{x^{(1)}(1), x^{(1)}(2), x^{(1)}(3), \ldots, x^{(1)}(n)\} \tag{2}$$

numerator: $x^{(1)}k = \sum_{i=0}^{k} x(i)$

The whitened differential equation (shadow equation) of GM(1,1)model is:

$$\overline{\varepsilon} = \frac{1}{n} \sum_{k=1}^{n} \varepsilon(k) \tag{3}$$

quorum:

a, u—The undetermined parameter, a reflects the parameter of system development;

u—The grey action quantity mined from the grey system reflects the relationship between data changes.

Applying the least square method, solving Formula (3), we can get:

$$A = \left(B^T B\right)^{-1} B^T Y = [a, u]^T \tag{4}$$

quorum:

$$B = \begin{bmatrix} -\frac{1}{2}(x^{(1)}1 + x^{(1)}2)1 \\ -\frac{1}{2}(x^{(1)}1 + x^{(1)}2)1 \\ \vdots \\ -\frac{1}{2}(x^{(1)}n - 1 + x^{(1)}n)1 \end{bmatrix} \quad B = \begin{bmatrix} x^{(1)}2 \\ x^{(1)}2 \\ \vdots \\ x^{(1)}n \end{bmatrix} \tag{5}$$

Then the solution of the time response function is:

$$\widehat{x}^{(1)}(k + 1) = \left[x^{(1)}1 - \frac{u}{a}\right]e^{-ak} + \frac{u}{a} \tag{6}$$

The fitting value of the initial data is:

$$\widehat{x}^{(0)}(k + 1) = \widehat{x}^{(1)}(k + 1) - \widehat{x}^{(1)}(k) \tag{7}$$

Based on the given information, a prediction model for the concrete durability atten-uation of silica fume concrete in the freeze-thaw + sulfate dry-wet cycle has been developed.

The residual of the test model is:

$$\varepsilon(k) = x^{(0)}(k) - \widehat{x}^{(1)}, \quad (k + 1)k = 1, 2, \cdots, n \tag{8}$$

The mean and variance of the original data sequence are:

$$\bar{x} = \frac{1}{n}\sum_{k=1}^{n} x^{(0)}(k) \tag{9}$$

$$p = P\{|\varepsilon(k) - \bar{\varepsilon}| < 0.6745S_1\} \tag{10}$$

The mean and variance of the data residual are:

$$\bar{\varepsilon} = \frac{1}{n}\sum_{k=1}^{n} \varepsilon(k) \tag{11}$$

$$S_2^2 = \frac{1}{n}\sum_{k=1}^{n} [\varepsilon(k) - \bar{\varepsilon}]^2 \tag{12}$$

Test mean difference ratio C:

$$C = S_2/S_1 \tag{13}$$

Small probability error P:

$$p = P\{|\varepsilon(k) - \bar{\varepsilon}| < 0.6745S_1\} \tag{14}$$

According to Table 4, The accuracy of the model can be evaluated based on the mean difference ratio C. It is expected that C should be less than 0.35 and should not exceed 0.65. Another index to assess the model's accuracy is the small error probability p, which should be greater than 0.95 and not less than 0.7. Therefore, based on these two indicators, the model accuracy can be classified into four level.

Table 4. Accuracy grade of grey prediction mode

Mean square error grade C	Small error probability p	Accuracy class
$C \leq 0.35$	$P \geq 0.95$	Grade 1(excellent)
$0.35 < C \leq 0.5$	$0.8 \leq p < 0.95$	Grade 2(good)
$0.5 < C \leq 0.65$	$0.7 \leq p < 0.8$	Grade 3(qualification)
$0.65 < C$	$p < 0.7$	Grade 4(below proof)

According to my country Standard for Test Methods for Long-term Performance and Durability of Ordinary Concrete" (GB/T50082-2009). The quick freezing test is evaluated based on the relative dynamic elastic modulus and mass loss rate, and the sulfate resistance test is evaluated based on the strength corrosion resistance coefficient and mass crrosion resistance coefficient. As can be seen from Figs. 1, 2 and 3, in the sodium sulfate erosion and freeze-thaw cycle alternating test, silica fume concrete was damaged the most, and the corrosion resistance coefficient (Kf) of compressive strength concrete dropped to 75%. Therefore, we use compressive strength as an indicator to predict the service life of silica fume concrete.

4.2 Prediction of Concrete Life Based on Compressive Strength

Based on the GM(1,1)model and the compressive strength data of silica fume concrete obtained from the KSD test, we have developed a prediction model for residual compressive strength. The accuracy grade of this model is presented in Table 5.

According to Table 5, the accuracy of predicting the failure behavior of concrete by compressive strength is high. This predictive accuracy can be utilized to forecast the lifespan of silica fume concrete. Figure 4 illustrates the failure time of silica fume concrete predicted by the GM(1,1)model for various dosages and tests.

The results shown in Fig. 4 indicate that, for the same duration of testing, the longer the damage time of concrete, the higher the ratio of dry-wet erosion of sodium sulfate to freeze-thaw cycle time. This demonstrates that the freeze-thaw damage at 20 °C is greater than that caused by a 5%solution of sodium sulfate at 80 °C. Additionally, the inclusion of silica fume has a positive impact on the service life of concrete, with the best effect observed at a 10%concentration. Compared to concrete without silica fume, the addition of silica fume prolongs the time for KSD1, KSD2, and KSD3 by 112 days, 96 days, and 64 days, respectively. This represents an increase of 124.4%, 85.7%, and 40.0%, respectively. Wu Hairong proposed dividing the freeze-thaw zone into D levels, and the average number of freeze-thaw cycles in the Qinghai Northwest Territories project area is 120 times [17]. Furthermore, the indoor fast freezing method specified in the 'ordinary concrete long-term performance and durability test method standard' (GB/T 50082-2009) is equivalent to 10–14 natural environment freeze-thaw cycles. The combination of sodium sulfate dry-wet erosion and freeze-thaw cycles has a synergistic effect, resulting in more accurate predictions of the concrete's lifespan in project areas. The concrete life of the Northwest Territories project is more accurate when considering 10 times the sulfate erosion and freeze-thaw environment. The results can be observed in Fig. 5.

Table 5. Qualitybased prediction model and accuracy evaluation

Experimentation	Silica fume content/%	Forecasting model	Difference ratio C	Small probability error p	Precise grade	Damage time /d
KSD1	0	$y = 15.89e^{(0.003997t)} + 16.88$	0.076	1	Grade 1 (excellent)	80
	5	$y = 25.67e^{(0.00246t)} + 26.66$	0.225	1	Grade 1 (excellent)	128
	10	$y = 39.11e^{(0.001619t)} + 40.1$	0.242	1	Grade 1 (excellent)	192
	15	$y = 28.95e^{(0.002218t)} + 29.95$	0.278	1	Grade 1 (excellent)	144
KSD2	0	$y = 22.7e^{(0.002833t)} + 23.69$	0.196	1	Grade 1 (excellent)	112
	5	$y = 29.11e^{(0.002193t)} + 30.1$	0.244	1	Grade 1 (excellent)	144
	10	$y = 40.27e^{(0.001582t)} + 41.26$	0.199	1	Grade 1 (excellent)	208
	15	$y = 32.95e^{(0.001943t)} + 33.94$	0.253	1	Grade 1 (excellent)	160
KSD3	0	$y = 35.5e^{(0.001813t)} + 36.5$	0.4844	0.83	Grade 2 (good)	160
	5	$y = 33.51e^{(0.00189t)} + 34.51$	0.1431	1	Grade 1 (excellent)	176
	10	$y = 44.36e^{(0.00143t)} + 45.36$	0.0404	1	Grade 1 (excellent)	224
	15	$y = 35.91e^{(0.00179t)} + 36.9$	0.3789	1	Grade 1 (excellent)	192

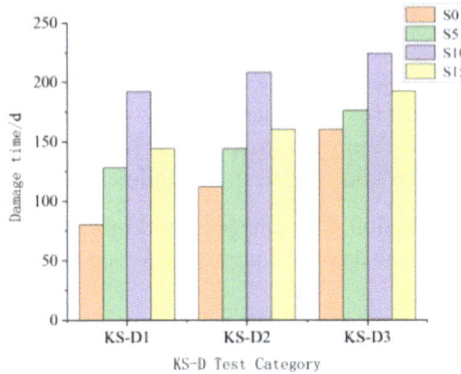

Fig. 4. Failure time diagram of silica fume concrete under KSD test

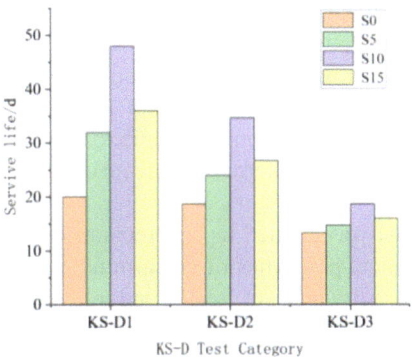

Fig. 5. Presents a service life diagram of silica ash concrete, illustrating the effects of dry and wet erosion as well as freeze-thaw on its durability.

The cycle time of three types of sodium sulfate salts is difficult to determine in the Northwest Territories area due to variations in precipitation, temperature, and altitude. To predict the service life of concrete, three different ratios of sodium sulfate dry-wet erosion and freeze-thaw cycles have been designed. Figure 5 shows that the increase in sulfate erosion time has a significant impact on the service life of concrete when the number of freeze-thaw cycles in the Qinghai area is determined to be 120 times. The service life of concrete without silica fume is 20 years, 18.7 years, and 13.3 years for freeze-thaw and dry-wet erosion ratios of 3:1, 1:1, and 1:3, respectively. On the other hand, the service life of silica fume concrete with the optimal durability, at a dosage of 10%, is 48 years, 34.7 years, and 18.7 years for the same ratios. These findings can provide valuable insights for predicting the life of silica fume concrete under different sodium sulfate erosion conditions in the Northwest Territories area.

5 Conclusion

In the context of the Garxi project, a sulfate sodium erosion-freeze-thaw cycle alternating test was conducted based on three different sulfate dry-wet erosion and freeze-thaw cycle time ratios. The grey system theory was used to establish the GM(1,1)prediction model for the mechanical properties under this test. The following conclusions were drawn:

(1) Under the same test duration, the damage caused by the 20°C freeze-thaw cycle to concrete is greater than the damage caused by 5%sodium sulfate solution(80°C)dry and wet erosion.

(2) In the sodium sulfate-freeze-thaw cycle alternating test, the addition of silica fume can reduce the loss rate of concrete quality, relative dynamic elastic modulus, and compressive strength. The optimal dosage is 10%. However, when the dosage exceeds 10%,the water retention and cohesion of concrete weaken, leading to the hardening of concrete in cement and causing significant defects. This ultimately reduces the frost and sulfate resistance of concrete.

(3) In the context of sulfate attack and freeze-thaw cycles, the compressive strength of silica fume concrete is initially compromised. To accurately predict the time at which the concrete gets damaged, the GM(1,1)model is employed based on the compressive strength. This prediction can be utilized to estimate the lifespan of silica fume concrete subjected to the alternating effects of sulfate attack and freeze-thaw cycles.

(4) The gray theory model was applied in the preparation of the GM(1,1) model prediction of the mechanical properties of silica fume concrete under the alternating action of sodium sulfate dry and wet erosion and freeze-thaw cycles. The prediction accuracy of the mechanical properties of strength and relative dynamic modulus is high. It can be utilized to predict the failure time of concrete residual compressive strength and relative dynamic elastic modulus in actual projects, but the prediction accuracy of the mass loss rate is poor and it cannot accurately predict the concrete mass failure time.

Acknowledgements. This research was funded by the Technology Funding Scheme of China Construction Second Engineering Bureau LTD. (2020ZX150002); National Natural Science Foundation of China (12262018); Special Funds for Guiding Local Scientific and Technological Development by The Central Government (22ZY1QA005).

References

1. Ho, L.S., Huynh, T.-P.: Long-term mechanical properties and durability of high-strength concrete containing high-volume local fly ash as a partial cement substitution. Results Eng. **18**, 101113 (2023). https://doi.org/10.1016/j.rineng.2023.101113
2. Babalu, R., Anil, A., Sudarshan, K., Amol, P.: Compressive strength, flexural strength, and durability of high-volume fly ash concrete. Innov. Infrastruct. Solut. **8**(5) (2023). https://doi.org/10.1007/s41062-023-01120-x

3. Peng, X., Shi, F., Yang, J., Yang, Q., Wang, H., Zhang, J.: Modification of construction waste derived recycled aggregate via CO2 curing to enhance corrosive freeze-thaw durability of concrete. J. Clean. Product. **405**, 137016 (2023). https://doi.org/10.1016/j.jclepro.2023.137016

4. Fiol, F., Revilla-Cuesta, V., Thomas, C., Manso, J.M.: Self-compacting concrete containing coarse recycled precast-concrete aggregate and its durability in marine-environment-related tests. Construct. Build. Mater. **377**, 131084 (2023). https://doi.org/10.1016/j.conbuildmat.2023.131084

5. Chen, H., Cao, Y., Liu, Y., Qin, Y., Xia, L.: Enhancing the durability of concrete in severely cold regions: mix proportion optimization based on machine learning. Construct. Build. Mater. **371** (2023)

6. Lei, G., et al.: Experimental and numerical investigations on damage evolution of concrete under sulfate attack and freeze-thaw cycles. J. Build. Eng. **71** (2023)

7. Chen, S., Ren, J., Ren, X., Li, Y.: Deterioration laws of concrete durability under the coupling action of salt erosion and drying–wetting cycles. Front. Mater. (2022)

8. Shakiba, M., Bazli, M., Karamloo, M., Mortazavi, S.M.R.: Bond-slip performance of GFRP and steel reinforced beams under wet-dry and freeze-thaw cycles: the effect of concrete type. Construct. Build. Mater. **342**, 127916 (2022). https://doi.org/10.1016/j.conbuildmat.2022.127916

9. Chen, D., Liu, S., Shen, J., Sun, G., Shi, J.: Experimental study and modelling of concrete carbonation under the coupling effect of freeze-thaw cycles and sustained loads. J. Build. Eng. **52**, 104390 (2022). https://doi.org/10.1016/j.jobe.2022.104390

10. Jaworska-Wędzińska, M., Jasińska, I.: Durability of mortars with fly ash subject to freezing and thawing cycles and sulfate attack. Materials **15**(1), 220 (2021). https://doi.org/10.3390/ma15010220

11. Zheng, Y., Yang, L., Guo, P., Yang, P.: Fatigue characteristics of prestressed concrete beam under freezing and thawing cycles. Adv. Civil Eng. (2020)

12. Hemmati, A., Arab, H.: Proposed model for stress-strain behavior of fly ash concrete under the freezing and thawing cycles. Periodica Polytech. Civil Eng. (2020). https://doi.org/10.3311/PPci.14805

13. Hakuzweyezu, T., Qiao, H., Lu, C., Twagirumukiza, J., Yang, B.:Life prediction of concrete mixed with NanoCaCO3 in semiimmersed corrosive environment. ACI Mater. J. **119**(5) (2022)

14. Manganelli, B.: Economic life prediction of concrete structure. Adv. Mater. Res. **3149**(919921) (2014)

15. Dinesh Vijay Kumar, M., Paulson, J.: Method of proportioning silica fume concrete. Int. J. Adv. Technol. **8**(3s) (2019)

16. Qin, Y., Guan, K., Kou, J., Ma, Y., Zhou, H., Zhang, X.: Durability evaluation and life prediction of fiber concrete with fly ash based on entropy weight method and grey theory. Construct. Build. Mater. **327**, 126918 (2022). https://doi.org/10.1016/j.conbuildmat.2022.126918

17. Tang, H.S., Mei, J.H., Chen, W., Li, D.W., Xue, S.T.: Uncertainty analysis in fatigue life prediction of concrete using evidence theory. Mater. Sci. Forum **866**, 25–30 (2016). https://doi.org/10.4028/www.scientific.net/MSF.866.25

18. Li, G.: Deterioration and service life prediction of concrete subjected to freeze–thaw cycles in Na_2SO_4 solution. Am. J. Civil Eng. **4**(3), 104 (2016). https://doi.org/10.11648/j.ajce.20160403.17

19. Özcan, F., Atiş, C.D., Karahan, O., Uncuoğlu, E., Tanyildizi, H.: Comparison of artificial neural network and fuzzy logic models for prediction of long-term compressive strength of silica fume concrete. Adv. Eng. Softw. **40**(9), 856–863 (2009). https://doi.org/10.1016/j.advengsoft.2009.01.005

20. Asad, A.O.: Production of high performance silica fume concrete. Am. J. Appl. Sci. **14**(11) (2017)
21. Farahani, H.Z., Farahani, A., Fakharian, P., Armaghani, D.J.: Experimental study on mechanical properties and durability of polymer silica fume concrete with vinyl ester resin. Materials **16**(2), 757 (2023). https://doi.org/10.3390/ma16020757
22. Ahmed, A.S., Yousry, M.M., Elshikh, W.E., Elemam, O.Y.: Influence of mixing-water magnetization method on the performance of silica fume concrete. Buildings **13**(1), 44 (2022). https://doi.org/10.3390/buildings13010044
23. Ahmed, S.H., et al.: Predicting compressive and splitting tensile strengths of silica fume concrete using M5P model tree algorithm. Materials **15**(15) (2022)
24. Atia, S.M., Abbas, W.A.: Effect of adding nano starch biopolymer on some properties of silica fume concrete. Key Eng. Mater. **911**, 145–150 (2022). https://doi.org/10.4028/p-2i42va
25. Bhatt, K., Singh, S.: Study on performance, durability and strength of silica fume concrete. J. Progress Civil Eng. **3**(8) (2021)
26. Science Applied Science. Recent Studies from Ataturk University Add New Data to Applied Science (Microstructural Analysis of Silica Fume Concrete with Scanning Electron Microscopy and XRay Diffraction). Science Letter (2020)
27. Golafshani, E.M., Behnood, A.: Estimating the optimal mix design of silica fume concrete using biogeography-based programming. Cement Concrete Composit. **96**, 95–105 (2019). https://doi.org/10.1016/j.cemconcomp.2018.11.005
28. Artificial Neural Networks. Reports Summarize Artificial Neural Networks Study Results from Purdue University(Predicting the compressive strength of silica fume concrete using hybrid artificial neural network with multiobjective grey wolves). Computers Networks & Communications (2018)
29. Fernandes, B., Khodeir, M., Perlot, C., Carré, H., Mindeguia, J.-C., La Borderie, C.: Durability of concrete made with recycled concrete aggregates after exposure to elevated temperatures. Mater. Struct. **56**(1) (2023). https://doi.org/10.1617/s11527-023-02111-1
30. Arasteh-Khoshbin, O., Seyedpour, S.M., Ricken, T.: The effect of Caspian Sea water on mechanical properties and durability of concrete containing rice husk ash, nano [Fotmula:see text], and nano[Formula:see text]. Sci. Reports **12**(1) (2022). https://doi.org/10.1038/s41598-022-24304-4
31. Sun, D., et al.: Effect of the moisture content of recycled aggregate on the mechanical performance and durability of concrete. Materials **15**(18) (2022)
32. Monkman, S., Hanmore, A., Thomas, M.: Sustainability and durability of concrete produced with CO2 beneficiated reclaimed water. Mater. Struct. **55**(7) (2022). https://doi.org/10.1617/s11527-022-02012-9
33. Corbu, O., Toma, I.-O.: Progress in sustainability and durability of concrete and mortar composites. Coatings **12**(7), 1024 (2022). https://doi.org/10.3390/coatings12071024
34. Zaki, S.I., Hodhod, O.A., Eid, M.F.: Evaluating the effect of using nano bentonite on strength and durability of concrete. Nano Hybrids Compos. **36**, 125–141 (2022). https://doi.org/10.4028/p-1qwl8w
35. Mingchang Hei, A., Fayou, X.J., Peng, W., Yin, C.: Study on durability of concrete under alkali-aggregate reaction. Geofluids **2022**, 1–14 (2022). https://doi.org/10.1155/2022/4223791
36. Da Silva, S.R., Andrade, J.J.O.: A review on the effect of mechanical properties and durability of concrete with Construction and Demolition Waste (CDW)and fly ash in the production of new cement concrete. Sustainability **14**(11) (2022)
37. Oikonomopoulou, K., Ioannou, S., Savva, P., Spanou, M., Nicolaides, D., Petrou, M.F.: Effect of mechanically treated recycled aggregates on the long term mechanical properties and durability of concrete. Materials **15**(8), 2871 (2022). https://doi.org/10.3390/ma15082871

38. Amer, O.A., Rangaraju, P., Rashidian-Dezfouli, H.: Effectiveness of binary and ternary blended cements of class C fly ash and ground glass fibers in improving the durability of concrete. J. Sustain. Cement-Based Mater. **11**(2), 127–136 (2022). https://doi.org/10.1080/21650373.2021.1899085

39. Guo, J., Guo, T., Zhang, S., Yan, L.: Experimental study on freezing and thawing cycles of shrinkage-compensating concrete with double expansive agents. Materials **13**(8), 1850 (2020). https://doi.org/10.3390/ma13081850

40. Ismail, M.K., Hassan, A.A.A.: Abrasion and impact resistance of concrete before and after exposure to freezing and thawing cycles. Construct. Build. Mater. **215**, 849–861 (2019). https://doi.org/10.1016/j.conbuildmat.2019.04.206

41. Ziaei-Nia, A., Tadayonfar, G.-R., Eskandari-Naddaf, H.: Effect of air entraining admixture on concrete under temperature changes in freeze and thaw cycles. Mater. Today: Proc. **5**(2), 6208–6216 (2018). https://doi.org/10.1016/j.matpr.2017.12.229

Research on Performance Evaluation of Different Dosages of Hot Recycled Asphalt Mixtures Based on High Temperature Shear Resistance

Jiahua You[1], Cong Xi[1], Zhaojie Zhang[2(✉)], Guohua Fu[1], and Huan Zhang[2]

[1] China Construction Eighth Bureau Development and Construction Co., Ltd., Qingdao 260000, China
[2] Shandong Transportation Institute, Jinan 250000, China
1165951186@qq.com

Abstract. Recycled asphalt pavement (RAP) is incorporated into AC-25C asphalt mixture at varying percentages—25%, 40%, and 60%—to assess the high-temperature performance through rutting and uniaxial penetration tests. The research reveals that as the RAP content increases, there is a significant improvement in the high-temperature performance of the asphalt mixture. Notably, at a 40% RAP content, the dynamic stability and penetration strength exhibit the most substantial enhancement. However, at a 60% RAP content, the high-temperature performance of the recycled mixture is markedly lower than at 40%, and the dynamic stability fails to meet the relevant requirements specified in JTG F40–2004. Furthermore, the study observes a notable increase in cohesion (C) as the hot recycled asphalt mixture's RAP content rises, leading to an extended stress retention stage compared to the conventional uniaxial penetration test process. Additionally, the stress retention stage becomes longer with the increasing RAP content.

Keywords: RAP hot recycled asphalt mixture · dynamic stability · shear strength · stress maintenance · cohesion C

1 Introduction

"Carbon peaking and carbon neutrality" is an important measure to address global climate change and promote sustainable development. We are committed to achieving this goal by accelerating the shift to renewable energy and promoting low-carbon transportation [1, 2]. The recycling and utilization technology of old road surfaces is one of the effective measures for energy conservation and emission reduction widely used in the process of highway construction both domestically and internationally. It improves the reuse rate of mineral resources and protects the development of new mineral resources. The high-value utilization of asphalt milling and recycled materials (RAP) for old road surfaces is still in the exploratory and experimental stage [3–7]. At present, high content

© The Author(s) 2024
A. Bieliatynskyi et al. (Eds.): CSTTE 2023, LNCE 603, pp. 30–37, 2024.
https://doi.org/10.1007/978-981-97-5814-2_3

hot recycled asphalt mixtures face challenges such as poor gradation stability and heating difficulties. The warm mix high content plant mix hot recycling technology is currently the mainstream research direction, with the majority being the addition of warm mix agents and mechanical foaming [8–11]. Research has shown that the high-temperature performance of hot recycled asphalt mixture increases with the increase of RAP content, but the upper limit of RAP content is unknown [12–14]. Due to the fact that hot recycled asphalt mixtures are mainly used in the middle and lower layers of pavement structures, this article uses AC-25C type mixtures to study the high-temperature performance of hot recycled asphalt mixtures with different RAP dosages. The high-temperature performance is evaluated through rutting tests and uniaxial penetration strength tests.

2 Raw Materials and Mix Design

2.1 Raw Materials

Due to the good high-temperature performance of modified asphalt, in order to reduce the impact of new asphalt on the performance of recycled mixtures, Qilu Petrochemical 70-A grade road petroleum asphalt was used in this study; Select typical limestone as raw materials: 20-30 mm, 10-20 mm, 5-10 mm, 3-5 mm, 0-3 mm; RAP uses a certain highway to recover old road surfaces. Considering the stability of the high content recycled mixture gradation, RAP is broken into three specifications: 10-20 mm, 5-10 mm, and 0-5 mm. The aggregates, asphalt, mineral powder, and other materials used in this study have been tested to meet the relevant technical requirements of the JTG F40–2004 specification. The RAP asphalt content from coarse to fine is 3.0%, 3.4%, and 7.1%, respectively.

2.2 Mix Ratio

The Marshall design method was used for the mix proportion design of the mixture used in this study. In order to achieve the high value utilization of RAP, in the mix design process, the amount of milling and planing materials in each stage should be as close as possible to the proportion of each stage in the secondary crushing and screening, to avoid excessive or insufficient situations. The content of milling material exceeds 30%, and the performance of old asphalt in the milling material is improved by adding a regenerant. The regenerant content accounts for 6% of the total asphalt content. During the mixing process, the regenerant and milling material are mixed first, and new aggregates are added for mixing to extend the dry mixing time. The wet mixing time for spraying asphalt should also be extended. This study compared the mix proportions of 25%, 40%, and 60% with 0%, as shown in Table 1. The grading curve is shown in Fig. 1, and the test specimens were formed according to the above ratio for experimentation. The experimental results are shown in Table 2.

(1) Due to the aging of RAP asphalt, the optimal asphalt content of plant mixed hot recycled asphalt mixture is usually higher than that of conventional asphalt mixture;

(2) With the increase of milling material content, the stability of asphalt mixture gradually decreases and the flow value gradually increases. The asphalt mastic content

in hot recycled asphalt mixture is relatively high compared to conventional mixture, and the limited heating temperature of milling material results in the mixture not being able to achieve the uniformity of conventional mixture, resulting in strong rheological properties of recycled asphalt mixture, and the specimen still maintains a certain cohesion after instability.

Table 1. Mix proportions of hot recycled asphalt mixtures with different dosages

Grading of mixture	Limestone					Reclaimed Asphalt Pavement(RAP)			Mineral powder
	20-30 mm	10-20 mm	5-10 mm	3–5 mm	0–3 mm	10–20 mm	5-10 mm	0-5 mm	
AC-25–0%	15	36	21	4	21	0	0	0	3
AC-25–25%	18	28	16	0	12	10	5	10	1
AC-25–40%	20	21	12	0	6	20	10	10	1
AC-25–60%	21	14	4	0	0	30	20	10	1

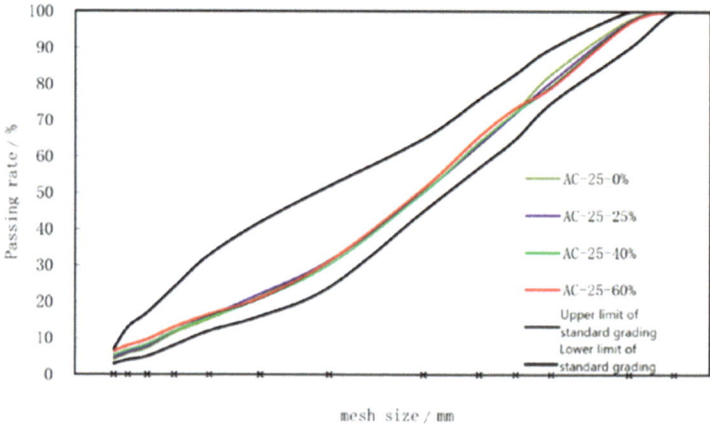

Fig. 1. Grading Curve of Hot Regenerated Asphalt Mixture with Different Dosages

3 High Temperature Performance Evaluation

3.1 High Temperature Anti Rutting Performance

This article uses rutting test to study the high-temperature resistance to rutting performance of asphalt mixtures. The rutting test is the most commonly used test method for evaluating the high-temperature performance of asphalt mixtures in China. The rutting test mainly studies the deformation of 45–60 min of rutting, analyzes the relationship between deformation and rolling times, and uses correlation equations to compare the rutting performance (Fig. 2).

Table 2. Volume Index of Hot Regenerated Asphalt Mixtures with Different Dosages

Type of mixture	Newly added asphalt content/%	Total asphalt content/%	Mixing regenerant	Gross volume relative density	Theoretical maximum relative density	Void rate/%	Stability/kN	Stream value/mm
AC-25–0%	4.0	4.0	no	2.433	2.542	4.3	11.8	2.82
AC-25–25%	3.0	4.2	no	2.447	2.558	4.3	9.8	2.96
AC-25–40%	2.6	4.2	yes	2.449	2.565	4.5	8.9	3.41
AC-25–60%	1.9	4.2	yes	2.452	2.565	4.4	8.2	4.58

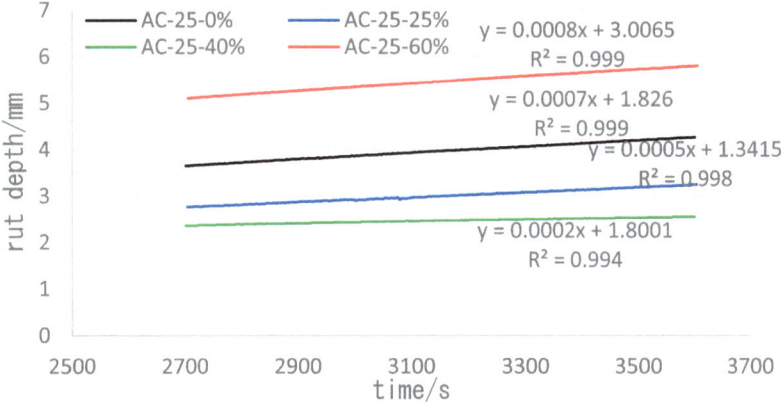

Fig. 2. Rutting deformation of mixtures with different dosages (45–60 min)

Table 3. Rutting Test Results of Mixtures with Different Admixtures

Type of mixture	Rutting deformation/mm		Dynamic stability/(time/mm)	Dynamic stability-rutting deformation equation(45 ~ 60min)(y = kx + b)	Correlation R^2
	45min	60min			
AC-25–0%	3.66	4.27	1033	y = 0.0007x + 1.826	0.999
AC-25–25%	2.77	3.25	1313	y = 0.0005x + 1.3415	0.998
AC-25–40%	2.36	2.55	3316	y = 0.0002x + 1.8001	0.994
AC-25–60%	5.10	5.80	900	y = 0.0008x + 3.0065	0.999

From Table 3, it can be seen that:

(1) The rutting deformation of plant mixed hot recycled asphalt mixture and conventional asphalt mixture tends to increase linearly within the range of 45–60 min, and the correlation R2 reaches above 0.99;

(2) According to the dynamic stability rutting deformation Eq. (45–60 min) of different types of mixtures, it can be seen that with the increase of RAP content, the k value first increases and then decreases, and the dynamic stability of 40% mixture reaches its highest. After the fusion of old asphalt and new asphalt in the milling material, it is equivalent to reducing the asphalt grade. When the content is 25%, the effect is not significant. When the content is 40%, the dynamic stability increases exponentially. When the content exceeds 60%, the dispersion uniformity of the milling material in the mixture decreases, resulting in a deviation between the actual grading and the theoretical grading, and a decrease in dynamic stability.

3.2 High Temperature Shear Failure Resistance

This article uses uniaxial penetration test to study the high-temperature shear failure performance, which was first proposed by renowned domestic road expert Professor Sun Lijun. During the test, the stress distribution of the test piece is consistent with that of the road surface. The data obtained from the uniaxial penetration test to determine the shear strength of the test piece is more practical. In the analysis of this article, the extreme point of axial pressure is used as the calculation point for the failure strength of the specimen.

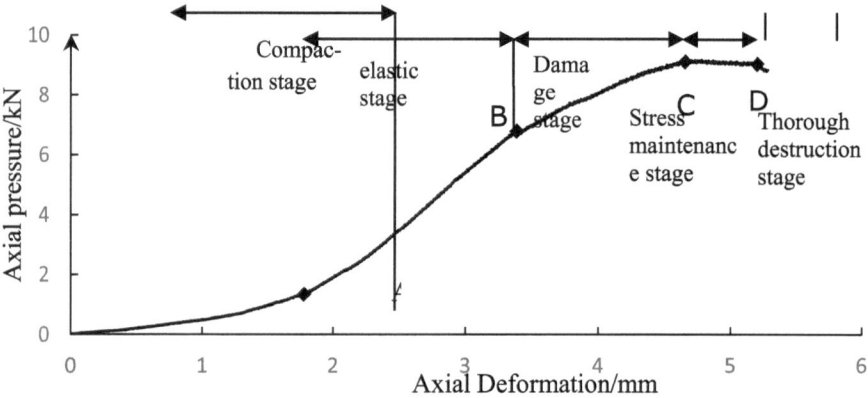

Fig. 3. Axial deformation pressure curve of uniaxial penetration test

The uniaxial penetration test specimens of asphalt mixtures have gone through five stages, including compaction stage (OA), elastic stage (AB), damage stage (BC), stress retention stage (CD), and complete failure (after point D) (Fig. 3). Due to the relatively high content of asphalt mastic in recycled asphalt mixture and the high cohesion C, there was no significant decrease in stress after specimen damage, and the mixture exhibited rheological characteristics.

With the increase of RAP content, the penetration strength of asphalt mixture first increases and then decreases. The shear strength of mixture with 40% content increases most significantly. When the RAP content reaches 60%, the penetration strength decreases rapidly. The penetration test results are consistent with the rutting

Table 4. Rutting deformation of different admixtures (45–60 min)

Type of mixture	Penetration depth/mm	Penetration strength/kN	Shear strength/MPa
AC-25–0%	3.239	4.207	0.536
AC-25–25%	4.330	4.624	0.589
AC-25–40%	4.650	9.186	1.170
AC-25–60%	5.601	4.636	0.590

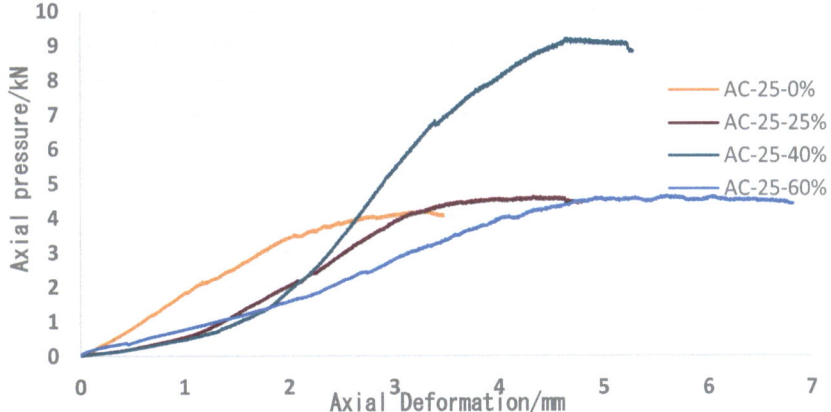

Fig. 4. Axial deformation pressure curve of uniaxial penetration test for mixtures with different dosages

test results. Although the penetration strength increases with the addition of RAP, the extreme penetration depth continues to increase (Table 4).

From Fig. 4, it can be seen that: The compression deformation stage of asphalt mixture without RAP addition is very short, and the damage stage is closely connected with the cracking and failure stage, without obvious stress retention stage. After adding milling materials, the mixture exhibits obvious compression deformation, and the variation of compression deformation with the dosage is consistent with the penetration strength; as the dosage increases, the cohesion C of the mixture increases, and the stress holding stage and axial deformation of the asphalt mixture are prolonged. The axial deformation is consistent with the results of the Marshall flow test.

4   Conclusion

By conducting rutting tests and uniaxial permeability tests, 25%, 40%, and 60% RAP are added to the AC-25C asphalt mixture to comprehensively evaluate the high-temperature performance of the recycled mixture. Based on the results and discussions presented above, the conclusions are obtained as below:

(1) Rutting tests and uniaxial penetration tests show that the addition of RAP asphalt mixture can significantly improve its high-temperature stability and shear strength, but

when the RAP content exceeds 40%, there is a significant downward trend, but it is still better than the asphalt mixture without RAP. When the milling material content exceeds 60%, the dynamic stability of the mixture does not meet the technical requirements of JTG F40–2004.

(2) Due to the presence of milling materials, the cohesive force C of hot recycled asphalt mixture increases, and the specimen still maintains a certain strength after instability. As the penetration depth continues to increase, the higher the RAP content, the longer the maintenance time of the mixture's penetration strength. Therefore, the uniaxial penetration test of hot recycled asphalt mixture is different from conventional mixture, which is divided into compaction stage, elastic stage, and damage stage. There are five stages: stress maintenance stage and complete failure stage.

(3) The problem faced by high-content hot recycled asphalt mixture is gradation stability. The milling process, secondary crushing and grading, storage, and other refined pre-treatment of the milling material are the foundation of RAP's high-value application, and ensuring the stability of the mixture gradation is the key.

From the perspective of high-temperature stability and shear strength, the optimal dosage of RAP is 40%. To ensure the grading stability of high-content RAP mixtures, it is crucial to study the RAP crushing process, which is also a prerequisite for the high-value utilization of RAP. Factory mixed hot regeneration technology can be vigorously promoted in future road maintenance.

Shortcomings and Prospects: This study only evaluates the high-temperature performance of hot recycled asphalt mixtures with different dosages. In the next step of research, the durability of the mixture can be comprehensively evaluated through its water resistance, low-temperature performance, dynamic modulus, and other properties, in order to further promote the high-value utilization of plant mixed hot recycled asphalt mixtures.

References

1. Li, H., Peng, X., Zhang, J.: Assessment and sensitivity analysis of carbon emissions during the life cycle of highways. Highw. Eng. **46**, 132–138 (2021)
2. Jie, P.: Research on energy-saving and emission reduction strategies in the context of low-carbon urban environment. Heilongjiang Environ. Bull., 114–116 (2023)
3. Chao, Z., Jia, X., Wei, B.: Research on technical indicators for classification and standardization of asphalt mixture recycled materials (RAP). Fujian Constr. Technol., 80–84 (2023)
4. Wang, L., Chen, H.: Evaluation of temperature effect performance of geothermal recycled asphalt mixture. Transp. Technol., 137–141+146 (2023)
5. Li Dingzhu; Fu Rong; Zhang Guiming; Time wins; Yang Mingsheng. Design of Mix Proportion for RAP Plant Mixed Hot Recycled Asphalt Mixture with High Content and No Regenerative Agent. Highway Transportation Technology (Application Technology Edition), 146–149 (2020)
6. Wang, B.: Optimization Design and Performance Study of Plant Mix Hot Recycled Mixtures. Shandong University, Jinan (2018)
7. Wang, D., Wang, Z., Li, N., Yuan, S., Li, J. Study on the effect of RAP Pre-treatment process on the road performance of recycled asphalt mixture. Henan Sci., 342–349 (2023)

8. Liang, B., Zhang, H., Liang, Y., Wang, X., Zheng, J. Research progress on the application of warm mixing technology in asphalt. J. Transpor. Eng., 1–24 (2023)
9. Kusam, A., et al.: Laboratory evaluation of workability and moisture susceptibility of warm-mix asphalt mixtures containing recycled asphalt pavements. J. Mater. Civ. Eng., 04016276 (2017)
10. You, Z., et al.: Preliminary Laboratory Evaluation of Methanol Foamed Warm Mix Asphalt Binders and Mixtures. J. Mater. Civ. Eng., 06017017 (2017)
11. Duan, C.: Research on key technologies for thermal regeneration of modified SMA asphalt mixture. Chongqing Jiaotong Univ. (2019)
12. Huang, T., Yin, Y., Lv, J., Li, J.: Study on the deformation resistance of warm mixed recycled modified asphalt with different RAP dosage. J. Guangdong Univ. Technol., 103–110 (2019)
13. Yang, H.: Experimental study on warm mix recycled asphalt mixture. Shandong Trans. Technol., 101–103+127 (2019)
14. Wang, J.: Research on the effect of RAP content on the road performance of recycled asphalt mixture. Trans. World., 17–19 (2023)

Research on Preparation Technology and Properties of Grouting Materials for Prefabricated Buildings

Jian Hu[1], Xiaohong Gao[2(✉)], Zengyin Wen[1], Guojiao Wen[2], and Miao Zhang[2]

[1] The Fourth Engineering Company of CCCC Second Harbor Engineering Co., Ltd.,
Anhui Wuhu 241000, China
wenzengyin@ccccltd.cn
[2] CCCC Wuhan Zhixing International Engineering Consulting Co., Ltd., Wuhan 430074, China
1137761047@qq.com

Abstract. Sleeve grouting materials are important to connect steel bars in prefabricated buildings since their quality and properties not only affect the stability, safety, and seismic capacity of buildings but also influence the operation of buildings. Quality issues in grouting materials can potentially result in joint damage and embrittlement, which may compromise the safety and service life of buildings, thereby leading to an increase in maintenance costs. Consequently, appropriate grouting materials should be carefully selected according to the operational characteristics of the buildings. This study conducted experiments to analyze the properties of grouting materials under different cement-sand ratios, with the cement-sand ratio of 1.0 selected as the base grout of grouting material. Subsequently, the orthogonal test method was employed to investigate the correlations between the properties of grouting materials and varying amounts of water reducers, expansion agents, and silica fume. Through this analysis, the optimal mixing amounts were determined as 15.7 g, 1.2 g, and 42 g for water reducers, expansion agents, and silica fume, respectivelys. Finally, based on the foregoing conclusions, a high-performance sleeve grouting material with a 3 h compressive strength of 40 MPa was successfully developed by adding sulphoaluminate and lithium carbonate. This study provides some references for the production and application of sleeve grouting materials.

Keywords: prefabricated buildings · Grouting materials · Admixtures · High performance

1 Introduction

Sleeve grouting materials, as an important part of prefabricated buildings, have been widely used with the rapid development of prefabricated buildings in recent years. Concurrently, the widespread construction of high-rise and super high-rise buildings also imposes higher requirements on the performance of grouting materials. However, the existing sleeve grouting material cannot fully meet the construction needs of high-rise

A. Bieliatynskyi et al. (Eds.): CSTTE 2023, LNCE 603, pp. 38–50, 2024.
https://doi.org/10.1007/978-981-97-5814-2_4

prefabricated buildings, and there are still certain performance defects in terms of fluidity, early strength and expansion. Therefore, it is of great significance to develop sleeve grouting material that meets the construction requirements and has good performance to improve the quality of construction, save resources, protect the environment and promote the development of prefabricated buildings.

By the present juncture, an extensive body of research has been dedicated to the investigation of sleeve grouting materials. In order to improve the performances of sleeve grouting materials at low temperatures, Lu et al. [1] studied the effects of different types of early strength agents on sleeve grouting materials; Sun et al. [2] studied the effects of five different components on the compressive strength, fluidity, and expansion rate of the sleeve grouting materials, and analyzed their microscopic appearance with an electron microscope; Li et al. [3] prepared a grouting material with low cost and good performance based on ordinary Portland cement through an orthogonal test; Hu et al. developed a cement-based grouting material [4] to improve the performance of the current grouting materials. The results show that the 28d compressive strength and 30-min fluidity of the grouting materials have been improved compared with the ordinary ones. Xiong et al. [5] designed 402 grouting material specimens in the study of the compressive strength of sleeve grouting materials, analyzing the relationship between their shape, size, and water-material ratio and the compressive strength, and establishing the conversion formula between the compressive strength of the standard specimen and the test one; Huang et al. [6] developed a sleeve grouting material with a good performance by mixing sulphoaluminate cement, Portland cement, and gypsum ternary cementitious materials, and analyzed the effects of water-binder ratio, silica fume content, and other factors on the performance of the grouting materials; When studying the composition of grouting materials for prefabricated buildings; Yazan Alrefaei et al. [7] studied the effect of capsules on the flowability, compressive strength, and porosity of grouting materials through experiments. The results showed that adding no more than 3% of capsules to grouting materials can meet the requirements of engineering for flowability and compressive strength; Kiarash Koushfar et al. [8] explore the feasibility of grouted joint sleeve connections (GSSC) using carbon fiber reinforced polymer (CFRP) and glass fiber reinforced polymer (GFRP) sheets; Xu et al. [9] considered the effects of quartz sand gradation, sand-cement ratio, water-cement ratio, admixtures and other factors on the performance of grouting materials, determining the ideal mix proportion; Yang et al. [10] analyzed the influence of water-binder ratio, sand-cement ratio, defoamer, silica fume, and expansion agent on the fluidity and compressive strength of sleeve grouting materials through experimental methods. The results indicate that the sand-cement ratio has a greater impact on the performance of grouting materials, and the addition of silica fume could improve the overall compressive strength. However, the performance of sleeve grouting material is the result of the interaction of multiple components, and the best mix proportion cannot be accurately obtained only through the analysis of a single component. Zhu Wen et al. [11] established the optimal ratio ternary composite material system of sulfonated aluminate cement, Portland cement, and gypsum through orthogonal experiments.

Consequently, based on the existing research results, a progressive test is adopted in this study to analyze the influence of cement sand ratio, water reducer, silica fume, sulphoaluminate, and lithium carbonate content on the performance of sleeve grouting materials. The best admixture content obtained from the former group of tests is used as the initial mix proportion of the next group of tests. The orthogonal test method can determine the influence of different materials on the fluidity, compressive strength and vertical expansion rate of sleeve grouting material. By employing this method, the optimal mixture ratio of the grouting materials can be continuously corrected, and the grouting material with excellent properties can be found. This study introduces novel ideas and methods for the preparation of new sleeve grouting materials for prefabricated construction.

2 Properties and Composition of Sleeve Grouting Materials

2.1 Technical Index of Sleeve Grouting Materials

The properties of sleeve grouting materials for steel bar connection have a great influence on the connection effect of prefabricated components. Reasonable preparation and excellent performance of grouting materials can not only ensure the stability and reliability of the connection, but also ensure the safety of the whole structure of prefabricated buildings, and at the same time, it can also assist in reducing the difficulty of construction. The technical requirements for the property indexes of sleeve grouting material are specified in Sleeve Grouting Materials for Steel Bar Connections [12], mainly including fluidity, compressive strength, and vertical expansion rate. The specific index requirements are shown in Table 1.

Table 1. Technical indexes of grouting material for steel bar sleeve connection.

Test items		Property indexes
Fluidity (mm)	Initial	≥ 300
	30 min	≥ 260
Compressive strength (MPa)	1 d	≥ 35
	3 d	≥ 60
	28 d	≥ 85
Vertical expansion rate (%)	3 h	0.02–0.2
	Difference between 24 h and 3 h	0.02–0.4

2.2 Selection of the Components of Grouting Materials

The different components and contents of the sleeve grouting materials directly affect their properties. Therefore, it becomes crucial to select raw materials that have a positive impact on the performance of the grouting materials while being easily accessible and preparable. Traditional grouting materials are mainly composed of cementitious materials, aggregates, and some admixtures, among which, the common admixtures are water reducers, expansion agents, and defoamers. Considering the strength of quartz sand is better than that of river sand, the former is used as a fine aggregate in the experimental preparation process. In order to address the problem of bleeding and segregation caused by water reducers, cellulose ether is added to increase viscosity and thickening.

3   Study on the Preparation of Base Grout

The main components of the grouting materials are cementitious materials and sand, and the base grout with excellent performance directly affects the overall characteristics of the grouting material, so it is necessary to determine the appropriate cement-sand ratio before the test, and on this basis, to study the effects of admixtures and their contents on the performance of grouting materials by adding different contents of admixtures.

3.1   Mix Proportion Design of Base Grout of Grouting Materials

When preparing the base grout of the grouting material, the content of the admixture shall be kept unchanged, and the change law of the base grout performance with the cement-sand ratio shall be studied by constantly adjusting the proportion of cementitious materials and sand, and then the appropriate cement-sand ratio shall be selected. The design scheme for preparing the base grout in this experiment is shown in Table 2 below.

Table 2. The design scheme of base slurry preparation

Mix ratio number	1	2	3	4	5
Water-binder ratio	0.28	0.28	0.28	0.28	0.28
Cement-sand ratio	0.7	0.8	0.9	1.0	1.2
Water reducer agent/g	10.6	11.2	11.8	12.6	14.1
Cement/g	804	846	895	950	1071
Fly ash/g	50	55	60	65	70
Silica fume/g	35	35	35	35	35
Quartz sand/g	1270	1170	1100	1050	980
Expansion agent/g	0.8	0.8	0.8	0.8	0.8
Cellulose ether/g	0.05	0.05	0.05	0.05	0.05
Defoamer	1	1	1	1	1

3.2   Properties of Base Grout of Grouting Materials

The fluidity, compressive strength, and expansion rate of the sleeve grouting materials obtained through the test are shown in Fig. 1 according to the above base grout preparation and design method.

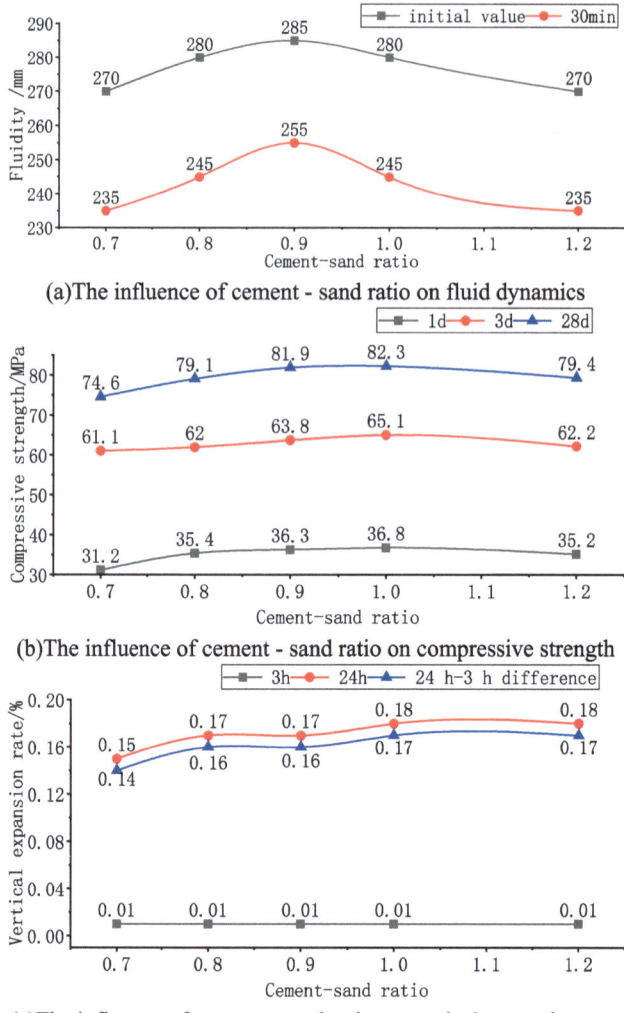

(a)The influence of cement - sand ratio on fluid dynamics

(b)The influence of cement - sand ratio on compressive strength

(c)The influence of cement - sand ratio on vertical expansion rate

Fig. 1. Test value of basic slurry performance of sleeve grouting materials

The figure above demonstrates a similarity between the changes in fluidity and compressive strength of the base grout in sleeve grouting materials with respect to the cement-sand ratio. Specifically, both properties exhibit a pattern of initial increase followed by subsequent decrease. However, from the experimental results, the fluidity reaches the peak value when the cement-sand ratio is about 0.9, while the compressive strength reaches the peak value when the cement-sand ratio is about 1.0. The 3 h vertical expansion rate increases with the cement-sand ratio, but the change range is limited. Overall, the fluidity and vertical expansion rate of the base grout cannot meet the requirements of the specification, and the 1 d compressive strength under the conditions of 28 d and Mix ratio 1 is also lower than the strength requirements. Therefore, considering the three indicators, the cement-sand ratio of 1.0 is selected as the mix proportion of the base grout.

4 Study on Preparation Technology of Sleeve Grouting Materials

4.1 Influence of the Content of Water Reducers on Sleeve Grouting Materials

Water reducer can be used as an important additive in sleeve grouting materials because it can enhance the hydration reaction of cement, and improve the fluidity and workability of cement stone materials.

Based on the mixing scheme of the base grout of sleeve grouting materials under the condition of the cement-sand ratio of 1.0, the variation law of the performance of sleeve grouting materials with the content of water reducer is studied by changing the percentage of water reducer in the total mass of cementitious materials. In this test, the dosages of water reducer are 10.5 g, 13.6 g, 15.7 g, 17.8 g, and 20 g respectively, accounting for 1.0%, 1.3%, 1.5%, 1.7%, and 1.9%. The experimental results are shown in Fig. 2 below.

The experimental results reveal that the effect of the water reducer on the vertical expansion rate of grouting material is not obvious, which cannot meet the requirements of the specification but can improve its fluidity and compressive strength. As the content of water reducer increases, the initial values of fluidity and compressive strength exhibit an initial ascent followed by a subsequent decline. The peak values are attained at approximately 1.5% water reducer content. Meanwhile, under this condition, the vertical expansion rate of 3 h and 24 h is also the maximum. Therefore, based on the above conditions, the optimal dosage of water reducer in this test is 15.7 g.

4.2 Influence of Expansion Agent Content on Sleeve Grouting Material

Based on the test results of the water reducer, the influence of the content of the expansion agent on the fluidity, compressive strength, and vertical expansion rate of sleeve grouting materials can be studied by adjusting the content of the expansion agent. During the test, the contents of other components remain unchanged, and the contents of the expansion agent are set to 0.4 g, 0.8 g, 1.2 g, 1.6 g, and 2.0 g. The test results are shown in Fig. 3 below.

It can be seen from the above figure that the expansion agent can remarkably improve the performance of the sleeve grouting materials. The vertical expansion rates of 3 h and 24 h demonstrate a significant increase as the content of the expansion agent increases. However, when the content of the expansion agent increases to 1.6 g, the difference between the vertical expansion rate of 24 h and 3 h exceeds the technical specification and does not meet the requirements. The fluidity and compressive strength indicators meet the technical requirements, and the overall fluctuation range is small. Therefore, the above data reveals that the dosage of the expansion agent is 1.2 g, that is, the dosage of 0.114% is the best.

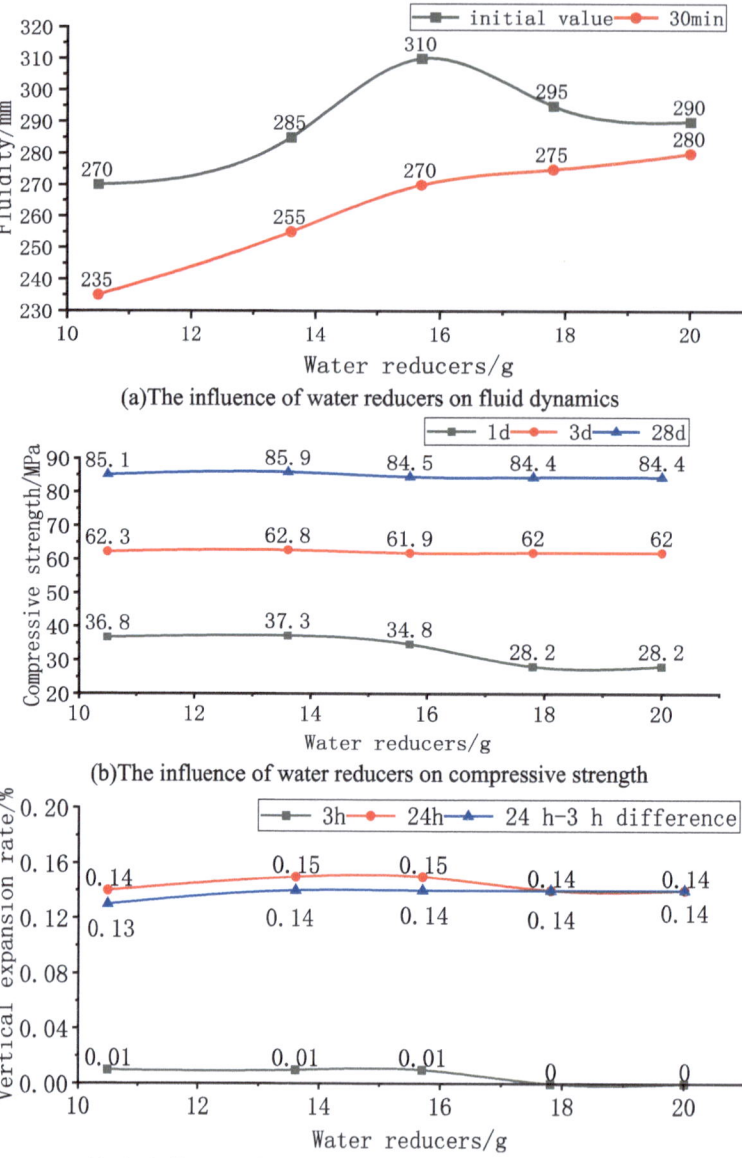

Fig. 2. Effect of the content of water reducer on properties of sleeve grouting materials

4.3 Influence of Silica Fume Content on Sleeve Grouting Material

Given the lightweight nature, high fineness, and low porosity of silica fume, the addition of a suitable quantity of this material to the sleeve grouting mixture is advantageous for enhancing both the overall strength and compactness of the slurry. Moreover, the active substances on the surface of the silica fume will make the electrostatic repulsion

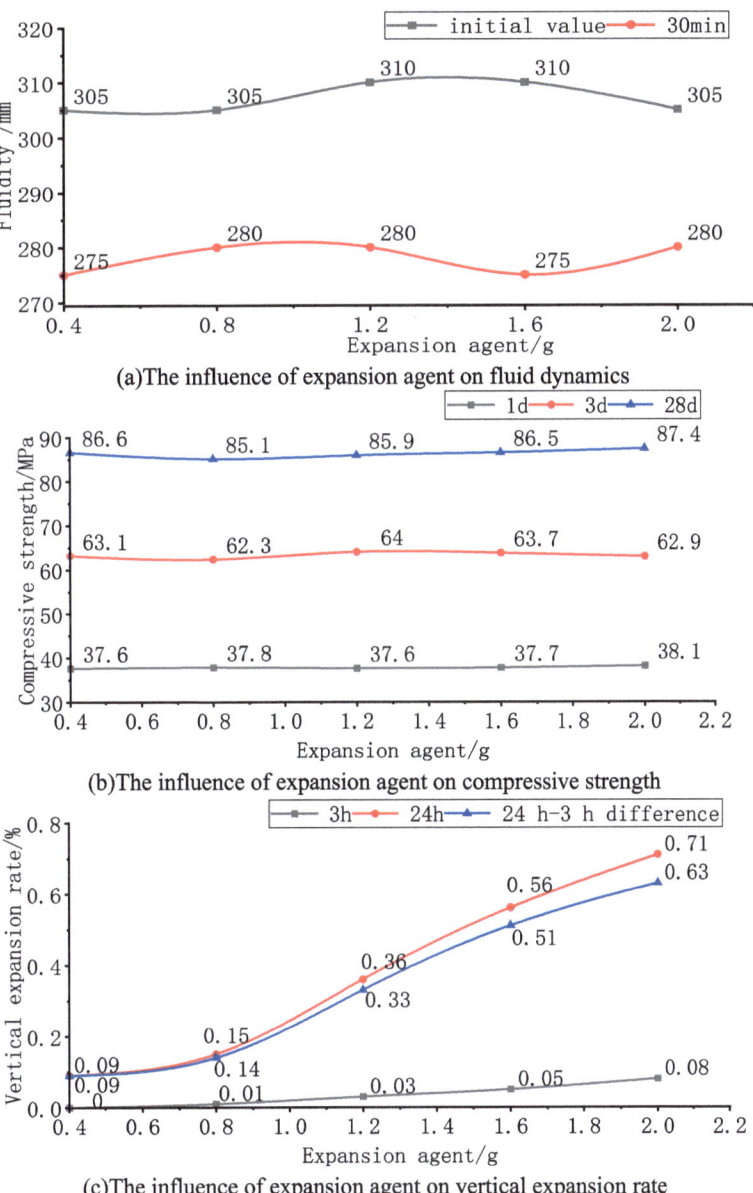

(a)The influence of expansion agent on fluid dynamics

(b)The influence of expansion agent on compressive strength

(c)The influence of expansion agent on vertical expansion rate

Fig. 3. Influence of expansion agent content on the performance of sleeve grouting materials

between particles greater than the cohesion, which can disperse cement particles, play a role in lubrication, and then improve the fluidity of slurry. In this test, the content of the expansion agent is kept at 1.2 g, and the content of the silica fume was set at 0 g, 21 g, 42 g, 63 g, and 84 g respectively, to explore the change in the properties of grouting materials. The experimental results are in Fig. 4 below.

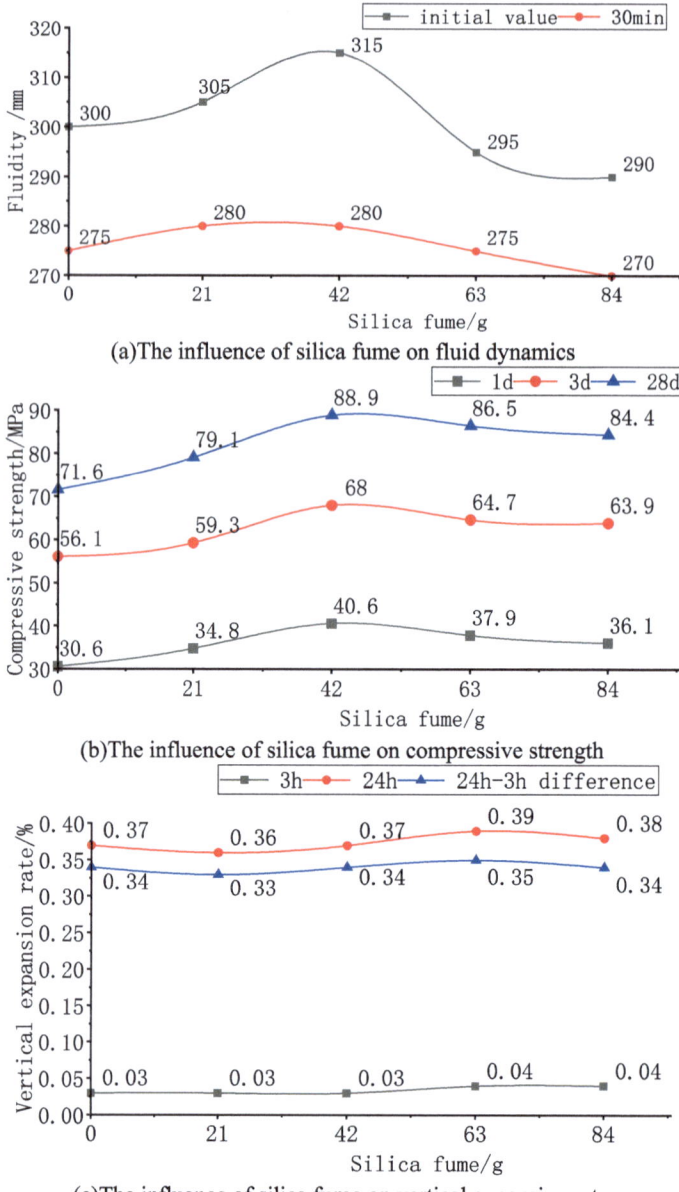

(a)The influence of silica fume on fluid dynamics

(b)The influence of silica fume on compressive strength

(c)The influence of silica fume on vertical expansion rate

Fig. 4. Effect of the content of silica fume on properties of sleeve grouting materials

The results show that with the increase of the content of silica fume, the fluidity and compressive strength of the sleeve grouting material increase first and then decrease, and reach the maximum when the silica fume content is about 42g. It is also found that the fluidity is sensitive to the content of silica fume. When the content of silica fume is low, the fluidity of the slurry can be improved by increasing content, which is related to the characteristics of the surface active substances of silica fume. However, with the further increase of the silica fume content, the fluidity of the slurry will be significantly reduced. Through this experiment, the optimum silica fume content is determined to be 42 g.

4.4 Effect of Sulphoaluminate and Lithium Carbonate Content on Sleeve Grouting Material

Based on the above experiments, the sleeve grouting material can be obtained when the cement-sand ratio is 1.0, the content of the water reducer is 15.7 g, the content of the expansion agent is 1.2 g and the content of silica fume is 42 g. Moreover, the fluidity, compressive strength, and vertical expansion rate of the prepared slurry can meet the technical requirements. To further develop new materials with better performance, based on the above basic experimental ratio, a better sleeve grouting material is obtained by mixing ordinary Portland cement with sulphoaluminate cement and using lithium carbonate as an early strength agent. The ratio of sulphoaluminate to lithium carbonate and the corresponding experimental results are shown in Table 3 and Fig. 5, respectively.

Table 3. Experimental proportioning of high-performance sleeve grouting material.

Mix ratio number	1	2	3	4	5
Silicate cement/%	80	90	90	90	90
Sulphoaluminate cement/%	20	10	10	10	10
Percentage of lithium carbonate in sulphoaluminate cement/%	0	0	0.03	0.04	0.05

Based on the experimental results of the first and second groups, when the sulphoaluminate cement is directly mixed with ordinary Portland cement without adding an early strength agent, the fluidity of the grouting material will be reduced. Moreover, the reduction will be more obvious with the increased sulphoaluminate content while the early strength demonstrates an improvement. Therefore, based on the experimental ratio of the second group, the performance of the grouting material can be further optimized by adding the early strength agent lithium carbonate.

It can be seen from the test results of the third, fourth, and fifth groups that the fluidity of the grouting material gradually decreases with the increased content of lithium carbonate. Moreover, the fluidity of the fifth group does not meet the requirements of Sleeve Grouting Materials for Steel Bar Connection (JG/T408–2019). In contrast, the fluidity of the third group satisfies the requirements, but its 3 h compressive strength fails to exceed 40 MPa. Generally speaking, the compressive strength is improved after adding an early strength agent, but the change of early strength tends to be stable when the content of the early strength agent is more than 0.04%. Based on the above analysis, the best proportion of sulphoaluminate and lithium carbonate can be determined as the fourth group.

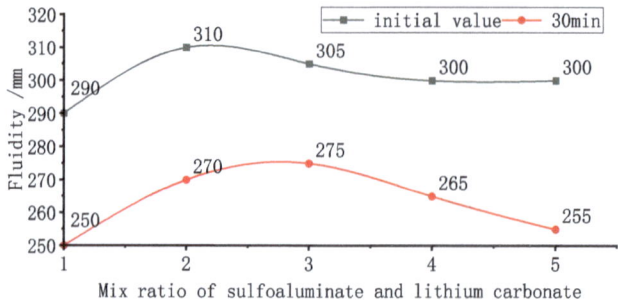

(a)The influence of sulfoaluminate and lithium carbonate on fluid dynamics

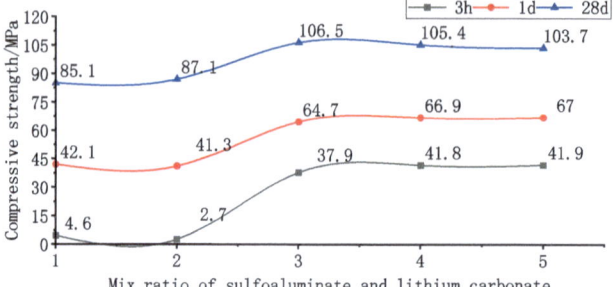

(b)The influence of sulfoaluminate and lithium carbonate on compressive strength

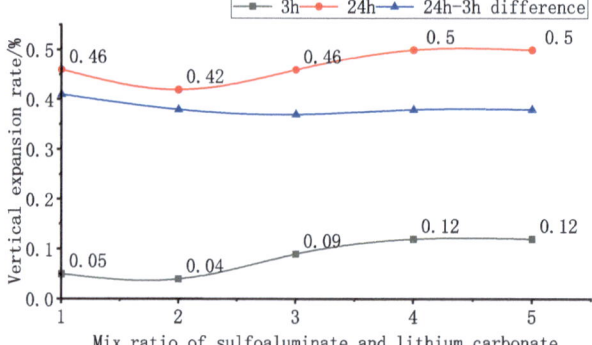

(c)The influence of sulfoaluminate and lithium carbonate on vertical expansion rate

Fig. 5. Effect of the content of sulphoaluminate and lithium carbonate on sleeve grouting material

5 Conclusion

In this paper, the effects of cement-sand ratio, superplasticizer, silica fume, sulfoaluminate and lithium carbonate contents on the fluidity, compressive strength and vertical expansion rate of sleeve grouting material are analyzed sequentially by using the orthogonal test method, which can not only improve the performance of grouting material while reducing the amount of material, but also save costs, and provide a reference for the preparation of high-performance grouting material, which is of great significance for promoting the development of prefabricated buildings. Based on the experimental results and the research of this paper, the following conclusions can be drawn:

1. With the increase of the cement-sand ratio, the fluidity and strength of the sleeve grouting materials increase first and then decrease, and the corresponding cement-sand ratio is close when they reach their peak values; water reducer can play a role in improving the compressive strength and fluidity of the sleeve grouting materials, but has little effect on the expansion. Notably, the dosage of the water reducer should be within the appropriate range, and excessive addition of water reducer will lead to a decrease in compressive strength and poor fluidity of grout. The optimal cement-sand ratio is 1.0, and the dosage of water reducer is 1.5%.

2. The expansion agent can improve the compressive strength, fluidity, and expansibility of the sleeve grouting material to a certain extent. However, when its content exceeds 0.114%, the properties of the grouting material no longer meet the requirements of the standard specification. Additionally, within a certain range of content, silica fume plays a role in lubricating the sleeve grouting material, and the compressive strength reaches a peak value when the silica fume content is 4%.

3. Both lithium carbonate and sulphoaluminate cement can accelerate the reaction rate of ordinary Portland cement and improve its early strength. Based on ordinary sleeve grouting materials, 10% of sulphoaluminate is added, and the content of lithium carbonate in sulphoaluminate is set to 0.04%. Based on the above ratio, the high-performance sleeve grouting material with a 3 h compressive strength of more than 40 MPa is obtained.

References

1. Yang. Y., Lu. X.F. Liu J.T.: Experimental study on properties of low temperature sleeve grouting material. New Build. Mater. **47**(10), 49–52 (2020). https://kns.cnki.net/kcms/detail/detail.aspx?Dbname=CJFD2020&filename=XXJZ202010012&dbcode=CJFD
2. Sun. X.W., et al.: Study on the influence of composition factors on the properties of sleeve grouting material for steel bar connection. Concrete. (07), 142–146 (2021). 150 https://kns.cnki.net/kcms/detail/detail.aspx?dbname=CJFD2021&filename=HLTF202107039&dbcode=CJFD
3. Li. X.Y., Mi. Y., Ding. H.: Study on sleeve grouting material for steel bar connection in prefabricated building. New Build. Mater. **49**(12), 75–78 (2022). 150https://kns.cnki.net/kcms/detail/detail.aspx?Dbname=CJFD2022&filename=XXJZ202212016&dbcode=CJFD
4. Hu. Z.G., Li. W., Wang. C.J., Song. T.W., Ding. X.P., Han. Y.D.: Comparative study on related properties of cementitious grout and reinforcement sleeve grouting material. Ind. Constr. **52**(08), 201–207 (2022). https://kns.cnki.net/kcms/detail/detail.aspx?dbname=DKFX2022&filename=GYJZ202208030&dbcode=DKFX.10.13204/j.gyjzg22030604
5. Xiong. Y., Li. J.H., Sun. B., Mao. S.Y.: Strength and influencing factors of sleeve grouting material for prefabricated building. J. Build. Mater. **22**(02), 272–277 (2019). https://kns.cnki.net/kcms/detail/detail.aspx?Dbname=CJFD2019&filename=JZCX201902018&dbcode=CJFD
6. Huang. S.M., Zhu. W., Wang. Y.Y.: Study on preparation and properties of high-performance reinforcement sleeve grouting material for prefabricated concrete structures. New Build. Mater. **46**(06), 10–13 (2019). https://kns.cnki.net/kcms/detail/detail.aspx?dbname=CJFD2019&filename=XXJZ201906003&dbcode=CJFD

7. Liu, Y.L., Wang, Y.S., Fang, G., Alrefaei, Y., Dong, B., Xing, F.: A preliminary study on capsule-based self-healing grouting materials for grouted splice sleeve connection. Constr. Build. Mater. **170**, 418–423 (2018). https://doi.org/10.1016/j.conbuildmat.2018.03.088

8. Koushfar, K., et al.: Behavior of grouted splice sleeve connection using FRP sheet. Eng. Struct. **296**, 116898 (2023)

9. Xu, C.W., Zeng, W., Ma, S.F., Chen, Y.: Study on composition of grouting material for prefabricated building. Concrete **07**, 133–137 (2016)

10. Yang. H.M., Ye. Q.J., Hu. M.M., Wu. C.P., Sun. X.W.: Study on preparation and properties of sleeve grouting material for steel bar connection. Concrete. (12), 177–180 (2019). https://kns.cnki.net/kcms/detail/detail.aspx?Dbname=CJFD2019&filename=HLTF201912048&dbcode=CJFD

11. Zhu, W., Huang, S., Zhang, S., Wu, J., Luo, K., Tang, M.: Study on the preparation of cementitious grout for grout sleeve splicing of rebars based on a ternary composite material system. Adv. Eng. Technol. Res. **1**(1), 80 (2022)

12. Jiang. Q.J., Wu. H.J., Qian. G.L., Zhang. Y.M., Sha. J.F., Li. C.G.: Introduction to revision and special test research of JG/T 408—2013 sleeve grouting material for steel bar connection. Concr. Cem. Prod. (11), 78–81 (2019). https://kns.cnki.net/kcms/detail/detail.aspx?Dbname=CJFD2019&filename=HNTW201911020&dbcode=CJFD.10.19761/j.1000-4637.2019.11.078.04

Study on the Optimization of Proportioning and Mechanism of Polymer Composite Repair Mortar Based on Dual Objectives of Shrinkage and Mechanical Performance

Kai Li[1], Gang Zhao[2], Ming Jiang[2], Kuixiang Guo[2], Yu Cui[2], and Jinsong Wang[1(✉)]

[1] School of Civil Engineering, South China University, Hengyang 421001, China
xhwjs@163.com
[2] Shanghai Urban Construction Design and Research Institute (Group) Co., Ltd.,
Shanghai 200000, China
{zhaogang,jiangming,guokuixiang,cuiyu}@sucdri.com

Abstract. This study aims to optimize the proportion of polymer composite repair mortar, with the objectives of enhancing its mechanical properties and reducing shrinkage, to ensure its application value in non-excavation repair construction of underground sewage pipelines. By taking the dosage of Hydroxypropyl Methyl Cellulose (HPMC), Cement-based Crystalline Capillary Waterproofing (CCCW), and Polypropylene Fiber (PPF) as key influencing factors, the Response Surface Methodology (RSM) was employed to design and optimize the proportions for 28-day compressive strength and 7-day shrinkage. The study not only analyzed the impact of the interaction of these factors on the material properties but also conducted microstructure analysis through SEM and XRD to reveal the reinforcement mechanism. The optimization results indicate that the optimal dosages of HPMC, CCCW, and PPF are 0.306%, 0.423%, respectively, achieving the predicted optimal performance, i.e., 28-day compressive strength of 73.6 MPa and 7-day shrinkage of 133.98 μm, with experimental verification errors of 2.3% and 4.5%, respectively. The achievements of this study provide important theoretical basis and practical guidance for developing high-performance, low-shrinkage repair materials, demonstrating the integration of traditional civil engineering practices with modern scientific methodologies, and contributing to the field of construction materials.

Keywords: Geopolymer · Response Surface Methodology · Hydroxypropyl Methylcellulose · Cement-Based Crystalline Capillary Waterproofing · Polypropylene Fiber

1   Introduction

With the rapid development and accelerated urbanization process in our country, the volume of urban sewage treatment and collection has continued to increase. As of 2022, the total length of urban drainage pipelines has exceeded 914,000 kilometers [1]. However, approximately 30% of these pipelines were built before the year 2000, mainly

A. Bieliatynskyi et al. (Eds.): CSTTE 2023, LNCE 603, pp. 51–61, 2024.
https://doi.org/10.1007/978-981-97-5814-2_5

serving as urban sewage mains, and are now nearing the end of their service life. These pipelines suffer from defects in materials and structures, compounded by the erosion of sewage, geological factors, and environmental influences, leading to aging and leakage issues. Consequently, the actual sewage treatment rate is only between 40% to 60%, with approximately 40% of sewage being discharged untreated directly into natural water bodies [2]. This underscores the urgency of pipeline repair, necessitating breakthrough research.

Traditional methods of pipeline network repair mainly involve excavation and replacement, which not only affect traffic and the environment but also incur high costs and construction difficulties. In contrast, non-excavation repair techniques address defects through inspection well entrances, reducing their impact on urban traffic and the environment, while being cost-effective and safe [3]. Currently, non-excavation repairs primarily use cement-based materials, but their high energy consumption and susceptibility to corrosion limit their application scope. With the increasing awareness of environmental protection, there is an urgent need to seek environmentally friendly and low-carbon alternative materials [4]. Polymer-based composite mortar, due to its excellent performance, abundant raw materials, and potential for utilizing solid waste, has become a research hotspot.

Due to the harsh working environment of sewage pipelines, higher standards are demanded for the mechanical performance and durability of the repair materials used on them. Additionally, it is difficult to achieve basic maintenance conditions within sewage pipelines, which necessitates repair materials with characteristics of high early strength and low shrinkage rate. Research has shown that modifying polymer mortar with fibers and cellulose ethers can effectively enhance its strength and durability while reducing shrinkage rate [5].

2 Experiment

2.1 Raw Materials

Kaolin (MK), obtained from Shanxi Chaopai Calcined Kaolin Co., Ltd., carries the model number K1100 and possesses a fineness of 800 mesh. The cementitious crystalline waterproofing material (CCCW), a gray powdered Xypex admixture, is supplied by Beijing Chengrong Waterproofing Materials Co., Ltd. As shown in Fig. 1 and Table 1, X-ray diffraction (XRD) and X-ray fluorescence (XRF) analyses indicate kaolin mainly comprises amorphous Al_2O_3 and SiO_2, plus minor TiO_2. CCCW contains calcium and silicon predominantly as tricalcium silicate, along with MgO and traces of $Mg(OH)_2$ due to moisture absorption.

To ensure the stability and reproducibility of multiple experiments, standard sand from Xiamen ISO Standard Sand Co., Ltd. Was selected as the aggregate. The particle size range is 0.08 mm to 2 mm, with a silica content of over 98%, and a mud content of less than or equal to 0.18%.

The alkali activator used is a laboratory-prepared water glass solution, combining previous laboratory research findings, utilizing commercial liquid sodium silicate solution, adjusted with deionized water and NaOH to a modulus of 1.2, and a concentration of 43.5%.

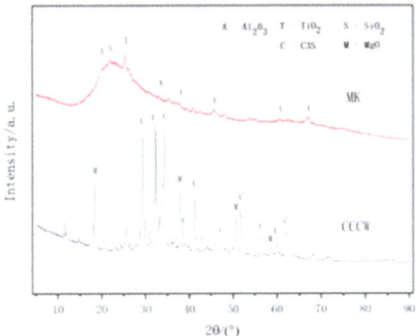

Fig. 1. XRD spectra of MK and CCCW

Table 1. Main chemical composition of Metakaolin and CCCW

Material	CaO	SiO2	Al2O3	TiO2	MgO	Na2O
Metakaolin	0.084	45.788	51.687	1.276	0.173	0.116
CCCW	52.925	15.876	4.925	0.814	15.366	4.146

Hydroxypropyl methylcellulose (HPMC), a common water-retaining agent in building mortar, can slow down the rate of water loss in mortar and reduce the likelihood of drying cracks. It is purchased from Hebei Weiteng Building Materials Co., Ltd., with a molecular weight (degree of polymerization) of 200,000, a fineness of 80 to 100 mesh, and a density of 1.26 to $1.31 g/cm^3$.

For use in non-excavation repair spraying operations of pipelines, the goal was to find a flexible and finer diameter fiber. Polypropylene fiber (PPF) meets these requirements. It is produced by Shanghai Chenqi Chemical Technology Co., Ltd., with an industrial grade polypropylene fiber (PP), length of 6mm, fiber diameter of $31\mu m$, and a Young's modulus of ≥ 3.5 Gpa.Experimental design.

2.2   Response Surface Method

The experimental design adopted in this study is Response Surface Methodology (RSM), which is implemented using the Design-Expert software as a platform, employing the Box-Behnken design method (abbreviated as BBD) within the software for experimental design.

The HPMC dosage, CCCW dosage, and PFF dosage are selected as independent variables, represented respectively as A1, A2, and A3. The compressive strength at 28 days and 7-day shrinkage of the mortar are the response values, denoted as Y1 and Y2, respectively. A three-factor three-level experiment is conducted, with each factor corresponding to high, medium, and low levels encoded as + 1, 0, and -1. Specifically, -1 represents the low level, 0 represents the center point, and + 1 represents the high level. The specific values are shown in Table 2.

Table 2. Coding and level of independent variables

Code value	Factor	Code lever (%)		
		-1	0	1
A1	HPMC	0.2	0.4	0.6
A2	CCCW	0.6	0.8	1
A3	PPF	0.4	0.6	0.8

3 Result

3.1 Experimental Plan and Results

According to the design of Box-Behnken experiment scheme, there are 17 groups of experiment points, each group contains 3 shrinkage tests and 6 compressive tests. The average value of experimental data is taken as the experimental result. After the experiment, the measured data of compressive and compressive values of specimens are obtained, as shown in Table 3.

3.2 Analysis of Variance

Utilizing DesignExpert software, regression model variance analyses were conducted for the 28-day compressive strength, 28-day flexural strength, and 7-day shrinkage values, with the analysis results shown in Table 4.

In response surface methodology, the significance testing of parameters is a crucial step to evaluate their impact on the response variable. The p-value serves as a key metric in this assessment, where values greater than 0.05 are considered not significant. On the other hand, the Lack of Fit test assesses whether the model adequately accounts for the random errors in the data. A larger p-value in this test indicates a lack of significance, suggesting a lower probability of errors in the model.

From Table 4, it's observed that for the 28-day compressive strength and shrinkage values model, all p-values are below 0.0001, signifying a highly significant model with excellent fitting accuracy. Moreover, the Lack of Fit test's p-value exceeds 0.05, indicating that the lack of fit is not significant, hence the model is less likely to contain errors. According to Table 4, within the compressive strength model, the factors HPMC, CCCW, and PPF all show p-values less than 0.05, suggesting a significant influence on compressive strength. The order of impact on compressive strength from these factors is CCCW > HPMC > PPF. In the model for 7-day shrinkage values, the p-values for HPMC and PPF are below 0.0001, demonstrating their substantial effect on the mortar's 7-day shrinkage. Meanwhile, CCCW's p-value is 0.0483, closely approaching the threshold of non-significance at 0.05, indicating a minimal impact on 7-day shrinkage of mortar. The ranking of these factors' effects on 7-day shrinkage of mortar is HPMC > PPF > CCCW.

Table 3. Response Surface Experimental Design Plan and Results

Std	Factor (%)			Result	
	HPMC	CCCW	PPF	28d Compressive strength (Mpa)	7d Shrinkage value (μm)
1	0.2	0.6	0.6	75.08	188
2	0.6	0.6	0.6	74.33	206
3	0.2	1	0.6	73.38	236
4	0.6	1	0.6	70.62	155
5	0.2	0.8	0.4	74.1	171
6	0.6	0.8	0.4	71.57	155
7	0.2	0.8	0.8	74.58	170
8	0.6	0.8	0.8	72.88	116
9	0.4	0.6	0.4	71.03	147
10	0.4	1	0.4	69.73	132
11	0.4	0.6	0.8	73.35	118
12	0.4	1	0.8	70.77	114
13	0.4	0.8	0.6	74.62	129
14	0.4	0.8	0.6	74.98	130
15	0.4	0.8	0.6	75.12	126
16	0.4	0.8	0.6	75.35	123
17	0.4	0.6	0.8	75.17	127

28d Compressive strength.

Based on the variance analysis of the 28-day compressive strength, only the P-value of the interaction term AB, with $0.0303 < 0.05$, indicates that the interaction between HPMC and CCCW has a significant effect on the strength of the mortar. As shown in Fig. 2(a), at a mid-range dosage of PPF (0.6%), the compressive strength of the mortar initially increases and then decreases with the increase in CCCW dosage. This phenomenon is due to the unique action mechanism of the penetrating crystalline material: in the alkaline environment within the polymer mortar, the active substances in CCCW react in the mortar's micropores and cracks to form C-S-H gel and generate a large number of ettringite (Aft) crystals. These crystals interconnect to form a three-dimensional network structure, reinforcing and filling the voids[6]. With the increase in CCCW dosage, the excessive generation of Aft will instead cause damage to the N-(C)-S-H gel, leading to microcracks. When the damage effect outweighs the filling reinforcement effect, it results in an increase in internal defects and a decrease in compressive strength [6]. Meanwhile, with the increase in HPMC dosage, the compressive strength shows a gradual decrease trend. This is because HPMC reduces the surface tension of the solution, has a strong air-entraining effect, increases the porosity of the mortar dramatically, reduces compactness, and consequently decreases compressive strength [7].

Table 4. Analysis of variance table

Source	Y1		Y2	
	F-value	p-value	F-value	p-value
Model	44.96	< 0.0001	201.9	< 0.0001
A-HPMC	54.34	0.0002	208.46	< 0.0001
B-CCCW	78.29	< 0.0001	5.7	0.0483
C-PPF	24.06	0.0017	89.2	< 0.0001
AB	7.33	0.0303	231	< 0.0001
AC	1.25	0.3005	34.03	0.0006
BC	2.97	0.1284	2.85	0.1351
A^2	1.03	0.3442	886.22	< 0.0001
B^2	107.88	< 0.0001	192.12	< 0.0001
C^2	116.07	< 0.0001	179.25	< 0.0001
Lack of Fit	2.97	0.1604	1.97	0.2611

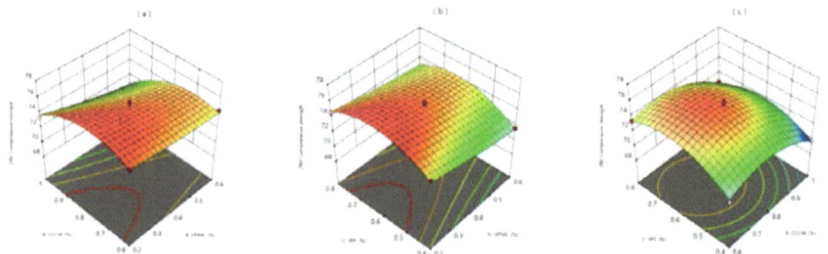

Fig. 2. The Effect of HPMC, CCCW, and PPF on 3d and 7d Strength

Figure 2 (b) and (c) reveal that with the increase in PPF, the strength of the mortar also shows an initial increase followed by a decrease trend. This is because PPF acts as a bridging agent during hydration, restricting the occurrence and propagation of cracks in the interfacial transition zone during loading. Moreover, the dense PPF fibers form a grid structure, creating a skeleton. After filling with N-(C)-S-H gel, they enhance the continuity of the material, reduce internal defects, and thus increase compressive strength [8]. However, excessive addition of polypropylene fibers leads to noticeable aggregation during the mixing stage, resulting in a large number of macropores and interconnected pores during the hardening stage, increasing the number of defects in the mortar and reducing compressive strength.

7d Shrinkage Value

According to the analysis of variance in Table 4, the P-value of the AB interaction term is less than 0.0001, the P-value of the AC interaction term is $0.0006 < 0.05$, indicating significance, while the P-value of the BC term is $0.1351 > 0.05$, indicating insignificance.

This proves that the AB and AC interactions have a significant impact on the 7-day shrinkage of the mortar. Analyzing the response surface plot of the AB interaction, Fig. 3 it can be observed that when the PPF dosage is at the middle level (0.6%), the response surface plot exhibits a cloak-shaped pattern with lower values in the middle and higher values around the edges. As the HPMC dosage increases, the shrinkage initially decreases and then increases. This is because the appropriate amount of HPMC forms a three-dimensional network structure and membrane structure inside the mortar, acting as both a skeleton and a filler, reducing the shrinkage of the mortar [9]. With increasing HPMC dosage, the internal pores of the mortar increase, leading to decreased structural stability and increased drying shrinkage. Similarly, with an increase in CCCW dosage, there is a trend of initially decreasing and then increasing 7-day shrinkage values. This is because the active substances in CCCW, such as C3S and MgO, generate expansive forces outwardly when forming ettringite and magnesium aluminate hydrate in the pores [10], providing compensation during the contraction period of the mortar and thus reducing shrinkage values.

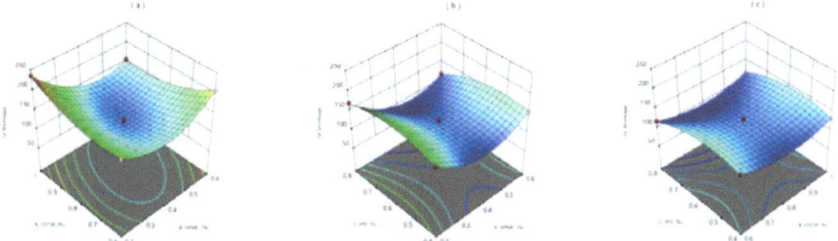

Fig. 3. Effects of HPMC, CCCW, and PPF on 7d shrinkage values

When CCCW is held constant at the middle level (0.8%), the response surface plot shows a noticeable downward slope with an increase in PPF dosage. This is because the addition of fibers forms a three-dimensional randomly distributed network support system in the mortar, improving the homogeneity and integrity of the mortar and reducing the occurrence of inherent cracks [11].

3.3 Best Mix Proportion

Using Design Expert software, optimization of the response surface model can be conducted. After optimization, it is determined that the optimal ratio exists for the effects of HPMC, CCCW, and PPF on the compressive strength and drying shrinkage of geopolymer repair mortar. The optimal ratios are when the HPMC content is 0.306%, CCCW content is 0.764%, and PPF content is 0.423%. Predicted values for compressive strength at 28 days and shrinkage at 7 days are 73.6 MPa and 133.98 μm, respectively. Experimental verification yielded measured values of 71.9 MPa for compressive strength at 28 days and 128 μm for shrinkage at 7 days. The errors were 2.3% and 4.5%, respectively, indicating that the use of response surface methodology results in higher accuracy in both optimization of the mix proportion and prediction.

4   Micro Analysis

4.1   Scanning Electron Microscopy (SEM) Analysis

Through scanning electron microscopy (SEM), it can be observed in Fig. 4(a) that the N(C)-A-S-H gel tightly wraps around the fibers, forming a skeleton-filling structure. This structure enhances the cohesion of the mortar, increases its overall integrity, and reduces the probability of defect formation. In Fig. 4(b), cracks can be seen in the mortar under stress or drying conditions. The addition of PPF effectively relieves some of the stress and restricts the formation and propagation of cracks.

However, the excessive addition of fibers can lead to some adverse effects. As shown in Fig. 4(c), the aggregation of fibers within the mortar is evident, with long and slender fibers intertwining, resulting in voids and defects. Additionally, the mixing process can reduce the flowability of the mortar, and the bubbles within the mortar are less likely to be broken during vibration, leading to an increase in porosity and a decrease in strength. The decrease in structural stability also affects the overall shrinkage of the mortar.

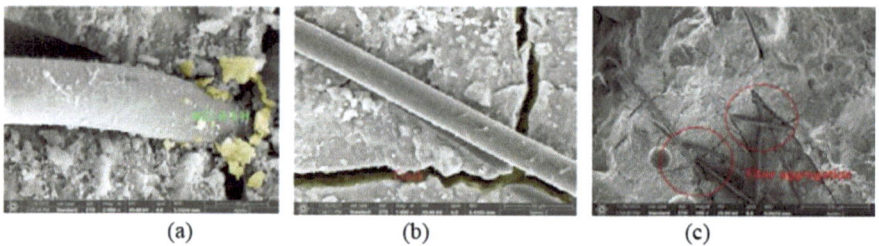

(a)   (b)   (c)

Fig. 4. SEM images of fiber doped samples

4.2   X-Ray Diffraction (XRD) Analysis

Figure 5 presents the XRD spectra of polymer mortar prepared with varying proportions of HPMC and CCCW compared to the blank polymer mortar. From the graph, a significant broad peak is observed in the diffraction angle range of 25° to 30°, characteristic of the amorphous phase of N(C)-A-S-H gel[12]. It is clearly observed that as the HPMC content increases, the diffraction peak around 22° originating from the amorphous $SiO2$ characteristic peak of the original kaolinite gradually disappears, while the broad peak of N(C)-A-S-H gel becomes more prominent. This indicates that the addition of HPMC promotes the formation of more N(C)-A-S-H gel in the polymer mortar. This is because water in the polymer does not directly react but rather provides a medium and space for the reaction of the polymer [13]. The addition of HPMC enhances the water retention of the mortar, delaying the rate of water loss, allowing more sufficient reaction time between the polymer precursors and activators. However, this does not lead to an increase in mechanical strength of the polymer mortar, which may be related to the introduction of more pores internally after adding HPMC.

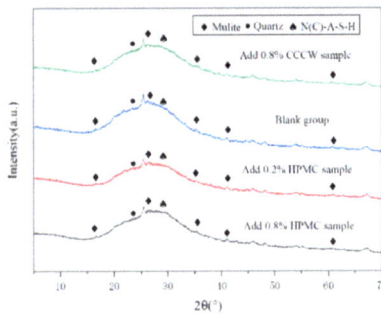

Fig. 5. XRD patterns of samples and blank groups with added HPMC and CCCW

Comparing the XRD spectra after adding CCCW, a slight enhancement is observed in the broad peak at the diffraction angle range of 25° to 30°. This is attributed to the formation of C-S-H gel and M-A-S-H gel from the tricalcium silicate and active magnesium in CCCW under alkaline conditions.

5 Conclusion

(1) By incorporating HPMC, CCCW, and PPF, modifications can be made to geopolymer-based repair mortar, and an optimal mix ratio exists. The Box-Behnken experimental design in response surface methodology can be used to optimize the mix ratio for different objectives accurately. The established model can predict experimental results and optimal mix ratios precisely.

(2) PPF and CCCW enhance the mechanical properties of geopolymer mortar and reduce shrinkage by forming a three-dimensional support system and generating substances such as ettringite (AFt) and C(M)A-S-H gel. HPMC improves the water retention of the mortar, allowing for the early formation of more N-A-S-H gel, reducing shrinkage. However, the special air-entraining effect of HPMC can introduce a large number of harmful pores to the mortar, adversely affecting its mechanical strength.

(3) After optimization through response surface methodology, the optimal dosages of HPMC, CCCW, and PPF for achieving the dual objectives of compressive strength and shrinkage are determined to be 0.306%, 0.764%, and 0.423%, respectively. Predicted values for compressive strength at 28 days and shrinkage at 7 days are 73.6 MPa and 133.98 μm. Experimental verification yields errors of 2.3% and 4.5%.

This study highlights the effectiveness of response surface methodology in developing high-performance, low-shrinkage repair materials and provides theoretical and practical insights into material formulation optimization. However, the study acknowledges some limitations, such as insufficient consideration of environmental impacts on material performance and the lack of discussion on long-term durability, which are crucial for practical applications. Future research should delve deeper into the environmental adaptability and long-term durability of materials to provide a more solid foundation for their application reliability.

Acknowledgements. 1. National Natural Science Foundation of China (No.42177074)

2. Tunnel Corporation Innovation Center Project: Research on Detection, Repair, and Operation & Maintenance Technology and Application of Urban Water Supply, Drainage, and Gas Pipeline Networks (Project Code: JK2023201A)

References

1. National Bureau of Statistics, 2022, China Statistical Yearbook. <https://www.stats.gov.cn/sj/ndsj/2022/indexch.htm>
2. Xu, Z.X., Xu, J., Jin, W., Yin, H.L., Li, H.Z.: The Challenges and Opportunities Facing Urban Black and Malodorous Water Bodies Treatment in China. Water Supply Drainage **55**, 1–5+77, https://doi.org/10.13789/j.cnki.wwe1964.2019.03.001 (2019)
3. Yang, Y.B.: Research on the corrosion resistance of ultra-high performance concrete to wastewater. Ind. Constr. **50**, 82–87 (2020). https://doi.org/10.13204/j.gyjz202004015
4. Wang, A.G.: Comparison of durability performance between alkali-activated materials and ordinary portland cement and concrete. Engineering **6**, 237–261 (2020)
5. Messan, A., Ienny, P., Nectoux, D.: Free and restrained early-age shrinkage of mortar: influence of glass fiber, cellulose ether and EVA (ethylene-vinyl acetate). Cement Concr. Compos. **33**, 402–410 (2011)
6. Yang, K.H., Xiao, H.B., Liu, J.M.: Influence and Mechanism Analysis of Cement-based Penetrating Crystalline Waterproofing Material on the Performance of Sulfate-Aluminate Cement Solidified Soil in Geotechnical Mechanics.**37**, 477–486, https://doi.org/10.16285/j.rsm.2016.02.021 (2016)
7. Pourchez, J., Ruot, B., Debayle, J., Pourchez, E., Grosseau, P.: Some aspects of cellulose ethers influence on water transport and porous structure of cement-based materials. Cem. Concr. Res. **40**, 242–252 (2010)
8. He, J., Zhu, M., Sang, G., Yu, S., He, J.: Effect of PVA latex powder and PP fiber on property of self-compacting alkali-activated slag repair mortar. Constr. Build. Mater. **408**, 133703 (2023)
9. Yang, W.X.: The influence of composite thickening agents on the performance of high-flow thin-layer mortar and its crack resistance mechanism. Silic. Bull. **42**, 1938–1949 (2023). https://doi.org/10.16552/j.cnki.issn1001-1625.20230505.001
10. Jin, F., Gu, K., Al-Tabbaa, A.: Strength and drying shrinkage of reactive MgO modified alkali-activated slag paste. Constr. Build. Mater. **51**, 395–404 (2014). https://doi.org/10.1016/j.conbuildmat.2013.10.081
11. Matalkah, F., Ababneh, A., Aqel, R.: Effect of fiber type and content on the mechanical properties and shrinkage characteristics of alkali-activated kaolin. Struct. Concr. **23**, 300–310 (2022)
12. Bernal, S.A., et al.: Gel nanostructure in alkali-activated binders based on slag and fly ash, and effects of accelerated carbonation. Cem. Concr. Res. **53**, 127–144 (2013). https://doi.org/10.1016/j.cemconres.2013.06.007
13. Weng, L., Sagoe-Crentsil, K.: Dissolution processes, hydrolysis and condensation reactions during geopolymer synthesis: part I—Low Si/Al ratio systems. J. Mater. Sci. **42**, 2997–3006 (2007)

Influence of Fly Ash and Basalt Fibers on the Properties of Recycled Pervious Concrete

Hailong Lou[1], Chenglong Ma[2(✉)], and Qiankun Hong[2]

[1] Hangzhou Municipal Construction Group, Hangzhou 310000, China
[2] Zhejiang Tongji Vocational College of Science and Technology, Hangzhou 311231, China
mcL1971@126.com

Abstract. As an environmentally friendly building material, recycled pervious concrete can not only alleviate the increasingly severe urban flooding and heat island effect, but also realize the resource utilization of construction waste. However, the porous nature of recycled pervious concrete leads to its low strength, and there is an urgent need to develop recycled pervious concrete with good mechanical and permeability properties. This paper proposes to mix fly ash and basalt fibers in the waste brick aggregate pervious concrete, and investigate the effects of mixing fly ash and basalt fibers on the properties of recycled brick pervious concrete by comparing the mechanical properties, water permeability and frost resistance of recycled brick pervious concrete. The results show that: with the increase of fiber admixture, the mechanical properties and frost resistance of recycled brick-mixed permeable concrete are improved, and the water permeability is decreased; fly ash is more obvious to improve the late strength of recycled brick-mixed permeable concrete, with the increase of fly ash admixture, the mechanical properties and frost resistance of recycled permeable concrete are improved, and the water permeability is decreased, and the excessive admixture of fly ash will greatly reduce the water permeability of recycled brick-mixed permeable concrete. Excessive incorporation of fly ash will greatly reduce the water permeability of recycled pervious concrete. Under the premise of better water permeability and mechanical properties, the optimal mix combination of 10% fly ash and 0.05% basalt fiber was selected on the basis of 85% recycled concrete aggregate and 15% brick aggregate as the mixed coarse aggregate.

Keywords: Pervious concrete · Fly ash · Basalt fiber · mechanical properties · water permeability · mixture ratio

1 Introduction

The rapid development of China's construction industry also brings a large amount of construction waste, which is dominated by brick and concrete type waste [1]. The use of construction brick waste as recycled aggregate, and the preparation of recycled permeable concrete, on the one hand, can effectively solve the problem of urban construction waste pollution, lack of natural sand and gravel aggregate resources, China's current

A. Bieliatynskyi et al. (Eds.): CSTTE 2023, LNCE 603, pp. 62–75, 2024.
https://doi.org/10.1007/978-981-97-5814-2_6

production of construction waste has reached 40% of the total production of munici-pal garbage, and with an average growth rate of 8% per year continues to grow [2], on the other hand, it can also alleviate the urban heat island effect and urban flooding phenomenon due to the intensive population activities, and bring significant benefits to the development of the building materials industry [1]. On the other hand, it can also alleviate the urban heat island effect and urban flooding caused by intensive popula-tion activities, and bring significant economic, social and environmental benefits for the development of building materials industry [3, 4]. However, due to the porous nature of pervious concrete and the thin bonding layer between aggregates, these proper-ties lead to its low mechanical proper-ties and poor durability, which greatly limits its further popularization and application [5–7].

Based on this, many scholars at home and abroad have carried out a large number of related studies around the preparation of pervious concrete with recycled aggregates, in which the addition of mineral admixtures, fiber admixtures, polymer-based admixtures and other external admixtures is considered to be one of the most effective ways to enhance the performance of pervious concrete [8–17]. Jul et al. [8] and Lutfur et al. [9] investigated the effects of fly ash, EVA emulsions, air-entraining agents and fibers on the permeability, compressive strength and freezing resistance of pervious concrete. The effects of fly ash, EVA latex, air-entraining agent and fiber on permeability, compressive strength and frost resistance of permeable concrete were investigated, and the results showed that the mechanical properties and frost resistance of permeable concrete with different admixtures were improved, but the degree of influence and the mechanism of action were different. Some researchers analyzed the mechanism to improve the mechan-ical properties and frost resistance of pervious concrete from macro and microscopic perspectives, and gave a reasonable range of mineral admixture mixing, and the results showed that mixing the appropriate amount of fly ash, silica fume, ultrafine mineral powder, nano-SiO2 and other admixtures in pervious concrete can improve the densifi-cation of the cementitious slurry, and improve the interfacial strength of the cementitious materials and aggregates as well as the microscopic pore structure[10–13].The results showed that the mixing of different aggregates can improve the mechanical properties of pervious concrete, but with different degrees of influence and mechanisms. Abdulka-der et al. [14] concluded that PET fiber composites have a significant modifying effect on the compressive strength, tensile strength and elastic modulus of pervious concrete. Mahapara et al. [15] pointed out that the addition of polymerized fibers in pervious concrete can better inhibit the drying shrinkage of pervious concrete, and the strength is increased in a certain range. Xu et al. [16] investigated the addition of different types, amounts and lengths of short-cut fibers (basalt fibers, polyvinyl alcohol fibers, glass fibers) in pervious concrete, and the results showed that the permeability coefficient was reduced more significantly, and the basalt fibers were optimal for the improvement of the strength of pervious concrete. Yu et al. [17] studied the modification effect of basalt fibers and carbon fibers on recycled aggregate permeable concrete under the condition of water-cement ratio of 0.3, and the results showed that the admixture of fibers did not have much effect on the porosity and permeability coefficient of the concrete, but it could improve its strength and abrasion resistance.

It can be seen that scholars at home and abroad have fully affirmed the role of adding external admixture materials to enhance the performance of pervious concrete, and have deeply analyzed the influence mechanism of various external admixture materials on the performance of pervious concrete. However, there is still a lack of thematic research on the modification of recycled brick-mixed pervious concrete, which does not match the scale of China's construction industry. Moreover, there is a lack of research on the modification of recycled brick-mixed pervious concrete by adding two different admixtures to improve the performance of pervious concrete, and the strengthening strategy of recycled brick-mixed pervious concrete needs to be studied in depth. Therefore, three questions are raised about the modification of recycled brick-mixed pervious concrete in China:

- Does the simultaneous addition of two different admixtures have the effect of further strengthening the performance of recycled brick-mixed pervious concrete?
- How do different admixtures affect the performance of recycled brick-mixed pervious concrete?
- What is the optimum admixture level for the mechanical and permeability properties of recycled brick-mix permeable concrete?

This paper proposes to mix fly ash and basalt fiber in the waste brick aggregate pervious concrete, through the use of different amounts of fly ash and basalt fiber to prepare recycled pervious concrete, to study its impact on the mechanical properties of recycled pervious concrete, durability and water permeability, and to derive the optimal mixing ratio, so that the mechanical properties and durability of recycled pervious concrete get the maximum improvement. This will provide experimental basis and technical reference for the application of recycled brick permeable concrete.

2 Materials and Methods

2.1 Materials

Recycled aggregate: the aggregate used in this experiment is selected from the construction waste generated by the demolition of a residential area in Hangzhou City, Zhejiang Province, through selective demolition and crushing, the collected concrete and brick mix crushing, sieving out the recycled concrete aggregate and recycled brick mix aggregate with a particle size of 9.5–13.2 mm for the test, the basic physical properties of recycled aggregate are shown in Table 1.

Table 1. Basic physical properties of recycled aggregates

Aggregate size (mm)	Bulk density (kg/m^3)	Apparent density (kg/m^3)	Water absorption (%)	Moisture content (%)	Crush index (%)
9.5 ~ 13.2	1226	1805	9.4	2.11	26

Cement: The selected strength grade is P-O 42.5 grade cement, its chemical composition and basic properties are shown in Table 2.

Table 2. Performance index of cement

Index	National standard	Test cement
Loss on burn (%)	≤ 5.0	2.98
Magnesium oxide (%)	≤ 5.0	4.32
Sulfur dioxide (%)	≤ 3.5	2.78
stability	eligible (voter etc.)	eligible (voter etc.)
Condensation time (h)	Initial setting ≥ 0.45 Final set ≤ 10.0	2.1 3.76
Compressive strength (MPa)	28 days ≥ 42.5	49.9
Heat of hydration (KJ/KG)	not have	a low fever (up 38 °C)
Alkali content (%)	≤ 5.0	0.44

Fly ash: The experiments were conducted with fly ash from Hangzhou Banshan Power Plant, which was determined to be class II fly ash, the main performance indexes are shown in Table 3, and the chemical composition of fly ash is shown in Table 4.

Table 3. Performance parameters of fly ash

factory owners	Hangzhou Mid-levels Power Plant	dates 2022.04.09	Based on criteria	GB1596–2005	hierarchy
norm	Fineness (0.45μm) 9.0%	Burning loss of volume: 4.36%	Sulfur dioxide content: 0.3%	water demand	category B

Table 4. Table of basic chemical composition content of fly ash (%)

ingredient	SiO_2	Al_2O_3	Fe_3O_4	MgO	CaO	SO_2	K_2O	Na_2O	heat loss
quantity contained	51.3	26.8	7.15	1.29	2.86	0.45	1.47	0.64	7.9

Fine aggregate: fine aggregate selection of particle size 0.35 ~ 0.5mm, fineness modulus 2.3 ~ 3.0 sand.

Fiber: basalt fiber basic properties shown in Table 5.

Water reducing agent: adopt high-efficiency carboxylic acid water reducing agent, and the dosage of water reducing agent is 1% of the dosage of cement.

Table 5. Basalt fiber basic properties

Length (mm)	Equivalent diameter (μm)	Density (kg/m³)	Elongation at break (%)	Tensile Strength (GPa)	Modulus of elasticity (GPa)
14.8	0.18	3050	3.4	4.7	95

2.2 Mixing Ratio

Selected aggregate size of 9.5–13.2 mm recycled concrete coarse aggregate and recycled brick mix coarse aggregate, the effective water-cement ratio is 0.25, i.e., the ratio of mixing water and cementitious materials, sand rate is 2%, the mixing amount of recycled brick mix aggregate is 15%, and the mixing amount of fly ash is 0, 10%, 20%, respectively. Basalt fiber dosage was 0.05%, 0.1%, 0.2% respectively. The concrete mix ratios are shown in Table 6. 85% recycled concrete aggregate and 15% recycled brick aggregate were used in the mix ratios, and it has been shown in the literature [18] that when the admixture of brick aggregate is greater than 15%, the mechanical properties and frost resistance of recycled pervious concrete will be significantly reduced.

Table 6. Concrete mixing ratio

Serial number	Recycled concrete aggregate (kg/m³)	Recycled brick mix coarse aggregate (kg/m³)	Cement (kg/m³)	Sand (kg/m³)	Mixing water consumption (kg/m³)	Aggregate adsorbed water (kg/m³)	Fly ash (kg/m³)	Basalt fiber (%)
F0B0	1370	193.78	566.8	32	142.7	87.35	0	0
F0B1	1370	193.78	566.8	32	142.7	87.35	0	0.05
F0B2	1370	193.78	566.8	32	142.7	87.35	0	0.1
F0B3	1370	193.78	566.8	32	142.7	87.35	0	0.2
F1B1	1370	193.78	510.12	32	142.7	87.35	56.68	0.05
F1B2	1370	193.78	510.12	32	142.7	87.35	56.68	0.1
F1B3	1370	193.78	510.12	32	142.7	87.35	56.68	0.2
F2B1	1370	193.78	453.44	32	142.7	87.35	113.36	0.05
F2B2	1370	193.78	453.44	32	142.7	87.35	113.36	0.1
F2B3	1370	193.78	453.44	32	142.7	87.35	113.36	0.2

Note: F0, F1, F2 represent 0, 10%, 20% of fly ash dosage respectively; B1, B2, B3 represent 0.05%, 0.1%, 0.2% of polypropylene fiber dosage respectively

2.3 Methods for Recycled Pervious Concrete

Mechanical properties

According to GB/T50081–2019 "Test Methods for Physical and Mechanical Properties of Concrete", the flexural strength and compressive strength of recycled pervious concrete are tested, the size of the sample is 100 mm × 100 mm × 100 mm cube, and the loading rate of the universal testing machine is 0.5 kN/s, and the test result is accurate to 0.1 MPa.

Permeability

The method used to determine the effective porosity of permeable concrete in this test is the drainage method [18, 19]. The length, width and height of the specimen were measured accurately with a steel ruler to calculate its volume, then the specimen was immersed in water, and its mass m1 in water was measured after no bubbles were produced on the surface; the specimen was dried in an oven at 60 °C for 24 h, and then placed to cool down to room temperature after removal, and its mass m2 was measured; finally, the porosity P of the specimen was calculated according to the following formula.

$$P = \left[1 - \frac{m_2 - m_1}{\rho V}\right] \times 100\% \tag{1}$$

where: P is the connected porosity, %; m1 is the mass of the specimen in water, g; m2 is the mass of the specimen after drying in the drying oven for 24 h, g; ρ is the density of water, g/cm3; V is the volume of the specimen, cm3.

The permeability coefficient of the specimen is determined according to the permeability meter test device in CJJ/T 253–2016 Technical Specification for Application of Pervious Concrete with Recycled Aggregate.

Frost Resistance

According to CJJ/T 253–2016 Technical Specification for Application of Pervious Concrete with Recycled Aggregate, anti-freezing test is carried out, the anti-freezing performance of recycled pervious concrete is measured by slow-freezing method, and the mass loss rate after 35 freeze-thaw cycles is recorded, and the mass loss rate is calculated as shown in Eq. (2):

$$K = \frac{m_i - m_j}{m_i} \times 100\% \tag{2}$$

In the formula, K is the mass loss rate after j times of freeze-thaw cycle, %; mi is the mass of the ith freeze-thaw cycle, kg; mj is the mass of the jth freeze-thaw cycle, kg; j > i.

3 Results and Discussion

The effects of different amounts of fly ash and basalt fiber on the water permeability, mechanical properties and frost resistance of recycled pervious concrete were investigated. In order to optimize the comprehensive performance of recycled pervious concrete, it is proposed to mix the two kinds of admixtures on the basis of preparing mixed aggregate pervious concrete to obtain an optimal combination of ratios, so that the comprehensive performance of recycled pervious concrete can be optimized.

3.1 Mechanical Properties

Figure 1 shows the test results of compressive strength and flexural strength of permeable concrete with different fly ash admixtures. As can be seen from Fig. 1, when the fly ash is the same dosage, the compressive strength and flexural strength gradually increase with the increase of fiber dosage. When the dosage of fly ash was 0, 10%, and 20%, respectively, along with the increase of basalt fiber dosage from 0.05% to 0.2%, the compressive strength increased by 5.8%, 9.0%, and 10.0%, and the flexural strength increased by 9.8%, 11.8%, and 12.3%. This is because the basalt fiber in the concrete mixing process, and the cementitious materials in full contact, the fiber like a needle interspersed in the aggregate and the cementitious materials, taking up part of the effective pore space, to improve the mechanical occlusion effect of the interface between the cement paste and the aggregate, can withstand a greater impact when subjected to the load stress, showing better mechanical properties, which, the flexural strength increased due to the crack-blocking effect of concrete, can effectively slow down the cracking of concrete, and can effectively slow down the cracking of concrete. Effect, which can effectively delay the formation and expansion of concrete microcracks [20, 21].

Through horizontal comparison, it can be found that in the case of the same amount of basalt fiber mixing, compressive strength and flexural strength with the increase in the amount of fly ash showing a downward trend, when the basalt fiber mixing amount was 0.05%, 0.1%, 0.2%, respectively, with the increase in the amount of fly ash mixing from 0 to 20%, the compressive strength decreased by 14.0%, 12.4%, 10.6%, flexural strength decreased by 18.7%, 21.8%, 21.8%, 21.8%, 21.8% and 21.8%, respectively. 18.7%, 21.8%, 22.3%. The main reason for this phenomenon is that due to the partial replacement of cement by fly ash, the number of hydration products of concrete in hardening is reduced, coupled with the delayed chemical reaction of the active component of fly ash particles, the interstices of the water film layer around the particles are not filled, and the structural densification is poor, which leads to a reduction in compressive strength and flexural strength.

Figure 2 shows the results of compressive strength and flexural strength tests for 28 and 56 d for some test groups. As can be seen from Fig. 2, the compressive strength and flexural strength increased with the increase of the curing time. And with the increase of fly ash dosage, the increase of compressive strength and flexural strength is more significant. This is consistent with the conclusion that fly ash admixture does not significantly enhance the early strength of concrete specimens, but significantly enhances the late strength of concrete, as described in the relevant literature [22, 23]. This is because the fly ash itself hydration reaction progresses slowly, not play the volcanic ash effect, in the maintenance of 28 d when only part of the fly ash completed hydration, the concrete still exists in part of the internal unhydrated fly ash; with the maintenance time up to 56 d, the active effect of the chemical reaction occurs, so that the concrete is more dense, the strength of the concrete is gradually improved.

3.2 Water Permeability

The water permeability of permeable concrete is related to the connecting porosity and water permeability coefficient, the larger the connecting porosity and water permeability

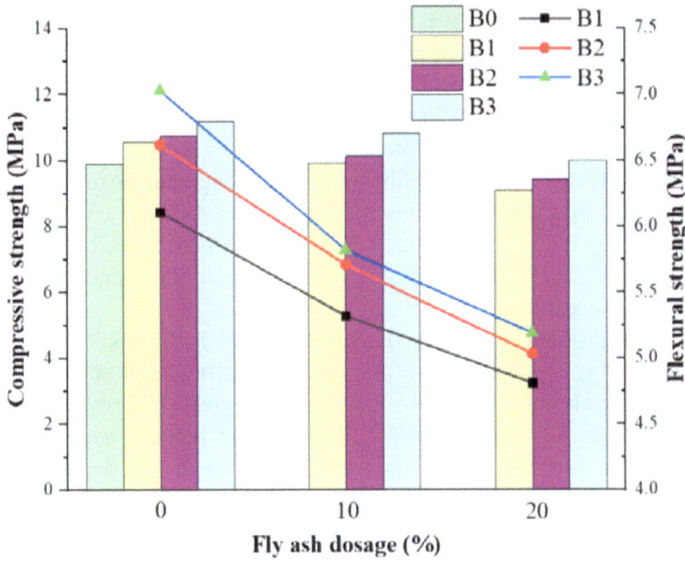

Fig. 1. Effect of different basalt fiber and fly ash dosage on compressive and flexural strengths

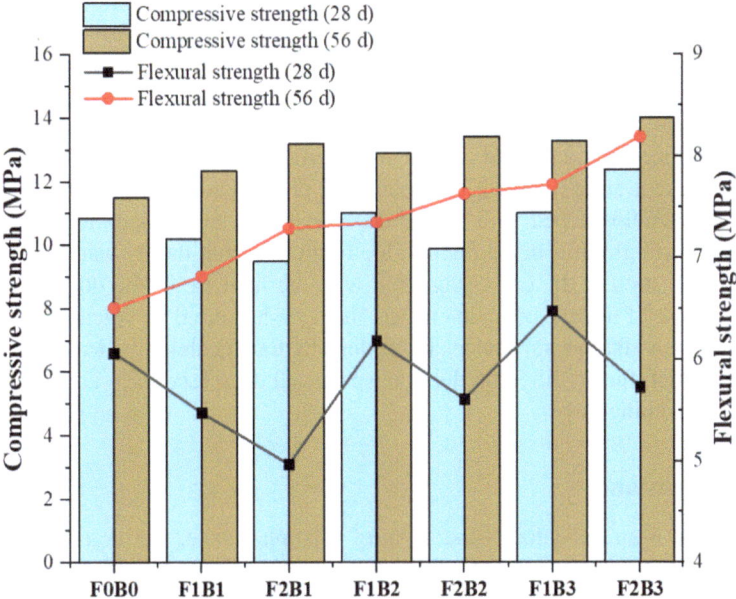

Fig. 2. Compressive and flexural strength tests of some test groups under different curing periods

coefficient, the stronger the water permeability of permeable concrete. Figure 3 shows the test results of connectivity porosity and water permeability coefficient of permeable concrete with different basalt fiber and fly ash admixtures. From Fig. 3, it can be seen that with the same amount of fly ash admixture, the connectivity porosity and water permeability coefficient have a significant decreasing trend with the admixture of basalt fibers. When the dosage of fly ash is 0, 10%, 20%, respectively, with the basalt fiber dosage increased from 0.05% to 0.2%, the connected porosity decreased by 2.8%, 2.9%, 3.3%, and the water permeability coefficient decreased by 9.8%, 11.8%, 12.3%, respectively. This is mainly due to the basalt fiber blending, along with the mixing of aggregate and cement slurry, the fiber will be wrapped by the cement slurry, the volume gradually increases, filling in the coarse aggregate, so that the pore space between the coarse aggregate gradually shrinks, which leads to the decrease of the connecting porosity and water permeability coefficient.

It can be found through horizontal comparison that when the basalt fiber doping is the same, the connecting porosity and water permeability coefficient show a slight decreasing trend. When the basalt fiber dosage was 0.05%, 0.1%, and 0.2%, respectively, with the increase of fly ash dosage from 0 to 20%, the connected porosity decreased by 0.83%, 1.12%, and 0.76%, respectively, and the permeability coefficient decreased by 4.66%, 3.27%, and 5.10%, respectively. It can be seen that the addition of fly ash has a slight effect on the water permeability of concrete in the early stage, which is because fly ash can reduce the friction between particles and particles, and the addition of fly ash in the appropriate range can improve the fluidity of the concrete mix, which will increase the thickness of the cement paste, however, in the early stage of the concrete fly ash has not yet been fully involved in the reaction, it will block some of the pores inside the concrete, which will cause a slight decrease in the connecting porosity and the water permeability. Slight decrease in water permeability properties.

Figure 4 shows the comparison of the connected porosity and permeability coefficient of some test groups at 28 and 56 d. It can be found that with the increase of the maintenance time, the connected porosity and permeability coefficient of the seven test groups decreased, among which, the connected porosity and permeability coefficient of the six test groups doped with fly ash decreased by a larger magnitude than that of the blank control group. This is mainly due to the fact that the fly ash did not participate in the reaction at the early stage, but with the increase of the dosage of fly ash and the increase of the age of maintenance, the fly ash exerted its volcanic ash activity and reacted with alkaline substances to produce hydrated calcium silicate and calcium aluminate, which further filled up the pore space and decreased the water permeability performance continuously.

3.3 Frost Resistance

Mass loss rate is an indicator of the merit of frost resistance, the larger the mass loss rate, the worse the frost resistance, the smaller the mass loss rate, the better the frost resistance [24].

Figure 5a reflects the effect of different dosages of fly ash and basalt fiber on frost resistance. When the dosage of fly ash was the same, the mass loss rate showed a decreasing trend with the increase of basalt fiber dosage, and the mass loss rates of the three groups of single-doped fibers were lower than those of the blank control group, and the mass loss rate gradually decreased with the increase of fiber dosage. When the fly ash dosage was 0, 10% and 20% respectively, the mass loss rate decreased by 22.4%,

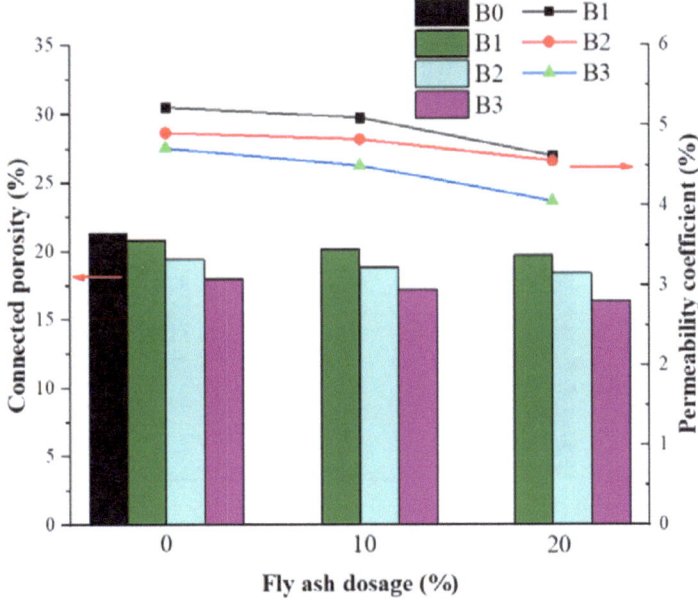

Fig. 3. Effect of different basalt fiber and fly ash dosage on connected porosity and water permeability coefficient

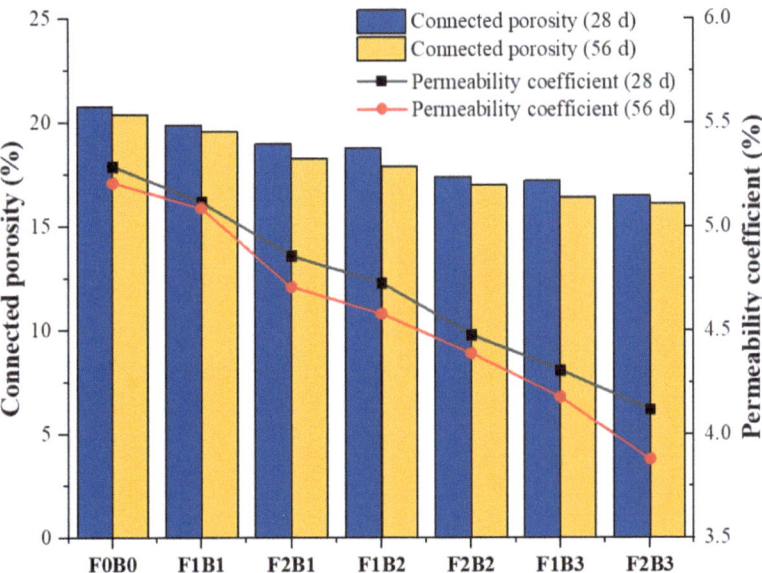

Fig. 4. Connected porosity and permeability coefficient tests for some test groups under different maintenance periods

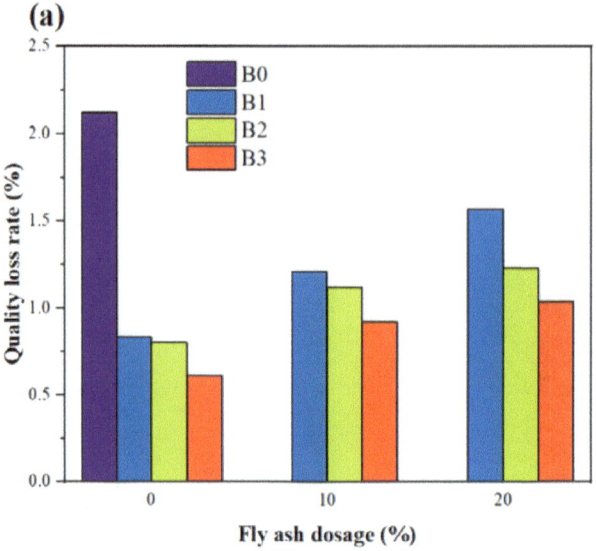

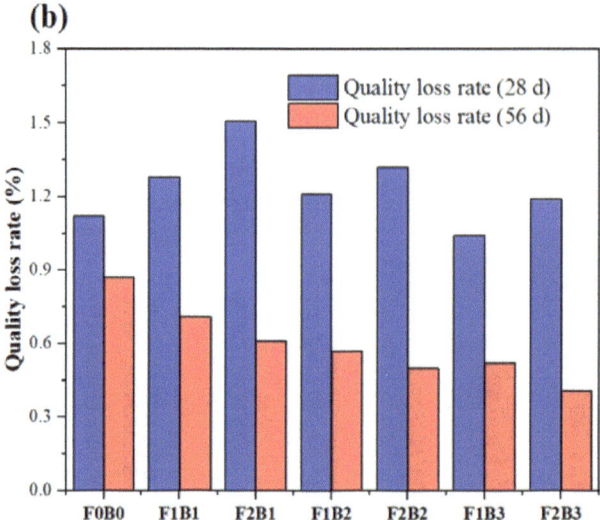

Fig. 5. Mass loss rate tests: (a) different fly ash dosage and basalt fiber dosage; (b) different test groups and conservation periods

25.3% and 29.5% with the increase of basalt fiber dosage from 0.05% to 0.2%. However, at the same basalt fiber dosage, the mass loss rate tended to increase with the increase of fly ash dosage, indicating the decrease in frost resistance of the specimens, which increased by 60.2%, 51.7%, and 46.1% with the increase of fly ash dosage from 0 to 20% when the basalt fiber dosage was 0.05%, 0.1%, and 0.2%, respectively. Figure 5b shows the comparison of mass loss rate of some test groups at 28 and 56 d. It can be seen from the figure that the mass loss rate of the same type of test group at 56 d decreased dramatically compared with that of the test group with fly ash doping at 28 d. The mass loss rate of the same type of test group at 56 d decreased dramatically. The above data show that the addition of basalt fibers enhances the frost resistance of concrete, and the higher the dosage, the better the frost resistance, the reason is consistent with the mechanism of the fiber to improve the mechanical properties, the basalt fibers play its crack-blocking effect, which can reduce the number and length of plastic cracks, and improve the integrity and continuity of the concrete material [25]. Overall, the addition of fly ash makes the concrete specimens become denser with the growth of the curing age, and the frost resistance gradually improves, and the mechanism of action is consistent with the previous section to improve the mechanical properties.

Comprehensive analysis of the above results shows that the mixing of fly ash and basalt fiber can enhance the mechanical properties of recycled pervious concrete, in which the addition of fly ash will improve the long-term mechanical properties and durability of recycled pervious concrete, but it will make its water permeability decline, and the larger the admixture, the more obvious the decline in the water permeability, therefore, the admixture amount of fly ash is selected as 10%; basalt fiber will make up for the shortcomings of the pre-activation of fly ash, which will enhance the overall permeability of concrete. The addition of basalt fiber will make up for the shortcomings of low activity of fly ash in the early stage, and improve the mechanical properties and frost resistance of permeable concrete, but it will affect the water permeability, so the dosage of basalt fiber is selected as 0.15%, and considering the comprehensive performance of regenerated permeable concrete, the combination of fly ash dosage of 10% and the dosage of basalt fiber of 0.15% is selected as the double admixture.

4 Conclusion

In this study, we investigated the effects of fly ash and basalt fibers on the mechanical properties, water permeability and frost resistance of recycled pervious concrete through compound mixing, and came to the following conclusions:

(1) The mixing of fly ash has a certain effect on the early strength and water permeability of recycled pervious concrete, and with the increase of the curing time, the mechanical properties and frost resistance of the specimens are greatly improved.

(2) The admixture of basalt fiber can significantly improve the frost resistance and mechanical properties of recycled pervious concrete, but the basalt fiber in the mixing process of pervious concrete, will be covered with a layer of cement paste, and the aggregates are more closely connected, reducing the permeable pores, resulting in a reduction in the permeability of recycled pervious concrete.

(3) After the test, it was found that the mechanical properties and water permeability of recycled pervious concrete were best achieved when the mixing amount of fly ash was 10% and the mixing amount of basalt fiber was 0.05%.

(4) There are many factors affecting the physical and mechanical properties of recycled pervious concrete by fiber. Its type, length-diameter ratio, shape characteristics and dosage will affect the performance of recycled pervious concrete. Therefore, the study of the influence mechanism of fiber on recy-cled pervious concrete based on multi-factor coordination is the focus of sub-sequent research.

Acknowledgements. This work was supported by the Foundation of Zhejiang Construction Research Project (2022K175).

References

1. Liu, P., Yao, S., Dong, X., et al.: Microstructure and performance analysis of pervious concrete with mineral admixtures. Bull. Chin. Ceram. Soc. **42**, 2504–2512 (2023)
2. Hu, M.C., et al.: Flood mitigation by permeable pavements in Chinese sponge city construction. Water (Switzerland), **10**, 172 (2018)
3. Xia, D., Li, X., Hu, J.: Experimental study on performance optimization of pervious recycled concrete based on orthogonal test. Bull. Chin. Ceram. Soc. **41**, 2748–2758 (2022)
4. Kumar, B.S., Sri-kanth, K.: Study on properties of pervious concrete with high-volume usage of supplementary cementitious materials as substitutes for cement. Asian J. Civ. Eng. **24**, 1997–2009 (2023)
5. Aoki, Y., Sri, R.R., Khabbaz, H.: Properties of pervious concrete containing fly ash. Road Mater. Pavement Des. **13**, 1–11 (2012)
6. Al-sallami, Z.H.A., Radi, Q.S., Marshdi, R.-A.-H., Mukheef: Effect of cement replacement by fly ash and epoxy on the properties of pervious concrete. Asian J. Civ. Eng. **21**, 49–58 (2020)
7. Xie, H., Li, L., Ng, P.-L., et al.: Effects of solid waste reutilization on performance of pervious concrete: a review. Sustainability (Switzerland) **15**, 6105 (2023)
8. Endawati, J., Utami, R., Rochaeti: The influence of fly ash and aggregates composition on pervious concrete characteristics. Mater. Sci. Forum **917**, 297–302 (2018)
9. Akand, L., Yang, M., Wang, X.: Effectiveness of chemical treatment on polypropylene fibers as reinforcement in pervious concrete. Constr. Build. Mater. **163**, 32–39 (2018)
10. Bright Singh, S., Murugan, M.: Effect of metakaolin on the properties of pervious concrete. Constr. Build. Mater. **346**, 128476 (2022)
11. Debnath, B., Sarkar, P.P.: Application of nano SiO2 in pervious concrete pavement using waste bricks as coarse aggregate. Arab. J. Sci. Eng. **47**, 12649–12669 (2022)
12. Xia, D., et al.: Frost resistance of polymer modified recycled pervious concrete based on orthogonal test. Shenyang Jianzhu Daxue Xuebao (Ziran Kexue Ban)/J. Shenyang Jianzhu Univ. (Natural Science), **39**, 707–715 (2023)
13. Jemimah Carmichael, M., Prince Arulraj, G., Meyyappan, P.L.: Effect of partial replacement of cement with nano fly ash on permeable concrete: a strength study. Mater. Today Proc. **43**, 2109–2116 (2020)
14. Al-Hadithi, A.I., Noaman, A.T., Mos-leh, W.K.: Mechanical properties and impact behavior of PET fiber rein-forced self-compacting concrete (SCC). Constr. Build. Mater. **224**, 111021 (2019)
15. Abbass, M., Singh, G.: Impact strength of rice husk ash and basalt fibre based sustainable geopolymer concrete in rigid pavements. Mater. Today Proc. **61**, 250–257 (2022)
16. Jun, X., Kang, A., Zhengguang, W., et al.: Effect of high-calcium basalt fiber on the workability, mechanical properties and micro-structure of slag-fly ash geopolymer grouting material. Constr. Build. Mater. **302**, 124089 (2021)
17. Haiying, Y., Meng, T., Zhao, Y., et al.: Effects of basalt fiber powder on mechanical properties and mi-crostructure of concrete. Case Stud. Constr. Mater. **17**, e01286 (2022)
18. Khankhaje, E., Kim, T., Jang, H.-O., et al.: Properties of pervious concrete incorporating fly ash as partial replacement of cement: a review. Dev. Built Environ. **14**, 100130 (2023)
19. Muthaiyan, U.M., Thirumalai, S.: Studies on the properties of pervious fly ash–cement concrete as a pavement material. Cogent Eng. **2017**(4), 1318802 (2017)

20. Ozel, B.F., Sakallı, Ş, Şahin, Y.: The effects of ag-gregate and fiber characteristics on the properties of pervious concrete. Constr. Build. Mater. **356**, 129294 (2022)
21. Jian, W., Pang, Q., Lv, Y., et al.: Research on the Mechanical and Physical Properties of Basalt Fiber-Reinforced Pervious Concrete. Materials **15**, 6527 (2022)
22. Nazeer, M., Kanish Kapoor, S.P., Singh: Strength, durability and microstructural investigations on per-vious concrete made with fly ash and silica fume as supplementary cementitious materials. J. Build. Eng. **69**, 106275 (2023)
23. Mahalingam, R., Mahalingam, S.-L.: Analysis of pervious concrete properties. Gradjevinar **68**, 493–501 (2016)
24. Mini, K.M., Hari, R.: Mechanical and durability properties of basalt-steel wool hybrid fibre reinforced pervious concrete – a box behnken approach. J. Build. Eng. **70**, 106307 (2023)
25. Huang, J., Luo, Z., Khan, M.-M.: Impact of aggregate type and size and mineral admixtures on the properties of pervious concrete: an experimental investigation. Constr. Build. Mater. **265**, 120759 (2020)

Study on Mechanical Performance Evolution Law of the Friction Pendulum Bearing Under the Influence of Friction Characteristics of Sliding Interface

Zhenhua Dong[1] ⓘ, Li Chen[2] ⓘ, Yi Li[3] ⓘ, and Zuohu Wang[2(✉)] ⓘ

[1] Ministry of Transport, Research Institute of Highway, Beijing 100088, China
[2] School of Civil and Transportation Engineering, Beijing University of Civil Engineering and Architecture, Beijing 100044, China
wangzuohu@bucea.edu.cn
[3] Guangdong Highway Construction Co., Ltd., Guangdong 510660, China

Abstract. At present, it is difficult to acquire the internal state of sliding friction bearing, so the assessment indexes and quantitative evaluation standard for service state of friction pendulum bearing (FPB) is lack. In order to propose the relationship between service state and mechanical performance, the key components of FPB affecting the seismic response is as the research subject. The 3D solid refinement finite element numerical model of FPB is established, and its rationality of and simulated results reliability is verified by experimental results. Combined the mechanical performance theory calculation models of FPB and friction characteristics of sliding interface, the simulated results of mechanical performance including effective stiffness, total energy consumption, effective period, effective damping ratio of FPB under the influence of sensitive parameters is analyzed. The results show that the compressive stress is the most sensitive factor. With the increase of compressive stress, the equivalent stiffness increases exponentially, but the equivalent period increases and the equivalent damping ratio decreases, which obviously deteriorates the seismic performance of the seismic isolation bridge. When the compressive stress is constant, the cumulative sliding distance and sliding velocity can obviously affect the mechanical performance of the bearing, but the variation range of the equivalent damping ratio is less than 10%. In addition, when the friction characteristics of the upper and lower sliding interfaces of the FPB are not consistent, the condition of the lower sliding interface has a significant impact on the mechanical properties of the FPB. The research results provide data and method support for further quantitative evaluation of the seismic performance for the serviced FPB and seismic isolation bridges.

Keywords: the serviced seismic isolation bridges · FPB · sliding interface friction characteristics · sensitive parameters · finite element numerical method

© The Author(s) 2024
A. Bieliatynskyi et al. (Eds.): CSTTE 2023, LNCE 603, pp. 76–88, 2024.
https://doi.org/10.1007/978-981-97-5814-2_7

1   Introduction

For the seismic isolation bridges, seismic isolation bearings are usually set in the bridge, and the natural vibration period of the structure is extended and the damping is increased to consume seismic energy and reduce the seismic response of the structure, while it also satisfies the functional requirements in normal servicing condition. Therefore, the rationality and reliability of the seismic isolation device are the key factors to meet the seismic requirements of the bridge. In order to timely and accurately identify the seismic risk sources and risk factors, and assess the seismic safety of seismic isolation bridges, the service station, function and performance level of seismic isolation devices are the main investigation and evaluation contents [1–5]. At present, the existing inspection standards on technical condition of bridges lack of the contents on isolation devices, in which the apparent state information is the main contents and the relevant quantitative indicators is lack [6–8].

The FPB is as the common typical seismic bearing in seismic isolation highway bridges in China. It uses the pendulum principle to realize the seismic absorption function by discharging seismic energy at the sliding interface friction, and realizes the seismic isolation function by extending the movement period of the beam through spherical swing. Researchers have carried out experimental and theoretical studies on the mechanical properties of different types of FPB [9–11], and the results show that the geometric configuration and friction characteristics of the sliding interface are the main factors affecting the function of seismic reduction and isolation, and are the key parameters for the seismic design and the design mechanical index of FPB. For the friction characteristics of the sliding interfaces, the sensitive parameters include surface pressure, sliding velocity, temperature, loading time, cumulative motion, roughness of the stainless steel surface, lubrication, wear, etc., and theoretical calculation models of the friction characteristics under the influence of different factors is proposed [12–17]. However, the ideal assumption method is usually used to determine the effective stiffness and friction parameters, which can result in great uncertainty of performance analysis results of the service seismic isolation devices and bridges.

In order to analyze the mechanical performance level under the effect of interface sliding friction state, it is further to improve the seismic safety of the serviced seismic isolation bridges, the FPB is as the research subject. Starting with theory calculation models of mechanical performance and interface friction characteristics, the testing and finite element numerical analysis methods are applied to analyze the mechanical performance of FPB under influence of different sensitive factors. It is to identify the correlation between service status and interface friction characteristics, mechanical performance, which is further to timely and accurately identify the seismic risk sources, risk factors and assess the service status.

2   Theoretical Calculation Model of FPB and Friction Characters of Sliding Interface

The load-displacement hysteresis curve for FPB is shown in Fig. 1.

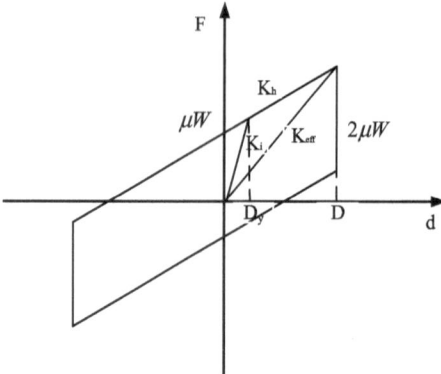

Fig. 1. The load-displacement hysteresis curve

The equivalent stiffness K_{eff} of the bearing is calculated as follows:

$$K_{eff} = \left(\frac{1}{R} + \frac{\mu}{D}\right)W \tag{1}$$

where: D is the design displacement (mm); R is the equivalent radius(mm); μ is the coefficient of dynamic friction (recommended value 0.05); W is the vertical load (kN).

The equivalent self-oscillation period T_e of the bearing is calculated as follows:

$$T_e = 2\pi\sqrt{\frac{W}{K_{eff}g}} \tag{2}$$

The equivalent damping ratio ξe of the bearing is calculated as follows:

$$\xi_e = \frac{E_D}{2\pi K_{eff}D^2} = \frac{4\mu WD}{2\pi K_{eff}D^2} \tag{3}$$

The isolation period T of the bearing is calculated as follows

$$T = 2\pi\sqrt{\frac{R}{g}} \tag{4}$$

Based on the above theoretical mechanical calculation model of FPB, and considering the time-varying nature of influencing factors of FPB, it can be seen that the main factors affecting its seismic performance are the sliding interface friction coefficient.

For the sensitive parameters of sliding interface friction characteristics and their theoretical calculation models, Mokha et al. [12], and Constantinou et al. [13] proposed different theoretical formulas for sliding friction coefficient with sliding speed and compressive stresses. In order to determine the sliding interface state under earthquake load, based on the classical Coulomb theory, researchers gave the discriminant conditions for the state of the sliding interface under the simple harmonic of acceleration time history, but the applicability of the theory needs to be further explored and analyzed.

During normal operational service, the sliding speed of the isolation sliding bearing is low, but it can produce certain wear under cyclic sliding displacement. A simple wear model is used to predict the wear of the bearing [14] as follows:

$$h_s = K_s \cdot P \cdot d \tag{5}$$

Where: h_s is the depth of wear; K_s is the wear coefficient; P is the bearing pressure; and d is the sliding distance. The wear coefficient of the sliding surface of FPB is approximately $5 \times 10^{-10} \mathrm{MPa}^{-1}$.

The sliding interface shows the wear phenomenon with the increase of cumulative distance, and then the correlation between the friction coefficient and the cumulative distance under a constant pressure is proposed [14]:

$$\begin{cases} \mu = A + B\ln(D + C) \\ \mu = A + B\ln(D_s + C) \end{cases} \tag{6}$$

Where: A, B, and C are constants; D is the cumulative movement distance; D_s is the critical point of cumulative movement distance.

3 Finite Element Analysis of Mechanical Properties Change Rule of FPB

3.1 Three-Dimensional Solid Finite Element Model of Bearing

According to the geometrical configuration parameters (Fig. 2 and Table 1) of FPB3000-ZX-e150-0 adopted for the test, Abaqus finite element software is used to establish the corresponding three-dimensional solid finite element model as shown in Fig. 3.

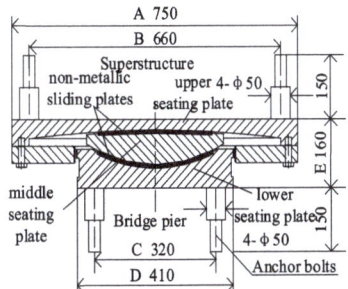

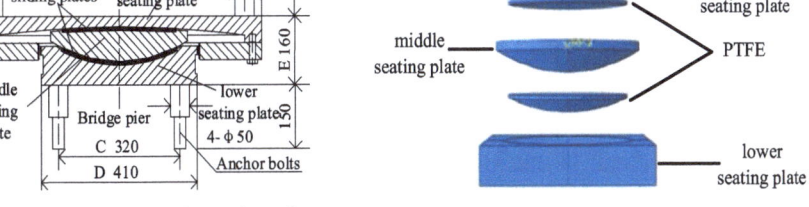

Fig. 2. Geometrical configuration of FPB

Fig. 3. Split diagram of FPB numerical model

In the finite element numerical model of the FPB, the property parameters of the materials used are shown in Table 2. Stainless steel seating plate adopts the double line stress-strain intrinsic model. Polytetrafluoroethylene (PTFE) adopts the ideal elastic-plastic model, and its stress-strain relationship curve is shown in Fig. 4. Face-to-face contact is used between upper seating plate and PTFE plate, middle seating plate and PTFE plate, with penalty function in the tangential direction and "hard" contact in the

Table 1. Geometric configuration specific parameters of FPB3000-ZX-e150-0

External dimensions			Anchor bolt positioning dimensions		Anchor bolts	
A	D	E	B	C	N	φ
750	410	160	660	320	4	50

Table 2. Material properties of FPB

Components	Material	Density (t/mm^3)	E (Mpa)	Poisson's ratio	Yield strength (N/mm^2)
upper seating plate	stainless steel	7.85×10^{-9}	210000	0.3	345
non-metallic upper sliding plates	PTFE	2.2×10^{-9}	280	0.42	30
middle seating plate	stainless steel	7.85×10^{-9}	210000	0.3	345
non-metallic lower sliding plates	PTFE	2.2×10^{-9}	280	0.42	30
lower seating plate	stainless steel	7.85×10^{-9}	210000	0.3	345

normal direction. The friction contaction element is applied to simulate the changing of friction characteristics. For middle seating plate and PTFE plate, the join connection unit is used.

$$\begin{cases} \sigma = E\varepsilon & |\varepsilon| \leq \varepsilon_S \\ \sigma = \sigma_S \operatorname{sgn}(\varepsilon) & |\varepsilon| > \varepsilon_S \end{cases} \tag{7}$$

According to the loading mode of quasi-static test, constant vertical load is applied on the top of the bearing, and horizontal unidirectional reciprocating displacement $d = dx \sin(2\pi ft)$ is applied. The horizontal displacement curve of FPB is shown in Fig. 5. The compressive stress is 2MPa, 4MPa, and 6MPa respectively.

3.2 Test Calibration of Three-Dimensional Solid Finite Element Numerical Model of Bearing

Test Scheme

The Double Spherical Seismic Isolation Bearing (DSSIB) is adopted in this test, and the model number FPB3000-ZX-e150-0. The design parameters of the test bearing are shown in Table 3.

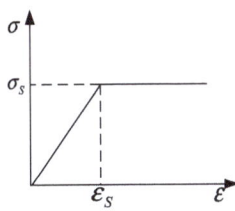

Fig. 4. Stress-strain relationship curve of PTFE

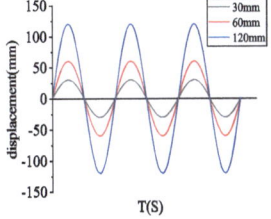

Fig. 5. Horizontal displacement loading curve

Table 3. Design parameters of the test bearing

Vertical capacity /kN	Horizontal force /kN	Horizontal ultimate capacity /kN	Turning angle /rad	Normal displacement /mm	Seismic displacement /mm	T/s
3000	600	750	0.02	± 50	± 150	3.5

Using the proposed Pseudo-static experimental to simulate the cycling laoding and cumulative movement displacement, the schematic loading diagram and the field loading test diagram of DSSIB are shown in Fig. 6. Under the constant vertical load, the displacement loading system was adopted in the horizontal one-way, and the loading rules follow $d = dx \sin(2\pi f t)$, with each level of horizontal displacement cyclic loading of 3 revolutions. The vertical compressive stresses are 2 MPa, 4 MPa and 6MPa, respectively, and the corresponding vertical loads were 884 kN, 1768 kN and 2652 kN, respectively, corresponding to test conditions P-1, P-2 and P-3, respectively. Throughout the process of the proposed static test of DSSIB, the horizontal displacements and the horizontal forces of the top plate under the bearing are measured and extracted.

Test and Simulation Results Comparison of the Mechanical Properties
The P1, P2, and P3 (corresponding to the vertical compressive stress 2 MPa, 4 MPa, and 6 MPa, respectively), the comparison between the test and simulation results of the fore -displacement hysteric curve is shown in Fig. 6. It shows that the overall shape of the hysteresis curve, residual displacement, lateral stiffness are basically the same, especially the horizontal displacement is less than 120 mm, and they all show the shape of "parallelogram", which is consistent with the theoretical hysteresis curve model (Fig. 1). To further validate the reliability of the simulation results under significant horizontal displacements, the calculated results of the equivalent period and equivalent damping ratio of the bearing under the horizontal displacement amplitude of 120 mm are compared, as shown in Table 4. As shown in Table 4, the error between the test and the simulated value of the equivalent period is less than 5%, and the maximum error of the equivalent damping ratio is 5.2%. It can be seen that the evolution law of the mechanical properties of FPB can be analyzed reliably by the three-dimensional solid finite element model of FPB.

 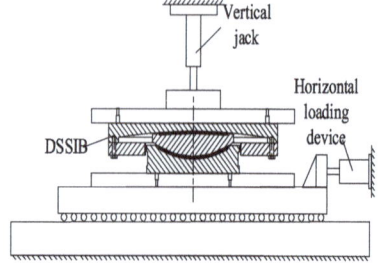

a) Schematic loading diagram   b) Field loading test diagram

Fig. 6. Comparison between hysteresis curve test results and simulation results

Table 4. Comparison of experimental results and simulation results of equivalent period and equivalent damping ratio

Case	Dd(mm)	μ	T(s)		Error	ξ_{eff}		Error
			Test	Simulation		Test	simulation	
P-1	120	0.049	2.42	2.33	3.7%	1.07	1.05	1.9%
P-2	120	0.034	2.63	2.53	3.8%	0.97	0.92	5.2%
P-3	120	0.025	2.75	2.70	1.8%	0.85	0.84	1.2%

3.3 Finite Element Analysis of the Sensitive Parameters of the Mechanical Properties of FPB

Taking into account the correlation of the sensitive parameters of its sliding interface state, the change law of the mechanical properties of FPB under the influence of factors such as compressive stress, sliding speed, cumulative moving distance and inconsistent sliding interface state is deeply analyzed, which provides data support for the performance evaluation of FPB in the whole life.

3.4 Compressive Stress

Based on the existing correlation between the friction coefficient and the compressive stress [15], the analysis cases of the friction coefficient of the sliding interface under different compressive stresses are shown in Table 5.

According to the calculation theory of basic mechanical properties of the FPB, combined with the force - displacement hysteresis curve of the bearing (Fig. 7a), the comparison results of mechanical properties of the bearing under different compressive stresses are shown in Table 6. As can be seen from Table 6, with the increase of compressive stress, the equivalent horizontal stiffness, total energy consumption and equivalent period gradually increase, in which the equivalent horizontal stiffness increases exponentially. However, the damping ratio decreases with the increase of compressive stress, and tends to be stable gradually.

Table 5. Analysis cases of compressive stress parameters

Case number		Compressive stress /MPa	Displacement amplitude /mm	$\mu_{up} = \mu_{down}$	Frequency /Hz	Sliding speed /mm/s	Number of cycles
1	p-1	1	120	0.064	0.01	4.8	3
2	p-2	2	120	0.041	0.01	4.8	3
3	p-3	3	120	0.032	0.01	4.8	3
4	p-4	4	120	0.026	0.01	4.8	3
5	p-5	5	120	0.023	0.01	4.8	3
6	p-6	6	120	0.020	0.01	4.8	3

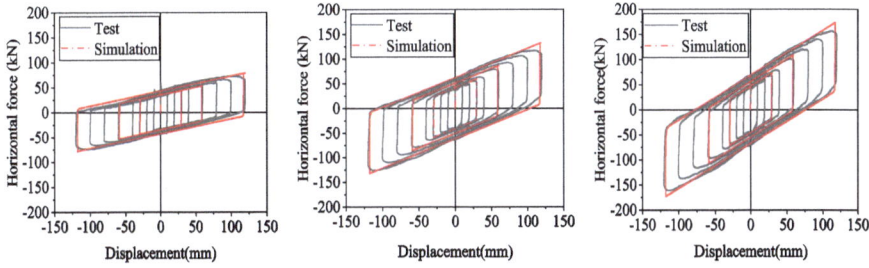

Fig. 7. Force-displacement curves companions of the bearing

Table 6. Comparison results of the mechanical properties of the bearing under different compressive stresses

Case	$\mu_{up} = \mu_{down}$	K_{eff}(N/mm)	T(s)	Total energy consumption (J)	ξ_{eff}
p-1	0.064	393.2	2.13	42262.2	0.40
p-2	0.041	610.4	2.42	54898.0	0.33
p-3	0.032	816.9	2.56	65960.6	0.30
p-4	0.026	990.9	2.68	75939.0	0.28
p-5	0.023	1180.6	2.75	88435.3	0.28
p-6	0.020	1347.3	2.82	99163.4	0.27

(2) Sliding speed

According to the correlation relationship between the friction coefficient and the sliding speed [12, 13], the friction coefficient of the sliding interface of the bearing under different sliding speeds is determined, and the corresponding analysis conditions are shown in Table 7.

Table 7. Analysis conditions of sliding speed parameters

Case number		Compressive stress /MPa	Dd /mm	$\mu_{up} = \mu_{down}$	Frequency/Hz	Sliding speed /mm/s	Number of cycles
1	vp-1	6(2652KN)	120	0.025	0.02	9.6	3
2	vp-2	6(2652KN)	120	0.029	0.04	19.2	3
3	vp-3	6(2652KN)	120	0.031	0.08	38.4	3
4	vp-4	6(2652KN)	120	0.032	0.16	76.8	3

According to the calculation theory of the basic mechanical properties of FPB, combined with the force - displacement hysteresis curve of the bearing (Fig. 7b), the calculation results of the mechanical properties of the bearing under different sliding speeds are shown in Table 8. As shown in Table 8, with the increase of sliding speed, the maximum variation ranges of equivalent stiffness, total energy consumption, equivalent damping ratio and equivalent period are 20.3%, 33.0%, 11.1% and 8.9%, respectively.

Table 8. Calculation results of the mechanical properties under different sliding speeds

Case	D_d(mm)	$\mu_{up} = \mu_{down}$	K_{eff}(N/mm)	T(s)	Total energy consumption (J)	ξ_{eff}
vp-1	120	0.025	1461.3	2.70	111177.2	0.28
vp-2	120	0.029	1552.5	2.62	122457.2	0.29
vp-3	120	0.031	1598.1	2.59	128652.9	0.30
vp-4	120	0.032	1621.0	2.57	131889.2	0.30

(3) Cumulative movement distance

The existing research results show that the friction coefficient gradually increases with the increase of cumulative movement distance, until it tends to be stable [14]. Assuming that the critical point of the cumulative movement distance is 1500 m, the friction coefficient correspond to 0.062 and 0.034 under the conditions of vertical compressive stress 2 MPa and 6 Mpa respectively, and the corresponding analyzed conditions are shown in Table 9.

Combined with the force-displacement hysteresis curve of the bearing (Fig. 8c), the calculation results of the mechanical properties of the bearing under different cumulative movement distances are shown in Table 10. As shown in Table 10, when the cumulative movement distance reaches the critical point, compared with p-2, the equivalent stiffness, the total energy dissipation value, the equivalent damping ratio increases by 26.2%, 49.2%, 18.7% respectively, and the equivalent period decreases by 11.2%; compared with p-6, the equivalent stiffness, the total energy dissipation value, the equivalent damping

Table 9. Analysis conditions of cumulative movement distance parameters

Case number		Compressive stress (Mpa)	D_d(mm)	$\mu_{up} = \mu_{down}$	Sliding speed (mm/s)	D(m)	Number of cycles
1	dp-1	2(884KN)	120	0.062	4.8	1500	3
2	dp-2	6(2652KN)	120	0.034	4.8	1500	3

ratio increases by 23.7%, 39.8%, 13.5%respectively, and the equivalent period decreases by 10.2%.It can be seen that the cumulative movement distance can affect its mechanical properties, especially in the case of smaller compressive stress.

Table 10. Calculation results of the mechanical properties of the bearing under different cumulative movement distances

Case	Compressive stress (MPa)	$\mu_{up} = \mu_{down}$	K_{eff}(N/mm)	T(s)	Total energy consumption (J)	ξ_{eff}
dp-1	2	0.062	770.6	2.15	81908.3	0.39
dp-2	6	0.034	1666.6	2.53	138638.8	0.31

(4) Non-consistent sliding interface state

DSSIB contains upper and lower sliding interfaces, two sliding interfaces, the upper and lower sliding interfaces are in different states. Therefore, the corresponding analysis conditions are formulated as shown in Table 11.

Table 11. Analysis conditions of different sliding interface condition parameters

Case number		Compressive stress (MPa)	μ_{up}	μ_{down}	Frequency (Hz)	Sliding speed (mm/s)	Number of cycles
1	mp-1	6	0.025	0.050	0.02	9.6	3
2	mp-2	6	0.025	0.100	0.02	9.6	3
3	mp-3	6	0.050	0.025	0.02	9.6	3
4	mp-4	6	0.100	0.025	0.02	9.6	3

Combined with the force-displacement hysteresis curve of the bearing (Fig. 8d), the calculation results of the mechanical properties of the bearing under the non-consistent interface condition are shown in Table 12. Compared with vp-1, with the increase of μ_{down}, the equivalent stiffness, the total energy dissipation, the equivalent damping ratio maximum increases by 13.8%, 37.7%, 11.7% respectively; but the equivalent period maximum decreases by 6.1%. With the increase of μ_{up}, the equivalent stiffness, the total energy dissipation, the equivalent damping ratio maximum increases by 103.2%, 247.7%, 58.1% respectively, and the equivalent period maximum decreases by 29.7%. This shows that the inconsistent sliding interface conditions can significantly affect the performance of FRP, especially for μ_{up}.

Table 12. Calculated results of the mechanical properties of the bearing under the condition of non-uniform sliding interface

Case	μ_{up}	μ_{down}	K_{eff}(N/mm)	T(s)	Total energy consumption (J)	ξ_{eff}
mp-1	0.025	0.05	1529.2	2.64	120001.8	0.29
mp-2	0.025	0.1	1663.6	2.53	141275.1	0.31
mp-3	0.05	0.025	1963.6	2.33	187968.9	0.35
mp-4	0.1	0.025	2969.5	1.90	356804.3	0.44

4 Conclusion

Starting with the interface friction characteristics and mechanical performance of FPB, the testing and finite element numerical analysis methods are applied to analyze the mechanical performance of FPB under influence of different sensitive factors. The conclusions are as follows:

(1) According to the geometric configuration and analysis essentials of the FPB, the corresponding three-dimensional solid refined finite element numerical model of the support is established. Based on the theoretical calculation model of the mechanical properties of the friction pendulum support, the maximum deviation of the simulation results and the test results of the horizontal force-displacement hysteresis curve, equivalent period and equivalent damping ratio of the friction pendulum support is 5.2%, which verifies the rationality and reliability of the three-dimensional solid refined finite element numerical model.

(2) Furthermore, the finite element numerical simulation method and the theoretical calculation formula of the interface friction coefficient are used to analyze the evolution of key mechanical parameters such as the equivalent stiffness, total energy dissipation, equivalent period and equivalent damping ratio of the FPB. The analysis results show that compressive stress is the most sensitive factor affecting the mechanical properties of FPB, with the increase of compressive stress, the equivalent horizontal stiffness increases exponentially, but the bearing damping ratio decreases. If the compressive stress is constant, the cumulative sliding distance and sliding velocity can obviously affect the performance of the FPB, but the variation range of the equivalent damping ratio is less than 10%. In addition, when the friction characteristics of the upper and lower sliding interfaces of the FPB are not consistent, the condition of the up sliding interface has a more significant impact on the mechanical properties.

In summary, the interface friction state can significantly affect the mechanical properties of FPB. According to the mechanical properties evolution law of the FPB under different interface friction states, it provided the data and theoretical basis for further qualitative evaluation of the service performance level of the serviced FPB, and lays a foundation for the evaluation of the seismic performance of the in-serviced seismic isolation bridge. In addition, in order to better clarify the applicability of the theoretical calculation model of friction coefficient, the verification and applicability of the theoretical model will be further studied in combination with the model test method.

Acknowledgment. The work was supported by the Central level public welfare research institutes basic research business fund project of China (Grant no. 2021-9083C).

References

1. Li, C., Zhang, P., Li, Y., Zhang, J.: Effects of friction pendulum bearing wear on seismic performance of long-span continuous girder bridge. J. Vibro Eng. **25**(3), 506–521 (2023). https://doi.org/10.21595/JVE.2022.22915
2. Zhong, J., Zhu, Y., Han, Q.: Impact of vertical ground motion on the statistical analysis of seismic demand for frictional isolated bridge in near-fault regions. Eng. Struct. **278**, 115512 (2023). https://doi.org/10.1016/J.ENGSTRUCT.2022.115512
3. Zhang, J., Ding, Y., Guan, X.: Overturning resistance of friction pendulum bearing-isolated structure subjected to impact. Int. J. Struct. Stab. Dyn. **22**(06), 2250072 (2022). https://doi.org/10.1142/S0219455422500729
4. JTG/T 2231-01-2020. Specifications for seismic design of highway bridges, Ministry of Transport of the People's Republic of China, Beijing
5. Chen, L., Dong, Z., Wang, Z.: Sensitivity factor analysis of seismic performance of seismically isolated girder bridges during the operation period. AIP Adv. **13**, 125124 (2023). https://doi.org/10.1063/5.0185971
6. JTG/T H21-2021. Standards for technical condition evaluation of highway bridges, Ministry of Transport of the People's Republic of China, Beijing
7. Song, L.L.: Bearing disease and prevention of prestressed concrete girder bridge. Urban Roads Bridges Flood Control (06): 208–213+230+23–24 (2021). https://doi.org/10.16799/j.cnki.csdqyfh.2021.06.054
8. Zhao, G., Ma, Y., Li, Y., Su, L., Zhou, F.: An experimental study on the behavior deterioration trend of friction pendulum bearings with corrosion time for offshore isolated bridges. IOP Conf. Ser. Earth Environ. Sci. **304**(4), 042025 (2019). https://doi.org/10.1088/1755-1315/304/4/042025
9. Chen, Z.W.: Study on mechanical properties of damping and isolation friction pendulum bearing under different parameters. Liaoning University of Technology, Jinzhou (2021). https://doi.org/10.27211/d.cnki.glngc.2021.000004
10. Li, X.D., Huo, J., Zhao, J.: Research on mechanical properties and application analysis of friction pendulum bearing. Build. Struct. **48**(19), 86–90 (2018). https://doi.org/10.19701/j.jzjg.2018.19.018
11. Han, Q., Liu, W.G., Du, X.L., et al.: Computational model and experimental validation of multi spherical sliding friction isolation bearings. China J. Highway Transp. **25**(5), 82–88 (2012). https://doi.org/10.19721/j.cnki.1001-7372.2012.05.013

12. Mokha, A., Constantinou, M., Reinhorn, A.: Teflon bearings in base isolation i: testing. J. Struct. Eng. **116**(2), 438–454 (1990). https://doi.org/10.1061/(ASCE)0733-9445(1990)116: 2(455)
13. Constantinou, M., Mokha, A., Reinhorn, A.: Teflon bearings in base isolation ii: modeling. J. Struct. Eng. **116**(2), 455–474 (1990). https://doi.org/10.1061/(ASCE)0733-9445(1990)116: 2(455)
14. Lin, L.L.: Numerical analysis of interface heat effect in sliding friction isolation bearing. Beijing University of Technology, Beijing (2015)
15. Campbell, T.I., Green, M.F., Koppens, N.C., et al.: Stress distribution at PTFE interface in cylindrical bearing. J. Bridg. Eng. **3**(4), 186–193 (1998). https://doi.org/10.1061/(ASC E)1084-0702(1998)3:4(186)
16. Braga, F., D'Amato, M., Gigliotti, R., Laguardia, R.: Numerical modelling of sliding isolators incorporating self-heating effects. Structures **46**, 1968–1980 (2022). https://doi.org/10.1016/ J.ISTRUC.2022.11.040
17. Constantinou, M.C., Caccese, J., Harris, H.G.: Frictional characteristics of Teflon–steel interfaces under dynamic conditions. Earthquake Eng. Struct. Dynam. **15**(6), 751–759 (1987). https://doi.org/10.1002/eqe.4290150607

Study on the Circumferential Mechanical Properties of Buried PE Pipes with Soil Erosion Void

Jinqiu Hu, Xuefeng Yan[✉], and Cong Zeng

Faculty of Engineering, China University of Geosciences – Wuhan, Wuhan 430074, Hubei, China
cugyanxf@163.com

Abstract. Soil erosion void is a common pipeline defect, which has a great impact on the stability and safety of pipeline operation, but the force characteristics for eroded pipes are still unclear. Buried PE pipes in municipal engineering are taken as the research object, and the calculation method of soil pressure at the bottom of the pipe with soil erosion void is proposed, and the influence of factors such as size and location of the void on the stress characteristics of buried PE pipes with soil erosion void is studied by using the numerical model. The results of the study show that the void will affect the structural stress of PE pipes, the size of the void is positively correlated with the influence degree and influence range of the pipe; the location of the void determines the influence location of the pipe.

Keywords: urban sewer system · PE pipe · soil erosion void · stress characteristics · numerical simulation

1 Introduction

Municipal pipe network is an important part of municipal infrastructure. However, due to lack of testing and maintenance, the pipes are commonly suffering from corrosion, leakage and other diseases. Erosion void (as Fig. 1) is caused by the pipe's own leakage, rainwater scouring, external disturbance and other factors, it makes the pipe structural stress change, which has a greater impact on the safety and stability of pipe operation, and the continued development of void may lead to secondary problems such as ground subsidence or landslides, posing a serious safety risk. For soil erosion void, there have been related repair techniques, such as grouting, polymer [1–3], but the force characteristics of erosion void are not known, resulting in the repair design is not based on evidence, and the operation is not prudent and may lead to the secondary damage of the pipe. Therefore, it is of great engineering significance to study the force distribution characteristics and mechanical response of eroded pipes to provide theoretical and data support for the design of pipe rehabilitation.

Scholars have carried out a lot of research on soil erosion void. Jiang L derived the erosion direction and erosion shape of the soil around the pipe through the phenomenon

A. Bieliatynskyi et al. (Eds.): CSTTE 2023, LNCE 603, pp. 89–100, 2024.
https://doi.org/10.1007/978-981-97-5814-2_8

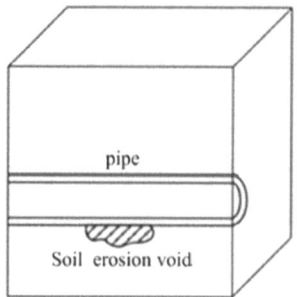

Fig. 1. Sketch of soil erosion void

of solid-liquid coupling around the pipe after seepage from a pressure pipe in a model test [4]. Li H J designed a model test to investigate the mechanism of pipe-soil interaction induced by the ground subsidence, and proposed a limit equilibrium solution for the soil pressure at the top of the pipe with ground settlement [5]. Tan Z found that erosion void had a large effect on the bending moment of the pipe and the degree of the effect would increase greatly when it exceeded 45° [6]. Meguid M A investigated the soil pressure redistribution around the pipe of the eroded pipes [7]. Peter J M found that the void produces a large change in the mechanical state of the buried pipe, but the grouting can effectively alleviate this phenomenon [8, 9]. At present, a large number of studies have been carried out on the quantitative analysis of the causes of soil erosion void and the load distribution around the pipe with soil erosion void, but most of the studies are focused on the axial force of the pipe, and are mainly aimed at the rigid pipe, there is a lack of circumferential force analysis of the flexible pipeline. The force and deformation characteristics of flexible pipes and rigid pipes are different, which leads to more uncertainty in parameter selection and difficulty in precise control of construction parameters when trenchless repairing and designing for flexible pipes with soil erosion void.

This paper takes PE pipe as the research object, adopts ABAQUS finite element software to establish the numerical model of buried PE pipe. Based on the test results, we analyze the influence the width and location of the erosion void on the annular force of the pipe under loading, provide theoretical and data support for rehabilitation design of eroded pipes.

2 Analysis of Force Characteristics of Eroded Pipes

2.1 Impact of Erosion Void on Buried Pipes

When void occurs, the pipe body above the void loses the support force of the soil, which destroys the original force equilibrium state of the pipe-soil, the soil pressure around the pipe is redistributed, and the soil pressure on the pipe body around the void is increased and the load concentration phenomenon occurs, as Fig. 2.

According to the equivalent load, the soil pressure in the void is transferred to void surroundings, and comparing with the force of the pipe when it is not eroded, the problem

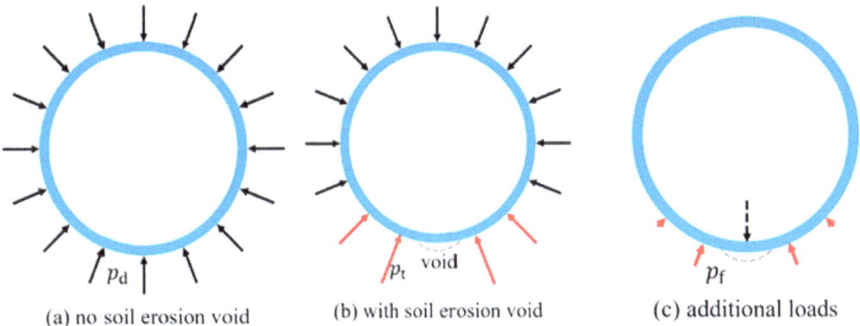

(a) no soil erosion void   (b) with soil erosion void   (c) additional loads

Fig. 2. Sketch of soil pressure around the pipe

of redistributing the soil pressure around the pipe with soil erosion void can be simplified to the problem of additional loading, and the formula is obtained:

$$p_t = p_d + p_f \qquad (1)$$

Where, p_t is the soil pressure on the bottom of the pipe with soil erosion void, kPa; p_d is the soil pressure on the bottom of the pipe without soil erosion void, kPa; p_f is the additional load on the bottom of the pipe with soil erosion void, kPa.

2.2 Calculation Method of Additional Load on Buried PE Pipe with Soil Erosion Void

This paper analyzes the soil pressure of buried PE pipe with soil erosion void on the basis of Spangler soil pressure model [10]. When the soil erosion void occurs, the soil pressure around the void will show obvious redistribution and load concentration phenomenon. Referring to the modified tunnel lining load-structure model under void conditions proposed by Ying [11], the Spangler soil pressure model is modified, and in order to simplify the model, the soil pressure model is obtained without considering that the soil erosion void of smaller width makes the pipe produce larger lateral deformation, which leads to larger changes of lateral soil pressure of the pipe, as Fig. 3.

Assuming that a soil erosion void with width l at the bottom of the pipe; x is the ratio of the influence range of the void on the soil pressure to the width of the void, and the length of the influence range of the void is xl, m; y is the ratio of the gradually increasing section of the additional load to the influence range of void; $p_{f,max}$ is the maximum additional load, kN/m^2; m is the length of the additional stress direction downward, m; p_m is the soil pressure at the top of the pipe, kN/m^2; p_0 is the horizontal soil pressure on the side of the pipe, kN/m^2; and p_d is the soil pressure at the bottom of the pipe at the initial stage, kN/m^2.

When the width of the soil erosion void is small, the soil pressure at the void is transferred to the surrounding of the void, from which it can be seen that the two shaded areas in the figure are equal; at the same time, according to the principle of similarity of triangles, the following formula can be introduced:

$$S_1 = S_2 \qquad (2)$$

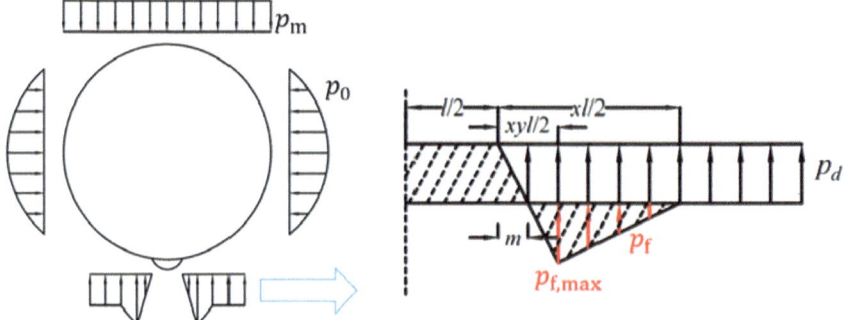

Fig. 3. Soil pressure model of buried PE pipe with soil erosion void

$$S_1 = p_{f,max}(xl/2 - m)/2 \tag{3}$$

$$S_2 = p_d(l + m)/2 \tag{4}$$

$$\frac{p_{f,max}}{p_d} = \frac{xyl/2 - m}{m} \tag{5}$$

The association leads to the equation:

$$p_{f,max} = p_d\left(\frac{2}{x} + y\right) \tag{6}$$

As a result, the formula for calculating the maximum additional load $p_{f,max}$ for buried PE pipe with soil erosion void is derived. However, the above formula only applies to the small width of erosion void, its influence range does not exceed or just reach the edge of the pipe base.

When the width of void is large, and the influence range of evacuation theory exceeds the edge of pipe base, i.e. $l + xl > D\sin\alpha$, the above model is no longer applicable. According to previous research, with the gradual increase of the width of the void, the influence range of the void also increases, the load concentration is gradually shifted to both sides, and the trend of the load concentration gradually disappears, and the additional stress tends to be evenly distributed.

Therefore, when $D\sin\alpha/(1 + x) < l < D\sin\alpha$, the buried PE pipe soil pressure model shown in Fig. 4.

Based on the equivalent loads the equation can be obtained:

$$p_f = \frac{p_d l}{(D\sin\alpha - l)} \tag{7}$$

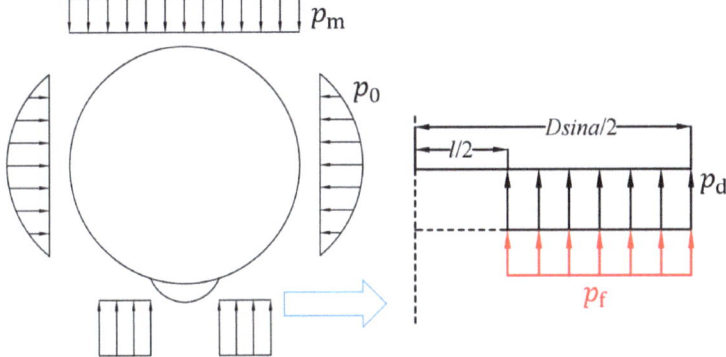

Fig. 4. Soil pressure model of pipe when the soil void grows

3   Finite Element Analysis

3.1   Finite Element Model

In order to further verify the analysis results, this paper uses ABAQUS finite element software to establish a numerical model of pipe-soil interaction of PE pipe with soil erosion void.

Since this study mainly analyzes the circumferential force of the pipe, it can be simplified to a plane strain problem and establish a two-dimensional model. The pipe material is 400 mm PE100 grade polyethylene pipe, which is commonly used in municipal projects. The width of the fill on the side of pipe is 3 times the radius of the pipe, and the depth of the pipe is set at 4 times radius. The size of the soil is 1.6 m × 1.6 m, and the outer diameter of the PE pipe is 0.4 m with a wall thickness of 19.1 mm. As shown in Fig. 5.

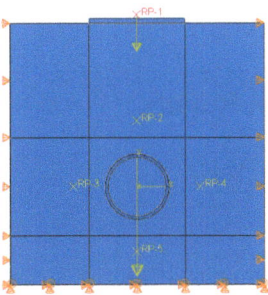

Fig. 5. 2D model

The density of PE pipe is $0.965 \text{ g} \cdot \text{cm}^{-3}$, elastic principal model. The density of soil is $0.965 \text{ g} \cdot \text{cm}^{-3}$, Moore Coulomb model, angle of internal friction is 37.8° and cohesion is taken as 0. The basic properties of the modeled materials are shown in Table 1.

The surface mutual contact is normal hard contact, tangential penalized contact, and the friction coefficient between PE pipe and sandy soil can be taken as 0.4. The boundary

Table 1. Material properties of the model

Material	Density/g · cm^{-3}	Modulus of elasticity/MPa	Poisson's ratio
Soil	1.700	15	0.3
PE pipe	0.965	1050	0.4

condition at the bottom of the model is set to be completely fixed, while the left and right sides are fixed to have the horizontal displacement as 0.

In this simulation, the analysis process includes four analysis steps, which are ground stress equilibrium, pipe activation, setting the soil erosion void around the pipe and loading. The model change function is used to make the soil at the void around the pipe in the unactivated state to set the soil erosion void around the pipe.

3.2 Simulation Program

Taking the erosion void width of 10 cm at the bottom of the pipe as the control group, and considering width and position of void, the simulation results are compared with the control group to study the influence law of different parameter characteristics on the force characteristics of buried PE pipes. The specific program is shown in Table 2.

Table 2. Numerical simulation program

Influencing factor	Value	Unit
width	0, 5, 10, 20	cm
position	bottom, side, top	\

4 Analysis of Simulation Results

4.1 Width of the Soil Erosion Void

(1) Pipe deformation

Table 3 shows the pipe deformation and vertical displacement of pipe top and bottom after 50 kN load is added to the model.

When the width of the void is 5 cm, the deformation of the pipe is the same as the deformation of the pipe without evacuation, which indicates that the degree of influence of the void on the pipe at this time is small, and it is not enough to change the deformation of the pipe. When the width of the void increases to 10 cm, the deformation of the pipe changes, the vertical deformation of the pipe decreases by 14.1%, and the lateral deformation decreases by 6.3%. When the width of the void increases to 20 cm, the deformation of the pipe changes more obviously, the vertical deformation of the

Table 3. Deformation of pipes

width of void	Vertical deformation	Horizontal deformation	Vertical displacement of top	Vertical displacement of bottom
0	−3.83 mm	3.18 mm	−5.38 mm	−1.55 mm
5 cm	−3.88 mm	3.21 mm	−5.44 mm	−1.56 mm
10 cm	−3.29 mm	2.98 mm	−5.53 mm	−2.24 mm
20 cm	−1.87 mm	2.25 mm	−5.94 mm	−4.07 mm

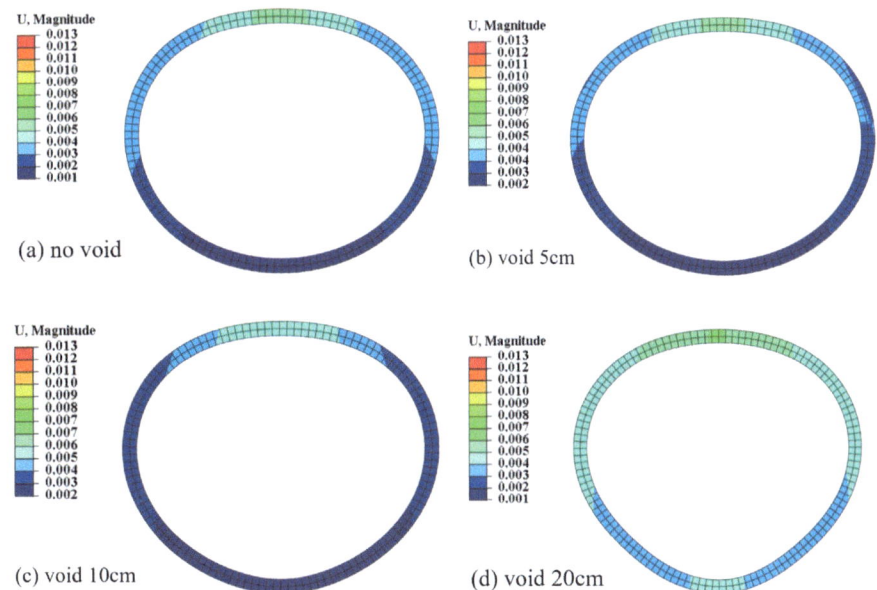

(a) no void   (b) void 5cm

(c) void 10cm   (d) void 20cm

Fig. 6. Sketch of pipe shape (deformation factor is 10)

pipe decreases by 51.2%, the lateral deformation decreases by 29.2%, and the lateral deformation is larger than the vertical deformation of the pipe.

When the soil erosion void at the bottom of pipe, the vertical displacements of the top and bottom of the pipe are increased. The vertical displacement of the bottom of the pipe increases by 0.01 mm, 0.69 mm and 2.52 mm with the increase of width of void. The vertical displacement of the top of the pipe is less affected by void, with the increment of 0.06 mm, 0.15 mm and 0.56 mm. Combined with the shape of the deformed pipe (Fig. 6), it can be found that the bottom of the pipe loses the support of the soil and produces a bulge to the bottom. Soil support part will produce a bulge to the bottom, and the void mainly affects the vertical displacement of the pipe bottom by affecting the vertical displacement of the pipe bottom, making the distance between the top of the pipe and the bottom of the pipe increase, which in turn leads to a reduction in the vertical deformation of the pipe.

(2) Pipe stress

Extract the Mises stress of the outer and inner walls of the buried pipe after the addition of 50 kN load, and draw a comparative diagram of the pipe stress (Fig. 7).

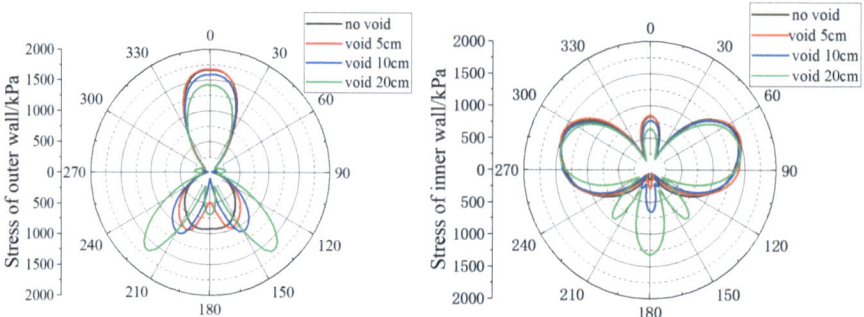

Fig. 7. Stress curve of pipe

The impact of the void on the pipe stress is mainly located in the bottom and waist of the pipe. The influence of void on the stress distribution of pipe is as follows: the stress on the outer wall of pipe bottom decreases, and the stress on the inner wall increases; the stress on the outer wall of pipe waist increases, and the stress on the inner wall slightly decreases. The maximum stress is 1671.51 kPa, located at the top outer wall of the pipe, when the void does not occur. When the width of 5cm of the void appeared, only the bottom and waist of the tube stress produces a small change, the other positions of the stress is not affected, and the location and value of the maximum stress remain unchanged. When the width of the void increases to 10 cm, the stress at the outer wall of pipe bottom and the inner wall of pipe waist decreases, and the stress at the inner wall of the pipe bottom and the outer wall of the pipe waist increases, but the maximum stress remains unchanged. When the width of the void is increased to 20 cm, the location of the maximum stress is shifted from the outer wall of the top of the pipe to the outer wall of the pipe waist, which is 1660.10 kPa.

(3) Vertical soil pressure at the bottom of the pipe

The initial vertical soil pressure at the bottom of the pipe in the model is plotted in Fig. 8. The vertical soil pressure around the void will produce the load concentration phenomenon, with the increase of the void, the load concentration is shifted to both sides, and the concentration trend is gradually reduced.

Under the action of soil gravity load, the average value of vertical soil pressure at the bottom of the pipe is 15.62 kPa when the void does not appear, and the maximum additional load is 12.64 kPa and 16.53 kPa when the width of the void is 5 cm and 10 cm respectively, and the maximum additional load calculated according to Eq. 6 is 12.51 kPa, with an error of 1.04% and 18.14 kPa, with an error of 8.87%. When the

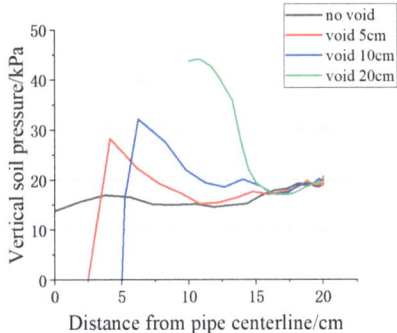

Fig. 8. Vertical soil pressure at the bottom of the pipe

width of void is 20 cm, the average additional load is 22.28 kPa, and the additional load calculated by Eq. (7) is 24.41 kPa, with an error of 9.56%.

There is a certain error between the simulation results and theoretical calculations, and the calculation error increases with the increase of the width of the void, which may not take into account the effect of the void on the width of the base of the pipe bottom. However, the above additional load calculation model for the bottom of eroded pipes can be used for reference.

4.2 Location of the Soil Erosion Void

(1) Pipe deformation

The deformation of the pipe after loading is shown in Table 4. When the void appears at the bottom or top of the pipe, the pipe deformation is reduced, but when the void appears at the top of the pipe, the pipe deformation is smaller. The pipe deformation increases when the void appears on the side of the pipe. Combined with the displacement of the pipe and the shape diagram of the pipe after deformation (Fig. 9), the void affects the pipe deformation by changing the displacement of pipe at the place where void is located, and the influence of void at different locations on the deformation of pipe is as follows: the pipe produces an obvious bulge towards the place where the void is located, while no obvious deformation occurs in other parts. In the case of the same width of the voids, the degree of influence of the location of the voids on the pipe deformation for the top of the pipe > side > bottom.

(2) Pipe stress

Figure 10 shows the comparison of pipe stresses after loading. Under external loading, the maximum stress of the pipe is located in the outer wall of the top of the pipe when there is no void, and the value is 1671.51 kPa. When the void is located in the bottom of the pipe, the maximum stress of the pipe basically remains unchanged. When the void is located in the left side of the pipe, the stress in the left inner wall of the pipe increases to 2409.87 kPa, an increase of 44.17%, which becomes the maximum stress of the pipe. When the void is located at the top of the pipe, the stress at the bottom and waist of the

Table 4. Deformation of pipes

Location of void	Vertical deformation	Horizontal deformation	Vertical displacement at top	Vertical displacement at bottom	Vertical displacement at left	Vertical displacement at right
no void	−3.83 mm	3.18 mm	−5.38 mm	−1.55 mm	−1.59 mm	1.59 mm
bottom	−3.29 mm	2.98 mm	−5.53 mm	−2.24 mm	−1.49 mm	1.49 mm
left	−4.09 mm	3.88 mm	−5.59 mm	−1.50 mm	−2.27 mm	1.51 mm
top	−2.55 mm	2.75 mm	−4.04 mm	−1.48 mm	−1.37 mm	1.37 mm

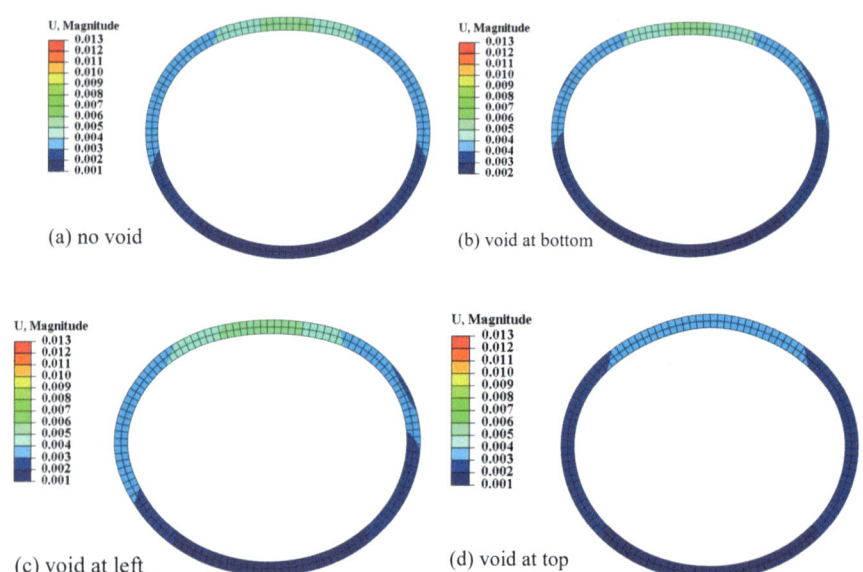

Fig. 9. Sketch of pipe shape (deformation factor is 10)

pipe remains unchanged, the stress at the outer wall of the pipe shoulder increases, the stress at the inner wall of the pipe shoulder and the top of the pipe decreases, and the maximum stress of the pipe decreases. It is hypothesized that the reason for this is that when there is a void at the top of the pipe, the top stress decreases due to the loss of part of the soil load at the top.

When the void at side, the scope and degree of influence on the pipe waist is greater than that on the pipe shoulder. In the case of the same width of voids, the degree of influence of the location of the voids on the stress of the pipe is side > bottom > top, but voids at top can cause ground collapse, its harm can't be ignored.

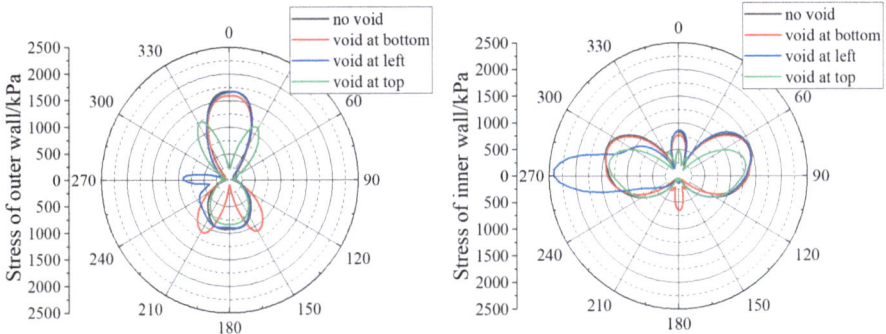

Fig. 10. Stress curve of pipe

5 Conclusions

This paper takes buried PE pipes in municipal engineering as the research object, establishes the calculation method of additional load at the bottom of the pipe with soil erosion void. Using ABAQUS to establish a 2D numerical model, the influence of the width and location of the void on the annular force characteristics of the buried PE pipe with soil erosion void was investigated by the control variable method.

(1) As a kind of pipeline disease, soil erosion void has a greater impact on the safety and stability of pipeline operation. The pipe at the void loses the restraining effect of soil, and load concentration effect around void results in changes in pipe structural stress, which has a greater impact on the safety and stability of pipeline operation.

(2) The scope and degree of influence of the void on the pipes are positively correlated with the width of the void, and when the void is small (less than 5 cm), the deformation of the pipe and the structural stress do not change significantly. With the increase of the width of the void, the bottom of the pipe produces a bulge downward, the location of the maximum strain and maximum stress of the pipe changes, and the value exceeds the initial maximum value, and the pipe bearing capacity is reduced.

(3) The influence of the location of the void on the deformation of the pipes is that the pipe bulges significantly to the place where the void is located, while no significant deformation occurs in other parts of the pipe. The degree of influence of the location of the void on the pipe stress is side > bottom > top.

References

1. Wang, F., Li, B., Fang, H.: Experimental and numerical study on polymer grouting repair of underground pipeline containing dehiscence and corrosion disease. Tunnel. Underground Eng. Disaster Prevent. Control **1**(03), 1–8 (2019)
2. Li, B., Fang, H., Zhai, K., et al.: Mechanical behavior of concrete pipes with erosion voids and the effectiveness evaluation of the polyurethane grouting. Tunn. Undergr. Space Technol. **129**, 104672 (2022)
3. Li, B., Fang, H., Yang, K., et al.: Impact of erosion voids and internal corrosion on concrete pipes under traffic loads. Tunn. Undergr. Space Technol. **130**, 104761 (2022)

4. Jiang, L., Zhang, B., Huang, S., et al.: Analysis of fluidized zone in transparent soil under jet induced by pipe leakage. Water Sci. Eng. **16**(2), 203–210 (2023)
5. Li, H.J., Zhu, H.H., Wu, H.Y., et al.: Experimental investigation on pipe-soil interaction due to ground subsidence via high-resolution fiber optic sensing. Tunn. Undergr. Space Technol. **127**, 104586 (2022)
6. Tan, Z., Moore, I.D.: Effect of backfill erosion on moments in buried rigid pipes (2007)
7. Meguid, M.A., Kamel, S.: A three-dimensional analysis of the effects of erosion voids on rigid pipes. Tunn. Undergr. Space Technol. **43**, 276–289 (2014)
8. Peter, J.M., Chapman, D., Moore, I.D., et al.: Impact of soil erosion voids on reinforced concrete pipe responses to surface loads. Tunn. Undergr. Space Technol. **82**, 111–124 (2018)
9. Peter, J.M., Moore, I.D.: Effects of erosion void on deteriorated metal culvert before and after repair with grouted slip liner. J. Pipeline Syst. Eng. Pract. **10**(4), 04019031 (2019)
10. Spangler, M.G., Shafer, G.E.: The structural design of flexible pipe culverts. In: Highway Research Board Proceedings, p. 17 (1938)
11. Ying, G., Zhang, D., Chen, L., et al.: Modification of load structure model in the presence of vault cavity. J. Civil Eng. **48**(S1), 181–185 (2015)

Research on Gyratory Compaction Characteristics of Low Void Modified Asphalt Concrete Materials

Qiang Chen, Wei Xu[✉], Yongjian Li, Jingxian Liang, Zizhan Du, Runan Zhang, and Jinhai Liu

South China University of Technology, Institute of Civil Engineering and Transportation, Guangzhou, Guangdong, China
{202121010542,xuweib,202121010542,202220107950,202121010559, 202321009858,202320107949}@scut.edu.cn

Abstract. The compaction characteristics of asphalt concrete materials have a decisive impact on the degree of field compaction. Therefore, the study of the compacting law of composite modified asphalt mixture has important guiding significance for the field construction of pavement. This study used a gyratory compactor (SGC) to analyze the compaction characteristics of composite modified asphalt concrete materials from the two dimensions of asphalt-aggregate ratio and gradation. The test added 10% PE modifier by mass of asphalt and 0.2% polymer fiber. The compaction characteristics of LA-10 (low-void asphalt concrete materials), SMA-10, and AC-10 gradations under asphalt-aggregate ratio of 7.5%, 8.0%, and 8.5% were studied and compared with the compaction situation of the commonly used EA-10 for steel bridge deck pavement. The results showed that with changes in the asphalt-aggregate ratio, the trend of the mixture voidage and compaction curve was consistent. The LA-10 mixture reached the target voidage range at an asphalt-aggregate ratio of 7.5%, with good workability, while the AC-10 and SMA-10 mixtures reached the target voidage range at an asphalt-aggregate ratio of 8.0%, and were easier to compact. The compaction density energy index was SMA-10 > AC-10 > LA-10 > EA-10, and the compaction ease of LA-10 was relatively close to that of epoxy asphalt mixture EA-10. According to the above compaction characteristics, LA-10 is suitable for the lower layer of pavement, and SMA-10 and AC-10 are suitable for the upper layer of pavement.

Keywords: steel bridge deck pavement · composite modified asphalt concrete · gyratory compaction · voidage · experimental research

1 Introduction

With the rapid advancement of the transportation industry, steel bridge decks play a crucial role in modern bridge structures and have a direct impact on driving safety and the service life of bridges. Modified asphalt concrete, recognized as high-performance pavement materials, have garnered widespread attention in steel bridge deck pavement

A. Bieliatynskyi et al. (Eds.): CSTTE 2023, LNCE 603, pp. 101–109, 2024.
https://doi.org/10.1007/978-981-97-5814-2_9

due to their exceptional crack resistance, durability, and high-temperature stability. Previous studies by Hao et al. [1], Liu et al. [2], and others have explored the application of high-elasticity modified asphalt concrete in steel bridge deck pavement, demonstrating significant improvements in temperature stability and fatigue performance. Additionally, Zhang et al. [3] investigated the influence of basalt fiber on epoxy asphalt concrete in cold regions, revealing that the addition of basalt fiber effectively enhanced flexibility and deformation characteristics of the epoxy asphalt concrete. However, the focus of research on modified asphalt concretes has primarily been on road performance, with relatively limited attention given to the compaction characteristics of construction. Additionally, there are technical challenges and difficulties associated with the application of steel bridge deck pavement.

Compaction plays a critical role in the construction process of modified asphalt concrete, and its effectiveness directly impacts the smoothness and density of the pavement [4], which ultimately determines the performance of the pavement. Modified asphalt concrete with added fibers exhibit significantly different compaction characteristics compared to regular asphalt concrete due to their unique composition and properties [5]. Therefore, conducting a comprehensive study on the compaction characteristics of composite modified asphalt concrete is highly significant for optimizing construction techniques and enhancing pavement quality.

Compared with Marcel compaction, the rotary compactor can better simulate the compaction process of pavement construction site by kneading, and obtain the density of asphalt concrete by controlling the compaction parameters. Jiu-peng Z et al. [6] studied the warm mix SBS asphalt concrete specimens through gyratory compaction and established the optimal compaction temperature. Zhang et al. [7] studied the relationship between the compaction characteristic parameters of the concrete and the composition of the asphalt concrete (gradation, asphalt content) and the molding temperature. The study showed that the medium gradation was easier to compact, and with the increase of the asphalt-aggregate ratio, the compactability increased linearly. Alexandros M et al. [8] studied the effect of compaction temperature on the compaction curve of the concrete.

This article aims to investigate the compaction characteristics of low-void modified asphalt concrete with added PE modifiers and polymer fibers through gyration compaction testing technology. It considers two factors: gradation type and asphalt-aggregate ratio. The experimental research and theoretical analysis aim to provide technical reference for improving the quality of modified asphalt concrete steel bridge deck pavement.

2 Materials and Test Methods

2.1 Materials

PG-76 Modified Asphalt

Based on previous research regarding the performance of pavement materials, The asphalt used in this study is PG-76 modified asphalt produced by Guangdong Xinyue Jiafu Asphalt Co., Ltd., and its performance indicators are shown in Table 1.

Table 1. Technical Indicators of PG-76 Modified Asphalt

Test Item	Unit	Test Result	Design or Technical Requirement
Penetration (25°C)	0.1 mm	52	40–60
Ductility (5°C)	cm	33	>20
Softening Point	°C	86.5	>75
Flash Point	°C	341	>230
Solubility	%	99.8	>99

Aggregate

The test aggregate is divided into three sizes: 0–3 mm, 3–5 mm, and 5–10 mm. We used aggregate produced by Zhongshan Aggregate Plant, with limestone powder from the same plant. Performance indicators can be found in Table 2.

Table 2. Technical Indicators for Coarse Aggregates

Test Item	Unit	Test Result	Design or Technical Requirement
Crushing Value	%	≤15	9.4
L.A. abrasion (%)	%	≤22	8.7
Apparent Relative Density	–	≥2.70	2.933
Water absorption	%	≤1.5	0.47
Adhesion	level	≥5	5

Modifier

In this study, we used PE-type modifiers and polymer fibers as modifiers.

2.2 Gradation

The concrete comprises three types of asphalt concrete: low void asphalt concrete LA-10 (with a 0.075 mm passing rate of 15%), SMA-10, and AC-10. The synthetic gradation of the concrete can be found in Table 3.

2.3 Research Plan

Modifier addition: Based on results from rut tests, beam bending tests, pull-out tests, etc., we found that adding a composite modifier with 10% asphalt mass fraction and 0.3% polymer fiber mass fraction to the concrete resulted in better road performance. The modifier addition method used in this test is dry addition.

Table 3. Gradation in this study

Log sieve size (mm)	13.2	9.5	4.75	2.36	1.18	0.6	0.3	0.15	0.075
LA-10	100	98.8	70.6	53.1	42.2	36.2	27.2	21.1	14.5
SMA-10	100	97.6	42.3	25.5	21.0	18.6	15.0	12.3	9.3
AC-10	100	98.7	67.6	43.2	32.2	26.2	17.2	11.6	6.0

To study the compaction of low void composite modified asphalt concrete, we selected LA-10, SMA-10, and AC-10 gradations as mentioned in 2.2. Road performance tests of the modified asphalt concrete showed that when the oil-stone ratio is 8.0, the concrete still has good high-temperature stability. Therefore, concrete with oil-stone ratios of 7.5%, 8.0%, and 8.5% were chosen, with a mixing temperature of 185 °C and a specimen preparation temperature of 185 °C.

2.4 Specimen Preparation

The gyratory compactor primarily simulates on-site compaction to form asphalt concrete. The PINE-AFG1 type of gyratory compactor was utilized in this experiment, following the standards AASHTO T312, ASTMD 6925-08, and EN 12697-31. The parameters set for this test include a vertical loading pressure of 600 kPa, a rotation speed of 3030r/min, and a rotation area inclination of 1.25°.

The gyratory compactor is capable of accurately recording the height changes of each compaction, thus obtaining the compaction curve of the asphalt concrete and understanding the compaction process of the specimen [9]. The compaction process generally consists of two stages: initial compaction during construction to achieve the design void ratio, and subsequent compaction under traffic load after open traffic to reach the limit void ratio and above. Therefore, the resulting compaction curve can reflect changes in compactness during both construction and operation.

When evaluating the compaction characteristics of asphalt concrete, two main indicators are relied upon: the Compaction Energy Index (CEI) during construction and Traffic Density Index (TDI) during traffic. CEI focuses on measuring the area under the compaction curve from initial compaction to reaching 92% density. A lower CEI value indicates easier workability during construction. It should be noted that when void ratio is less than 2%, it is considered close to its limit [10].

Through these indicators, a more comprehensive understanding can be gained regarding the compaction characteristics of asphalt concrete during both construction and use, ensuring quality and safety in road construction.3Test Results and Analysis.

3 Test Results and Analysis

3.1 Analysis of Modified Asphalt Concrete Void Ratio

As shown in Fig. 1, after 160 compaction cycles, the void ratios of the nine kinds of asphalt concrete were measured and all maintained relatively low levels.

It is noteworthy that for the LA-10 concrete, when the asphalt-aggregate ratio increased from 7.5% to 8.0%, the void ratio decreased by 16.8%, but this change was not significant. However, when the asphalt-aggregate ratio further increased to 8.5%, the void ratio remained almost unchanged. In contrast, the void ratio of the AC-10 concrete exhibited a different trend with changes in the asphalt-aggregate ratio. From 7.5% to 8.0%, the void ratio of AC-10 rapidly decreased by 42%, and then the decrease narrowed to 19% when the asphalt-aggregate ratio increased to 8.5%. Additionally, for the SMA-10 concrete, as the asphalt-aggregate ratio increased from 7.5% to 8.5%, the decrease in void ratio first dropped from 25% to 8.6%, indicating that the rate of decrease in void ratio gradually slowed down with increasing asphalt content.

For all three kinds of concretes, the void ratios showed relatively significant decreases when the asphalt-aggregate ratio increased to 8.0%, but the decreases became gradually smaller. Therefore, under the same compaction conditions, all three kinds of concrete exhibited relatively ideal compaction effects when the asphalt-aggregate ratio was 8.0%. Additionally, a comparison at the same asphalt-aggregate ratio revealed a trend of SMA-10 > AC-10 > LA-10 in terms of void ratio. When the asphalt-aggregate ratio exceeded 8.0%, the difference in void ratio between LA-10 and AC-10 gradually decreased, indicating similar performance characteristics.

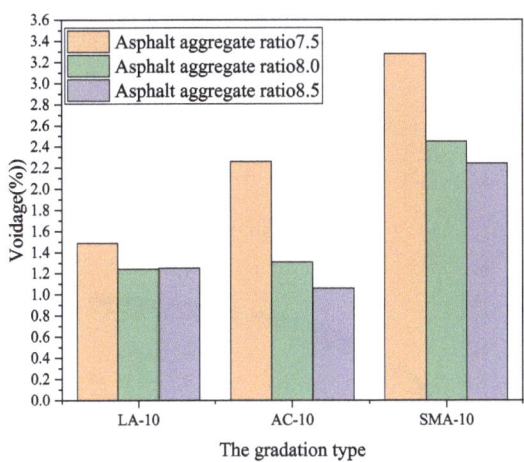

Fig. 1. The compaction void ratio of 9 types of concrete

3.2 Compaction Curves of Three Asphalt Concretes with Different Asphalt-Aggregate Ratios

The asphalt-aggregate ratio is a critical factor affecting the compaction performance of asphalt concrete. As the asphalt-aggregate ratio increases, the compaction rate of asphalt concrete typically accelerates, enabling them to reach the desired void ratio or compaction density faster. However, it is noteworthy that not all types of concrete exhibit significant improvements in compaction rate with increasing asphalt-aggregate ratios.

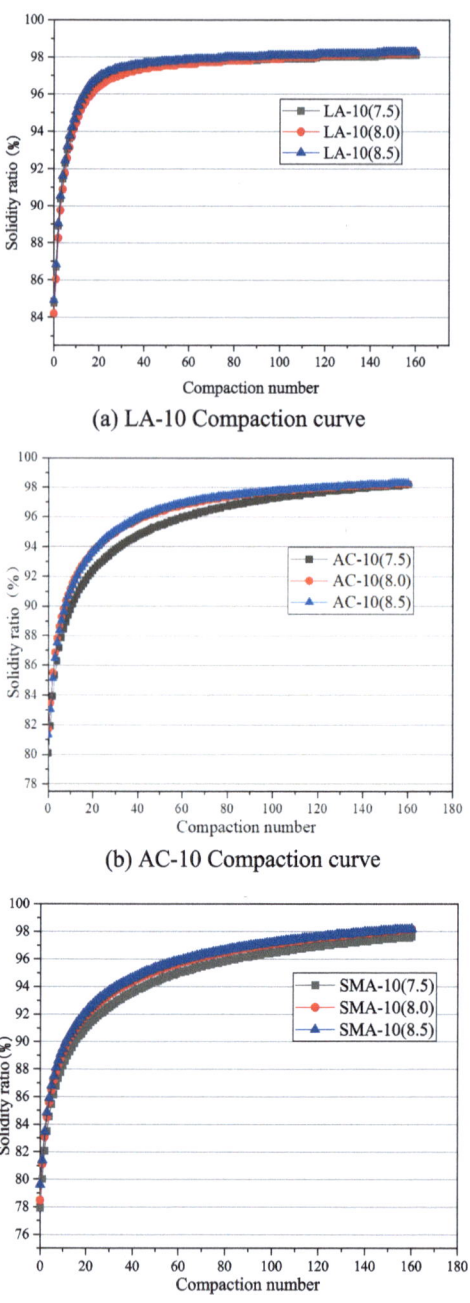

(a) LA-10 Compaction curve

(b) AC-10 Compaction curve

(c)SMA-10 compaction curve

Fig. 2. Compaction curves of concrete with different oil stone ratios

As shown in Fig. 2, compaction curves for three different graded concrete at asphalt-aggregate ratios of 7.5%, 8.0%, and 8.5% are presented. It can be clearly observed that as the asphalt-aggregate ratio gradually increases, the compaction performance of the asphalt concrete also improves. Further analysis of compaction curves for different graded concrete reveals the following:

For LA-10 grade, the compaction curves for the three asphalt-aggregate ratios are relatively close. Notably, when the asphalt-aggregate ratio exceeds 7.5%, there is no significant improvement in the compaction performance of the concrete, indicating that further increases in the asphalt-aggregate ratio have limited contributions to the compaction process.

For AC-10 grade, when the asphalt-aggregate ratio increases from 7.5% to 8.0%, the compaction rate of the concrete shows a noticeable improvement. However, when the asphalt-aggregate ratio further increases to 8.0% and 8.5%, the compaction curves become similar, and the increase in compaction rate is not significant, indicating that the compaction performance of the concrete has stabilized under these two asphalt-aggregate ratios.

Similarly, SMA-10 grade exhibits a similar trend to AC-10. During the increase of asphalt-aggregate ratio from 7.5% to 8.0%, the compaction rate increases slightly. However, when the asphalt-aggregate ratio exceeds 8.0%, the compaction curves become similar, indicating that the growth of compaction rate has leveled off.

Based on the above information, there are significant differences in the compaction performance of asphalt concrete with different graded types and asphalt-aggregate ratios. Comparing Fig. 1 with Fig. 2, it can be seen that for the same graded concrete, as the asphalt-aggregate ratio increases, the trend of increasing compaction rate is similar to the decreasing trend of void ratio. Therefore, there is a clear correlation between the void ratio and compaction curve. When selecting an appropriate asphalt-aggregate ratio, it is necessary to consider the graded type of the concrete and the compaction target to achieve the best compaction performance.

3.3   Compaction Characteristics of Asphalt Concrete with Different Grades

Based on the influence of asphalt-aggregate ratio on the void ratio and compaction characteristics of asphalt concrete discussed earlier, it is found that when the asphalt-aggregate ratio is 8.0%, the asphalt concrete can meet the requirement of a low void ratio while providing good construction ease. Therefore, this study compares the compaction curves of three graded mixtures with an asphalt-aggregate ratio of 8.0% to the commonly used epoxy asphalt concrete EA-10 for steel bridge deck pavement.

Figure 3 shows the compaction curves and their fitting curves for the four graded concrete. The compaction curves in the figure are relatively distinct, indicating significant differences in compaction characteristics among different graded concrete. It can be clearly seen from the figure that the compaction energy index (CEI) follows the order: SMA-10 > AC-10 > LA-10 > EA-10. Calculations reveal that the traffic density index (TDI98) is highest for AC-10 (172), followed by LA-10 (170), SMA-10 (164), and EA-10(28).

In summary, the LA-10 concrete has a CEI close to EA-10 and a low void ratio. Considering its high-temperature stability and other pavement performance characteristics,

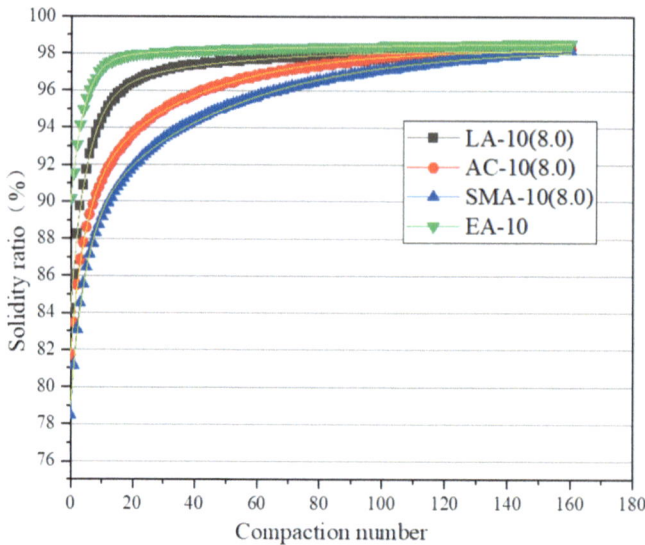

Fig. 3. Compaction curves of the four gradations

it is suitable for application in the lower layer of steel bridge deck pavement. On the other hand, SMA-10 has a higher CEI, making it difficult to compact during construction. However, it has a higher TDI98 and exhibits good high-temperature performance, making it suitable for the upper layer of steel bridge deck pavement.

4   Conclusion

In this study, the influence of three grading types of asphalt concrete (LA-10, AC-10, and SMA-10) on the compaction performance of asphalt concrete at oil-stone ratios of 7.5%, 8.0%, and 8.5% was investigated through gyratory compaction tests. The results showed that:

1) Both the grading and oil-stone ratio significantly impact the compaction performance of asphalt concrete, with a close relationship between void content and compaction curves.
2) For LA-10 graded concrete, little change in void content and similar compaction curves were observed when the oil-stone ratio exceeded 7.5%. Similarly, for AC-10 and SMA-10 graded concrete, changes in void content and compaction curves were not significant when the oil-stone ratio exceeded 8.0%.
3) The dense energy index ranking is as follows: SMA-10 > AC-10 > LA-10 > EA-10. During construction compaction, LA-10 exhibits similarities to epoxy asphalt concrete EA-10.

The comprehensive test results indicate that the low-void composite modified asphalt concrete designed in this study can achieve target voids under conventional construction conditions. However, this paper only tested specific gradations and oil stone ratios.

Future studies can further expand the scope of tests to more fully understand the effects of gradations and oil stone ratios on the compaction properties of asphalt concretes.

References

1. Hao, Z., Zhang, X., Sheng, X., et al.: Application research of high elastic modified asphalt in steel bridge deck pavement. J. Highway Transport. Res. Develop. **26**(04), 22–28 (2009). https://doi.org/10.3969/j.issn.1002-0268.2009.04.005
2. Liu, P., Hao, Z., Sheng, X., et al.: Performance analysis of high elastic modified asphalt concrete SMA10 for steel bridge deck pavement. J. Technol. Highway Transport **37**(06), 14–20 (2021). https://doi.org/10.13607/j.cnki.gljt.2021.06.003
3. Zhang, J.: The influence of basalt chopped fibers on the performance of epoxy asphalt and its mixture. J. Bull. Chin. Ceramic Soc. **39**(09), 3032–3039 (2020). https://doi.org/10.16552/j.cnki.issn1001-1625.20200819.001
4. Dai, W., Qian, G., Zhu, X., et al.: Research on void characteristics during compaction of asphalt mixtures. J. Construct. Build. Mater. 416135069 (2024). https://doi.org/10.1016/J.CONBUILDMAT.2024.135069
5. Qu, Y., Yu, D.: Research on construction technology of glass fiber modified asphalt concrete. J. Transport. Sci. Technol. (S2), 33–35 (2011). https://lib-cqvip-com.webvpn.scut.edu.cn/Qikan/Search/Index?from=index
6. Zhang, J., Pei, J., Xu, L., et al.: Gyratory compaction characteristic of SBS warm mixed asphalt mixture. J. Traffic Transport. Eng. **11**(1), 1–6(2011). https://doi.org/10.19818/j.cnki.1671-1637.2011.01.001
7. Zhang, Z., Bian, X., Du, Q., et al.: Study on factors effecting on compaction property of asphalt mixture. J. Wuhan Univ. Technol. **34**(6), 36–41 (2012). https://doi.org/10.3963/j.issn.1671-4431.2012.06.008
8. Alexandros, M., Tine, T., Stefan, V., et al.: Impact of the mastic phase and compaction temperature on the sigmoidal gyratory compaction curve of asphalt mixtures. J. Construct. Build. Mater. **391** (2023). https://doi.org/10.1016/j.conbuildmat.2023.131283
9. Li, L., Li, X., Zhong Z.: Analysis of influencing factors on compaction characteristics of asphalt mixtures. J. China J. Highway Transport (S1), 33–36+40 (2001). https://doi.org/10.19721/j.cnki.1001-7372.2001.s1.009
10. Smith, R.W.: Asphalt concrete mix properties related to pavement performance (with discussion). J. Assoc. Asphalt Paving Technol. Proc. **58** (1989). https://trid.trb.org/View/486929

Experimental and Simulation Study on the Stress Characteristics of Precast U-Shaped Beam Slab of Prestressed Concrete for High-Speed Railways

Binghe Zhang[1], Yicai Yang[1], Haijun Jiang[2], Zhangsheng Yue[2], and Shumin Wan[2(✉)]

[1] Qingdao Metro Line 8 Co., Ltd., Qingdao 266100, Shandong, China
zhjhc_bhxgs@qd-metro.com
[2] Qingdao Municipal Engineering Design and Research Institute Co., Ltd., Qingdao 266100, Shandong, China
449211215@qq.com

Abstract. The precast U-shaped beam slab of prestressed concrete is a crucial structural form in railway bridge construction, with its stress characteristics directly affecting the safety and stability of railway bridges. This study examines the stress characteristics of the slab track, including transverse bending, shear lag effects, and the stress state at the junction between the web and the bottom slab. It selects precast U-beams from actual field projects to investigate the transverse bending performance, shear lag effects, and stress conditions at the junctions through experimental studies. Additionally, by integrating the results of linear and nonlinear finite element analysis, it summarizes the stress characteristics of precast U-shaped beam slabs made of prestressed concrete. This provides references for the design and construction of precast U-shaped beam slabs of prestressed concrete.

Keywords: Rail Transit · U-shaped Beam · Slab Track · Transverse Bending · Stress Characteristics · Experimental Study · Finite Element Analysis

1 Introduction

Prestressed concrete, due to its excellent structural performance and economic benefits, has been widely applied in fields such as bridges and high-speed railways. Precast U-beams, as a new type of prestressed concrete structural element, are increasingly used in modern rail transit systems due to their convenience in construction, environmental friendliness, and superior structural performance. Urban rail transit elevated bridges widely employ precast U-shaped beams because their track slabs (slab tracks) are located at the bottom of the two webs, serving as a supporting structure. This arrangement's notable advantage is its low construction height. Compared to traditional upper-bearing beams such as box beams and I-beams, it offers benefits like lower construction height, high-quality factory prefabrication, and savings on building materials [1, 2]. With the

A. Bieliatynskyi et al. (Eds.): CSTTE 2023, LNCE 603, pp. 110–116, 2024.
https://doi.org/10.1007/978-981-97-5814-2_10

rapid development of urban rail transit in China, many cities extensively use precast U-shaped beams of prestressed concrete in elevated rail transit bridges [3]. Particularly, the post-tensioning method in the field of precast U-beams has developed a comprehensive design and construction system, with numerous corresponding scientific research achievements [4–6]. In recent years, the use and research of pre-tensioned precast U-beams have gradually increased [7, 8]. The adoption of a mixed tensioning method for precast U-shaped beams in Qingdao Metro Line 8 further promotes the widespread application of prestressed concrete precast U-beams in elevated rail transit bridges [9]. U-shaped beams, as a type of open thin-walled structure, exhibit significant spatial characteristics in their stress properties: longitudinal bending stresses are borne jointly by the side webs and bottom slab, designed according to full prestressed concrete structure standards; the slab track directly bears the train load and transfers it to the webs, showing transverse bending characteristics, designed according to reinforced concrete structure standards [10, 11]. However, a critical gap in existing research is the comprehensive understanding of the stress characteristics of the slab track, particularly in its connection to the webs. This gap is crucial as the connection area between the slab track and the webs is often prone to longitudinal cracking, a problem whose causes have not been sufficiently explored. Addressing this knowledge gap is essential for advancing the design and durability of prestressed concrete structures in urban rail transit systems.

Therefore, this study aims to thoroughly investigate the stress characteristics of prestressed concrete precast U-shaped beam slab tracks, with a specific focus on the transverse bending, shear lag effects, and the stress characteristics at the critical connection between the slab track and the webs. This study is based on the Qingdao Metro research project "Study on the Comprehensive Mechanical Properties of Precast U-Beams with Prestressed Concrete Using the Pre-Tensioning Method". It selects cast-in-place precast U-beams for the investigation of the slab track's transverse bending, shear lag effects, and the stress characteristics at the connection between the slab track and the webs (neck skew). Additionally, it incorporates finite element analysis to study the stress performance of the slab track.

2 Introduction to Slab Track Test and Finite Element Analysis

2.1 Testing Program

The selected U-shaped standard span is 30 m, with the overall appearance of the beam in a "U" shape and the webs designed in an arc shape, resulting in a calculated span of 28.7 m. The slab track is thickened to 0.4 m at the beam ends over a length of 1.2 m. At the mid-span section, the slab track is 4.08 m wide and 0.26 m thick; at the end sections, it is 4.68 m wide and 0.4 m thick; an increase in thickness is applied at the connection between the web and the slab track for the neck skew. The thickened sections at the ends are reinforced with tension steel bars at both the upper and lower edges, whereas the non-thickened sections are only reinforced at the lower edge, with the upper edge having structural reinforcement. The experiments on the U-beam slab track include static load tests for bending performance, shear lag effect tests, and stress state measurements at the neck skew.

The static load test for the slab track bending performance is a supplementary part of the main beam load test. The loading method utilizes a steel structure portal reaction frame in combination with a synchronized jack system for loading. The transverse loading points simulate train wheel pairs, with a spacing of 1.4 m between loading points, as shown in Fig. 1. In Fig. 2, the loading force P1 on the bottom slab simulates the wheel-rail load, with the maximum loading value being 1.2 times the train axle load. The loading force P2 on the web is the loading value after exceeding 1.2 times the designed load.

The experiment employs three testing methods: affixing strain gauges to the concrete surface, welding rebar meters to the reinforcement, and embedding strain gauges within the concrete. Strain gauges are placed on three test sections of the half-span U-beam: the mid-span section, the 1/4 span section, and the section 0.5 m from the support, as illustrated in Fig. 3. Rebar meters and concrete strain gauges are placed on three test sections on the right half-span: the mid-span section, the center loading face, and the side loading face, as depicted in Fig. 4.

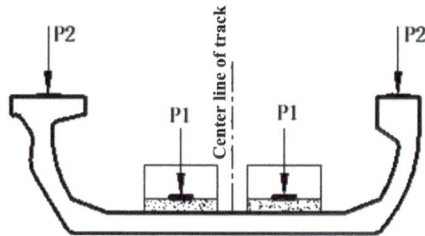

Fig. 1. Lateral Layout of Loading Points on the Reaction Frame

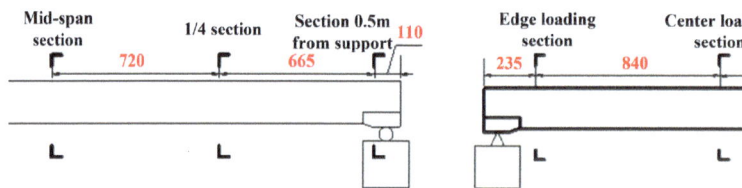

Fig. 2. Cross-Section for Testing Lateral Strain Gauges in Concrete (unit: cm)

Fig. 3. Test Section for Rebar Meter and Concrete Strain Gauge (unit: cm)

2.2 Finite Element Model of Slab Track

The finite element analysis employed both linear and nonlinear analyses to meticulously simulate the structural behavior of precast U-shaped beams under various load conditions. A spatial solid finite element model was constructed using ANSYS. In the model, Solid45 elements were used for concrete due to their ability to accurately simulate three-dimensional stress states in solid structures, which is essential for understanding the stress distribution within the beams. Additionally, Link8 elements were utilized for both prestressed strands and ordinary reinforcement bars because of their proficiency in modeling the behavior of slender structural members under tension, which is crucial for simulating the prestress applied to the beams. The ANSYS finite element models are shown in Figs. 4 and 5.

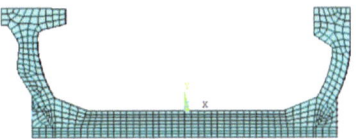

Fig. 4. ANSYS Solid Model   **Fig. 5.** Diagram of Element Division

3   Shear Lag Effect of Slab Track

Based on the ANSYS finite element analysis results, the distribution curves of the normal stress in the concrete of the slab track at the mid-span and the 1/4 span sections are drawn, as shown in Fig. 6.

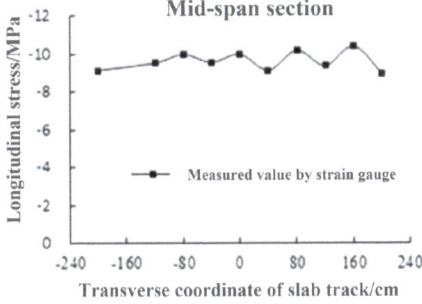

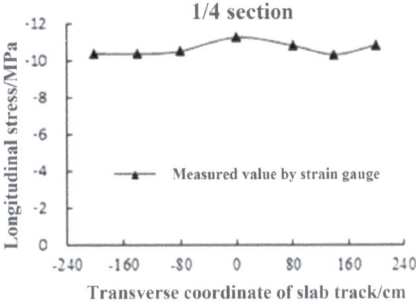

Fig. 6. Curve of longitudinal stress distribution in slab track concrete after prestress release

During the load test study, when the loading level K = 1.0, the strain increments measured by the embedded steel strand strain gauges in the slab track at the mid-span and the 1/4 span sections are converted into concrete stress. The distribution curves of the longitudinal normal stress in the concrete of the slab track are drawn, as illustrated in Fig. 7.

Based on the finite element analysis and experimental test results, the following conclusions can be drawn: As shown in Fig. 7, the finite element analysis of the longitudinal normal stress on the bottom plate closely matches the actual measurements at the mid-span, with differences observed at the quarter span section due to its location in the experimental loading area. According to the finite element analysis, taking into account the shear lag effect, it can be determined that the shear lag coefficient at the mid-span section is 1.09, and at the 1/4 span section, it is 1.08. Analysis of the experimental results shows that the distribution of longitudinal normal stress in the slab track does not vary significantly, suggesting that the impact of the shear lag effect can essentially be disregarded.

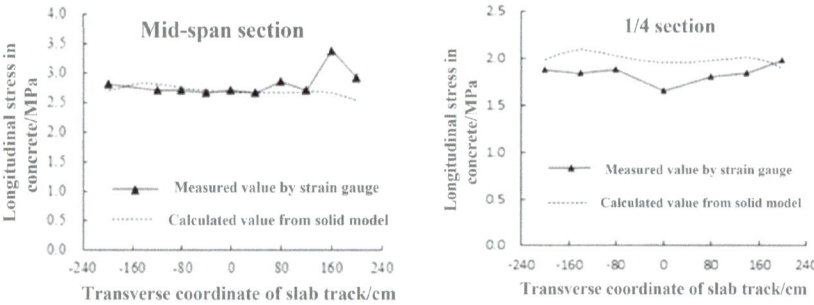

Fig. 7. Curve of longitudinal stress distribution in slab track concrete at load level K = 1.0

4   Analysis of Transverse Bending Mechanical Properties

The transverse bending normal stresses at the mid-span section and the center loading section (where stress concentration occurs) under load level K = 1.0 are extracted from the finite element model, as shown in Figs. 8 and 9.

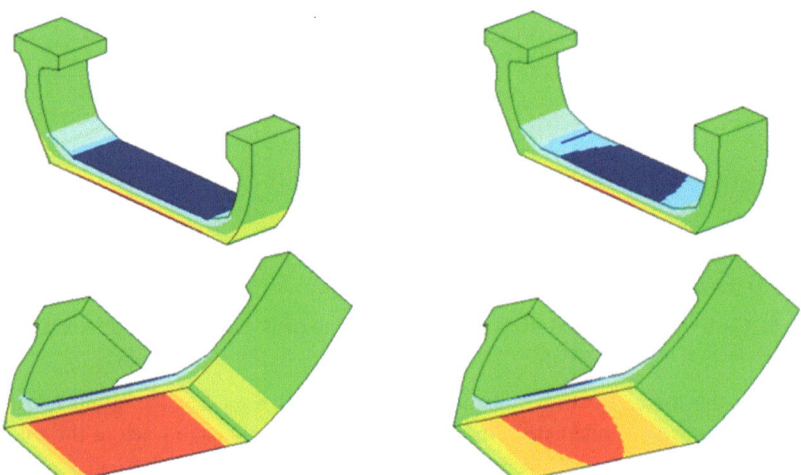

Fig. 8. Stress contour map of transverse stress at Mid-span section

Fig. 9. Stress contour map of transverse stress at 1/4 Section

From the linear elastic finite element analysis results shown in Figs. 8 and 9, it can be observed that even in sections with stress concentration, the maximum compressive stress in the upper concrete of the slab track is 4.0 MPa, which is significantly lower than the concrete's design compressive strength [12]. For non-loading sections, such as the mid-span and 1/4 sections, the maximum tensile stress in the lower concrete of the slab track is 1.88 MPa, not exceeding the limit tensile strength.

5   Conclusions

For the commonly used single-track 30 m standard span precast U-beams in rail transit, under the influence of live loads, the shear lag effect should be considered. The shear lag coefficient at the mid-span section is 1.09, and at the quarter-span section, the shear lag coefficient is 1.08.

Despite stress concentration in certain sections, the maximum compressive stress in the slab track's upper concrete remains well below the design strength, and the maximum tensile stress in non-loading sections does not exceed the concrete's tensile strength limit.

The precast U-shaped beam slabs of prestressed concrete exhibit distinct stress characteristics under various conditions, including transverse bending, shear lag effects, and stress concentrations at the web-to-bottom slab junctions. These findings are instrumental in understanding the structural behavior of railway bridge components under load.

Furthermore, the implications of our findings for the design, construction, and maintenance of railway bridges are significant. By integrating the shear lag effect and stress characteristics into design practices, engineers can enhance the durability and reliability of railway bridge components, potentially leading to more efficient material use and longer service lives.

References

1. Hu, K.Z., Jiang, X.Y., Lu, G.L.: Slot Beams, vol. 1987, pp. 1–10. China Railway Publishing, Beijing house (1987). https://doi.org/10.15935/j.cnki.jggcs.2012.01.002
2. Geng, D.M.: Research and application of integral track bed construction technology for urban rail transit u-beams. Railway Construct. Technol. **1**, 141–144 (2020). https://doi.org/10.3969/j.issn.1009-4539.2020.01.032
3. China urban rail transit association. Annual statistics and analysis report on urban rail transit. Urban Rail Transit **04**, 16–34 (2019). https://doi.org/10.14052/j.cnki.china.metros.2019.04.005
4. Paudel, S., Ganchai, T., Somnuk, T.: Numerical study on seismic performance improvement of composite wide beam-column interior joints. J. Build. Eng. **46**, 103637 (2022). https://doi.org/10.1016/j.jobe.2021.103637
5. Xue, W., Yang, X., Hu, X.: Full-scale tests of precast concrete beam-column connections with composite T-beams and cast-in-place columns subjected to cyclic loading. Struct. Concr. **21**(1), 169–183 (2020). https://doi.org/10.1002/suco.201800171
6. Song, X.D., Li, Q., Wu, D.J.: Prediction method of low-frequency noise in concrete bridges in rail transit. J. China Railway Soc. **40**(3), 126–131 (2018). https://doi.org/10.3969/j.issn.1001-8360.2018.03.019
7. Dawood, M.B., Taher, H.M.A.M.: Various methods for retrofitting prestressed concrete members: a critical review. Period. Eng. Natl. Sci. **9**(2), 657–666 (2021). https://doi.org/10.21533/pen.v9i2.1849
8. Yang, R., Yang, Y., Zhang, X., Wang, X.: An experimental study on secondary transfer performances of prestress after anchoring failure of steel wire strands. Metals **13**(8), 1489 (2023). https://doi.org/10.3390/met13081489
9. Wei, L.D.: Design of pre-stressed u-beams for Qingdao red island—Jiao Nan intercity rail transit. Railway Stand. Des. **61**(7), 101–107 (2017). https://doi.org/10.13238/j.issn.1004-2954.2017.07.023

10. Liu, M.L., Song, Y.M., Wu, D.J.: Study on the mechanical performance of precast u-beams in rail transit. Railway Stand. Des. **62**(3), 75–79 (2018). https://doi.org/10.13238/j.issn.1004-2954.201704100003
11. Wang C.B.: Study on the mechanical performance of pre-tensioned u-beams in rail transit. Shanghai University of Engineering Science (2019). https://doi.org/10.07-6344(2024)03-0248-03
12. Paudel, S., Pudasaini, A., Shrestha, R.K., Kharel, E.: Compressive strength of concrete material using machine learning techniques. Clean. Eng. Technol. **15**, 100661 (2023). https://doi.org/10.1016/j.clet.2023.100661

Study on Fatigue Damage Characteristics of a Cement-Stabilized Soil for the Composite Foundation of Immersed Tunnels

Xiang-qiu Wang$^{(\boxtimes)}$, Tian-jun Feng, and Guo-ping Lei

Department of Transportation, Civil Engineering and Architecture,
Foshan University, Foshan, China
tongji_wxq@163.com, guoping.lei@fosu.edu.cn

Abstract. Composite foundations involving cement-stabilized soils are often adopted for immersed tunnels. In this paper, the cumulative damage characteristics of cement-stabilized soil were studied considering long-term traffic vibration loads. In total, four sets of dynamic fatigue damage tests with different cement contents and stress levels were carried out by applying an asymmetric sine-wave cyclic dynamic load. The fatigue cumulative damage characteristics and damage failure modes of the cement-stabilized soil were analyzed based on Miner's linear elastic fatigue cumulative damage theory. The correlation between the cyclic loading stress levels and the fatigue life N_f of the cement-stabilized soil was obtained, revealing the fatigue damage developing characteristics of the cement-stabilized soil.

Keywords: Composite foundation of immersed tunnels · Cement-stabilized soil · Fatigue damage test · Damage parameter

1 Introduction

Immersed tunnels have been widely used as river-crossing channels for urban public transportation in major cities in China. However, the immersed tube tunnels are usually buried in soft grounds, which do not meet the requirements of foundation bearing capacity or settlement of the submerged tunnels. One of the solutions is to use a composite foundation, such as by applying deep mixing piles or jet grouting piles. Even so, fatigue damages and uneven settlements of the immersed tunnel are commonly seen owing to the long-term effects of traffic loadings, which significantly affect the waterproofing at the tunnel joints. Therefore, it is crucial to investigate the fatigue damage characteristics of cement-stabilized soils under such circumstances.

At present, a number of studies have been carried out on the fatigue damage characteristics of cement-stabilized soils. Jian Wenbin et al. [1] have studied the fatigue life of the cement reinforced soil using fatigue tests with asymmetric sine wave. Zhang Minxia et al. [2] have established the evolution equation for fatigue cumulative damage of cement soil based on uniaxial compression fatigue tests. Th. Tika [3] and Jiang Guolong et al. [4] have established the regression equation for the relationship between the tensile strength

© The Author(s) 2024
A. Bieliatynskyi et al. (Eds.): CSTTE 2023, LNCE 603, pp. 117–123, 2024.
https://doi.org/10.1007/978-981-97-5814-2_11

of cementitious soil and the frequency, amplitude and accumulated vibration times using cyclic loading tests. Chen and Tong [5] have analyzed the fatigue damage characteristics and fatigue life of the basalt fiber-cemented soil using uniaxial compression fatigue tests. Guo et al. [6] have studied the effect of freeze-thaw cycles on the fatigue resistance of cemented soil with and without fibers by performing unconfined fatigue tests. Zhao et al. [7] have analyzed the evolution rules of the compressive strength of cemented soil with freeze-thaw time, fatigue time, and loading amplitude through unconfined compression tests. Zhang zhen, Qinyong Ma, et al. [8–10] have analyzed the influences of confining pressure, static deviator stress, and replacement rate on the cumulative plastic strain of the cemented soil unit of a composite pile by conducting large-scale dynamic triaxial tests.

In conclusion, the majority of previous research on the fatigue damage properties of cement-stabilized soil was based on the results of unconfined uniaxial cyclic loading tests; little research was published using data from dynamic triaxial tests. Furthermore, the majority of previous studies concentrated on cemented soils for subgrades in highways, which may not be the same as those for immersed tunnel composite foundations. It is crucial to use a dynamic triaxial apparatus to expose the fatigue damage features of such cement-stabilized soil.

2 Dynamic Triaxial Fatigue Damage Tests of Cement-Stabilized Soil

2.1 Specimen Preparation

The cement-stabilized soil of the composite foundation of an immersed tunnel in the Pearl River Delta was taken as the research object. The testing material was prepared using three components: fine sand, silt, and cement. The particle size of the fine sand is not larger than 0.075 mm, and 325 ordinary Portland cement was used. The amount of silt was kept at 5% for all the tests, while the mixing ratio of cement varies from 10% to 25%, as seen in Table 1. A constant water-cement ratio of 0.5 was used when mixing the sample material. The specimen is cylindrical in shape, with a diameter of 39.1 mm and a height of 80 mm. The standard curing procedure was adopted for all the specimens.

Table 1. Mechanical and physical properties of the cement-stabilized soil

Test set	CM1	CM2	CM3	CM4
Mixing ratio of cement	10%	15%	20%	25%
Mixing ratio of silt	5%	5%	5%	5%
Elastic modulus (/MPa)	105.7	216.0	326.8	430.5
Uniaxial compressive strength (f_{cm}) (/MPa)	1.12	2.03	3.45	4.25

2.2 Fatigue Damage Tests

The fatigue damage tests of cement-reinforced soil were carried out using a dynamic triaxial test system with an electro-hydraulic servo. A dynamic load of the asymmetric sine-wave was adopted to simulate the effect of traffic loads. In each test set, four maximum amplitudes of the dynamic stress σ_{max}, namely $0.85 f_{cm}$, $0.8 f_{cm}$, $0.7 f_{cm}$, and $0.6 f_{cm}$, , were adopted to consider the effect of stress level, corresponding to the stress level $(S = \sigma_{max}/f_{cm})$ of $0.85, 0.8, 0.7,$ and $0.6,$ respectively (Table 2). The minimum stress amplitude was taken as $0.05 f_{cm}$, and the loading frequency was 2 Hz. Three repetitions were made for each test condition, i.e., there are 12 tests in each test set.

Table 2. Test configurations

Test set	Number of specimens	Maximum stress (σ_{max})	Minimum stress
CM1	$3 \times 4 = 12$	$0.85 f_{cm}, 0.80 f_{cm}, 0.7 f_{cm}, 0.6 f_{cm}$	$0.05 f_{cm}$
CM2	$3 \times 4 = 12$		
CM3	$3 \times 4 = 12$		
CM4	$3 \times 4 = 12$		

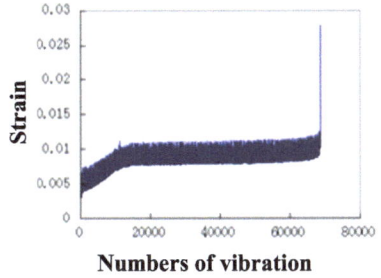

Numbers of vibration

(a) Dynamic strain vs vibration number

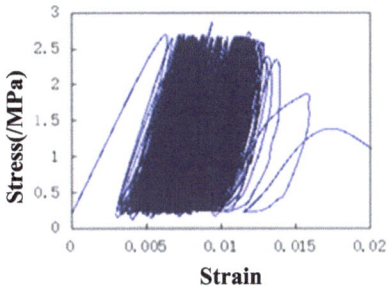

Strain

(b) Hysteretic curve

Fig. 1. Results of a fatigue test in CM3

The results of a typical test in CM3 are plotted in Fig. 1, showing the evolution of strain with vibration number, and the hysteretic stress-strain curve.

In the fatigue tests, mainly two typical specimen failure modes were observed, namely the combined tension-shear failure at the specimen ends and the combined compression-shear failure in the middle. The specific failure mode mainly depends on the initial damage state inside the specimen. When the initial damage of the specimen occurs at the ends, under the action of the axial dynamic load, cracks are first seen at the ends of the specimen along the axial direction. With the increase in load cycles, the axial cracks continue to accumulate and expand, resulting in a combined axial shear and tensile failure of the specimen. When the initial damage of the specimen is not obvious or the initial damage crack exists only in the middle zone of the specimen, under the axial dynamic load, an "X" type shear crack first appears in the middle zone, where the dynamic compressive stress also concentrates. With the increase in the number of load cycles, the damage in the middle zone continues to accumulate, finally giving rise to a combined compression-shear failure in the middle zone.

3 Analysis of the Fatigue Characteristics

Using the test results under different stress levels, the fatigue life (N) - stress level (S) relationship can be obtained for each test set, some of which are shown in Fig. 2. A certain dispersion can also be seen for some repeated test results; however, the general tendency is not affected. Overall, the lower the dynamic stress level, the longer the fatigue life. Similar S-lg$N$ curves were obtained for cemented soil with different compressive strengths. They are fitted by a second-degree polynomial, as shown in Table 3.

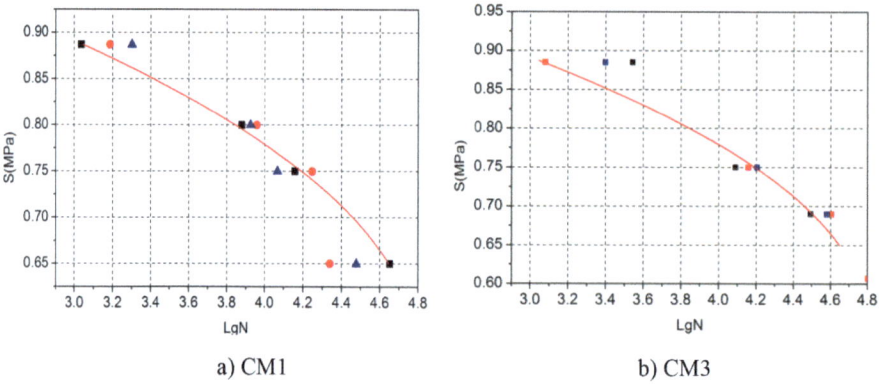

a) CM1   b) CM3

Fig. 2. S-lgN fatigue curves of the cemented soil

Table 3. Fitted results of the fatigue curves

Test set	Elastic modulus (/MPa)	f_{cm} (/MPa)	Fitting results	Correlation coefficient
CM1	107.23	1.12	$S = -0.083(\lg N)^2 + 0.459 \lg N + 0.263$	0.999
CM2	206.00	2.03	$S = -0.051(\lg N)^2 + 0.2271 \lg N + 0.685$	0.983
CM3	377.94	3.95	$S = -0.044(\lg N)^2 + 0.173 \lg N + 0.791$	0.988
CM4	541.66	5.01	$S = -0.046(\lg N)^2 + 0.236 \lg N + 0.612$	0.999

4   Analysis of Fatigue Damage Characteristics

In order to obtain the relationship between the number of load cycles and the damage parameter, according to the results of initial ultrasonic sound velocity tests, specimens with similar wave velocity were selected to perform acoustic damage tests. Based on the damage theory of elastic body, the damage parameter D can be calculated using measured elastic wave velocity as follows:

$$D = 1 - \frac{\hat{E}}{E_0} = 1 - \frac{\hat{v}^2}{v_0^2} \tag{1}$$

Where E_0 and $\hat{E}$ respectively represent the elastic modulus of the specimen before and after the damage, v_0 and $\hat{v}$ respectively represent the elastic wave velocity before and after the damage. According to Miner's linear fatigue cumulative damage theory, if the fatigue life of the material under a cyclic load with a stress level of S_i is N_{fi}, then the material damage caused by a load cycle is $1/N_{fi}$.

For a multi-stage fatigue test with varying stress amplitude, the cumulative damage of the material can be calculated as follows:

$$D = D_1 + D_2 + ... + D_n = \sum_{n=i} D_i = \frac{n_1}{N_{f1}} + \frac{n_2}{N_{f2}} + ... + \frac{n_i}{N_{fi}} = \sum_{n=i} \frac{n_i}{N_{fi}} \tag{2}$$

where D_i represents the part of cumulative damage when the loading stress level is S_i; n_i represents the cumulative vibration number under the stress level of S_i, N_{fi} is the ultimate cumulative vibration number when the stress level is kept constant at S_i.

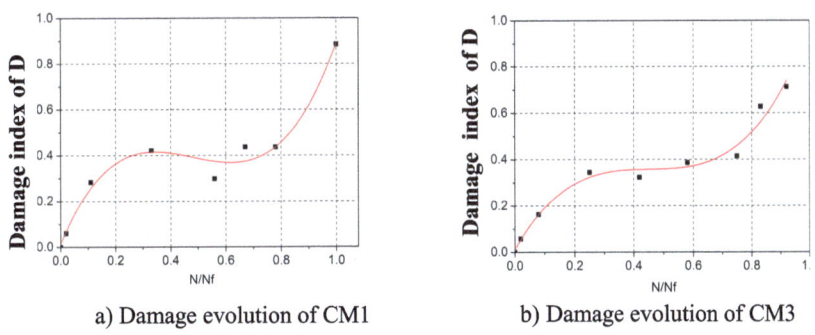

a) Damage evolution of CM1   b) Damage evolution of CM3

Fig. 3. Damage evolution curves of cemented soil with different compressive strengths

Applying the above definition of cumulative damage parameter D, the relationship of $D - N/N_f$ is obtained as shown in Fig. 3, where N is the cumulative vibration number. A similar dynamic cumulative damage process can be seen for the cement-stabilized soil with different compressive strengths. When N is low, the cumulative damage parameter D of the specimen increases with N due to the influence of initial damage. Afterwards, a relatively stable stage is reached, where D remains almost unchanged. As N increases further to exceed a critical value, due to the continuous expansion of the initial damage and the generation and development of new cracks, the cumulative damage parameter accelerates significantly with N, resulting in the macro-damage of the specimen and the ultimate life of N_f. At the same dynamic stress level, the cumulative ultimate vibration number N_f increases linearly with the uniaxial compressive strength f_{cm}.

5 Conclusions

(1) Under the long-term action of traffic loads, the cement-stabilized soil of the composite foundation for immersed tunnels has shown evident fatigue damage characteristics. Mainly, two typical modes of fatigue damage were observed: the combined tension-shear failure at the specimen ends and the combined compression-shear failure in the middle zone, depending on the initial damage location and damage degree of the cement-stabilized soil.

(2) A similar relationship between the cumulative damage parameter and the cumulative vibration number was obtained for cement-stabilized soil with different compressive strengths. It is characterized into three stages: when the cumulative vibration number N is low, the damage parameter D increases with N, which is followed by a relatively stable stage where D remains almost unchanged; at last, when N exceeds a certain critical value, the growth rate of damage parameter D accelerates significantly and eventually leads to the macroscopic failure.

Acknowledgement. The paper is supported by the Guangdong Basic and Applied Basic Research Foundation (2023A1515012085) and the Science and Technology Innovation Project of the Guangdong Provincial Department of Housing and Urban-Rural Development (2023-K33-071697).

References

1. Jian, W., Huang, C., Wu, W., et al.: Research on fatigue behavior cement-sodium-silicate grouted soil. Chin. J. Rock Mech. Eng. **2004**(11), 1949–1953 (2004)
2. Zhang, M., Xu, P., Jian, W.: Experimental study on the ultrasonic wave velocity of the cemented soil fatigue life under dynamic load. Indust. Build. **41**(S1), 728–730+733 (2011)
3. Tika, Th., Kallioglou, P., Koninis, G., Michaelidis, P., Efthimiou, M., Pitilakis, K.: Dynamic properties of cemented soils from Cyprus. Bull. Eng. Geol. Environ. **69**(2), 295–307 (2010). https://doi.org/10.1007/s10064-010-0262-6
4. Guolong, J., Sili, C., Junxiang, W., et al.: Experimental study on mechanical properties of cement-soil under cycle load. Chin. J. Underground Space Eng. **13**(S2), 524–528 (2017)
5. Feng, C., Sheng-hao, T.: Experimental study on fatigue properties of fiber cement soil. J. Shandong Agric. Univ. (Natl. Sci. Edn.) **50**(05), 815–820 (2019)
6. Guo Shaolong, L., Qun, L.Y.: Experimental study on fatigue resistance of basalt fiber reinforced cement soil under freeze-thaw condition. J. Water Resourc. Water Eng. **31**(01), 200–206 (2020)
7. Baichao, Z., Sili, C., Rui, H.: Experimental study on mechanical properties of soil and water under freeze-thaw cycles and fatigue loads. J. China Foreign Highway **41**(04), 362–365 (2021)

8. Zhen, Z., Wenqian, Z., Guanbao, Y., et al.: Deformation and long-term settlement calculation method of unit cell of soil cement column reinforced soft soil under cyclic loading. China J. Highway Transport **35**(11), 21–29 (2022)
9. Zhen, Z., Yong, C., Tianliang, Y., et al.: An experimental study of the dynamic characteristics of cement soils subjected to staged cyclic loading. Hydrogeol. Eng. Geol. **48**(02), 89–96 (2021)
10. Ma, Q., Gao, C.: Effect of basalt fiber on the dynamic mechanical properties of cement-soil in SHPB test. J. Mater. Civil Eng. **30**(8) (2018). https://doi.org/10.1061/(ASCE)MT.1943-5533.0002386

Study on Wave Velocity Under Different Uniaxial Loading Conditions of Weathered Granite

Bo Li, Wei Xiao, Jun Li, Bo Zhao, Hui Chen[✉], Chi Zhang, Qi Sun, and Youliang Bai

Northwest Institute of Nuclear Technology, Xi'an 710024, China
695789500@qq.com

Abstract. Rock strength is an important parameter for rock mass quality evaluation, engineering design, and construction. However, engineering excavation causes rock damage and decreases in rock mass strength. Rock damage is closely related to its acoustic properties during the fracture process. Therefore, in order to obtain the rock stress and damage characteristics of granite during the uniaxial loading failure process under different weathering conditions, rock samples with different weathering degrees were carried out using a triaxial testing machine and wave velocity acquisition equipment. Uniaxial loading and acoustic parameter testing experiments. The test results show that the weathering and integrity of rocks can be characterized by comprehensive parameters such as density, wave speed and wave speed ratio; the elastic deformation stage characteristics of different weathered rocks during uniaxial loading are very significant, and there is an inflection point effect in the rock strength and wave speed curves, but due to Due to the influence of micro-joints and cracks inside the rock, the wave speed value does not completely correspond to the strength of the rock. A new damage variable is established based on rock density and integrity, which can characterize the rock loading failure damage process and has a clearer physical meaning.

Keywords: uniaxial loading · weathering · stress · wave velocity · damage

1 Introduction

In the field of hydropower, transportation and other industries, construction disturbances (excavation, blasting, etc.) will inevitably cause damage to the surrounding rock, the rock belongs to the non-homogeneous initial damage brittle materials, its deformation and destruction process is accompanied by the emergence of local cracks and the expansion of the overall until the destruction of the damage is a cumulative process. Rock damage is a phenomenon that occurs both inside and on the surface at the same time, and the damage process is very complicated. Correctly describing the rock damage evolution phenomenon is a prerequisite for revealing the deformation and damage mechanism of rock, and at the same time, the weakening of the mechanical parameters of the surrounding rock leads to safety risks. Therefore, it is necessary to carry out research on quantitative assessment of rock damage.

A. Bieliatynskyi et al. (Eds.): CSTTE 2023, LNCE 603, pp. 124–137, 2024.
https://doi.org/10.1007/978-981-97-5814-2_12

Many scholars at home and abroad have carried out fruitful work in rock damage research. At present, rock damage research methods mainly include elastic modulus method, ultrasonic method, dissipated energy method, acoustic emission method, CT scanning method, resistivity method, and damage method, structural modeling method and others [1–17]. With the continuous development of deep learning technology, digital images have also become a hotspot in current rock damage research [18–26]. The acoustic wave testing method can reflect the internal structure information of rocks and is widely used in the field of rock mechanics. It is fast, non-destructive and efficient. Cai Z L, et al. [3] obtained the acoustic characteristics of the rock failure process through a uniaxial compression test of granite, and the results showed that the wave velocity it is more sensitive to changes in compressive stress, but it is still difficult to detect changes in local micro defects; Engelder et al. [27] conducted a study on changes in rock ultrasonic waves and strain relaxation, and the results showed that there is a correlation between the change amplitude of wave speed and the change amplitude of strain relaxation; Zhang et al. [28] conducted a full-process ultrasonic detection test on the damage and fracture evolution of cross-cracked rock mass. The test results showed that wave speed can better reflect the damage and fracture evolution process of fractured rock mass; Wang et al. [29] used acoustic emission CT imaging technology obtained the regional characteristics of wave velocity distribution inside the rock during the uniaxial loading process, and analyzed the high and low wave velocity zones and stress response characteristics; Dou et al. [30] proposed an impact hazard vibration wave CT inversion method for impact mine pressure disaster, which can invert the vibration wave speed distribution characteristics in the study area and realize the judgment of impact hazardous area.

From the current research results, rock damage has been the research hotspot of related technicians, rock damage has been carried out a variety of technical methods of research work, wave velocity, acoustic emission, etc., and the relationship between rock damage research has been continuous, but for different weathering degree of rock damage research is less, and the physical significance of damage indicators is not clear. Therefore, in this paper, by conducting uniaxial and wave velocity tests on granite under different weathering conditions, we obtain the characteristic stress and characteristic wave velocity values of the progressive damage process of granite, and establish the density-wave velocity relationship equation for granite with different weathering degrees. At the same time, this paper tries to adopt the density and integrity indexes that can reflect the damage characteristics of the rock body macroscopically to define the damage variables, and discusses the comparison with the rock damage indexes defined by wave velocity.

2 Experimental Design

2.1 Test Principle

Due to weathering, unloading and alteration, granite has the characteristics of inhomogeneity, discontinuity and anisotropy, its internal contains a large number of pores and microfractures, in the loading process of the rock volume generated by the compression and expansion, while the density change is a macroscopic reflection of the change in the volume of the rock, can be through the density of the characterization of the rock damage in the loading process. Rock integrity coefficient is a geological parameter related to the quality and mechanical strength of the rock body, and it is also a link from the rock specimen to the damage of the rock body, which can be obtained by the number of volume nodules of the rock body or the elastic wave velocity. For granite with different degrees of weathering, the macro-parameters of integrated density and integrity of the rock mass solve the problem of quantitative characterization of the damage parameters of the rock mass in the loading process, and at the same time, the change of wave velocity can reflect the rich information of the internal structure of the rock, so the wave velocity parameter of the rock sample in the loading process is obtained by using sonic testing technology.

The wave speed value can be calculated by the length of the rock sample and the time difference between the excitation probe and the receiving probe. The calculation formula is:

$$V_p = \frac{l}{t - t_0} \tag{1}$$

where V_p is compressional wave velocity, m/s; l is the distance between the corresponding acoustic transducers at both ends of the rock sample, m; t is the acoustic signal receiving time, s; t_0 is the acoustic signal excitation time, s.

2.2 Test Process

In accordance with the "Standard for Test Methods of Engineering Rock Masses" (GB/T50266-2013), the sample is a standard specimen of a cylinder with a diameter of 5 cm and a height of 10 cm. The end parallelism and axis deflection do not exceed 0.25°. Unqualified samples are removed and dried, and data such as sample size and quality are obtained.

Before the uniaxial test, density and axial wave velocity tests were carried out on the prepared rock samples. The acoustic wave testing system during uniaxial loading uses a compression-resistant acoustic wave transducer. During the test, the acoustic wave transducers are placed at both ends of the rock sample. The acoustic wave testing arrangement is shown in Fig. 1a. To carry out rock uniaxial testing, the testing machine adopts a 3000 kN fully digitally controlled electro-hydraulic servo rock rigid triaxial testing machine, which can obtain physical and mechanical parameters such as rock strain, mechanical strength and compressional wave velocity during uniaxial loading. The test process is shown in Fig. 1b.

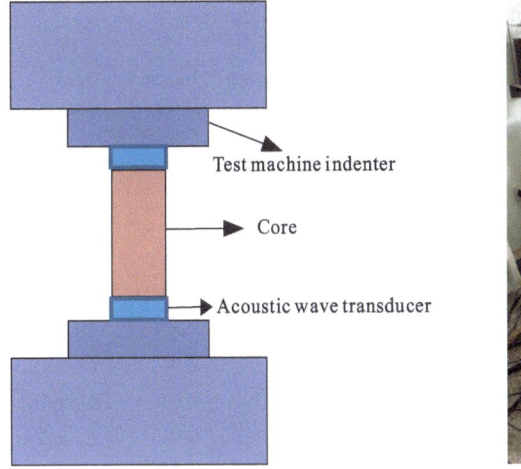

a. Sonic test probe layout

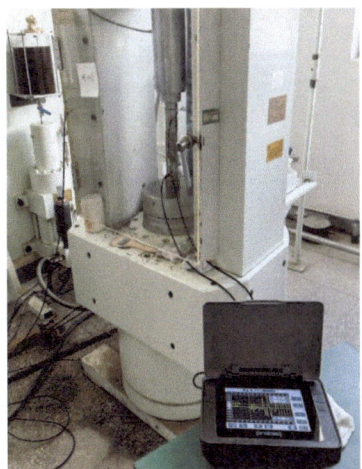

b. Rock uniaxial loading and wave velocity test

Fig. 1. Rock uniaxial test layout and test experiments

3   Test Data Statistics

3.1   Basic Physical and Mechanical Parameters of Rock Samples

Obtain the density value of the rock sample based on the geometric parameters and weight of the rock sample. At the same time, a wave velocity test was conducted on the unweathered granite rock sample. The initial compressional wave velocity $V_{P0} = 6000$ m/s. Combining the measured wave speed values of each sample, the weathering degree and integrity index can be obtained. The rock sample density, compressional wave velocity, shear wave speed, longitudinal to transverse wave ratio, weathering degree and integrity index (K_{v0}) are detailed in Table 1. Table 1 V_P represents the compressional wave velocity, V_S represents the shear wave speed, and V_{P0} is the initial wave speed of fresh granite.

The wave speed ratio of sample H3 is 0.4–0.6, indicating strong weathering, the remaining samples are all greater than 0.85, and are unweathered to slightly weathered.

Judging from the integrity index, the K_v value of the integrity index of the H3 sample is less than 0.35, and the integrity is broken. At the same time, its density value is small, the wave speed is low, and the wave speed ratio data is abnormal than other samples; the H4-2 sample is relatively complete, can be judged by the wave speed value, wave speed ratio and integrity index, but the density value parameter is difficult to evaluate the integrity; the other rock samples are all complete and have a good correspondence with the wave speed parameters.

From the perspective of longitudinal and transverse wave speeds, the higher the degree of weathering, the lower the rock longitudinal and transverse wave speeds. The ratio of H3 transverse waves to the transverse waves of other rock blocks is about 1/3, and the ratio of H3 compressional wave to the compressional waves of other rock blocks is about 1/2. The longitudinal and transverse waves can reflects changes in the physical

Table 1. Statistical table of test parameters of rock samples

Sample number	Density (g/cm^3)	Compressional velovity (m/s)	Transverse wave velocity (m/s)	V_P/V_S	V_P/V_{P0}	Weathering degree	Intactness index of rock mass (K_{v0})	Rock integrity
H1-1	2.6799	5250	3300	1.5910	0.88	Slightly weathered	0.77	Complete
H1-2	2.6870	5548	3480	1.5943	0.92	Unweathered	0.86	Complete
H1-3	2.6875	5473	3803	1.4391	0.91	Unweathered	0.83	Complete
H1-4	2.6770	5469	3311	1.6518	0.91	Unweathered	0.83	Complete
H2-1	2.6258	5714	3600	1.5872	0.95	Unweathered	0.91	Complete
H2-2	2.6286	5889	3269	1.8015	0.98	Unweathered	0.96	Complete
H2-3	2.6100	5806	3411	1.7021	0.97	Unweathered	0.94	Complete
H3-1	2.3461	2560	1080	2.3704	*0.43*	strongly weathered	*0.18*	Broken
H3-2	2.3689	2477	1150	2.1539	*0.41*	strongly weathered	*0.17*	Complete
H3-3	2.4016	3000	1630	1.8405	*0.50*	strongly weathered	*0.25*	Complete
H4-1	2.5424	5523	3644	1.5156	0.92	Unweathered	0.85	Complete
H4-2	2.5561	5102	3115	1.6379	0.85	Slightly weathered	0.72	Relatively complete
H4-3	2.5889	5423	3396	1.5969	0.90	Unweathered	0.82	Complete
H4-4	2.5954	5558	3700	1.5022	0.93	Unweathered	0.86	Complete

properties of rocks; from the perspective of longitudinal and transverse wave ratios, H3 samples are all greater than 1.8, while other samples are less than 1.7, reflecting that the weathering degree and integrity of H3 samples are worse than other samples.

It can be seen from the wave speed data that both the shear wave speed and the longitudinal to shear wave ratio can reflect the changes in the physical properties of the rock. Therefore, the comprehensive density, wave speed and wave speed ratio can better evaluate the weathering and integrity of the rock.

3.2 Uniaxial Loading Test Data

Patterns of Change in Intensity

Rock stress and strain data can be obtained through uniaxial loading tests, and H1-2, H2-1, H3-2, and H4-2 stress-strain relationship curves are selected, as shown in Fig. 2 for details.

In a typical stress-strain relationship curve, the rock sample failure process can be divided into a compression section, an elastic deformation section, a stable crack expansion section, an unstable rupture stage (yield stage), and a rupture to residual strength section [31]. While the pressure-dense section in Fig. 2a and b is not significant,

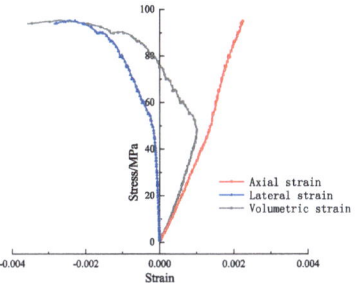
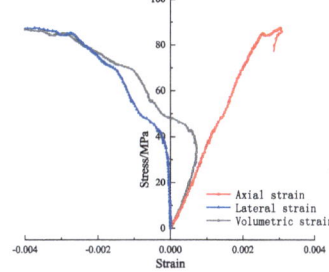

a H1-2 uniaxial compression test stress-strain curve

b H2-1 uniaxial compression test stress-strain curve

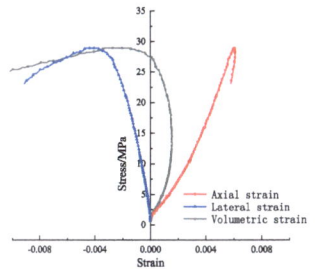
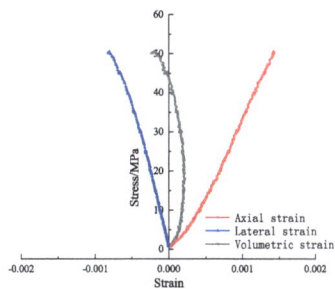

c H3-2 uniaxial compression test stress-strain curve

d H4-2 uniaxial compression test stress-strain curve

Fig. 2. Stress-strain curves of uniaxial compression test

the pressure-densified section can be seen in Fig. 2c and d, and the elastic deformation section is very significant overall. Due to the direct destruction of the rock sample in Fig. 2a, c, and d, the yield stage is not visible, Fig. 2b It can be seen that there is a yield section.

It can be seen from Fig. 2 that the granite rock sample is significantly brittle under uniaxial action, and the linear elastic stage is obvious. The H3-2 rock sample is strongly weathered and broken. Figure 2c shows that the degree of weathering and completeness do not affect the development of the elastic segment during uniaxial loading; the H4-2 rock sample is slightly weathered and relatively complete, but the peak intensity is low, indicating that the wave speed value It does not completely correspond to the strength of rock, mainly because micro-joints and cracks control the strength of rock blocks.

Wave Speed Change Rules

During the uniaxial loading process, a compression-resistant acoustic wave transducer was used to obtain acoustic wave test data. The wave speed-stress curves of specimens H1-2, H2-1, H3-2, and H4-2 are shown in Fig. 3. In order to better compare and analyze the above sample data, the wave speed is normalized, as shown in Fig. 4.

It can be seen from Figs. 3 and 4 that as the uniaxial loading process continues, the wave speed values of samples H3-2 and H4-2 show an increasing trend with the increase

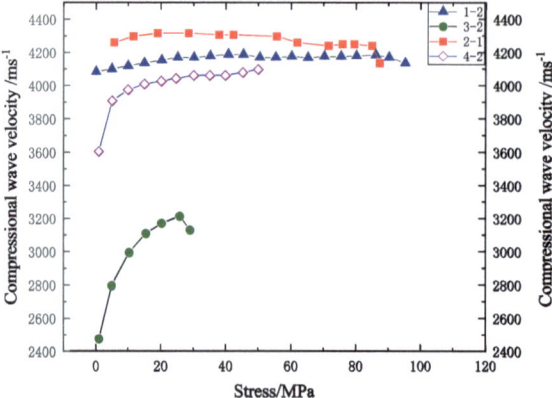

Fig. 3. Wave speed-stress curves of uniaxial compression test

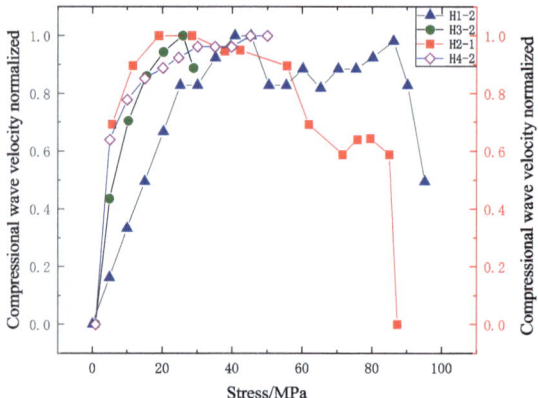

Fig. 4. Relationship curves between wave speed normalization and stress in uniaxial compression test

of stress. The increase first increases sharply and then develops steadily, and the higher the weathering, the higher the stress. The increase is more obvious in the early stage; the degree of weathering of samples H1-2 and H2-1 is the same, both are unweathered. Among them, sample H1-2 shows a trend of increasing first and then stabilizing, while sample H2-1 increases first and then decreases, but the overall wave speed of the two tends to be stable, indicating that the lower the weathering degree of the rock, the smaller the wave speed value fluctuates during the uniaxial loading process.

It can be seen from Fig. 3 that the higher the wave speed value, the greater the peak stress. There is a positive correlation between wave speed and peak stress. The peak wave speed-stress was fitted. The fitting results are shown in Fig. 5. It can be seen from Fig. 5 that when the wave speed is less than 4000 m/s, the stress level is low and the change amplitude is small. When the wave speed is greater than 4000 m/s, the stress increases sharply, indicating that there is an inflection point effect in the rock strength and wave speed curve. The shape of the stress -wave speed curve in this article is the same as the relationship curve between rock strength and wave speed in the literature of Wang et al. [29] and Zhao et al. [32] indicating that the relationship between stress and compressional wave velocity during rock loading can be established through an exponential function.

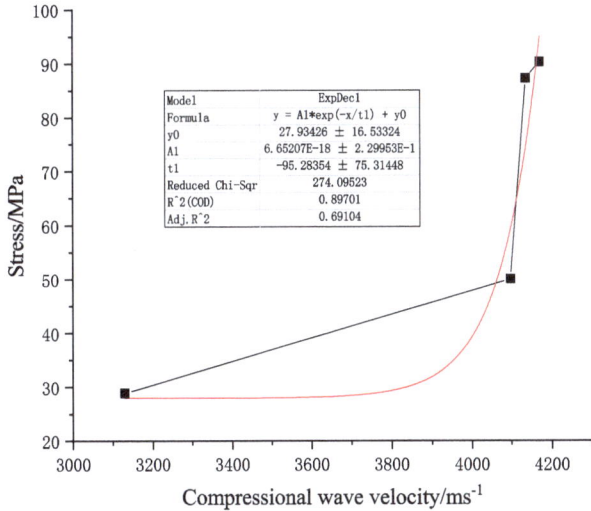

Fig. 5. Peak stress- compressional wave velocity curve of uniaxial compression test

4   Quantitative Analysis of Rock Damage

Rock damage variables can be selected from micro and macro perspectives [33]. The micro perspective mainly determines the damage benchmark through crack density, such as the GK model [34] and TCK model [35]; the macro perspective is mainly based on rock physical and mechanical properties are used as damage benchmarks, such as elastic modulus method, ultrasonic wave velocity method, energy method, CT number method and acoustic emission cumulative number method, etc.

The internal damage of rock under external force will cause the change of acoustic parameters, which can characterize the change of rock mechanical properties and internal structure, and has the characteristics of non-destructive, therefore, it is more widely used. Jin et al. [36] proposed a method to define the damage variables based on the wave impedance of rocks and gave the expression for defining the damage variables; in order to study the effect of static stress on the propagation of rock stress waves, Jin et al. [37] carried out small disturbance stress wave propagation tests on a long specimen of red sandstone to obtain the propagation and attenuation characteristics of the stress waves under different static stress conditions; Jia et al. [38] used a real-time ultrasonic acoustic parameter prediction method to predict the propagation and attenuation characteristics of stress waves under different static stress conditions. Taking the fine structural characteristics of rocks as the entry point and damage mechanics as the theoretical basis, Li Bo [39] obtained the critical state parameters of rock damage using CT image processing technology, damage theory analysis, fractal theory analysis, and improved crack strain model method. Therefore, this study tries to establish the damage variables by rock physical and mechanical parameters, and based on the wave velocity definition of the damage variables of the previous researchers, it is proposed to use the density and integrity index to define the damage variables.

4.1 Define Damage Variables Based on Wave Speed

The change in rock wave speed can effectively reflect the damage of rock before and after loading. The wave speed expression of the damage variable is:

$$D = 1 - (\frac{V_p}{V_{p0}})^2 \tag{2}$$

where D is the damage variable; V_p is the compressional wave velocity during rock loading, m/s; V_{p0} is the initial compressional wave velocity of fresh rock, m/s.

The initial compressional wave velocity of fresh granite is 6000 m/s. Taking sample H1-2 as an example to calculate the damage change of the rock sample under uniaxial loading of the granite sample, the damage variable changes with stress curve is shown in Fig. 6.

It can be seen from Fig. 6 that under uniaxial loading conditions, the damage variables of granite do not increase linearly with the increase of stress. Instead, they first decrease, then become dynamically stable and finally increase sharply. This shows that as the stress increases, the damage variables of rock first decrease due to crack closure. Small, the damage variables change dynamically within a small range during the stable crack expansion stage, and the damage variables increase sharply during the unstable expansion of the crack to the failure stage. Generally speaking, the change pattern of damage variables is consistent with the rock loading failure process.

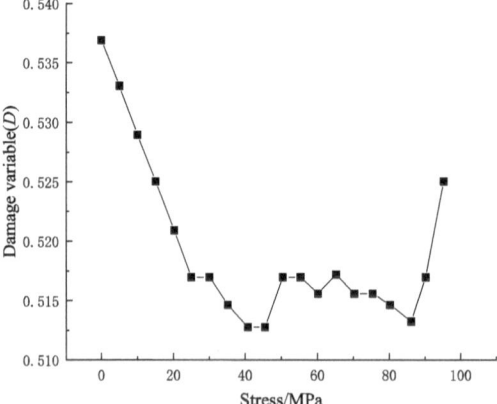

Fig. 6. Relationship curve between damage variable D and stress in uniaxial compression test

4.2   Define Damage Variables Based on Density and Integrity

The density of rock will change during the loading process. In order to establish the density-wave speed relationship, combined with previous research results, a relationship curve between rock density and wave speed in Table 1 was established and fitted (the red line is the fitting curve). See Fig. 7 for details. The fitting relationship between density and wave speed is:

$$\rho = 0.80303 V_{\mathrm{P}}^{0.13745} \tag{3}$$

Where ρ is rock density, g/cm^3; V_{p} is rock compressional wave velocity, m/s.

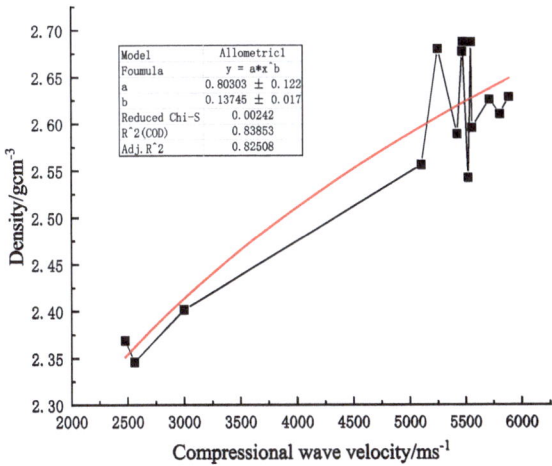

Fig. 7. Fitting curve of rock density and wave speed

Rock integrity can be characterized by the rock mass integrity row index K_{v}, which quantitatively reflects the integrity condition. The relationship expressions of comprehensive density and integrity index definition of damage variables and integrity index are respectively:

$$D' = \frac{\rho K_{\mathrm{V}}}{\rho_0 K_{\mathrm{V0}}} \tag{4}$$

$$K_{\mathrm{v}} = \left(\frac{V_p}{V_{p0}}\right)^2 \tag{5}$$

where ρ is rock density, g/cm^3; ρ_0 is initial density of rock, 2.7 g/cm^3; K_{v0} is initial rock integrity index, ranging from 0 to 1; V_{p0} is initial wave speed of rock, 6000 m/s.

Taking sample H1-2 as an example, the change curve of damage variables with stress under rock uniaxial loading is shown in Fig. 8.

Comparing Figs. 6 and 8, since the relationship between the rock density and rock wave speed parameters in the damage variable D' has been established, from a morphological point of view, the shape of the damage variable D' and the damage variable

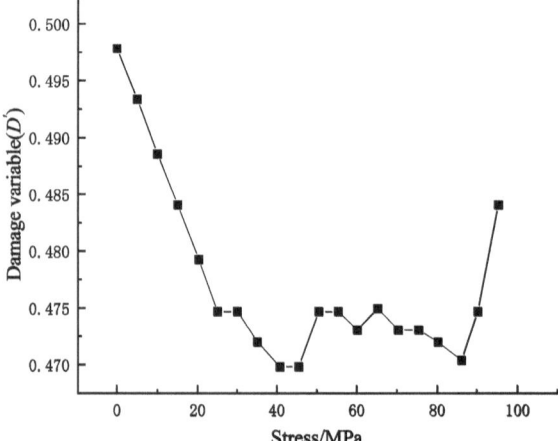

Fig. 8. Relationship curve between damage variable D' *and stress* in uniaxial compression test

D with the stress change curve are consistent; from the damage perspective From the numerical perspective of the variables, the D' value is slightly smaller than the D value, and the change is small; from the definition parameter index, the physical meaning of the damage variable D' based on density and integrity index is more clear, and the damage variable values are equivalent, indicating that the damage variable The definition is feasible.

5   Conclusion

By conducting uniaxial loading and compressional wave velocity tests on granite with different weathering degrees, the conclusions are as follows:

(1) Density, wave velocity and wave velocity ratio can better evaluate the degree of rock weathering and integrity, meanwhile, considering the change of density of rock in the loading process, the exponential function relationship between density and wave velocity is established, and the rock damage indexes of density and integrity are constructed, which provides a new idea for the study of rock damage evolution.

(2) Granite rock samples in the uniaxial action of the rock brittleness is significant, the line elasticity stage is obvious, but the degree of weathering and integrity does not affect the development of the elastic section of the uniaxial loading; uniaxial loading process, the wave velocity value with the stress of the first increase in the trend of smooth development, and the higher the weathering of the increase in the first period of time is more pronounced, and the strength of the rock and the wave velocity curves there is a point of inflexion effect.

(3) Based on the density and integrity of the definition of damage variables, the physical significance of the damage variables is clear, and can better characterize the damage process in the process of uniaxial loading of rocks, but the density and integrity of the indicators did not take into account the characteristics of the nodule production and filler, the author will follow up from the indoor experiments, numerical analyses,

theoretical calculations, etc., to carry out systematic research on the effect of nodules on the indicators, and to establish a quantitative relationship between the number of groups of nodules, the production of nodules, the filler, etc., and the indicators, to further expand the damage model. The relationship between the number of nodule groups, yield and filling materials and the damage indexes will be established to further expand the application scope of the damage model.

References

1. Lemaitre, J.: Evaluation of dissipation and damage in metals submitted to dynamic loading. Mech. Behav. Mater. **76**, 540–549 (1972)
2. Kemeny, J., Cook, N.G.W.: Effective moduli, non-linear deformation and strength of a cracked elastic solid. Int. J. Rock Mech. Min. Sci. Geomechan. Abst. **23**, 107–118 (1986). https://doi.org/10.1016/0148-9062(86)90337-2
3. Cai, Z.L., Liu, K., Wu, M.B., et al.: The research on acoustic properties of granite under uniaxial compression condition. Rock Soil Mech. **7**(2), 27–36 (1986)
4. Ren, J.X., Ge, X.R.: Study of rock meso-damage evolution law and its constitutive model under uniaxial compression loading. Chin. J. Rock Mech. Eng. **20**, 2–12 (2001). https://doi.org/10.3321/j.issn:1000-6915.2001.04.001
5. Chen, Y.S., Li, N., Han, X., et al.: Research on crack developing process in non-interpenetrated crack media by using CT. Chin. J. Rock Mech. Eng. **24**, 2665–2670 (2005). https://doi.org/10.3321/j.issn:1000-6915.2005.15.011
6. Xie, H.P., Ju, Y., Li, L.Y., et al.: Energy mechanism of deformation and failure of rock masses. Chin. J. Rock Mech. Eng. **27**, 1729–1740 (2008). https://doi.org/10.3321/j.issn:1000-6915.2008.09.001
7. Li, B., Wu, R.J., Gao, W.C.: A modified damage softening constitutive model for rock. China Earthquake Eng. J. **38**, 783–786 (2016). https://doi.org/10.3969/j.issn.1000-0844.2016.05.0783
8. Zhang, G.K., Li, H.B., Wang, M.Y., et al.: Comparative study on damage characterization and damage evolution of rock under uniaxial compression. Chin. J. Geotechn. Eng. **41**, 1074–1082 (2019). https://doi.org/10.11779/CJGE201906011
9. Wang, W., Li, Z., Chen, M.M., et al.: Characteristics of infrared radiation variation coefficient of rock under uniaxial loading. Energy Energy Conserv. **11–12** (2021). https://doi.org/10.16643/j.cnki.14-1360/td.2021.05.003
10. Liang, M.C., Miao, S.J., Cai, M.F., et al.: A damage constitutive model of rock with consideration of dilatation and post peak shape of the stress-strain curve. Chin. J. Rock Mech. Eng. **40**, 2392–2401 (2021). https://doi.org/10.13722/j.cnki.jrme.20 21.0107
11. Zhao, Y.G., Huang, L.Q., Li, X.B.: Identification of stages before and after damage strength and peak strength using acoustic emission tests. Chin. J. Geotechn. Eng. **44**, 1908–1916 (2022). https://doi.org/10.11779/CJGE202210017
12. Lei, M.F., Zhao, C.Y., Zeng, C., et al.: Rock damage calculation method based on the damage release energy. Chin. J. Rock Mech. Eng. **41**, 3210–3218 (2022). https://doi.org/10.13722/j.cnki.jrme.2022.0065
13. Qiao, J.Y., Liu, D.Q., Guo, Y.P.: Study on iterated damage and swelling evolution of rock. China Min. Magaz. **31**, 12–23 (2022). https://doi.org/10.12075/j.issn.1004-4051.2022.09.023
14. Liu, D.Q., Guo, Y.P., Li, J.Y., et al.: Experimental study on damage and failure energy evolution of brittle rocks under uniaxial compression. J. Eng. Geol. **31**, 843–853 (2022). https://doi.org/10.13544/j.cnki.jeg.2022-0799

15. Li, S.N., Xiao, J., Li, Y., et al.: A new damage constitutive model of rock considering microscopic crack growth. Chin. J. Rock Mech. Eng. **42**, 640–648 (2023). https://doi.org/10.13722/j.cnki.jrme.2022.0364
16. Duan, M.K., Jiang, C.B., Guo, X.W., et al.: Experimental study on mechanical and damage characteristics of coal under cyclic true triaxial loading. Chin. J. Rock Mech. Eng. **40**, 1110–1118 (2022). https://doi.org/10.13722/j.cnki.jrme.2020.0916
17. Liu, D.Q., Guo, Y.P., Li, J.Y., et al.: Damage constitutive model for layered yellow sandstone based on dissipative energy evolution and its verification. Chin. J. Eng. **1–22** (2023). https://doi.org/10.13374/j.issn2095-9389.2023.06.18.002
18. Li, X.F.: Research on rock fracturing and fragmentation subject to intensive impact loading. Chin. J. Rock Mech. Eng. **40**, 432 (2021). https://doi.org/10.13722/j.cnki.jrme.2020.0832
19. Fan, J., Zhu, X., Hu, J.W., et al.: Experimental study on crack propagation and damage monitoring of sandstone using three-dimensional digital image correlation technology. Rock Soil Mech. **43**, 1009–1019 (2022). https://doi.org/10.16285/j.rsm.2021.1132
20. Zhang, Q.H., Chen, C., Yuan, L., et al.: Early and intelligent recognition of dynamic cracks during damage of complex fractured rock masses based on DIC and YOLO algorithms. J. China Coal Soc. **47**, 1208–1219 (2022). https://doi.org/10.13225/j.cnki.jccs.XR21.1744
21. Zhang, Q.H., Yuan, L., Fang, Z.Y., et al.: Failure law and early warning method of precast fractured rock specimen based on multi monitoring information fusion. J. Min. Safety Eng. **39**, 797–807 (2022). https://doi.org/10.13545/j.cnki.jmse.2021.0065
22. Qi, X.Y., Wang, S.W., Yang, Z., et al.: Composite rock test and damage model based on digital image. Sci. Technol. Eng. **22**, 13450–13459 (2022). https://doi.org/10.3969/j.issn.1671-1815.2022.30.038
23. Cheng, B., Li, D.R.: Full-field dynamic measurement method for fatigue cracks based on decorrelation DIC. Chin. J. Theor. Appl. Mech. **54**, 1040–1050 (2022). https://doi.org/10.6052/0459-1879-21-650
24. Zhou, Y., Cheng, Y.T.: Non-contact structural displacement measurement based on digital image correlation method. J. Hunan Univ. (Natl. Sci.) **48**, 1–9 (2021). https://doi.org/10.16339/j.cnki.hdxbzkb.2021.05.001
25. Li, J., Liu, C., Liu, H.M., et al.: Study on meso-damage mechanism of shale reservoir rock based on digital cores. Chin. J. Rock Mech. Eng. **41**, 1103–1113 (2022). https://doi.org/10.13722/j.cnki.jrme.2021.0269
26. Zhang, J.Y., Han, F., Jiang, B.H., et al.: Peridynamic modeling through micro-CT images for failure simulation of composite microstructure. Acta Mech. Solida Sin. **43**, 143–157 (2022). https://doi.org/10.19636/j.cnki.cjsm42-1250/o3.2021.036
27. Engelder, T., Plumb, R.: Changes in in situ ultrasonic properties of rock on strain relaxation. RockMech. Min Sci. Geomech. Abstr. **21**, 75–82 (1997). https://doi.org/10.1016/0148-9062(84)91175-6
28. Zhang, X.J., Lin, Q.J., Xiu-li, S., et al.: Quantitative ultrasound prediction on damage and fracture evolution of fractured rock mass. J. Min. Safety Eng. **24**, 378–383 (2017). https://doi.org/10.13545/j.cnki.jmse.2017.02.026
29. Wang, C.B., Cao, A.Y., Jing, G.C., et al.: Evolution characteristics of rock fracture under uniaxial loading by combining acoustic emission and CT imaging. Chin. J. Rock Mech. Eng. **35**, 2044–2053 (2016). https://doi.org/10.13722/j.cnki.jrme.2015.1735
30. Dou, L.M., Jiang, Y.D., Cao, A.Y., et al.: Monitoring and pre-warning of rockburst hazard with technology of stress field and wave field in underground coalmines. Chin. J. Rock Mech. Eng. **36**, 29–37 (2017). https://doi.org/10.13722/j.cnki.jrme.2016.0756
31. Li, S.J.: Experimental study on the failure characteristics of rock under different confining pressures and it's grouting reinforcement. Jiangxi University of Science and Technology, Ganzhou (2020)

32. Zhao, M.J., Xu, R.: The rock damage and strength study based on ultrasonic velocity. Chin. J. Geotechn. Eng. **22**, 720–722 (2000). https://doi.org/10.3321/j.issn:1000-4548.2000.06.018
33. Sun, C.S.: The determination method of elastic modulus of deep damage rock mass. Northeastern University, Shenyang (2013). https://doi.org/10.7666/d.J0118531
34. Grady, D.E., KIPP, M.E.: Continuum modelling of explosive fracture in oil shale. Int. J. Rock Mech. Min. Sci. Geomech. Abst. **17**, 147–157 (1980). https://doi.org/10.1016/0148-9062(80)91361-3
35. Taylor, L.M., Chen, E.P., Kuszmaul, J.S.: Microcrack-induced damage accumulation in brittle rock under dynamic loading. Comput. Methods Appl. Mech. Eng. **55**, 301–320 (1986). https://doi.org/10.1016/0045-7825(86)90057-5
36. Jin, J.F., Li, X.B., Yin, Z.Q., et al.: A method for defining rock damage variable by wave impedance under cyclic impact loadings. Rock and Soil Mechanics **32**(1385–1393), 1410 (2011). https://doi.org/10.3969/j.issn.1000-7598.2011.05.017
37. Jin, J.F., Cheng, Y., Chang, X.X., et al.: Experimental study on stress wave propagation characteristics in red sandstone under axial static stress. Chin. J. Rock Mech. Eng. **36**, 1939–1950 (2017). https://doi.org/10.13722/j.cnki.jrme.2016.1458
38. Jia, P., Zhu, P.C., Li, B., et al.: Characteristics of real-time ultrasonic wave during uniaxial compression of rock. J. Central South Univ. (Sci. Technol.) **53**, 3967–3977 (2022). https://doi.org/10.11817/j.issn.1672-7207.2022.10.017
39. Li, B.: Study on crital information of rock identification and strength variation fragmentation process. Inner Mongolia Univ. Sci. Technol. (2017). https://doi.org/10.7666/d.D01250324

Study on Dynamic Porosity and Dynamic Characteristics of Unsaturated Soils

Jiaxin Zhao[✉]

College of Mechanical Engineering, Xi'an University of Science and Technology, Xi'an 710054,
Shaanxi, China
15002960558@163.com

Abstract. Under dynamic load, the porosity of unsaturated soil will change correspondingly, affecting the research accuracy of soil. Therefore, a dynamic response model of unsaturated soil considering dynamic porosity is established based on mixture theory to reveal the dynamic characteristics of unsaturated soils fully. The Comsol Multiphysics PDE is used to analyze the porosity change and dynamic response characteristics of two-dimensional unsaturated soil under vertical harmonic load. The numerical results show that the difference in the upper boundary is an essential factor affecting the dynamic characteristics of the subgrade. Compared with the upper boundary impervious to water and air, the pressure of the liquid and gas phases is zero, and the change of porosity and vertical displacement is more prominent. When considering the dynamic porosity, the porosity decreases under the load, and the resistance becomes more prominent when the soil skeleton is compressed, so the vertical displacement decreases, resulting in a corresponding decrease in liquid and gas phase pressure.

Keyword: Unsaturated soil · Dynamic porosity · Mixture theory · Comsol Multiphysics PDE

1 Introduction

The problem of dynamic response of unsaturated soils under dynamic loading has been widely studied in many fields, such as earthquake engineering, oilfield exploration, and geophysics.

Currently, in studying the dynamic properties of unsaturated soils, most are based on the equivalent fluid model [1] and the mixture theory [2] as the theoretical basis. Yang [3] studied the propagation of elastic waves in unsaturated porous media using the equivalent fluid model. Yuanqiang Cai [4] investigated the significant influence of saturation change on elastic wave propagation in saturated and unsaturated porous media based on the equivalent fluid model, indicating that the change of saturation should be emphasized in the study of soil dynamics. However, the equivalent fluid model considers the liquid and the gas as the same fluid, which shows that this kind of model does not belong to the three-phase medium model, and the equivalent fluid only applies to the highly saturated soil. Therefore, the application range of the equivalent fluid model is limited.

A. Bieliatynskyi et al. (Eds.): CSTTE 2023, LNCE 603, pp. 138–152, 2024.
https://doi.org/10.1007/978-981-97-5814-2_13

Mixture theory can effectively study complex porous media's deformation and overall motion and helpfully describe unsaturated soil's mechanical properties. Therefore, Hu Yayuan [5] introduced the isotropic linear elastic equation of unsaturated double medium based on the mixture theory to analyze the complex multi-physical field coupling effect of geotechnical soil; Lu [6] established the phase field model of porous medium in the freezing situation based on the mixture theory.

Porosity directly reflects the degree of compactness of the soil and is an important parameter affecting the fluid transport and permeability properties within unsaturated soils. Zhou Fengxi [7] used the theory of mixtures to give the dynamic control equations of unsaturated porous elastic media, studied the effect of unsaturated soil porosity, shear modulus, and other factors on the fluctuation response of the soil and carried out parametric analysis, Zhou Song et al. [8] studied the nature of heat conduction in unsaturated bentonite soil and analyzed the effect of porosity and other factors on the effective heat conduction characteristics of the soil.

The soil skeleton is bound to be deformed in unsaturated soil under the action of dynamic load, which in turn leads to a corresponding change in soil porosity [9]. However, the above studies regarded the porosity as a constant, ignoring the dynamic shift in porosity under dynamic loading and the effect on the dynamic properties of unsaturated soil, and the change of porosity will lead to a series of changes in the mechanical properties of soil permeability. Failure to consider dynamic porosity can lead to a decrease in model prediction accuracy, limiting the understanding of soil deformation mechanisms and affecting long-term performance assessment. However, some studies have concluded that soil permeability does not change with the change in porosity [10]. Therefore, this paper does not consider the influence of the change of porosity on permeability.

In summary, the mixture theory can characterize the complex interactions between multiphase substances and deal with complex media's ontological relationships. Therefore, this paper takes the mixture theory as the theoretical basis, establishes the dynamic response equation of unsaturated soil considering the dynamic porosity, and numerically solves it through the Comsol Multiphysics PDE module to analyze the dynamic characteristics of two-dimensional unsaturated soil under the action of vertically concentrated sinusoidal loading. Studies using dynamic modelling of unsaturated soils incorporating dynamic porosity can provide more accurate predictions of vertical displacements and pore water pressures. This approach takes into account the changes in the pore structure of the soil body under dynamic loading, which enables a more realistic simulation of the response behaviour of the soil body and provides critical guidance for engineering design.

2   Equations Governing the Dynamics of Unsaturated Soils Considering Dynamic Porosity

The correctness of the mixture theory that can be used to solve the skeleton deformation problem of unsaturated soils has been proved theoretically [13], and it has been proved by a large number of experiments that the motion of liquids and gases in the pores of unsaturated soils can be described by Darcy's law. [19].

In order to effectively establish the dynamic control equation of unsaturated soil considering dynamic porosity, the following assumptions are made:

(1) The three phases of unsaturated soil have the same temperature and thermal effects are not taken into account;
(2) There is a clear demarcation line between the three phases of unsaturated soil;
(3) The unsaturated soil is a homogeneous isotropic material;
(4) The transformation of mass, internal energy and momentum between the phases is not taken into account.

2.1 Ontological Relationships for Soil Skeleton Deformation

The effective stress obtained based on the deformation work is [11]

$$\bar{T}_{ij} = T_{ij} - [S_r p_l + (1 - S_r)p_g]I \tag{1}$$

where $\bar{T}_{ij}$ and T_{ij} are the practical and total stresses of the unsaturated soil, respectively; S_r is the degree of saturation of the unsaturated soil; p_l and p_g are the liquid-phase and gas-phase pressures, respectively; and I is the unit tensor.

Relationship between each of the corresponding variables and their displacements in unsaturated soils [12]:

$$\varepsilon_a = -\frac{[\nabla X_a + (\nabla X_a)^T]}{2} \tag{2}$$

where: when $a = s, l, g, X_a$ is the displacement of the solid, liquid, and gas phases, respectively.

The ontological relationship for the unsaturated soil skeleton is [12]

$$\bar{T}_{ij} = \lambda e I + 2\mu\varepsilon_s \tag{3}$$

where $e = tr\varepsilon_s$. λ and μ are Lame constants for the unsaturated soil skeleton.

2.2 Equations of Control for Unsaturated Soil Dynamics

The volume fractions of each phase of the unsaturated soil are:

$$\begin{aligned} \Phi_s &= 1 - n & (a) \\ \Phi_l &= nS_r & (b) \\ \Phi_g &= n(1 - S_r) & (c) \end{aligned} \tag{4}$$

where Φs, Φl, and Φg are the volume fractions of the solid, liquid, and gas phases, respectively, n is the porosity of the unsaturated soil.

The liquid-phase and vapor-phase pressures of unsaturated soils are [13]:

$$\begin{aligned} p_l &= -\lambda_{cl}\nabla \cdot X_s + M_{ll}\zeta_l + M_{lg}\zeta_g & (a) \\ p_g &= -\lambda_{cg}\nabla \cdot X_s + M_{gl}\zeta_l + M_{gg}\zeta_g & (b) \end{aligned} \tag{5}$$

Where: $\lambda_{cv} = (\Lambda_{sv} + \Lambda_{lv} + \Lambda_{gv})/\alpha_v$, Λ is the equilibrium interaction force, Φ_a^+ is the volume fraction of the phases in the unsaturated soil at static equilibrium, ζ is the volume increment of the phases in the unsaturated soil, U_v are the displacements of the liquid phase and the gaseous phase concerning the solid phase, respectively. $a, b = s, l,$ g. $v = l, g$.

Equation (5a) with Eq. (5a) can be simplified to:

$$
\begin{aligned}
-\nabla \cdot U_l &= c_{l1}p_l + c_{l2}p_g + c_{l3}\nabla \cdot X_s \qquad (a)\\
-\nabla \cdot U_g &= c_{g1}p_l + c_{g2}p_g + c_{g3}\nabla \cdot X_s \qquad (b)
\end{aligned}
\tag{6}
$$

Where: $\Delta c = M_{ll}M_{gg} - M_{gl}M_{lg}$, $c_{l1} = M_{gg}/\Delta c$, $c_{l2} = c_{g1} = -M_{lg}/\Delta c$, $c_{g2} = M_{ll}/\Delta c$, $c_{l3} = (M_{gg}\lambda_{cl} - M_{lg}\lambda_{cg})/\Delta c$, $c_{g3} = (M_{ll}\lambda_{cg} - M_{gl}\lambda_{cl})/\Delta c$.

The equations governing the dynamics of unsaturated soils are [13]

$$
m_l\frac{\partial^2 U_l}{\partial t^2} + \gamma_l\frac{\partial^2 X_s}{\partial t^2} = -\nabla p_l - \frac{\eta_l}{K_{ll}}\frac{\partial U_l}{\partial t} - \frac{\eta_g}{K_{lg}}\frac{\partial U_g}{\partial t} + \gamma_l b_l \qquad (a)
$$

$$
m_g\frac{\partial^2 U_g}{\partial t^2} + \gamma_g\frac{\partial^2 X_s}{\partial t^2} = -\nabla p_g - \frac{\eta_l}{K_{gl}}\frac{\partial U_l}{\partial t} - \frac{\eta_g}{K_{gg}}\frac{\partial U_g}{\partial t} + \gamma_g b_g \qquad (b)
\tag{7}
$$

$$
\rho\frac{\partial^2 X_s}{\partial t^2} + \sum_v \gamma_v\frac{\partial^2 U_v}{\partial t^2} = \nabla \cdot (-T_{ij}) + \rho b \qquad (c)
$$

Where: γ_l and γ_g are the actual densities of the liquid phase and gas phase in unsaturated soil, respectively. $m_l = \gamma_l/\Phi_l$, $m_g = \gamma_g/\Phi_g$, ρ is the relative density of unsaturated soil, $\rho = \Phi_s\gamma_s + \Phi_g\gamma_g + \Phi_l\gamma_l$; η_l and η_g are the viscosity coefficients of the liquid phase and the gas phase, respectively. K_{ll}, K_{lg}, K_{gl} and K_{gg} are the permeability of the liquid phase to the solid phase, the permeability of the gas phase to the liquid phase, the permeability of the liquid phase to the gas phase, and the permeability of the gas phase to the solid phase, respectively; and b_l, b_g and b are the external body forces of the liquid phase, the gas phase, and the solid phase, respectively.

Since the liquid phase and gas phase in unsaturated soil can flow freely and stably, the effect of gas-phase pressure gradient on the liquid-phase flow is negligible compared to the liquid-phase pressure gradient on the liquid-phase flow; similarly, the effect of liquid-phase pressure gradient on the gas-phase flow is negligible compared to the gas-phase pressure gradient on the gas-phase flow of the deterministic role of the gas-phase flow, which can be obtained from the $K_{lg} \to \infty$, $K_{gl} \to \infty$.. Equations (4b) and (4c) are obtained by bringing them into Eqs. (7a) and (7b), respectively:

$$
\frac{\gamma_l}{nS_r}\frac{\partial^2 U_l}{\partial t^2} + \gamma_l\frac{\partial^2 X_s}{\partial t^2} = -\nabla p_l - B_{ll}\frac{\partial U_l}{\partial t} + \gamma_l b_l \qquad (a)
$$

$$
\frac{\gamma_g}{n(1 - S_r)}\frac{\partial^2 U_g}{\partial t^2} + \gamma_g\frac{\partial^2 X_s}{\partial t^2} = -\nabla p_g - B_{gg}\frac{\partial U_g}{\partial t} + \gamma_g b_g \qquad (b)
\tag{8}
$$

where $Bll = \eta_l/Kll$, $Bgg = \eta_g/Kgg$.

Substituting Eqs. (1), (2), (3), (4a), (4b) & (4c) into Eq. (7c) gives:

$$
[\gamma_s + n(\gamma_g - \gamma_s) + nS_r(\gamma_l - \gamma_g)]\frac{\partial^2 X_s}{\partial t^2} + \gamma_l\frac{\partial^2 U_l}{\partial t^2} + \gamma_g\frac{\partial^2 U_g}{\partial t^2}
$$

$$
= \mu\nabla^2 X_s + (\lambda + \mu)\nabla\nabla \cdot X_s - S_r\nabla p_l - (1 - S_r)\nabla p_g + [\gamma_s + n(\gamma_g - \gamma_s) + nS_r(\gamma_l - \gamma_g)]b
\tag{8c}
$$

Equations (8a), (8b) and (8c) are the equations governing the dynamics of unsaturated soils expressed in terms of solid-phase displacement X_s, liquid-phase pressure p_l, and gas-phase pressure p_g.

2.3 Dynamic Porosity Modeling of Unsaturated Soils

The mass conservation equation for solid phase media, i.e., the equation for porosity with time, is [14]:

$$\frac{d(1-n)}{dt} + \frac{1-n}{\rho_s}\frac{d\rho_s}{dt} + (1-n)\nabla \cdot v_s = 0 \tag{9}$$

Where ρ_s is the relative density of the solid phase, ρ_s is the absolute velocity of the solid phase.

The material derivative of the solid phase in unsaturated soils can be simplified as:

$$\frac{d(\cdot)}{dt} \approx \frac{\partial(\cdot)}{\partial t} \tag{10}$$

The relationship between the absolute velocity of the solid phase and the bulk strain in unsaturated soils is [15]:

$$\nabla \cdot v_s = \frac{de}{dt} \approx \frac{\partial e}{\partial t} \tag{11}$$

The solid phase density in unsaturated soils can be expressed as [16]:

$$\frac{1-n}{\rho_s}\frac{d\rho_s}{dt} = -(1-\alpha^1)\frac{de}{dt} + \frac{\alpha^1-n}{K_s}(\chi_l\frac{dp_l}{dt}+\chi_g\frac{dp_g}{dt}) - (\alpha^1-n)\beta_T\frac{dT}{dt} \tag{12}$$

Where χ_l and χ_g are Bishop's effective stress coefficients, β_T is the coefficient of thermal expansion, T is the temperature, $\alpha^1 \equiv 1 - K/K_s$. K is the bulk compression modulus of the solid phase under drained conditions, K_s is the bulk compression modulus of the solid phase.

From Eqs. (9)–(12), the equation for the change in porosity with time can be derived as:

$$\frac{\partial n}{\partial t} = (\alpha^1 - n)(\frac{\partial e}{\partial t}+\frac{\chi_l}{K_s}\frac{\partial p_l}{\partial t} + \frac{\chi_g}{K_s}\frac{\partial p_g}{\partial t} - \beta_T\frac{\partial T}{\partial t}) \tag{13}$$

Since thermal effects are not taken into account in this paper and Eq. (1) is used as the effective stress for unsaturated soils, Eq. (13) can be simplified as:

$$\frac{\partial n}{\partial t} = (\alpha^1 - n)(\frac{\partial e}{\partial t} + \frac{S_r}{K_s}\frac{\partial p_l}{\partial t} + \frac{1-S_r}{K_s}\frac{\partial p_g}{\partial t}) \tag{14}$$

When the change in porosity is not significant, Eq. (14) can be obtained by integrating over time:

$$n=n_0+(\alpha^1 - n_0)(e + \frac{S_r}{K_s}p_l + \frac{1-S_r}{K_s}p_g) \tag{15}$$

where n_0 is the initial porosity of the unsaturated soil.

Equations (8a), (8b), (8c), and (15) are the equations governing the dynamics of unsaturated soils considering dynamic porosity.

3   Based on Comsol Multiphysics PDE Model Construction

3.1   Generalised Transformations of the Equations Governing the Dynamics of Unsaturated Soils Considering Dynamic Porosity

The general form of the system of partial differential equations of generalized type is:

$$e_a \frac{\partial^2 u}{\partial t^2} + d_a \frac{\partial u}{\partial t} + \nabla \cdot \Gamma = f \tag{16}$$

where a is the mass factor, da is the damping factor, Γ is the conserved flux, and f is the source term.

Equation (8a), Eq. (8b), and Eq. (8c) are generalized and transformed in a two-dimensional cartesian coordinate system according to the solution form provided by the Comsol Multiphysics PDE module:

According to the form of solving the system of partial differential equations of generalized type in Comsol Multiphysics PDE, it is necessary to rewrite Eq. (6) and Eq. (15) into the form of matrices that satisfy Eq. (16) and the corresponding matrices are thus obtained as:

$$d_a = \begin{pmatrix} 0 & 0 & 0 & 0 & 0 & 0 & 0 & 0 & 0 \\ 0 & 0 & 0 & 0 & 0 & 0 & 0 & 0 & 0 \\ 0 & 0 & B_{ll} & 0 & 0 & 0 & 0 & 0 & 0 \\ 0 & 0 & 0 & B_{ll} & 0 & 0 & 0 & 0 & 0 \\ 0 & 0 & 0 & 0 & B_{gg} & 0 & 0 & 0 & 0 \\ 0 & 0 & 0 & 0 & 0 & B_{gg} & 0 & 0 & 0 \\ 0 & 0 & 0 & 0 & 0 & 0 & 0 & 0 & 0 \\ 0 & 0 & 0 & 0 & 0 & 0 & 0 & 0 & 0 \\ 0 & 0 & 0 & 0 & 0 & 0 & 0 & 0 & 0 \end{pmatrix} \tag{17}$$

$$u = \begin{bmatrix} U_x & U_y & U_{lx} & U_{ly} & U_{gx} & U_{gy} & p_l & p_g & n \end{bmatrix}^T \tag{18}$$

$$\Gamma = \begin{bmatrix} -\nabla(\mu U_x) & -\nabla(\mu U_y) & 0 & 0 & 0 & 0 & -U_l & -U_g & 0 \end{bmatrix}^T \tag{19}$$

$$e_a = \begin{pmatrix} \gamma_s + n(\gamma_g - \gamma_s) + nS_r(\gamma_l - \gamma_g) & 0 & \gamma_l & 0 & \gamma_g & 0 & 0 & 0 & 0 \\ 0 & \rho & 0 & \gamma_l & 0 & \gamma_g & 0 & 0 & 0 \\ \gamma_l & 0 & \frac{\gamma_l}{nS_r} & 0 & 0 & 0 & 0 & 0 & 0 \\ 0 & \gamma_l & 0 & \frac{\gamma_l}{nS_r} & 0 & 0 & 0 & 0 & 0 \\ \gamma_g & 0 & 0 & 0 & \frac{\gamma_g}{n(1-S_r)} & 0 & 0 & 0 & 0 \\ 0 & \gamma_g & 0 & 0 & 0 & \frac{\gamma_g}{n(1-S_r)} & 0 & 0 & 0 \\ 0 & 0 & 0 & 0 & 0 & 0 & 0 & 0 & 0 \\ 0 & 0 & 0 & 0 & 0 & 0 & 0 & 0 & 0 \\ 0 & 0 & 0 & 0 & 0 & 0 & 0 & 0 & 0 \end{pmatrix} \tag{20}$$

$$
f = \left(
\begin{array}{c}
-S_r\frac{\partial p_l}{\partial x} - (1-S_r)\frac{\partial p_g}{\partial x} + \gamma_s + [n(\gamma_g - \gamma_s) + nS_r(\gamma_l - \gamma_g)b_x] + (\lambda + \mu)(\frac{\partial^2 U_x}{\partial x^2} + \frac{\partial^2 U_y}{\partial x \partial y}) \\
-S_r\frac{\partial p_l}{\partial y} - (1-S_r)\frac{\partial p_g}{\partial y} + \gamma_s + [n(\gamma_g - \gamma_s) + nS_r(\gamma_l - \gamma_g)b_y] + (\lambda + \mu)(\frac{\partial^2 U_x}{\partial x \partial y} + \frac{\partial^2 U_y}{\partial y^2}) \\
-\frac{\partial p_l}{\partial x} + \gamma_l b_{lx} \\
-\frac{\partial p_l}{\partial y} + \gamma_l b_{ly} \\
-\frac{\partial p_g}{\partial x} + \gamma_g b_{gx} \\
-\frac{\partial p_g}{\partial y} + \gamma_g b_{gy} \\
c_{l1}p_l + c_{l2}p_g + c_{l3}\nabla \cdot X_s \\
c_{g1}p_l + c_{g2}p_g + c_{g3}\nabla \cdot X_s \\
n_0 + (\alpha^1 - n_0)(e + \frac{S_r}{K_s}p_l + \frac{1-S_r}{K_s}p_g) - n
\end{array}
\right) \tag{21}
$$

Where U_x and U_y are the displacements of the solid phase of unsaturated soil in the horizontal and vertical directions, respectively. U_{lx} and U_{ly} are the displacements of the liquid phase compared to the solid phase in the horizontal and vertical directions, respectively. U_{gx} and U_{gy} are the displacements of the gaseous phase compared to the solid phase in the horizontal and vertical directions, respectively. b_x, b_{lx}, and b_{gx} are the external body forces of the solid, liquid, and gaseous phases of unsaturated soil in the horizontal direction, respectively, and b_y, by and b_{gy} are the external body forces of the solid, liquid and gas phases in the vertical direction, respectively.

3.2 Geometrical Modeling of Two-Dimensional Unsaturated Soils

To study the dynamic response characteristics of two-dimensional unsaturated soil considering dynamic porosity under vertical concentrated harmonic loading, a two-dimensional unsaturated soil model is established in Comsol Multiphysics PDE module, and the geometrical model is set to have a height and a width of 2×10^4 m, with the center of the upper top surface of the model as the origin of the co-ordinates, the positive half-axis of the y-axis vertically downward, and the positive half-axis of the x-axis horizontally to the right. The axis is vertically downward, the x-axis is horizontal to the right, and the top surface of the soil body is permeable, air permeable, and impermeable; impermeable two different boundary conditions: the bottom surface boundary is fixed and impermeable, impermeable boundary conditions, the left and right boundaries of the two along the x-direction is zero, and both are impermeable, impermeable boundary conditions (Fig. 1).

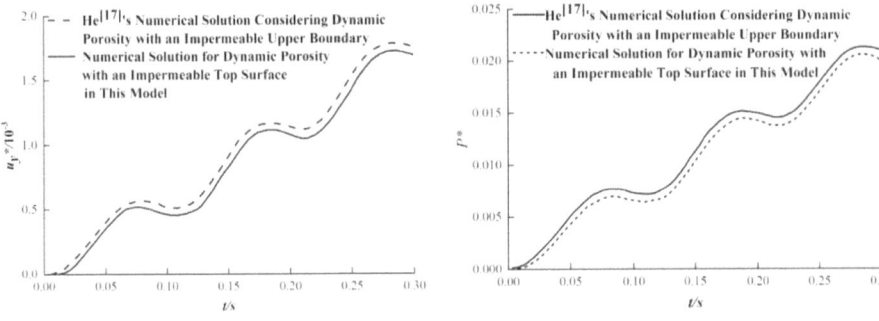

(a) Variation pattern of dimensionless vertical displacement with time

(b) Variation of dimensionless liquid phase pressure with time

Fig. 1. Comparison of the results of this paper with those of He Wenhai [17] considering dynamic porosity

4 Simulation and Analysis

4.1 Model Verification

To demonstrate the reliability of the research methodology in this paper, The dynamic equations of unsaturated soil considering dynamic porosity were degraded to the dynamic governing equations of saturated soil considering dynamic porosity, and the results were compared with the calculations in the literature [17]. The degraded soil contains only a liquid phase, saturation degree Sr = 1, and the mechanical parameter about the gas phase in the expression is 0. The geometric model is set to have impermeable boundary conditions at the top surface, bottom surface, and left and proper boundaries, and the dimensionless mechanical parameter, the geometric model, the magnitude of the excitation and its imposition are the same as that in the literature [17], and Comsol Multiphysics PDE solves the equation. The patterns of dimensionless vertical displacement and dimensionless liquid phase pressure with time are shown in Figs. 2a and b.

As can be seen from Fig. 2, when the constructed model degraded saturated soil dynamic control equations, the results of this paper are basically in agreement with the calculation results of literature [17] and literature [18]. Therefore, the reliability of the research method in this paper can be illustrated.

In this paper, considering the actual situation of the project, a group of artificial fill as unsaturated soil is selected for research, and according to the relevant engineering specifications as well as engineering experience, to determine its specific mechanical parameters as shown in Table 1 [20], and the harmonic load f(t) acted vertically on the soil at the center of the upper top surface:

$$f(t) = A_0 \sin(2\pi \omega t) \tag{22}$$

where A_0 is the load amplitude, $A0 = 1 \times 10^6$ Pa; ω is the harmonic load frequency, in this paper $\omega = 10$ Hz.

Table 1. Mechanical parameters of unsaturated soils

Parameter	Parameter value	Unit of parameter	Parameter	Parameter value	Unit of parameter
S_r	0.5	—	ρ	1970.3	kg · m^{-3}
λ	6×10^7	Pa	m_l	6666.7	kg · m^{-3}
μ	1×10^9	Pa	m_g	13.3	kg · m^{-3}
M_{ll}	9.3×10^6	Pa	γ_l	1000	kg · m^{-3}
M_{gg}	8×10^4	Pa	γ_g	2	kg · m^{-3}
M_{gl}	5×10^5	Pa	b_{lx}	1.5×10^{-3}	N
M_{lg}	5×10^5	Pa	b_{ly}	100	N
λ_{cl}	9.5×10^7	Pa	b_{gx}	2×10^{-3}	N
λ_{cg}	1×10^5	Pa	b_{gy}	5	N
K_s	4.46×10^7	Pa	b_x	100	N
B_{ll}	6×10^7	N · S · m^{-4}	b_y	2×10^{-5}	N
B_{gg}	4200	N · S · m^{-4}	α^l	1	—
n_0	0.3	—			

4.2 Changing Law of Unsaturated Soil Porosity Under Different Boundary Conditions

Unsaturated soil in the lower bottom surface and the left and right boundaries are impermeable, impermeable premise, when the upper top surface is permeable, permeable and impermeable, impermeable two different boundary conditions, under the action of harmonic load $f(t)$, the rule of change of the change of porosity at the origin of the coordinates of the law of change with time is shown in Fig. 3

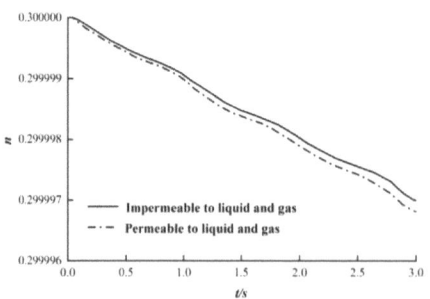

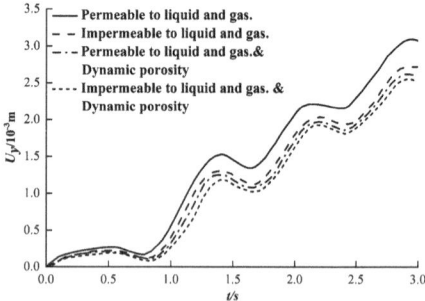

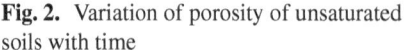

Fig. 2. Variation of porosity of unsaturated soils with time

Fig. 3. Variation of vertical displacement of unsaturated soil with time

As can be seen in Fig. 3, in the process of harmonic loading, with the change of time, unsaturated soil in the upper top surface permeable, permeable and impermeable, impermeable with two different boundary conditions, the porosity is fluctuating decreasing trend. The change rule of porosity in the first cycle ($0 < t < 0.75$ s) during the harmonic loading is thus analyzed. When $0 < t < 0.375$ s, the soil skeleton is continuously compressed by the vertical downward load, resulting in a gradual decrease in porosity; when 0.375 s $< t < 0.5625$ s, the load is vertically upward along the y-axis, but due to the existence of inertia, the soil skeleton is still continuously compressed, and porosity continues to decrease; when $t = 0.5625$ s, the inertia disappears, and the porosity reaches the minimum value, when 0.5625 s $< t < 0.75$ s, the load is still vertically upward along the y-axis, and the soil skeleton is compressed to a gradually decreasing degree, and the porosity shows an increasing trend. During the subsequent action of harmonic load, the rule of change of porosity is the same as that of the first cycle. Therefore, the porosity of unsaturated soil shows a fluctuating decrease.

It can also be seen from Fig. 3 that the porosity of unsaturated soil changes more under the boundary conditions of water permeability and air permeability at the upper top surface. This is because when the upper top surface is impermeable, the liquid and gas phases in the pores of the soil can not be freely discharged from the upper top surface, and the pressure of the liquid and gas phases in the pores resists the deformation of the soil body under the action of load. Under the boundary condition of water and air permeability on the upper top surface, the liquid and gas phases in the soil pore space can be discharged freely. The load borne by the liquid and gas phases will be borne by its skeleton, and the soil skeleton is continuously compressed. Therefore, the porosity of unsaturated soil changes more.

4.3 Effect of Dynamic Porosity on the Dynamic Response of Unsaturated Soils

Effect of Dynamic Porosity on Vertical Displacement of Unsaturated Soils
Unsaturated soil in the lower bottom surface and the left and right boundaries are impermeable, impermeable premise, when the upper top surface is permeable, permeable and impermeable, impermeable two different boundary conditions, under the action of harmonic load f(t), the change rule of the vertical displacement of the soil body at the origin of the coordinates to time is shown in Fig. 4.

As shown in Fig. 4, under the effect of harmonic loading, with time, the vertical displacement of unsaturated soil exhibits a fluctuating increase under two different boundary conditions at the top surface: permeable to water and air and impermeable to both, regardless of whether the dynamic porosity is considered or not. Taking the vertical displacement of soil during the first cycle of harmonic loading as an example for analysis, the load moves vertically downward along the y-axis in the 1st half of the cycle, continuously compressing the soil skeleton, resulting in an increase in vertical soil displacement. During the 1/2 to 3/4 period of the cycle, the load moves vertically upward along the y-axis. Still, due to inertia, the soil skeleton continues to be compressed, increasing vertical soil displacement. In the final 1/4 of the cycle, as the inertia dissipates and the load continues to move vertically upward along the y-axis, the degree of compression on the soil skeleton decreases, causing a decrease in vertical soil displacement. In subsequent cycles of harmonic load application, the pattern of change in the soil's vertical displacement is the same as in the first cycle. Therefore, regardless of whether dynamic porosity is considered, the vertical displacement of unsaturated soil shows a fluctuating increase.

As seen in Fig. 4, under the boundary conditions allowing for permeability and aeration at the top surface, the vertical displacement of unsaturated soil is more significant than that under non-permeable and non-aerated conditions at the same surface, regardless of the consideration of dynamic porosity. This occurs because, under the non-permeable and non-aerated conditions at the top surface, the liquid and gas phases within the soil pores cannot freely escape, being retained inside the soil body. When the soil is subjected to external loads, the pressure from the liquid and gas phases in the soil pores can resist the deformation of the soil. In contrast, under the conditions allowing for permeability and aeration at the top surface, the liquid and gas phases within the soil pores can freely escape, leading to a zero pressure increment under the load. Therefore, the vertical displacement of the unsaturated soil is more significant.

Furthermore, as also observed in Fig. 4, dynamic porosity significantly affects the vertical displacement of unsaturated soil, primarily manifesting as smaller vertical displacement under the same top surface boundary conditions when dynamic porosity is considered. This is because, under harmonic load, the porosity dynamically decreases over time (as shown in Fig. 3), enhancing the interaction between the soil skeleton and the liquid and gas phases in the pores. The resistance increases as the soil skeleton is compressed under load, resulting in a smaller vertical displacement of the unsaturated soil.

Effect of Dynamic Porosity on Pore Liquid Phase Pressure in Unsaturated Soils. Unsaturated Soil in the Lower

Unsaturated soil in the lower bottom surface and the left and right boundaries are impermeable, impermeable premise, when the upper top surface is permeable, permeable and impermeable, impermeable two different boundary conditions, respectively, in the harmonic load $f(t)$ under the action of the liquid-phase pressure in the pore space of the unsaturated soil at the origin of the coordinates of the law of change over time is shown in Fig. 5.

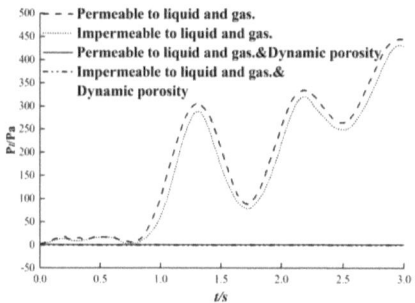
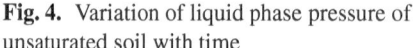
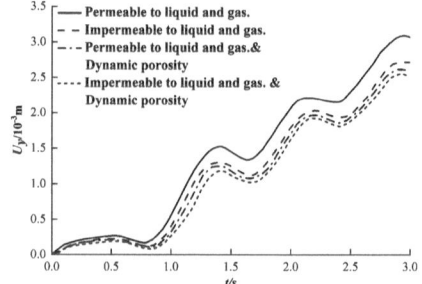

Fig. 4. Variation of liquid phase pressure of unsaturated soil with time

Fig. 5. Variation of gas phase pressure of unsaturated soil with time

As can be seen from Fig. 5, under the action of harmonic loading, with the growth of time, when the upper top surface of the soil body is impermeable and impermeable boundary condition, the pressure of the liquid phase in the pores of the soil body shows a fluctuating growth, no matter whether the dynamic porosity is considered or not. This is because, under the boundary conditions of the impermeable and impermeable upper top surface, the liquid phase in the soil pore space cannot be discharged freely from the upper top surface and is retained inside the soil body, and under the action of sinusoidal load, the soil body is fluctuating compression, which leads to fluctuating compression of the liquid phase in the pore space, and the liquid-phase pressure shows a fluctuating growth tendency. Under the boundary condition of water permeability and air permeability on the upper top surface, the pore liquid phase pressure increment is zero, regardless of whether dynamic porosity is considered. The main reason is that when the upper top surface is water-permeable and air-permeable, the liquid phase in the pore space can be discharged freely, and the liquid phase pressure change is not affected by the vertical deformation of the soil. Hence, the liquid phase pressure increment in the pore space of the unsaturated soil is zero.

Consideration of dynamic porosity has a specific effect on the liquid phase pressure in the pores of unsaturated soils, as can be seen from Fig. 5; under the boundary conditions of the impermeable and impermeable upper top surface, the liquid phase pressure in the pores of the soil is relatively tiny when dynamic porosity is considered. This is because compared to not considering dynamic porosity, in the presence of dynamic porosity, as previously mentioned, the porosity decreases, and the soil skeleton is compressed when the resistance increases. Hence, the vertical displacement of the soil body is smaller (as shown in Fig. 4), and the degree of compression becomes smaller, resulting in the degree of compression of the pore liquid phase being smaller; therefore, the unsaturated soil pore liquid phase pressure is more minor.

Effect of Dynamic Porosity on Pore Gas Phase Pressure in Unsaturated Soils
Unsaturated soil in the lower bottom surface and the left and right boundaries are impermeable, impermeable premise, when the upper top surface of the soil body is permeable, permeable and impermeable, impermeable two different boundary conditions, in the harmonic load $f(t)$, in the origin of the coordinates of the soil body pore space in the law of change of the law of the gas-phase pressure with time is shown in Fig. 6.

As can be seen from Fig. 6, under the action of harmonic loading, with the change of time, when the boundary of the upper top surface of the soil body is impermeable, the pressure of the gas phase in the pore space of the soil body shows a fluctuating growth, regardless of whether the dynamic porosity is considered or not. As mentioned above, under the boundary condition that the upper top surface is impermeable and impermeable, the gas phase in the pores of the soil body cannot be discharged freely and is retained inside the soil body, and under the action of harmonic load, the soil body is compressed by fluctuation, which leads to the fluctuation compression of the gas phase in the pores as well, and therefore, the pore gas-phase pressure shows a fluctuating growth. Under the boundary conditions of water permeability and air permeability at the upper top surface of the unsaturated soil, the increment of gas-phase pressure in the pores of the soil body is zero, regardless of whether dynamic porosity is considered. This is because the gas-phase in the pores of the soil body can be discharged freely, and the vertical deformation of the soil body doesn't affect the change of the gas-phase pressure in the pores.

It can also be seen from Fig. 6 that under the boundary conditions of the impermeable and impermeable upper top surface, the soil pore gas-phase pressure is smaller when dynamic porosity is considered. As mentioned above, when dynamic porosity is considered, under harmonic loading, the porosity decreases, and the soil body movement is affected by resistance. The soil body is compressed to a smaller extent, which causes the degree of compression of the gas phase in the soil pores to be smaller, so the soil pore gas-phase pressure is smaller when dynamic porosity is considered.

From Figs. 5 and 6, it can be seen that when the unsaturated soil is impermeable and impermeable on the upper top surface, the fluctuation amplitude of the liquid phase pressure in the soil pore space is much larger than the fluctuation amplitude of the gas phase pressure. This is because the liquid-phase pressure is much larger than the gas-phase pressure under harmonic loading, which indicates that the load borne by the liquid-phase is much larger than the gas-phase. The degree of compression is also more significant, so the amplitude of liquid-phase pressure fluctuation is much larger than that of gas-phase pressure fluctuation.

From Fig. 4, 5 and 6, it can be seen that the dynamic porosity does not affect the change rule of unsaturated soil dynamic response. Still, it affects the intensity size of its dynamic response. In the study of the dynamic response of unsaturated soil, full consideration of dynamic porosity can improve its accuracy.

5 Conclusion

This paper takes mixture theory as the theoretical basis, establishes the expression of the dynamic response of unsaturated soil considering dynamic porosity, and numerically solves and analyses the two-dimensional unsaturated soil based on Comsol Multiphysics PDE module, and the results show that:

(1) The porosity, vertical displacement, liquid-phase pressure, and gas-phase pressure of unsaturated soil are directly affected by the boundary conditions of the upper top surface. Compared with the boundary conditions of the impermeable and impermeable upper top surface, when the upper top surface of the soil is permeable and permeable, the liquid phase and gas phase in the pore space can be discharged freely from the upper top surface. The pressure of the liquid and gas phases in the pore space is zero, and the soil porosity and vertical displacement change are more significant.

(2) In the study of the dynamics of unsaturated soil, whether to consider the dynamic porosity only on the soil body response strength size does not affect the change rule of its dynamic response. When considering dynamic porosity, harmonic loading, the soil porosity decreases, the soil skeleton and pore liquid phase, gas phase interaction between the enhancement of the soil skeleton is compressed resistance becomes larger; the soil vertical displacement becomes smaller, the liquid phase and gas phase is compressed to a lesser extent, the liquid phase and the gas phase pressure is also reduced accordingly. This shows that considering dynamic porosity helps to reduce research errors.

The dynamic model of unsaturated soils considering dynamic porosity is closer to the behaviour of real soils, which also leads to more accurate mechanical models and analytical methods for better prediction of the nonlinear characteristics of soils under complex loading conditions.

In view of the lack of directly comparable experimental results, this paper validates the established mechanical model and numerical simulation method by using the results of related literature, based on which numerical simulations of the dynamic response of two-dimensional unsaturated soils excited by harmonic loading under different conditions are carried out. It provides a basis for the next work to be carried out by scholars who study the dynamic response characteristics of unsaturated porous media considering dynamic porosity.

References

1. Domenico, S.N.: Effect of water saturation on seismic reflectivity of sand reservoirs encased in shale. Geophysicists **39**(6), 759–769 (1974). https://doi.org/10.1190/1.1440464
2. Bowen, R.M.: Incompressible porous media models by use of the theory of mixtures. Int. J. Eng. Sci. **18**, 1129–1148 (1980). https://doi.org/10.1016/0020-7225(80)90114-7
3. Yang, J.: Saturation effects of soils on ground motion at free surface due to incident SV waves. J. Eng. Mech. **128**(12), 1295–1303 (2002). https://doi.org/10.1061/(ASCE)0733-9399(2002)128:12(1295)
4. Cai, Y., Li, B.: Effect of saturation variation on reflection and transmission of elastic waves on unsaturated sandstone surface. Chin. J. Rock Mech. Eng. **25**(03), 520–527 (2006). https://doi.org/10.3321/j.issn:1000-6915.2006.03.013
5. Hu, Y.: Elastoplastic model for saturated rock based on mixture theory. Chin. J. Geotechn. Eng. **42**(12), 2161–2169 (2020)
6. Lu, J.F., Tan, Y.P., Wang, J.H.: A phase field model for the freezing saturated porous medium. Int. J. Eng. Sci. **49**(8), 768–780 (2011). https://doi.org/10.11779/CJGE202012001
7. Zhou, F., Zhai, R., Cai, Y.: Analysis of elastodynamic response in unsaturated porous media. J. Basic Sci. Eng. **30**(02), 407–420 (2022). https://doi.org/10.16058/j.issn.1005-0930.2022.02.013
8. Zhou, S., Chen, Y., Zhang, Q., et al.: A model for effective thermal conductivity of unsaturated bentonite. Rock Soil Mech. **35**(04), 1041–1048+1055 (2014). https://doi.org/10.16285/j.rsm.2014.04.033
9. Liu, G.B., Xie, K.H., Zhang, R.Y.: Model of nonlinear coupled thermo-hydro-elastody-Namics response for a saturated poroelastic medium. Sci. China Ser. E: Technol. Sci. **52**(8), 2373–2383 (2009). https://doi.org/10.1007/s11431-008-0220-8
10. Palmer, I.: Permeability changes in coal analytical modeling. Int. J. Coal Geol. **1**(77), 119–126 (2009). https://doi.org/10.1016/j.coal.2008.09.006
11. Zhao, C.G., Liu, Y., Gao, F.P.: Work and energy equations and the principle of generalized effective stress for unsaturated soils. Int. J. Numer. Anal. Meth. Geomech. **34**(9), 920–936 (2010). https://doi.org/10.1002/nag.839
12. Chen, Z.: On basic theories of unsaturated and special soils. Chin. J. Geotechn. Eng. **36**(02), 201–272 (2014). https://doi.org/10.11779/CJGE201402001
13. Huang, Y., Zhang, Y.: Constitutive relation of unsaturated soil by use of the mixture theory(ll)—Linear constitutive equation and field equation. Appl. Math. Mech. **02**, 124–137 (2003). https://kns.cnki.net/kcms2/article/abstract?v=M7N75Hb03FUxztrhu41Qix7C9WOYa8Jasi2hmUQjDzOw9sxi67l3gCK9ZQXOgHnjEnx4fImD9wgHZL3uKfHBGbwIldsDeIenxfk43qMfNw1r_PZY1aUqlEjaRMNTwpA3&uniplatform=NZKPT&language=CHS

14. Yan, J., Wei, Y., Cai, H., et al.: Research of hydro-thermo-mechanical coupling model of saturated-unsaturated porous media. J. Hydraul. Eng. **45**(S2), 152–160 (2014). https://doi.org/10.13243/j.cnki.slxb.2014.S2.025

15. Bear, J., Bachmat, Y.: Introduction to Modeling of Transport Phenomena in Porous MediA, vol. 553. Theory & Applications of Transport in Porous Media (1990). https://link.springer.com/content/pdf/bbm:978-94-009-1926-6/1?pdf=chapter%20toc

16. Rutqvist, J., Börgesson, L., Chijimatsu, M., et al.: Thermo hydromechanics of partially saturated geological media: governing equations and formulation of four finite element models. Int. J. Rock Mech. Min. Sci. **38**(1), 105–127 (2001). https://doi.org/10.1016/S1365-1609(00)00068-X

17. He, W., Wang, T.: Dynamic porosity and related dynamic response characteristic of two-dimensional saturated soil. Rock Soil Mech. **41**(08), 2703–2711 (2020). https://doi.org/10.16285/j.rsm.2019.1671

18. Chen, J.: Time domain fundamental solution to Biot's complete equations of dynamic poroelasticity. Part I: Two-dimensional solution. Int. J. Solids Struct. **31**(10), 1447–1490 (1994). https://doi.org/10.1016/0020-7683(94)90186-4

19. Fredlund, D.G., Rahardjo, H.: Geotechnics of unsaturated soils. Translated by Chen Zhongyi, Zhang Zaiming and Chen Guijiong, pp. 126–146. China Construction Industry Press, Beijing (1997). https://doi.org/10.11779/CJGE201402001

20. Feng, J.: Experimental research on unsaturated engineering characteristics of clayey foundation soil in Hefei area. Southwest Jiaotong University (2016)

Triaxial Experimental Study on Strength Characteristics of Saturated Soft Soil Under Different Shear Strain Rates

Yongjian Liu[1], Lan Luo[1(✉)], You Zhang[2], Mingyang Lai[3], and Yangpan Fu[4]

[1] School of Guangzhou Institute of Science and Technology, Guangzhuo 510040, Guangdong, China
799935169@qq.com
[2] Guangzhou Engineering Contractor Group Co. Ltd., Guangzhuo 510310, Guangdong, China
[3] Guangzhou Urban Planning and Design Survey Research Institute Co.Ltd., Guangzhuo 510040, Guangdong, China
[4] School of Civil Engineering, Guangzhou University, Guangzhuo 510040, Guangdong, China

Abstract. The China Nansha District of Guangzhou City is located at the geographical center of the Pearl River estuary and the Greater Bay Area of Guangdong, Hong Kong, and Macau, is characterized by widely distributed saturated soft soil layers with poor engineering properties. Although many scholars have explored the influence of loading rate on the mechanical properties of soft soil, research on the variation of undrained shear strength of saturated soft soil under different shear strain rates is still limited. This article conducts an in-depth analysis of the mechanical behavior of saturated soft soil under different strain rates through true triaxial consolidated undrained shear tests. Under load, these layers exhibit a significant rate-dependent mechanical behavior. In order to reveal the influence of shear strain rate on the mechanical properties of highly saturated soft soils based on SPAX-2000 test system, a series of true triaxial consolidation undrained shear tests were carried out under different confining pressures and different strain rates, variation rules of undrained shear strength of saturated soft and the pore water pressure with strain rate were analyzed. The experimental results indicate that the strain-rate softening exists in saturated soft soil. Under the same consolidation conditions, the relative growth rate of shear strength is a monotonic increasing function of the logarithm of shear strain rate, and the smaller the intermediate principle stress, the more significant the effect of shear strain rate on shear strength. During the shearing process (The strain-rate range from 10^{-6}/s to 10^{-2}/s), the undrained shear strength of saturated soft soil increases with the growth of shear strain rate, following an exponential variation pattern. In the early stages of loading, the increase in pore water pressure and the maximum pore water pressure are significantly affected by the strain rate. Throughout the testing process, the variation of pore water pressure exhibits some fluctuation and hysteresis. Different from strongly structured clay, the strength of saturated soft soil undergoes a progressive change with the shear strain rate, without showing a distinct critical rate of transition.

Keywords: soft soil · shear strain rate · undrained shear strength · true triaxial test · strain rate softening

© The Author(s) 2024
A. Bieliatynskyi et al. (Eds.): CSTTE 2023, LNCE 603, pp. 153–167, 2024.
https://doi.org/10.1007/978-981-97-5814-2_14

1 Introduction

Due to the high porosity, high water content, high clay content, low permeability, and rheological properties of soft soil, the strength of soft soil is related to deformation and time, exhibiting certain mechanical effects of loading rate under the action of loads [1, 2]. As early as the 1930s, Buisman [3] pointed out that the influence of loading rate cannot be ignored in the study of constitutive relationships and strength characteristics of soft soil. However, there is usually a significant difference between the loading rate in practical engineering applications and the loading rate in indoor tests. If the influence of loading rate is not considered, it may lead to instability or excessive settlement of certain saturated soft soil foundations during construction or after construction. [4, 5].

Domestic and foreign scholars have conducted a series of exploratory studies on loading rate and its impact on soil mechanical response characteristics [5–13]. Diaz Rodrigues et al. studied the relationship between strength and strain rate of Mexico cohesive soil, indicating that the peak value of the stress-strain curve increases with the increase of strain rate, and found the phenomenon of "strain rate softening" [6]. Cai Yu, Kong Lingwei, and others [7] confirmed the influence of soil structure on soil mechanical properties by studying the strain rate of Zhanjiang clay. Zhu Qiyin et al. conducted a systematic study on the normalization of the loading rate effect of soft soil under uniaxial and triaxial stress conditions. Graham [11] summarized 15 types of clay and found that the average increase rate of undrained shear strength is 10%; Gao Yanbin [4] used an anisotropic elastic-plastic constitutive model to calculate the strain rate parameter ρ Between 7.2% and 12.2%, while foreign test results ρ ranges from 5% to 23% [12–21], with an average of 12%. The above literature mostly studies the loading rate under certain specific conditions (certain consolidation state, stress history, and test methods); studying soil properties involves structural clay, structural loess, soft clay with general moisture content, and remolded soil; in terms of experimental methods, conventional triaxial apparatus, direct shear apparatus, and ring shear apparatus are mainly used for testing; sowever, there are still few reports on the research of true triaxial tests, and there are few research results on the strain rate of saturated soft soil [22–27]. Previous studies have shown that strain rate has an impact on the undrained strength, pre consolidation pressure, pore water pressure, yield pressure, and shear dilation (shear dilation) characteristics of soil. The strain rate effect of clay with different characteristics varies greatly. Therefore, in-depth research is still needed for soft soil in different regions.

This article takes common saturated soft soil in the Nansha area of Guangzhou as the research object, and conducts true triaxial consolidated undrained shear tests with different confining pressure and strain rates. Based on the comparison of other cohesive soils in the literature, the variation law of undrained shear strength and pore water pressure with strain rate is analyzed, aiming to provide a certain basis for the engineering characteristics of saturated soft soil and the study of elastic-plastic constitutive models of saturated soft soil.

2 Test Soil Samples and Test Plan

2.1 Test Instruments

The SPAX-2000 improved true triaxial testing instrument produced by GCTS company in the United States was used to conduct triaxial consolidated undrained shear tests on saturated soft soil. The main functions and parameters of the testing system are shown in Table 1. This experiment used the SPAX-2000 static and dynamic true triaxial tester from GCTS company in the United States. The testing system consists of six major components: pressure chamber, rigid loading actuator, SCON digital servo controller and acquisition system, confining/back pressure volume controller, advanced servo software, and variant constant pressure hydraulic source. The testing system can simulate the real triaxial stress state. This means that the stress conditions experienced by the soil sample during the experiment are closer to the actual engineering conditions. In addition, the system can provide more accurate test results, which is crucial for capturing the subtle changes in saturated soft soil under different strain rates.

Table 1. Main functions and indicators of SPAX-2000

Major function	Performance index
Static and dynamic vertical loading functions	Static maximum axial pressure of 25 kN, static maximum axial pressure of 20 kN, maximum frequency of 20 Hz, maximum speed of 100 mm/s
Three way simultaneous dynamic function	Three directional forces can be independently controlled, with a maximum principal stress of 5 MPa, a maximum intermediate principle stress of 5 MPa, and a minimum principal stress of 2 MPa
Stress/strain control can switch functions at will	Using control and post-processing software to automatically control soil sample saturation, consolidation, static and dynamic loading, etc.
P/q path control function	Maximum dynamic frequency 20 Hz, confining pressure dynamic frequency 10 Hz, bidirectional coupling loading frequency 10 Hz

2.2 Test Soil Sample

Nansha District of Guangzhou City is located at the southernmost end of Guangzhou City, on the west bank of Humen Waterway of the Pearl River, where Xijiang River, Beijiang River and Dongjiang River converge. The test soil sample was taken from a soft soil foundation treatment project site in Nansha District, Guangzhou City, and was a high water content marine saturated soft soil. As Nansha is located at the mouth of the

the Pearl River, due to its unique geographical environment and geological conditions, it forms a deep soft soil sedimentary layer. The burial depth of this batch of soil samples is about 6.0–8.0 m, with a soft soil porosity ratio of 1.94, a natural moisture content of 74.72%, a weight of 15.61 kN/m^3, a liquid limit of 67.23%, a plastic limit of 37.98%, and a clay content of 41.0% less than 0.005mm. Sampling and wax sealing were carried out in situ according to the requirements of geotechnical tests.

2.3   Test Scheme

In this experiment, three types of confining pressures (100 kPa, 150 kPa, 300 kPa) were applied to the undisturbed saturated soft soil samples; three types of intermediate principle stresses (120 kPa, 175 kPa, and 230 kPa); three types of pore water pressures (30 kPa, 60 kPa, and 90 kPa); under the same consolidation stress ratio $K_c = 1.28$, triaxial consolidated undrained shear tests were conducted at five shear rates (10^{-6}/s, 0.8×10^{-5}/s, 10^{-4}/s, 0.5×10^{-3}/s, 10^{-2}/s). Based on the second set of test conditions (confining pressure σ_3 of 150 kPa, intermediate principle stress σ_2 of 175 kPa, and axial strain of 10%), the true triaxial consolidated undrained shear tests with different strain rates were mainly conducted, supplemented by the comparison of results from other test conditions. A total of 38 specimens were subjected to CU tests to obtain the influence of strain rate on the mechanical properties of high moisture content marine saturated soft soil.

Test steps: Preparation of specimens (rectangular specimens with dimensions of 50 mm × 50 mm × 120 mm) → Vacuum saturation → Backpressure saturation (pore water pressure coefficient B ≥ 0.95) → Sample consolidation (set consolidation program by GCTS-CATS software: through isobaric consolidation and eccentric consolidation) → Conduct CU tests at the set strain rate until the soil sample reaches the set axial strain value or the specimen fails.

3   Testing Results and Analysis

3.1   Characteristics of Stress-Strain Relationship Curves Under Different Strain Rates

Due to the structural and rheological properties of saturated soft soil, there is a significant correlation between the stress-strain strength relationship of the soil and the loading rate. To study the effect of shear strain rate on the stress-strain relationship curve, the specimens were vertically sheared at the same consolidation ratio ($K_c = 1.28$), different confining pressures (100 kPa, 150 kPa, and 200 kP), and different intermediate principle stresses (120 kPa, 175 kPa, and 230 kPa) under three consolidation pressure conditions. The stress-strain relationship curves were obtained by dividing them into different shear strain rates (10^{-6}/s, 0.8×10^{-5}/s, 10^{-4}/s, 0.5×10^{-3}/s, and 10^{-2}/s) until the axial strain reached 10%, as shown in Figs. 1 and 2.

Figure 1 shows the stress-strain relationship curves of specimens under three different strain rates: the first group ($\sigma_2 = 100$ kPa, $\sigma_2 = 120$ kPa), the second group ($\sigma_3 = 150$ kPa, $\sigma_2 = 175$ kPa), and the third group (σ_3 A $= 200$ kPa, $\sigma_2 = 230$ kPa). Figure 2 shows the

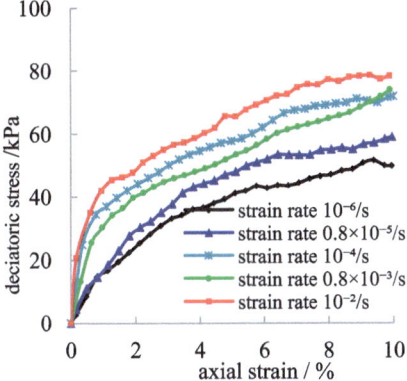

(a) Consolidation pressure (σ_3=150kPa, σ_2=175kPa)

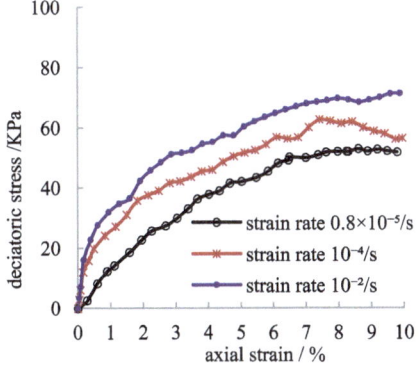

(b) Consolidation pressure (σ_3=100kPa, σ_2=120kPa)

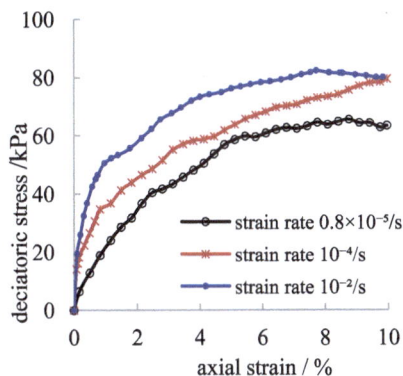

(c) Consolidation pressure (σ_3=200kPa, σ_2=230kPa)

Fig. 1. Comparison curves of deviation stress - strain rate - axial strain

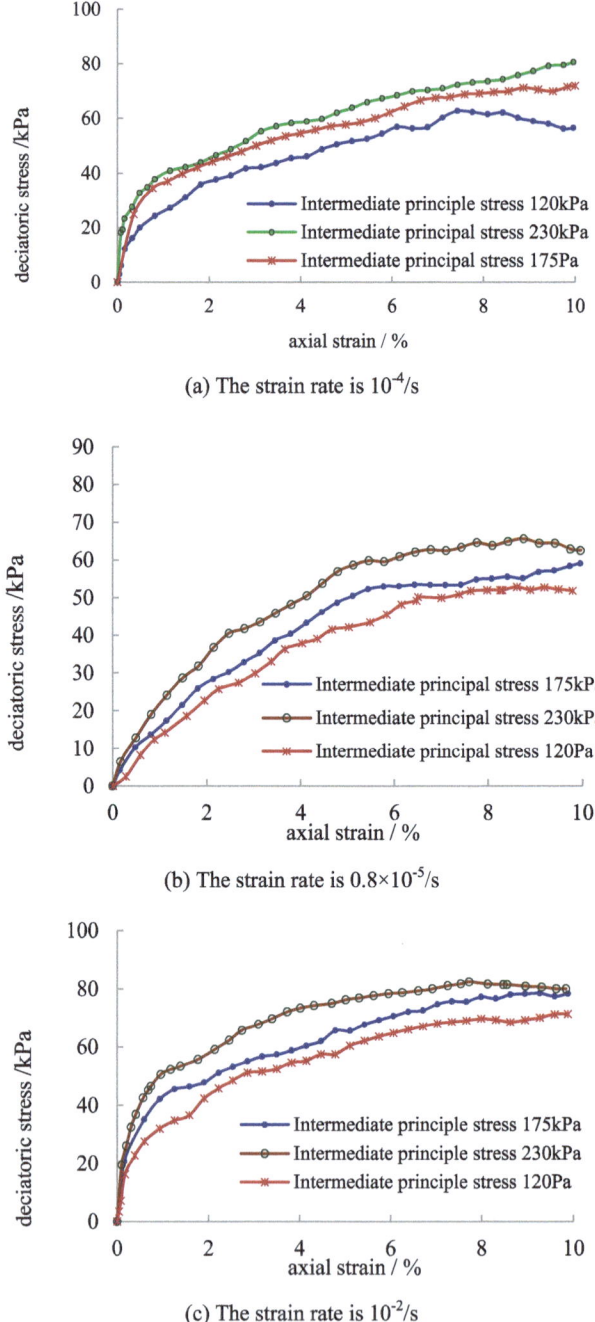

(a) The strain rate is 10^{-4}/s

(b) The strain rate is 0.8×10^{-5}/s

(c) The strain rate is 10^{-2}/s

Fig. 2. Comparison curves of deviation stress - intermediate principle stress - axial strain

stress-strain relationship curves under different vertical strain rates. From the graph, it can be seen that when the strain is less than 1%, the deviatoric stress of each specimen during the CU test increases sharply, and the faster the strain rate, the faster its growth rate; when the axial strain is between 1% and 8%, the deviatoric stress still increases rapidly; when the strain is greater than 8%, there are differences in the trend of deviatoric stress changes among various soil samples, with some samples maintaining a continuous growth trend and some samples experiencing strain softening phenomenon; the peak stress increases with the increase of strain rate. The shear strength of soil increases with the increase of shear strain rate under the same confining pressure; strain softening has a strain rate effect; the faster the shear rate under the same confining pressure, the more obvious the strain softening phenomenon. The strain softening phenomenon under lower intermediate principle stress is more obvious than that under higher intermediate principle stress.

Compared with reference [7], the stress-strain relationship curve of Zhanjiang strongly structured clay (with a water content of 48.6% and sensitivity of 5–7) in conventional triaxial CU tests has two significant characteristics: (1) the phenomenon of strain rate softening in strongly structured soil is significant; (2) under the same confining pressure, the peak shear stress of CU decreases first and then increases with the increase of shear strain rate, and there is a clear critical rate turning point. Reference [10] found through conventional triaxial CU and UU tests that Dalian saturated remolded soil (with a moisture content of 29%) undergoes strain strengthening at low shear rates and strain softening at high shear rates. The soft soil in Tianjin Binhai New Area (with a water content of 50.6%) [24] has a critical shear rate under low confining pressure conditions, and the critical shear rate disappears under high confining pressure conditions. The experimental results show that the saturated soft soil in Nansha has the basic characteristics of general clay, but its strain rate effect is different from that of strong structured soil and remolded soil. Strong structured soft soil has obvious critical rate turning points [7, 8], and the strength of saturated soft soil gradually changes with shear strain rate, without any obvious critical rate turning points. The differences in strain rate effects are related to factors such as the genesis, high moisture content, low permeability, and structural characteristics of saturated soft soil [23, 26].

In summary, "strain softening" in indoor tests refers to the phenomenon where the stress of a soil sample decreases as the strain increases after reaching its peak strength; during the process of strain softening, the rate of strain growth accelerates with the increase of stress. Through the comparison of this experiment and soil properties, it is shown that there are three main reasons for the above phenomenon: (1) the hardening or softening phenomenon during the experiment depends first on the physical properties of the soil, such as the obvious softening phenomenon of saturated soft soil at a certain moisture content, internal structure, and physical state; (2) secondly, it is closely related to the stress state and test conditions of the soil sample. When the confining pressure and intermediate principle stress are small, due to small lateral constraints and large vertical deformation, the strain growth rate accelerates, the load growth rate relatively decreases, and the softening phenomenon becomes more obvious; (3) the strain softening of saturated soft soil has a strain rate effect. When the strain rate is high, due to the poor permeability of saturated soft soil, stress superposition occurs in the sample, and the softening phenomenon becomes more obvious.

3.2 The Effect of Shear Strain Rate on Undrained Strength

The undrained shear strength is an important indicator in engineering design. The method for determining the undrained strength Su of saturated soft soil in this article takes 10% axial strain when the stress-strain curve of the soil sample shows strain hardening, i.e. no peak value appears ε_1 corresponds to the difference in principal stress $(\sigma_1 - \sigma_3)$ Half of it; when the stress-strain curve shows a softening form, take half of the peak principal stress difference q_k.

In order to quantitatively analyze the effect of strain rate on undrained shear strength, three sets of relationship curves between peak (or maximum) shear stress q_k and strain rate $\dot{\varepsilon}$ under test conditions were plotted in semi logarithmic coordinates, as shown in Fig. 3. It can be seen that $q_k - \dot{\varepsilon}$ shows a linear distribution in semi logarithmic coordinates, and the regression equations of strength strain rate under the three test conditions are (1), (2), and (3), respectively. (The three experimental conditions include: the first group $(\sigma_3 = 100$ kPa, $\sigma_2 = 120$ kPa), second group $(\sigma_3 = 150$ kPa, $\sigma_2 = 175$ kPa) and the third group $(\sigma_3 = 200$ kPa, $\sigma_2 = 230$ kPa)

$$q_k = 3.390\ln(\dot{\varepsilon}) + 94.101 \tag{1}$$

$$q_k = 3.062\ln(\dot{\varepsilon}) + 96.476 \tag{2}$$

$$q_k = 2.236\ln(\dot{\varepsilon}) + 90.825 \tag{3}$$

The slope of the regression line can reflect the rate of change of shear strength with strain rate; when the slope of the straight line is larger, the undrained shear strength increases faster with the rate. As shown in Fig. 3, the undrained shear strength of soil is a monotonic increasing function of strain rate. At the same consolidation ratio, the slope of the first group (confining pressure 120 kPa) is the highest ($k = 3.390$), and the strength increases the fastest with strain rate. The growth rate of strength in the three experimental groups is in the following order: the first group > the second group > the third group, indicating that the strain rate has a more significant impact on strength under low confining pressure.

The degree of influence of shear strain rate on undrained strength is represented by the strain rate parameter η. When the strain rate is increased by 10 times (one order of magnitude), the growth rate of undrained shear strength [4] is defined as follows:

$$\eta = [\Delta S_u/S_{u0}]/\Delta lg\dot{\varepsilon} \tag{4}$$

In the formula: ΔS_u represents the strength increment; S_{u0} is a reference undrained strength under the same consolidation conditions, and the sample with the smallest strain rate in each group of this test is used as the reference strength value; $\Delta lg\dot{\varepsilon}$ is the axial strain rate.

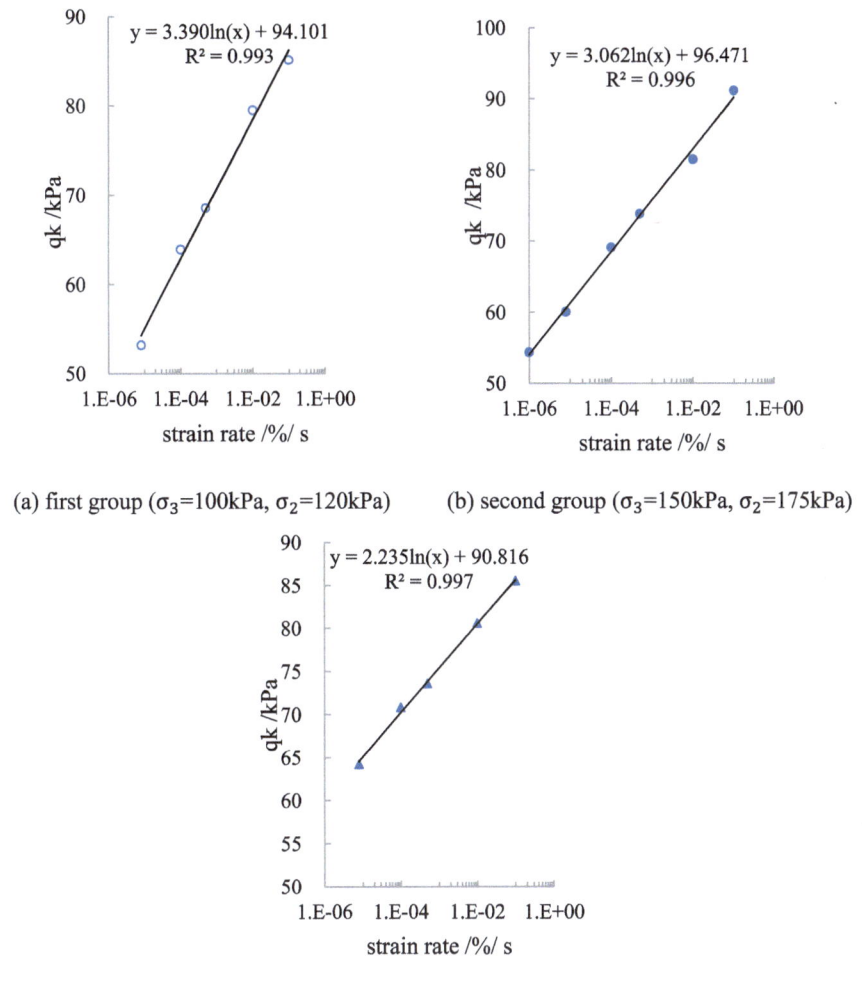

(a) first group (σ_3=100kPa, σ_2=120kPa)   (b) second group (σ_3=150kPa, σ_2=175kPa)

(c) third group (σ_3=200kPa,σ_2=230kPa)

Fig. 3. The peak shear stress -strain curves

To analyze the variation characteristics of shear strength with loading rate, Fodil [12] proposed an exponential strain rate equation based on Graham, and some scholars proposed a logarithmic equation [4, 7, 15]. Using the second set of experiments as an example, analyze the adaptability of exponential and logarithmic rate equations to saturated soft soil in Nansha.

Exponential rate equation:

$$\eta_{e1} = \frac{q_{jk}/q_0 - 1}{\lg(\dot{\varepsilon}_i/\dot{\varepsilon})} \tag{5}$$

$$\eta_{e2} = \frac{q_{jk}/q_0 - 1}{\lg\left(\frac{\dot{\varepsilon}_i}{\dot{\varepsilon}} + 1\right)} \tag{6}$$

Logarithmic rate equation:

$$\eta_{L1}=\frac{\lg(q_{ik}/q_0)}{\lg(\dot{\varepsilon}_i/\varepsilon)} \tag{7}$$

$$\eta_{L2}=\frac{\lg(\frac{q_{ik}}{q_0})}{\lg(\frac{\dot{\varepsilon}_i}{\dot{\varepsilon}}+1)} \tag{8}$$

In the formula, η_{e1} and η_{e2} are exponential rate equation parameters, η_{L1} and η_{L2} are logarithmic rate equation parameters, $\dot{\varepsilon}_i$ and $\dot{\varepsilon}_0$ are strain rate and strain rate of the reference sample group, respectively; q_{ik} and q_0 are the peak shear stress and reference peak shear stress of a certain sample, respectively.

The occupation fitting in semi logarithmic coordinates is good, and the regression coefficients R^2 of the regression equation are all high. Among them, the fitting degree between the two logarithmic equations and the distance between points is slightly poor, with R^2 below 0.920; the regression coefficients R^2 of the two exponential rate equations are both greater than 0.956, indicating that the exponential equation can better reflect the characteristics of the undrained strength of saturated soft soil changing with shear strain rate, and Eq. (5) is more concise in expression. As shown in Fig. 4, when the rate parameters of saturated soft soil in Nansha are $\eta_{e1}= 0.1305$ and $\eta_{e2}= 0.1319$, and the strain rate increases by 10 times, the undrained shear strength growth rate is 13.12%.

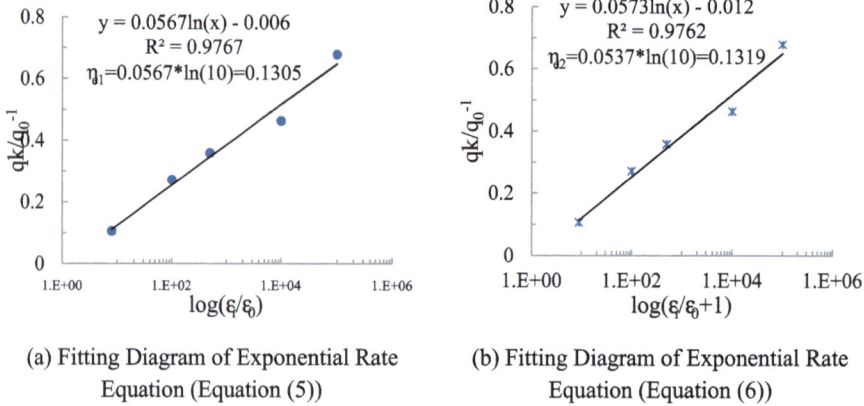

(a) Fitting Diagram of Exponential Rate Equation (Equation (5))

(b) Fitting Diagram of Exponential Rate Equation (Equation (6))

Fig. 4. Comparison curves of rate formulation for undrained shear strength of saturated soft soil

Research has shown that under the same consolidation conditions, the relative growth rate of shear strength is a monotonic increasing function of the logarithm of shear strain rate, and the smaller the intermediate principle stress, the more significant the effect of shear strain rate on shear strength. In addition, due to the structural strength of undisturbed saturated soft soil, there is a slight dispersion in the test results near the yield stress level. The magnitude of the rate parameter is closely related to the internal factors of soil formation, material composition, structure, and physical properties, as well as factors such as consolidation state, stress history, and testing methods.

3.3   The Effect of Shear Strain Rate on Pore Water Pressure

As shown in Fig. 5, the relationship curves of ultra static pore water pressure u, strain rate $\dot{\varepsilon}$, and axial strain ε_1 under three consolidation conditions and different shear rates show that the strain rate has a significant impact in the initial loading stage, but has no significant impact on pore water pressure in the later stage. In the early stage of shearing, the strain rate is low (10^{-6}/s and 0.8×10^{-5}/s), and the sample has a higher rate of pore water pressure growth. The excess pore water pressure rises rapidly, and this trend gradually disappears with increasing strain; after reaching a certain axial strain (about 3% to 7%), the pore water pressure of samples with high shear rates gradually increases compared to samples with low shear rates. Under the conditions of this experiment (axial strain $\varepsilon_1 \leq 10\%$ at the end of the experiment), the final excess pore water pressure stabilized at a certain value. In addition, the influence of intermediate principle stress on soil strength is significant, while its effect on pore water pressure is not significant.

Another noteworthy phenomenon is the fluctuation and hysteresis of pore water pressure. This is because the saturated soft soil has a flocculent structure, with particles mostly in the form of flakes or columns. Under the action of loads, the soil properties and structure are adjusted, causing the clay particles in the soil to transition from "edge-edge" contact and "edge-surface" contact to "surface-surface" contact form. The structure tends to homogenize, and the soil structure gradually transitions to reshaped soil properties. In the initial stage of applying load, the internal stress of the soil is concentrated, and the non-uniform distribution of stress causes the pore water in the sample to flow from the middle to both ends; in addition, the bonding effect of clay particles and poor vertical permeability have a certain inhibitory effect on pore water flow, and the pore water pressure measured by the pore pressure sensor at the bottom of the specimen is low, with a time lag. In addition, due to the influence of multiple factors, the variation of pore water pressure is not an increasing function of strain rate, resulting in slight fluctuations.

According to reference [7], the peak pore water pressure of Zhanjiang strongly structured clay obtained at different shear rates under low confining pressure is basically equal; under high confining pressure, the effect of shear rate on pore water pressure gradually becomes apparent. Crawford [15] found in his study on Leda clay that shear rate has a significant impact on pore water pressure, while Guy Lefebvre's study on Quebec clay found that pore water pressure is independent of shear rate during triaxial testing. The comparison results show that the properties and test conditions of cohesive soil are different, and the influence of shear rate on pore water pressure during the triaxial consolidated undrained shear test process varies greatly.

In summary, we have obtained similar conclusions as previous studies on the mechanical effects of shear strain rate in saturated soft soil through true triaxial tests, but there are also new insights. There are still significant differences in the characteristics and rate parameters of undrained shear strength of high moisture saturated soft soil with increasing shear strain rate, as well as the variation law of pore water pressure compared to soil types such as strong structured clay, remolded soil, and general low moisture clay in literature. This is related to the causes of saturated soft soil, high water content, high clay content, high liquid limit, medium sensitivity, flocculation structure, rheological properties, and true triaxial test conditions, which are different from previous research results and have certain representativeness.

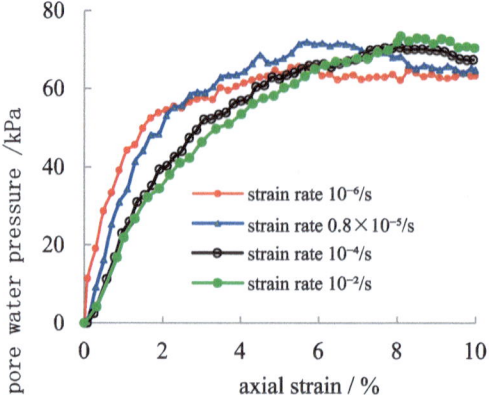

(a)   Consolidation pressure (σ_3=150kPa, σ_2=175kPa)

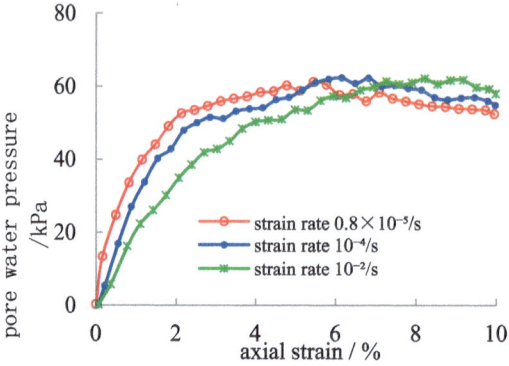

(b) Consolidation pressure (σ_3=100kPa, σ_2=120kPa)

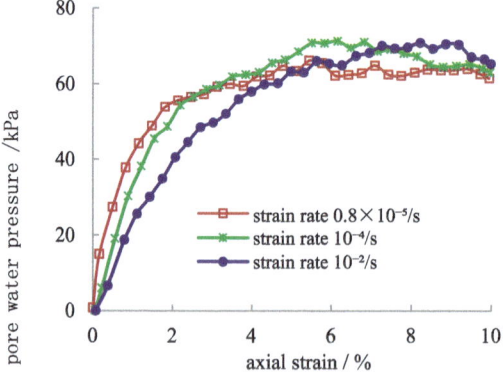

(c) Consolidation pressure (σ_3=200kPa, σ_2=230kPa)

Fig. 5. Comparison curve of pore pressure-axial strain

4   Conclusion

(1) Under the conditions of this experiment, the undrained shear strength of saturated soft soil in Nansha follows an exponential equation with strain rate. The exponential rate parameters η_{e1} are 0.1305 and η_{e2} are 0.1319. When the strain rate increases by 10 times, the undrained shear strength growth rate is 13.12%.

(2) Under the same consolidation conditions, in the early stage of shear, the excess pore water pressure of samples with lower strain rates increases rapidly, and this trend gradually disappears with increasing strain; the ultra static pore water pressure is influenced by multiple factors and is not an increasing function of strain rate.

(3) High moisture content soft soil has the basic characteristics of general clay, but it is different from strong structured clay and remolded soil. The strength of saturated soft soil gradually changes with strain rate, and there is no obvious critical rate turning point. The difference in strain rate effect is related to factors such as high moisture content, low permeability, and structural characteristics of soft soil.

(4) The strain softening of saturated soft soil has a strain rate effect; the faster the shear rate under the same confining pressure, the more obvious the strain softening phenomenon. As the consolidation pressure increases, the influence of strain rate on the undrained shear strength of soil gradually weakens. The strain softening phenomenon under lower intermediate principle stress is more obvious than that under higher intermediate principle stress.

(5) The conclusion of this article is suitable for saturated soft soil with similar physical properties and the same test conditions (load conditions and shear strain rate) as the soil used in this experiment. For other soils with a natural moisture content far greater than or less than 74.72%, if the moisture content of the soil sample decreases significantly after exposure to sunlight, the significant changes in soil physical properties will inevitably cause corresponding changes in the mechanical properties (shear characteristics) of the soil. Therefore, in the construction of soft soil engineering, it is necessary to reasonably control the loading rate and closely pay attention to the mechanical effects of shear strain rate to ensure the safety, stability, and economic rationality of engineering buildings.

(6) This article investigates the mechanical response of saturated soft soil under different shear strain rates, but the limited number of samples may affect the universality of the results. Future research can further validate these findings by increasing sample size and considering saturated soft soil types in different regions.

Acknowledgments. National Natural Science Foundation of China (52078142), Guangdong Provincial Natural Science Foundation (2021A1515011691,2022A151011047), Guangdong Higher Education Association 's 14th Five-Year Plan 2023 Higher Education Research Project (23GZD14), Guangdong Province Education and Teaching Reform Engineering Construction Project (Guangdong Education Gaohan [2023] No.4).

References

1. Li, Z.M.: Reinforcement and quality control of soft soil foundation. China Construction Industry Press, Beijing (2011)
2. Liu, Y.J., Fu, N., Qian, X.M., Lin, H.: Experiment study on dynamic response and variation of bound water of soft clay under impact loading. Chin. J. Underground Space Eng. **12**(2), 322–329 (2017)
3. Buisman, A.S.: Results of long duration settlement tests. In: Proceedings of 1st International Conference on Soil Mechanics and Foundation Engineering. Cambridge, vol. 1, pp. 103–107 (1936)
4. Gao, Y.B., Wang, Z.W.: Effect of strain rate on undrained shear strength of clays. Chin. J. Rock Mech. Eng. **24**(Supp. 2), 5779–5783 (2005)
5. Liu, Y.J., Li, Z.M., Liang, S.H.: Neural network model on the relationship between engineering properties and microstructure of soft soils. Chin. J. Underground Space Eng. **9**(4), 777–782 (2013)
6. Díaz-Rodríguez, J.A., Martínez-Vasquez, J.J., Santamarina, J.C.: Strain-rate effects in mexico city soil. J. Geotechn. Geoenviron. Eng. **135**, 300–305 (2009)
7. Cai, Y., Kong, L.W., Guo, A.G., et al.: Effects of shear strain rate on mechanical behavior of Zhanjiang strong structured clay. Rock Soil Mech. **27**(8), 1235–1240 (2006)
8. Zhu, Q.Y., Yin, Z.Y., Zhu, J.G., Wang, J.H.: Progress and trend of experimental investigation on rate-dependent behavior of soft clays. Rock Soil Mech. **35**(1), 7–19 (2014)
9. Qi, J.F., Luan, M.T., Nie, Y.: Experimental study of shear and strength behavior of saturated clay. J. Dalian Univ. Technol. **48**(4), 551–556 (2008)
10. Shi, B.T., Zhang, Y., Hu, L.: The saturated remolded clay strength under different direct shear rates. Coal Geol. Explor. **45**(2):90–95 (2017)
11. Graham, J., Crooks, J.H., Bell, A.L.: Time effects on the stress-strain behaviour of natural soft clays. Geotechnique **33**, 327–340 (1983)
12. Fodil, A., Aloulou, W., Hicher, P.Y.: Viscoplastic behavior of soft clay. Geotechique **47**(3), 581–591 (1997)
13. Yang, G.B., Xu, S.L., Wang, D.K., Ding, F.: Research on the stability of soft soil foundation under different loading rates and foundation treatment methods. Energy Environ. Protect. **11**, 273–277 (2022)
14. Rowe, R.K., Hinchberger, S.D.: The significance of rate effects in modelling the Sackville test embankment. Can. Geotech. J. **35**, 500–516 (1998)
15. Crawford, C.B.: The influence of rate of strain on effective stresses in sensitive clay. ASTM Spec. Tech. Publ. **254**, 36–48 (1960)
16. Yin, J., Cheng, C.: Comparison of strain-rate dependent stress-strain behavior from k o - consolidated compression and extension tests on natural hong kong marine deposits. Mar. Georesour. Geotechnol. **24**, 119–147 (2006)
17. Yin, Z., Chang, C.S., Karstunen, M., Hicher, P.: An anisotropic elastic-viscoplastic model for soft clays. Int. J. Solids Struct. **47**, 665–677 (2010)
18. Vaid, Y.P., Campanella, R.G.: Time-dependent behavior of undisturbed clay. J. Geotechn. Geoenviron. Eng. **103**, 693–709 (1977)
19. Hinchberger, S.D., Rowe, R.K.: Evaluation of the predictive ability of two elastic-viscoplastic constitutive models. Can. Geotech. J. **42**, 1675–1694 (2005)
20. Sheahan, T.C., Ladd, C.C., Germaine, J.T.: Rate-dependent undrained shear behavior of saturated clay. J. Geotechn. Eng. **122**, 99–108 (1996)
21. Wang, Q.S., Qiu, S.N., Zheng, H., Zhang, R.T.: Undrained shear strength prediction of clays using liquidity index. Acta Geotech. **1–18** (2023). https://doi.org/10.1007/s11440-023-021 07-9

22. Dan, H.B., Wang, L.Z.: Strain-rate dependent behaviors of K0 consolidated clays. Chin. J. Geotechn. Eng. **30**(5), 718–725 (2008)
23. Chen, B., Sun, D., Lv, H.: Experimental study on compressive characteristics of marine soft soil. Geotechn. Mech. (02), 381–388 (2013)
24. Yang, A.W., Lei, B.: Experiment study on effect of shearing rate on mechanical behavior of structured soft dredger fill. J. Eng. Geol. **23**(1), 7–12 (2015)
25. Yuan, W., Wang, J.: One-dimensional consolidation analysis of saturated soft clay based on water content variation. KSCE J. Civ. Eng. **25**, 107–113 (2021)
26. Lu, B.L., Wang, J.W., Cai, D.G.: Experimental study of influence of shear rate on clay undrained strength. Railw. Eng. **1**, 73–75 (2013)
27. Zhang, C., Wang, W.S., Cao, G.S.: Analysis on the influence of shear velocity on tailing sand shear strength under different stress. Chin. J. Underground Space Eng. **12**(Supp. 2), 463–469 (2016)

Analysis of Bridge Bearing Capacity and Tunnel Stability

Determination and Application of Excavation Footage in the Shallow-Buried Biased Section of the North-South Road Tunnel in Kyrgyzstan

Lu Zhang[1], Hao Feng[1], Yunfeng Du[1], Jian Tao[1], and Kangbao Lun[2(✉)]

[1] CCCC Construction Group Co., Ltd., Beijing 100011, China
[2] Shaanxi Key Laboratory of Geotechnical and Underground Space Engineering/Xi'an University of Architecture and Technology, Xi'an 710000, Shaanxi, China
489656917@qq.com

Abstract. When constructing tunnels in soft rock, improper excavation footage selection can cause instability and collapse. Using Xie Jiajie's shallow tunnel surrounding rock pressure model, this study expands from 2D to 3D. A formula is deduced to calculate the excavation footage in soft rock, ensuring stability. This formula is used to analyze the excavation footage of Kyrgyzstan's north-south Yueling tunnel using the step method, and combined with numerical simulation analysis, the optimal excavation footage is obtained, which provides a reference for tunnel construction.

Keywords: Tunnel engineering · Excavation footage · Numerical simulation

1 Introduction

The step method is widely used for grade V surrounding rock in tunnel excavation [1, 2]. Improper selection of excavation footage can lead to problems like face instability and vault collapse [3–5]. Therefore, selecting a suitable excavation footage is crucial for safe tunnel construction [6, 7]. Shi Xianhuo et al. [8] used silo theory to analyze the excavation footage and found that reserved core soil enhances face stability; Cai Jun et al. [9] obtained a reasonable excavation footage for the Yangjiaping tunnel considering the dilatancy angle. Li Hui et al. [10] derived a calculation formula for shallow rock tunnels' excavation footage, showing that it is primarily determined by vault and face stability.

Drawing from the rock pressure calculation model for the Xie Jiajie tunnel, this study extends its 2D framework into a comprehensive 3D model. It establishes an excavation footage formula tailored for shallow-buried soft rock tunnels, prioritizing face stability. The adapted formula is applied in assessing the excavation progress within the north-south Yueling tunnel in Kyrgyzstan. Utilizing the step method in conjunction with numerical simulations, it aims to ascertain the most efficient excavation footage for optimal tunneling outcomes. Compare and analyze the results with existing literature's formula calculation results, verifying the rationality of the formula to calculate tunnel excavation footage.

© The Author(s) 2024
A. Bieliatynskyi et al. (Eds.): CSTTE 2023, LNCE 603, pp. 171–178, 2024.
https://doi.org/10.1007/978-981-97-5814-2_15

2 Determination Method of Tunnel Excavation Footage

2.1 Calculation Model

We introduce a computation model for excavating soft rock strata in tunnels: ① The rock and soil mass exhibit an inclined plane as the fracture surface, inclined at an angle to the horizontal. ② As the overlying rock settles, compressed by surrounding rock, it induces decline in triangular masses on both sides. The sliding mass moves under resistance of undisturbed rock on both sides and the rock ahead of tunnel head. Refer to Fig. 1) and c for details.

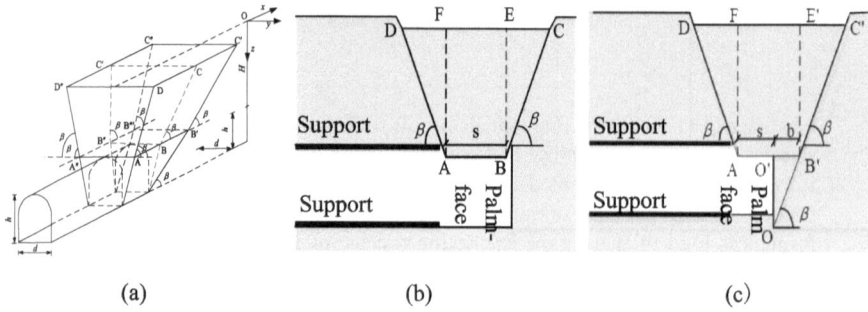

(a)   (b)   (c)

Fig. 1. Calculation model of excavation footage.

2.2 Determination of Footage Based on Vault Stability

Post-excavation, it is imperative that the upper section of the tunnel's surrounding rock remains stable within the specified loosening range, avoiding collapse. The self-weight of the surrounding rock should not exceed the four-side frictional force. To establish equilibrium in the extreme scenario, an equilibrium equation is formulated. The calculation diagram is shown in Fig. 2.

$$W_1 - 2T_1 \sin \theta - 2T_2 \sin \theta = 0 \tag{1}$$

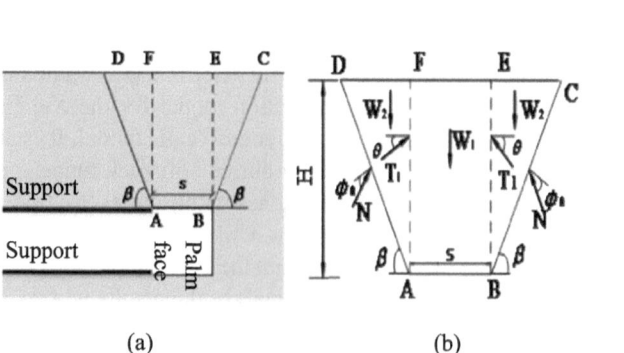

(a)   (b)

Fig. 2. Calculation model.

where, W_1 is the self weight of the overburden rock mass ABEF on the tunnel top.

$$W_1 = sdH\gamma \tag{2}$$

$2T_1 \sin \theta$ is the friction force of the soil on both sides when the rock mass ABEF slides, and $2T_2 \sin \theta$ is the friction force on the other two sides. The solution is shown in Fig. 3.

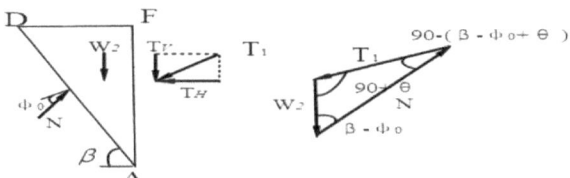

Fig. 3. Calculation model.

The self weight of the triangular prism is:

$$\frac{1}{2}\gamma \times \overline{AF} \times \overline{DF} \times s = \frac{1}{2}\gamma H^2 \frac{1}{\tan \beta} d \tag{3}$$

According to the sine theorem:

$$\frac{T_1}{\sin(\beta - \phi_0)} = \frac{W_2}{\sin[90^0 - (\beta - \phi_0 + \theta)]}$$

Bring Eq. (3) into the above equation and simplify it to obtain:

$$T_1 = \frac{1}{2}\gamma H^2 \frac{\lambda}{\cos \theta} d \tag{4}$$

Similarly:

$$T_2 = \frac{1}{2}\gamma H^2 \frac{\lambda}{\cos \theta} s \tag{5}$$

Where,

$$\tan \beta = \tan \phi_0 + \sqrt{\frac{(\tan^2 \phi_0 + 1)\tan \phi_0}{\tan \phi_0 - \tan \theta}} \tag{6}$$

$$\lambda = \frac{\tan \beta - \tan \phi_0}{\tan \beta[1 + \tan \beta(\tan \phi_0 - \tan \theta) + \tan \phi_0 \tan \theta]} \tag{7}$$

Bring Eqs. (2) (4) (5) into the Eq. (1):

$$s = \frac{dH\lambda \tan \theta}{d - H\lambda \tan \theta} \tag{8}$$

Where, s is the excavation footage; d is the width or diameter of the tunnel; H is the covering thickness of soil layer; ϕ_0 is the friction angle.

3 Determination of Excavation Footage of Tunnel Portal Section of North South Road in Kyrgyzstan

3.1 Determination of Excavation Footage Based on Arch Crown Stability

The North-South ridge crossing tunnel in Kyrgyzstan is 3750 m long, and the main tunnel of tunnel section 3-A is 1850 m long. The service pilot tunnel spans from the initial chainage of K431+90 to the final chainage of K450+40, covering a length of 1850 m. The construction of the primary tunnel involves the upper and lower bench method, facilitating an excavation width of 12.6 m. The long pipe shed section at the portal of the service pilot tunnel is constructed by the bench method, and the full section method is adopted in other places, with the excavation width of 5.2 m. When calculating excavation footage, select physical and mechanical parameters for rock mass corresponding to grade V, as shown in Table 1, and use them for calculation.

Table 1. Calculation parameters.

Severe γ (kN/m^3)	Cohesion c (kPa)	Friction angle ϕ_0 (°)	Lateral pressure coefficient K_a
20	200	27	0.57
17	50	20	0.38

According to the requirements of highway tunnel code, the equivalent load height is 7.5 m. From Eq. (6) the excavation footage of the main tunnel under the two sets of parameters is S = 1.5 m.

3.2 Numerical Simulation Analysis Based on Excavation Footage Optimization

To determine the optimal excavation footage, three values of s (1.0 m, 1.2 m, and 1.5 m) were simulated numerically, and the displacement fields were compared and analyzed. Based on prior tunnel mechanics experience, the model's total width is 96.2 m, with a left boundary-to-main tunnel distance of 36 m, a clear distance between the two tunnels of 9 m, a right boundary-to-pilot tunnel distance of 33 m, and a height direction 20 m below the inverted arch. The tunnel's rock is homogenous, behaving elastically and plastically by Mohr-Coulomb criterion. Shotcrete and anchor bolts also follow homogeneous, elastic principles. The finite element model is shown in Fig. 4.

Surface Subsidence. Through numerical simulation, it is found that the surface settlement under the conditions of surrounding rock ①, surrounding rock ② and surrounding rock ③ is consistent with the vertical displacement nephogram, and the maximum settlement occurs above the arch crown of the main tunnel. The maximum settlement under the three surrounding rock conditions is 2.5 mm, 4.3 mm and 5.3 mm respectively. The surface subsidence results are shown in Fig. 5.

Vault Settlement and Peripheral Convergence. The maximum vault subsidence under the three surrounding rock conditions is respectively 3.4 mm, 5.5 mm, 6.4 mm. The

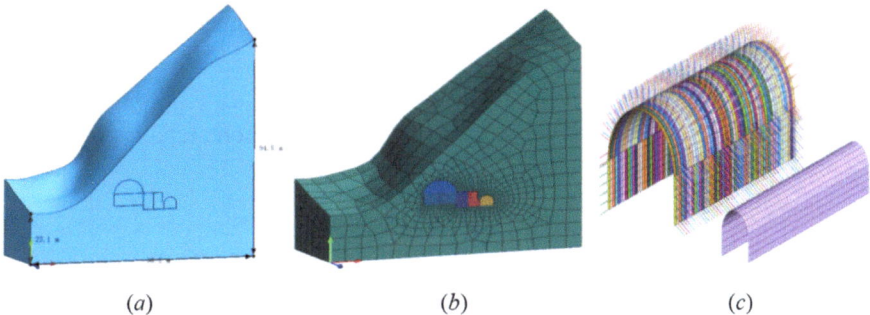

(a)   *(b)*   *(c)*

Fig. 4. Finite element model diagram.

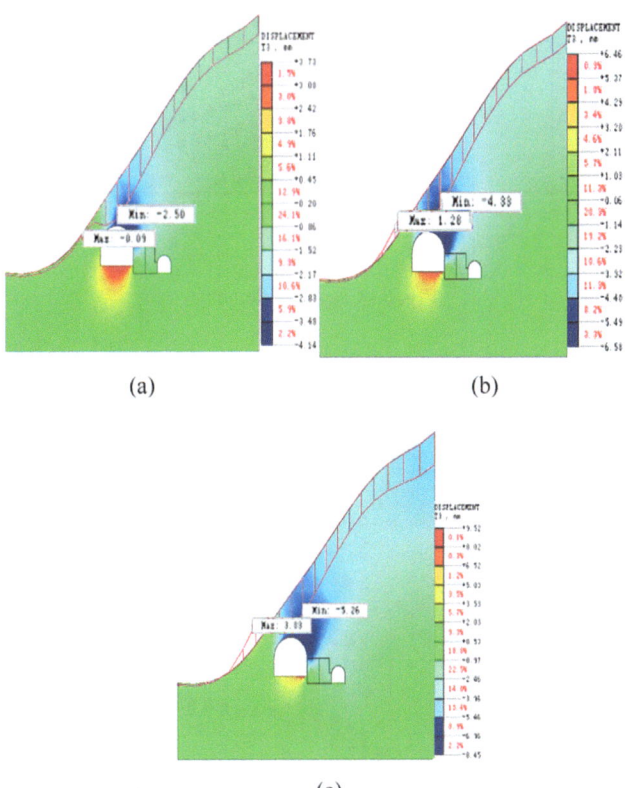

(a)   (b)

(c)

Fig. 5. The max surface settlement and displacement nephogram under three kinds of surrounding rock.

maximum peripheral convergence under the three surrounding rock conditions is 4.8 mm, 10.8 mm and 15.7 mm respectively. According to the above analysis, under the same surrounding rock, the surface settlement, vault settlement and peripheral convergence increase with the increase of excavation footage. Therefore, it is recommended to set the excavation footage to s = 1.5 m. The vault settlement and peripheral convergence are shown in Figs. 6 and 7.

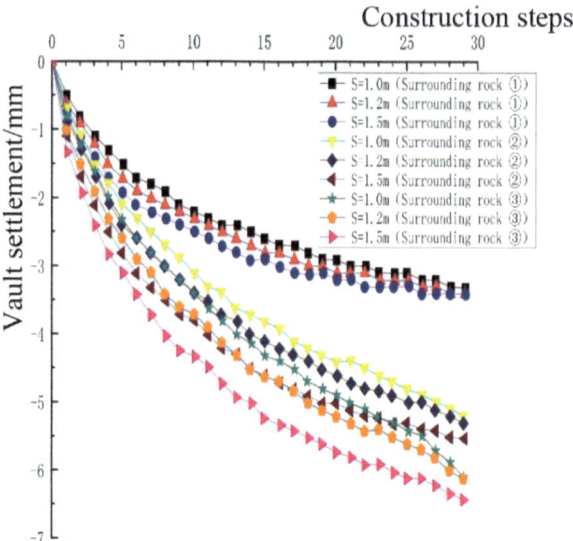

Fig. 6. Crown settlement.

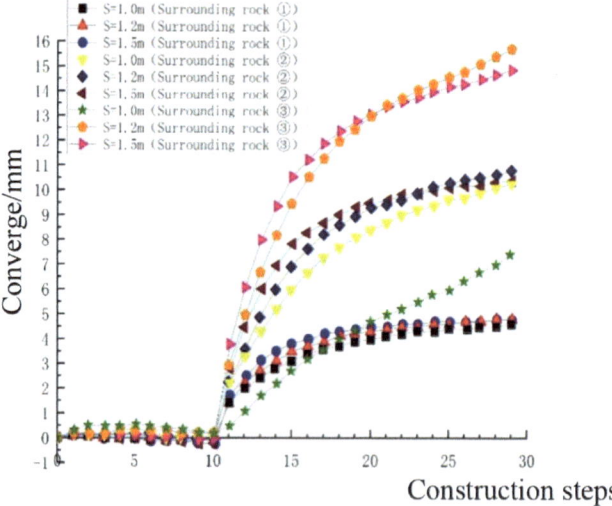

Fig. 7. Peripheral convergence.

4   Conclusion

The formula for calculating the excavation footage of a shallow buried tunnel with eccentric pressure is established. This formula is applied to the North-South crossing tunnel in Kyrgyzstan using the bench method, and the optimal excavation footage is obtained through numerical simulation analysis.

(1) The analysis indicates that the excavation footage of the tunnel is influenced by both the stability of the vault and face. Outcomes highlight sensitivity to cohesion and friction angle. Emphasize precise determination in engineering computations.
(2) The excavation footage of shallow tunnels in soft rock strata is directly related to the action of the tunnel face, increasing linearly with it.
(3) According to the construction characteristics of North-South mountain crossing tunnel in Kyrgyzstan, the optimal excavation footage is s = 1.5 m through numerical simulation analysis.

Acknowledgments. The financial support for this work was provided by the Innovation Capability Support Plan of Shaanxi Province-Innovation team (2020TD-005).

References

1. Xu, J., He, C., Zhou, Y., et al.: The research for tunnel excavation optimization in broken soft rock based on benching tunneling method. Hydrogeol. Eng. Geol. **2013**(40), 42–48 (2013)
2. Hui, M.A., Jie, L.I., Qiaofeng, Z.U.O., et al.: Micro bench method for tunnelling with fine blasting and its engineering application. J. Railway Eng. Soc. **29**(1), 57–61 (2012)
3. Kirsch, A.: Experimental investigation of the face stability of shallow tunnels in sand. Acta Geotechn. **5**(1), 43–62 (2010)
4. Kamata, H., Masimo, H.: Centrifuge model test of tunnel face reinforcement by bolting. Tunnell. Underground Space Technol. **18**(2), 205–212 (2003)
5. Ibrahim, O.: A new approach for estimating the transverse surface settlement curve for twin tunnels in shallow and soft soils. Environ. Earth Sci. **72**(72), 2357–2367 (2014)
6. Song, Z., Zhang, D., Qu, J., et al.: Study on the deformation control of PBA con-struction method in pressure water sand stratum. J. Xi'an Univ. Arch. Tech. (Natl. Sci. Edn.) **47**(1), 33–38 (2015)
7. Song, Z., Yang, T.T., et al.: Experimental investigation and numerical simulation of surrounding rock creep for deep mining tunnels. J. South. African Inst. Min. Metallurgy **116**(12), 1181–1188 (2016)
8. Shi, X., Dai, Y., Guo, J.: Calculation and analysis on excavation eycle length of shallow portal sections of tunnels located in lalus deposits: case study on zhaojiawu tunnel on Mazhao Highway in Yunnan, China. Tunnel Construct. **2015**(8), 787–791 (2015)
9. Cai, J., Ye, H., Lei, T., et al.: Optimization of tunnel excavation footage in jointed rock mass considering dilatancy angle. Rock Soil Mech. **37**(S1), 639–644+650 (2016)
10. Li, H., Tian, X., Song, Z., et al.: Study on calculation method of digging length for digging length for shallow tunnel based on Xie Jiajie's surrounding rock pressure Formula. Xi'an Univ. Arch. Tech. (Natl. Sci. Edn.) **50**(5), 662–667 (2018)

Study on the Stability of Rock Walls in Shallow-Buried Tunnels with Small Clearances in Layered Rock Masses

Xiaojun He[(✉)]

School of Civil Engineering, Lanzhou Jiaotong University, Lanzhou 730070, People's Republic of China
2029501373@qq.com

Abstract. In the engineering of tunnels with small clearances, the structural features of rock layers play a crucial role in the stability of the central rock walls. This study employed the Discrete Element Method software (UDEC) to simulate the mechanical response of tunnels with small clearances under various rock layer dip angles, both in unfortified conditions and when reinforced with shotcrete. The research thoroughly examined the deformation patterns of the central rock walls, the development process of the plastic zones, and the dynamic changes in the stress distribution along the rock layer interfaces. The findings indicate that the deformation of the central rock walls is most significant when the rock layer dip angle is around 60°. Timely reinforcement with shotcrete effectively restricts the expansion of the plastic zones. Additionally, the characteristics of the rock layer interfaces play a decisive role in the stability of the central rock walls, with a higher risk of slip failure under certain dip angles.

Keywords: Layered Rock Mass · Tunnel with Small Clearance · Central Rock Wall · Discrete Element Simulation

1 Introduction

As China's major projects like the "Belt and Road Initiative" continue to advance, tunnel construction faces unprecedented challenges in dealing with geological diversity and complexity [1]. Particularly in layered rock masses, the presence of jointed rock bodies can lead to complex bedding plane dislocation and anisotropy, which, if not properly addressed, may cause tunnel collapse and instability [2]. Small-clearance tunnels, due to their flexibility in terrain and route selection, are widely used in both short tunnels and the entrance sections of long tunnels. Although current research has covered the failure modes and mechanical properties of layered rock masses, as well as the impacts of construction responses in small-clearance tunnels, studies using the Discrete Element Method (DEM) to analyze the effects of structural planes on the stability of rock walls in small-clearance tunnels are still limited. This study aims to address this gap by exploring the impact of rock layer dip angles on rock wall stability in small-clearance tunnels, thus providing theoretical support for similar projects.

The utilization of DEM is particularly advantageous for this study due to its ability to simulate the complex behaviors of discontinuities and jointed rock masses, essential

A. Bieliatynskyi et al. (Eds.): CSTTE 2023, LNCE 603, pp. 179–185, 2024.
https://doi.org/10.1007/978-981-97-5814-2_16

for understanding the stability challenges in small-clearance tunnel environments. This approach offers a more comprehensive understanding of the structural plane impacts on tunnel stability, making a significant contribution to the field.

2 Establishment of the Numerical Model

This study employed the classic Discrete Element software UDEC to construct plane strain models of shallow-buried small-clearance tunnels with different rock layer dip angles β, both with and without support measures such as shotcrete. To eliminate boundary effects [3], the model dimensions were set at 100 m × 80 m, with displacement constraints applied to the left, right, and bottom boundaries, and the top boundary simulating the natural ground surface without constraints. The model features a tunnel buried at a depth of 20 m, with a net distance between the left and right tunnels of 0.5 times the tunnel span, and a rock layer thickness of 1.5 m. The schematic diagram of the calculation model is shown in Fig. 1.

The choice of UDEC is grounded in its proficiency for simulating the non-continuity and complex behaviors of rock mass, particularly beneficial for studying stress redistribution and failure processes in small-clearance tunnels.

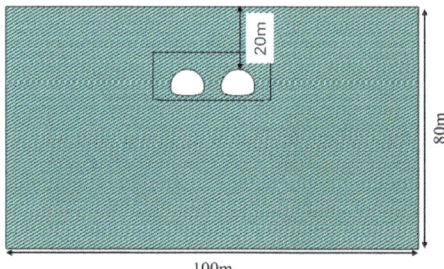

Fig. 1. Schematic diagram of the calculation model

Due to the difficulty in obtaining joint interface parameters through experiments, this study relied on rock mass parameters and geological conditions from field surveys, referencing the methods of Weiyuan Zhou [4] and B Sainsbury [5] for selecting structural plane parameters. The physical and mechanical parameters are as follows, with both the surrounding rock and rock layers modeled using the Mohr-Coulomb criterion, and the initial support modeled using an elastic criterion. The table of physical and mechanical parameters is shown in Table 1:

Table 1. Physical and Mechanical Parameters

	Density /kg·m³	Elastic Modulus /GPa	Poisson's Ratio	Normal Stiffness /GPa·m⁻¹	Shear Stiffness /GPa·m⁻¹	Friction Angle /°	Cohesion /MPa	Tensile Strength /MPa
Rock	2250	1.48	0.32			25	0.135	0.08
Joint				6	0.9	17	0.07	0
Support	2200	23.5	0.2					

3 Results Analysis

3.1 Displacement Analysis

Since the deformation of the central rock wall mainly causes horizontal convergence of the tunnel, this study focuses on the displacement in the thickness direction (i.e., the X-direction in the model) of the central rock wall as a reference to analyze the deformation distribution characteristics under different rock layer dip angles. The displacement cloud diagrams for 0°, 45°, and 90° scenarios are shown in Fig. 2.

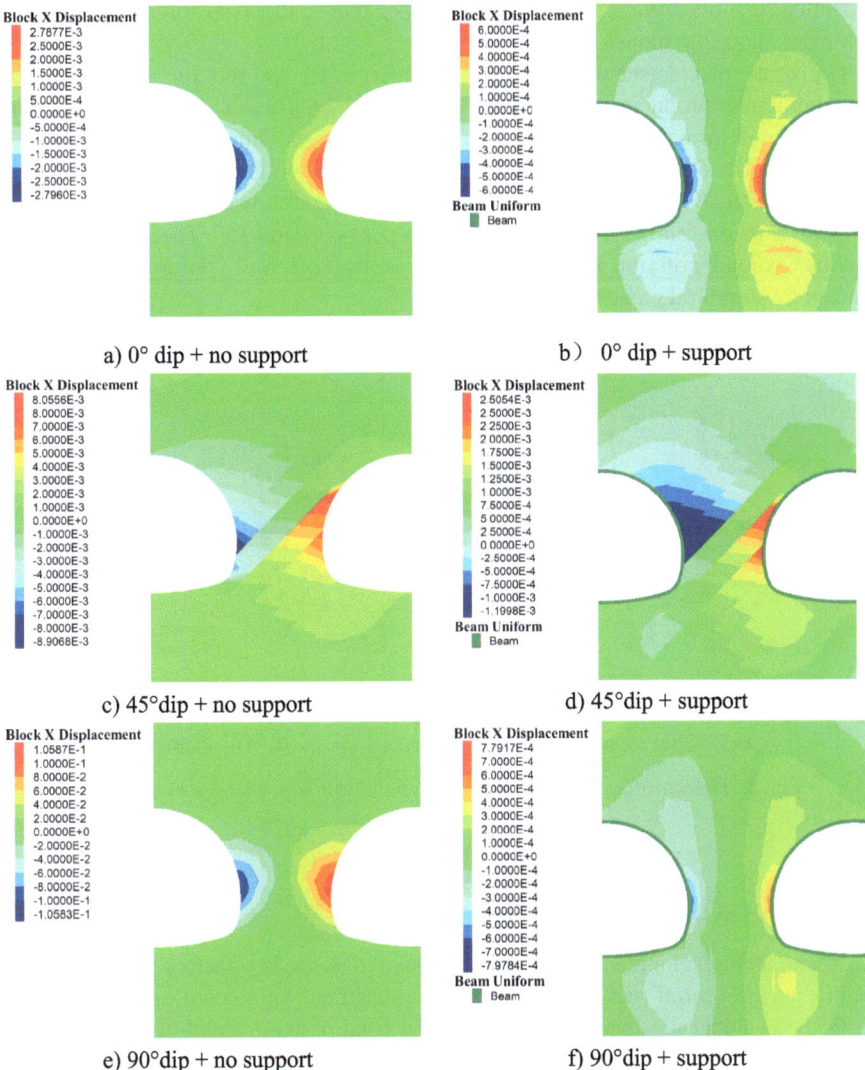

a) 0° dip + no support   b) 0° dip + support

c) 45°dip + no support   d) 45°dip + support

e) 90°dip + no support   f) 90°dip + support

Fig. 2. Deformation cloud diagram in the thickness direction of the central rock wall

Figure 2 illustrates that variations in rock layer dip angle significantly influence the deformation distribution of central rock walls in small-clearance tunnels. Without support, at a 0° dip angle, maximum displacements of 2.78 mm occur symmetrically near the arch waist on both sides of the wall. With increasing angles, displacement peaks shift following the excavation-formed sharply angled rock blocks. For example, at 60°, the left side shows -17 mm displacement near the right arch foot, and the right side 15 mm near the left arch shoulder. At 90°, the rock mass above the tunnel collapses, causing the central rock wall to bend and bulge longitudinally, with an 83 mm peak displacement at the arch waist. With shotcrete support, deformation patterns resemble those without support but are less pronounced due to the support's restraining effect. This effect is more significant at high dip angles due to the axial force exerted by the rock mass above the tunnel.

Horizontal displacements at the arch shoulder, waist, and foot on both tunnel sides were monitored, producing deformation curves (Fig. 3). These curves reveal that without support, deformation at the three positions fluctuates with the dip angle: initially increasing, then decreasing, and finally increasing again. The most notable increase occurs between 50° and 55°, peaking at 118 mm at the arch waist. With shotcrete support, deformation follows a similar pattern but peaks around 55°. These findings align with the Coulomb criterion, suggesting that the central rock wall's deformation is mainly affected by structural plane slip failure.

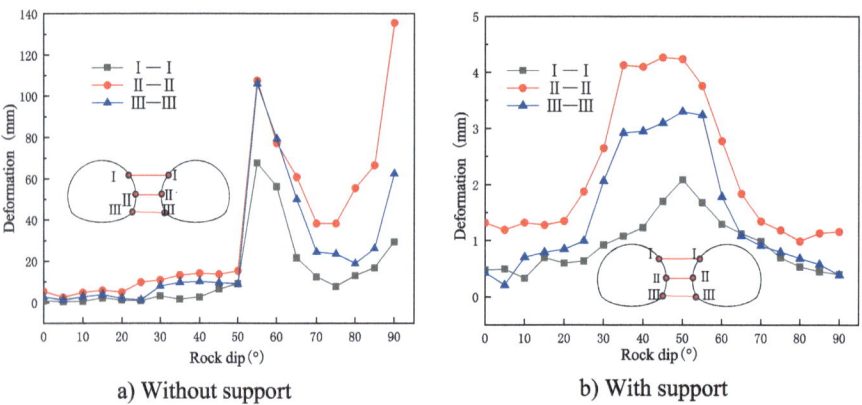

a) Without support   b) With support

Fig. 3. Deformation curves of the central rock wall with changes in dip angle

3.2 Analysis of the Plastic Zone Range

The shape and development trend of the plastic zone are key factors in assessing the stability of the central rock wall. Figure 4 demonstrates significant differences in the plastic zones under unsupported and supported conditions. In small-clearance tunnels, the plastic zones primarily appear around the central rock wall and the lateral arch waist of the tunnel. In the unsupported condition, the plastic zone penetrates through the central rock wall, while shotcrete reinforcement inhibits the development of the plastic zone, preventing penetration. Simultaneously, irrespective of shotcrete application, the distribution of the plastic zones in the surrounding rock noticeably shifts following the dip angle of the rock layers: initially, at a 0° dip angle, the plastic zones appear at the tunnel arch waist. As the dip angle increases, the plastic zones gradually incline in the direction of the increasing dip angle, with the left side of the central rock wall shifting towards the arch foot, and the right side towards the arch shoulder. At a 90° angle, the plastic zones again present a symmetrical distribution.

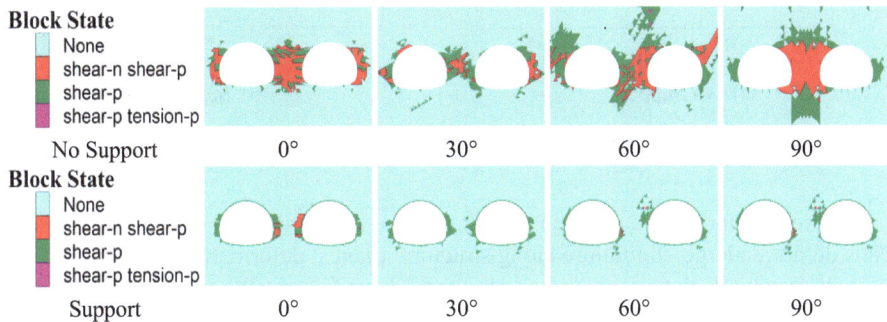

Fig. 4. Distribution of plastic zones under different rock dip angles

In summary, without support, the plastic zones penetrate the center of the central rock wall, indicating that the wall is subjected to extreme stress, posing a direct threat to the stability of the tunnel. After reinforcement, the plastic zones are confined to a smaller area and do not cross the central rock wall, signifying the significant effectiveness of shotcrete reinforcement in preventing the development of plastic zones. Further analysis shows that as the rock layer dip angle increases, the plastic zones exhibit a characteristic inclination along the direction of the dip angle, consistent with theoretical analyses of the influence of rock layer angles on shear stress distribution.

3.3   Analysis of Stress on Structural Planes

The stress distribution on structural planes is crucial for predicting rock mass behavior. Changes in stress on these planes, especially sudden changes at certain critical angles, provide early warning signs of potential slip surfaces. This study analyzed the patterns of normal and shear stress changes on three structural planes in the central rock wall area as the rock layer dip angle varies. As shown in Fig. 5, in the unsupported condition, the normal stress on structural planes generally decreases with an increasing dip angle, while shear stress is zero at a 0° angle, then generally increases and then decreases with increasing dip angle, returning to zero at 90°. Notably, at around 55°, both normal and shear stresses on the structural planes undergo a sudden decrease. Under shotcrete reinforcement, the trends in normal and shear stresses on structural planes are generally similar to those in the unsupported condition, but do not show a sudden decrease around 55°.

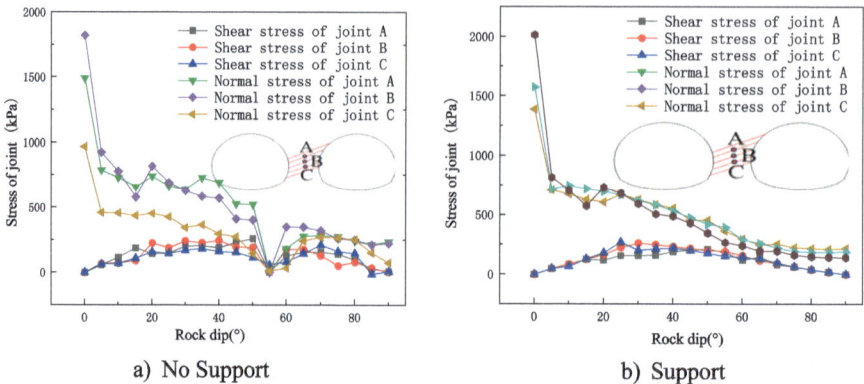

a) No Support   b) Support

Fig. 5. Variation curve of plane stress with dip angle of rock formation

The results indicate that when rock layers are horizontal or vertical, central rock walls do not undergo slip failure along structural planes; deformation primarily occurs through structural plane tension or rock mass shear failure. When rock layers are at certain angles, the central rock walls tend to slide along structural planes, particularly at around 55°. If support is not timely applied, the central rock walls are at risk of slip failure.

4   Conclusion

This paper investigates the mechanics of deformation and slip failure in the central rock walls of small-clearance tunnels through numerical simulation, leading to the following conclusions:

When the rock layer structure interfaces with the tunnel profile to form sharp-angled rock blocks, large deformations occur in these sharp-angled areas. As the dip angle of the rock layer increases, the deformation of the central rock wall first increases and then decreases. The range of dip angles at which the maximum deformation occurs is consistent with the range of angles at which slip failure occurs along structural planes.

In the absence of support, the plastic zone spans the entire central rock wall. When support is applied promptly after excavation, the extent of the plastic zone is significantly reduced. As the dip angle of the rock layer increases, the plastic zone gradually shifts in the direction of the increasing angle.

The stability of the central rock wall is primarily influenced by the rock layer structural planes. As the dip angle of the rock layer increases, the shear stress on the structural planes first increases and then decreases. When the rock layer dip angle is in a specific range, the central rock wall tends to slip.

Importantly, these insights are crucial for designing and maintaining small-clearance tunnels, particularly in managing rock layer angles and construction support. This research also guides future tunnel mechanics studies in complex geological settings.

References

1. Editorial Board of China Journal of Highway and Transport. Overview of academic research in Chinese transportation tunnel engineering 2022. China J. Highway Transp. **35**(04). 1–40 (2022). https://doi.org/10.19721/j.cnki.1001-7372.2022.04.001
2. De Silva, G.P.D., Ranjith, P.G., Perera, M.S.A., et al.: Effect of bedding planes, their orientation and clay depositions on effective re-injection of produced brine into clay rich deep sandstone formations: implications for deep earth energy extraction. Appl. Energy **161**, 24–40 (2016). https://doi.org/10.1016/j.apenergy.2015.09.079
3. Li, T.-Z., Dias, D.: Journal of Central South University **26**(7), 1735–1746 (2019). https://doi.org/10.1007/s11771-019-4129-0
4. Zhou, W.Y., Yang, Y.Y.: Study on the determination of mechanical parameters of jointed rock masses. J. Geotechn. Eng. **14**(5), 1–11 (1992). https://doi.org/10.3724/SP.J.1235.2011.00398
5. Sainsbury, B., Pierce, M., Ivars, D.M.: Simulation of rock mass strength anisotropy and scale effects using a Ubiquitous Joint Rock Mass (UJRM) model. In: Proceedings First International FLAC/DEM Symposium on Numerical Modeling. Minneapolis (2008). https://doi.org/10.1116/j.asw.2013.100753

Study on Seismic Response Mechanism of Continuous Rigid Frame Composite Girder Bridge of High-Speed Rail

Gangyi Liang[1], Zunwen Liu[1,2(✉)], Xingjing Li[1], and Hong Song[1]

[1] School of Civil Engineering, Lanzhou Jiaotong University, Lanzhou 730070, China
liuzunwen@lzjtu.edu.cn
[2] Key Laboratory of Road and Bridge and Underground Engineering of Gansu Province,
Lanzhou Jiaotong University, Lanzhou 730070, China

Abstract. In order to explore the earthquake response mechanism of continuous rigid frame composite girder bridge of high-speed railway, based on a rigid frame bridge of high-speed railway in northwest China and combined with the characteristics of CRTS type I double block non-ballast track structure, three continuous rigid frame composite girder bridge models were established, which considered the constraints of track and subsequent structure, only considering the constraints of track and no considering the constraints of track. The seismic response of structural systems is studied by using response spectrum method. The results show that the track constraint effect reduces the natural vibration period of the structure, and the subsequent structure has a inhibition effect on the seismic response of the track system, which is 4.18%. At the top beam joint of the transition pier, the change of pier height has a significant effect on the seismic response of the track. With the increase of pier height, the maximum rail stress of model 1 and model 2 increases by 50% and 65%, respectively. Under the constraint of track, when the height of rigid frame pier is 28% different from that of simple beam pier, the seismic internal forces of the two piers are equal. When the height of rigid frame pier is greater than 28%, the transition pier is the main energy-consuming rod, and when the height of rigid frame pier is less than 28%, the rigid frame pier is the main energy-consuming rod. Therefore, for bridge types with different pier heights, corresponding design reinforcement measures should be taken at the bottom of the transition pier and rigid frame pier respectively.

Keywords: High-speed railway bridge · Double block ballastless track bridge · Continuous rigid frame composite beam bridge · Earthquake response mechanism

1 Introduction

China is a country with frequent earthquakes [1, 2]. Meanwhile, the development time of rigid frame Bridges of high-speed railway in China is late, and there are many rigid frame Bridges of high-speed railway across gullies in western China, many of which are

A. Bieliatynskyi et al. (Eds.): CSTTE 2023, LNCE 603, pp. 186–196, 2024.
https://doi.org/10.1007/978-981-97-5814-2_17

located in seismic zones with high activity [3–5]. Therefore, according to the dynamic behavior characteristics of rail-bridge system of high-speed railway under earthquake action, establishing the integrated model of high-speed railway continuous rigid frame composite beam bridge and studying the response law of key components of the bridge are of great significance for the optimization design and reinforcement maintenance of practical engineering.

Scholars at home and abroad have studied the integrated model of high-speed railway line bridge and the seismic response of rigid frame bridge under earthquake. Lai Z [6] established a refined finite element model of rail-bridge considering interlayer hysteretic characteristics, and carried out nonlinear dynamic response analysis of rail-bridge system under typical near-field earthquakes. Yu M [7] established a rail-bridge model of a simply-supported beam bridge, and discussed the changes of various near-fault pulse ground motions to the dynamic response of the rail-bridge system. Taking the rigid frame continuous beam bridge as an example, Liang Y [8] took rigid structure-continuous beam Bridges as an example to study the response mechanism of Bridges under large earthquakes based on time-varying vulnerability. Wei J [9] established a finite element model of abutment-approach bridge-rigid frame continuous beam bridge to explore the influence of structural parameters on collision effects at expansion joints and seismic response of bridge structures. Zhao J [10] studied the vulnerability of multi-span high-pier continuous rigid frame Bridges under near-field rotating seismic waves, relying on five-span high-pier continuous rigid frame Bridges.

There are few studies on the seismic response mechanism of rigid frame Bridges of high-speed railway considering track constraints. Therefore, based on the characteristics of CRTS I double-block ballastless track structure, this paper establishes three continuous rigid frame composite girder bridge models, which consider track and subsequent structure constraints, only consider track constraints and do not consider track constraints, to compare and study the key parts of structural system damage under earthquakes, in order to provide theoretical basis for the maintenance and reinforcement of similar Bridges and the study of seismic design optimization.

2 Project Overview and Research Model

2.1 Project Overview

The main bridge is a (48 + 80 + 48) m double-line rigid frame box beam bridge, the beam body material is C50 concrete, the roof width is 13.4 m, the floor width is 5.5 m, the beam height at the rigid frame pier is 7.4 m, the beam height at the mid-span section is 2.4 m, and the bottom line is a quadratic parabola. The rigid frame pier is a thin-wall pier spanning a gully, 40 m high, 6m long section, 3 m wide, 0.6 m thick, and made of C40 concrete. The approach bridge is a simple beam bridge with 12m pier height, 6m section length and 2.5 m width. The material is C35 concrete. HRB335 steel bars with a diameter of 20mm are used for the stirrup and longitudinal reinforcement of the pier. The concrete strength of the track plate and the base plate is C40, and the section dimensions are 2.8×0.26 m and 3.4×0.175 m, respectively.

2.2 Track System Simulation

The track system of the model adopts CRTS type I double block ballastless track, which is composed of base and groove (or convex), isolation layer, track plate, double block sleeper, rail and fasteners. The beam body is provided with embedded steel bars, which are used to connect with the track base plate. The road bed plate and the base plate are isolated by geotextile. The groove is arranged around the elastic plate with high stiffness [11]. The track structure is shown in Fig. 1.

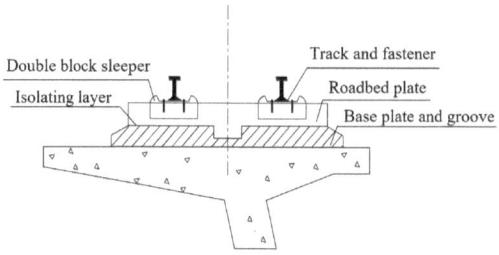

Fig. 1. CRTS I double block ballastless track structure diagram

Ignoring the longitudinal resistance effect of the track fastener, the rail, double-block sleeper and track bed plate are integrated into one section. The integration method is to convert the section of the rail into the section of concrete according to the elastic modulus ratio of steel and concrete materials to ensure the constant stiffness and the total mass is maintained by adjusting the bulk density of the material. The converted rail section size is 0.176×0.44 m, the concrete density is 2.5 t/m^3, the unit mass of the rail is 60 kg/m, the equivalent mass of the integrated section of the track is 2.1t/m, and the integrated rail is used to describe the rail system. A rigid arm is used to simulate the connection between the embedded steel bar and concrete between the base plate and the main beam. The friction stiffness of the isolation layer between the track plate and the track plate and the stiffness of the groove gasket are superimposed and simulated by linear spring. The successor structure is simulated by linear spring.

2.3 The Establishment of Research Model

Combined with the characteristics of CRTS I double-block ballastless track structure, the integrated model of the cable bridge was established. Basin rubber supports were selected for the support, which was established with the master-slave constraint. The stiffness values of each component were shown in Table 1. Bridge and track system are simulated by beam element, rigid pier and beam body are consolidated and connected by rigid arm. Six linear springs are applied to the bottom of the cap to simulate the foundation stiffness, and the spring parameters are calculated by m method. The pier adopts elastoplastic fiber cross section. The concrete constitutive model is Mander model, and the reinforcement constitutive model is Menegotto-pinto model.

A high-speed railway bridge model of simple supported beam bridge+rigid frame bridge+simple supported beam bridge (5×32 m + 48 m + 80 m + 48 m + 5×32

Table 1. Stiffness of each component

Unit	Insulating friction layer	Fluted gasket	Successor structure
stiffness	9.5 kN/m	1.8×10^5 kN/m	7.72×10^4 kN/m

m) was established by using MIDAS, which includes three models considering track and subsequent structure constraints, only track constraints and no track constraints, as shown in Fig. 2 and Fig. 3. There are 2,724 beam units in the integrated model, including 2,004 beam units for the track system, 496 beam units for the main beam, 224 beam units for the bridge pier, and 1,368 spring units.

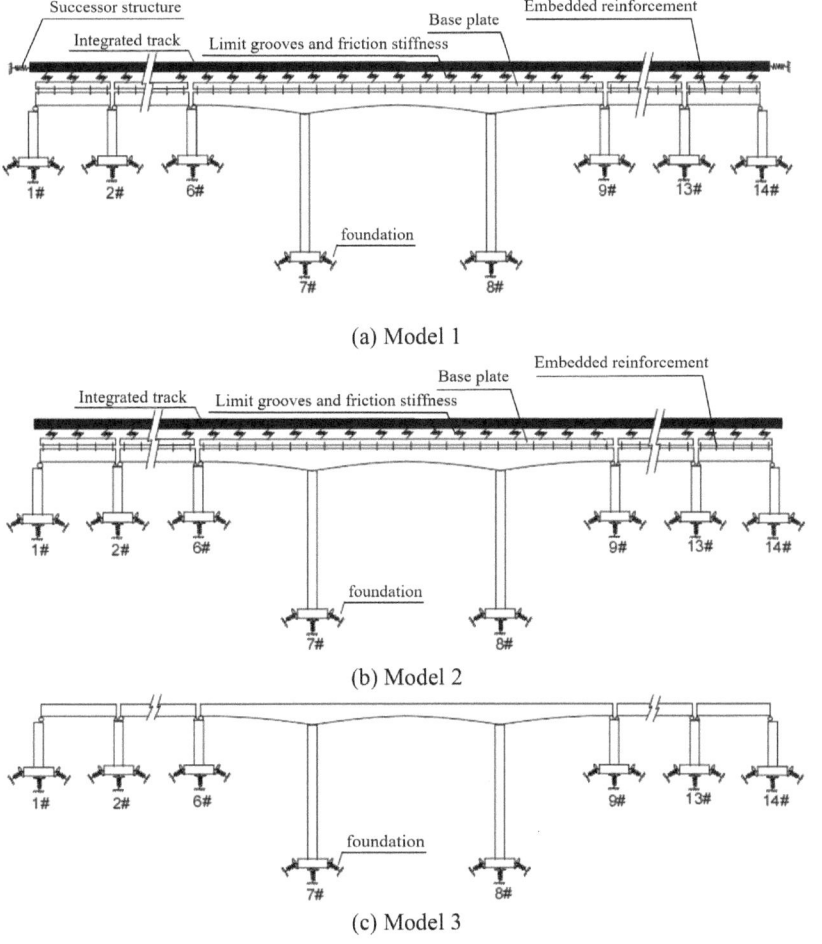

(a) Model 1

(b) Model 2

(c) Model 3

Fig. 2. Finite element calculation model

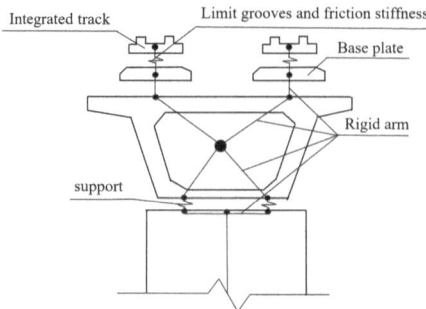

Fig. 3. Structure diagram of bridge cross section

2.4 Introduction of Study Methods

Response spectrum method is an equivalent static method based on linear elastic analysis of a single degree of freedom system under uniform seismic excitation. The response spectrum refers to the spectrum curve of different single-degree-of-freedom systems whose period is horizontal coordinate and whose maximum relative bit, maximum relative velocity and maximum absolute acceleration are vertical coordinate. The response spectrum can be used to quickly calculate the peak value of structural response, which is simple in concept and calculation and suitable for engineering applications. It is the first calculation method for seismic design of common span Bridges in current national norms.

2.5 Study Condition

In order to study the seismic response mechanism of structural system, the study conditions are defined by adjusting the height of rigid frame pier. Working condition 1: the pier of rigid frame bridge is 40m, which is the pier height of rigid frame bridge in actual engineering; Working condition 2: Take half of the pier height of rigid frame bridge, that is, 20 m; Working condition 3: Taking the height of rigid frame pier 17 m, the internal force at the bottom of rigid frame pier under earthquake tends to be equal to that of simply-supported beam bridge, but this is not an accurate value, and the accurate value will be determined by calculation in this paper. Working condition 4: The height of rigid frame pier and simple supported beam pier is equal, which is 12 m.

3 Dynamic Characteristics of Different Bridge Types

The modal analysis of the three bridge models was carried out under working condition 1. The calculation results of the first 5 natural vibration periods of the structural system are shown in Table 2, and Fig. 4 is the corresponding vibration pattern diagram under working condition 1.

Table 2. The first five natural vibration periods of the three models

Order of mode	Model 1	Model 2	Model 3
	Natural period/s	Natural period/s	Natural period/s
1	1.169	1.169	1.426
2	0.524	1.095	1.426
3	0.511	0.532	1.159
4	0.462	0.511	0.486
5	0.298	0.462	0.451

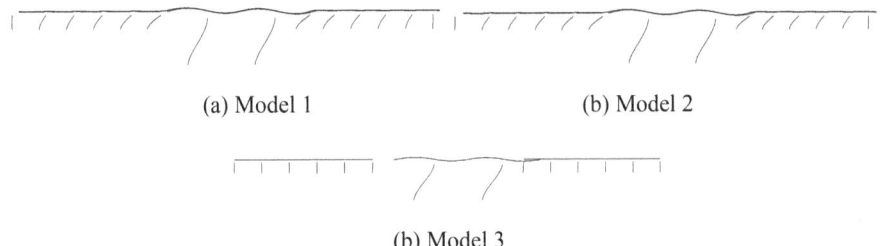

(a) Model 1   (b) Model 2

(b) Model 3

Fig. 4. The mode of each model is described in the working condition

It can be seen from Table 2 that the restraint effect of the track and the subsequent structure reduces the natural vibration period of the structure. According to the calculation of the first 5 orders of natural vibration period, the average natural vibration period of model 1 is reduced by 35.2%, and that of model 2 is reduced by 17.55%. It can be seen that the restraint effect of the subsequent structure has a certain inhibition effect on the natural vibration period of the structure, with the inhibition effect reaching 17.66%. As can be seen from Fig. 4, this is because the track constraints make the separate vibration of the rigid frame bridge and the simply-supported beam bridge become the overall vibration of the entire combined bridge, resulting in shortened natural vibration period of the bridge.

4   Study on Seismic Response of Different Bridge Types

The seismic responses of the three models under four working conditions were compared by the response spectrum method. Bridge type B, regional characteristic period 0.45 s, site type I, fortification intensity 7 (0.15 g) degree, seismic code E2, damping ratio 0.05. The peak seismic acceleration of horizontal design is 0.39 g, the first 200 vibration modes are selected, and the participating mass of vibration modes is more than 95%. The combination of vibration modes adopts CQC method.

4.1 Analysis of Bridge Pier Bottom Internal Force

The seismic internal forces at the bottom of the pier of the three models under four working conditions are shown in Fig. 5.

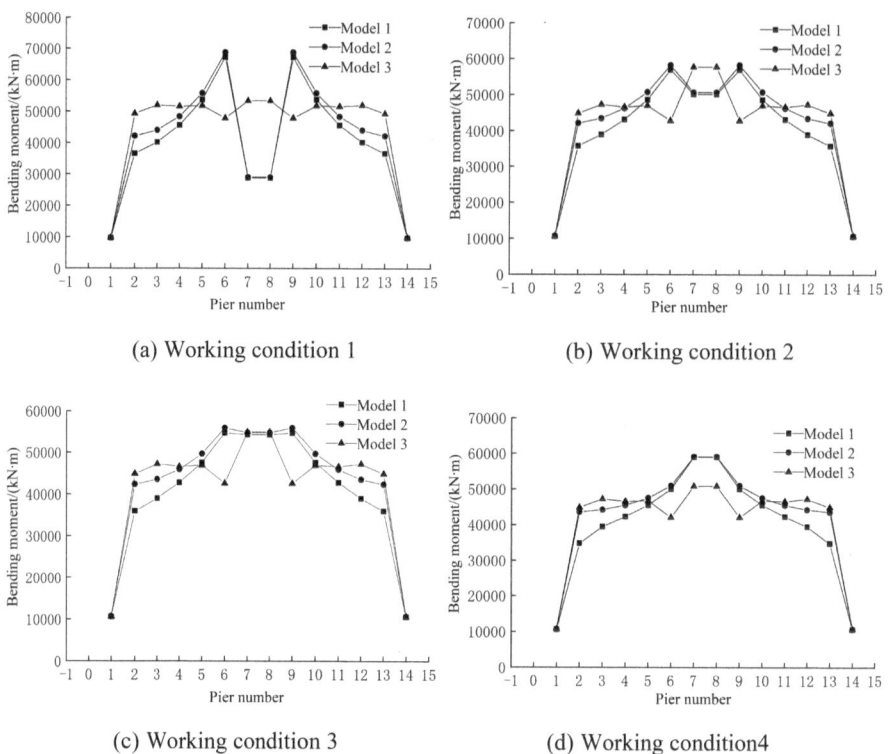

(a) Working condition 1

(b) Working condition 2

(c) Working condition 3

(d) Working condition4

Fig. 5. Distribution of internal forces at the bottom of pier of bridge system

It can be seen from Fig. 5 that under different working conditions, track constraints change the internal force behavior of composite girder Bridges. When the height difference between rigid frame pier and simply supported beam pier is too large, transition pier is the main energy dissipating member. The seismic responses of the two types of piers tend to be equal, and further studies are needed to determine their exact values.

4.2 Analysis of Influence of Rigid Frame Pier Height on Structural System

Figure 6 shows the variation of the internal forces at the bottom of the three models with the change of the height of the rigid frame pier, the transition pier and the rigid frame pier. The internal forces at the bottom of the rigid frame pier and the transition pier of each model are plotted, and the intersection point is the place where the internal forces at the bottom of the pier are equal, so that the height of the rigid frame pier at this time can be determined.

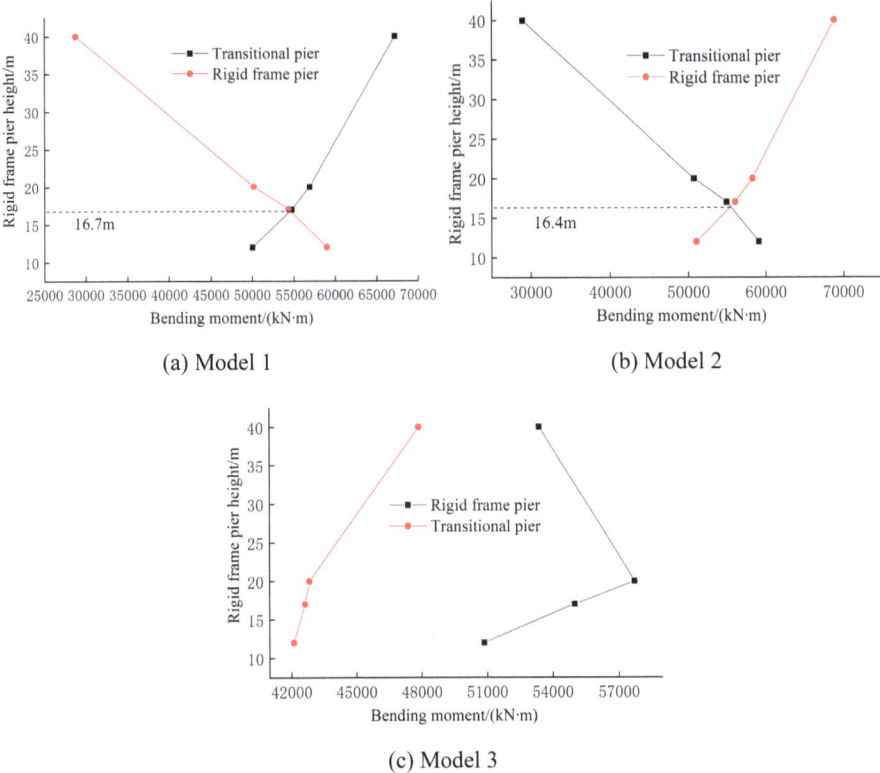

(a) Model 1   (b) Model 2

(c) Model 3

Fig. 6. Diagram of internal force change of model pier bottom when pier height changes

It can be seen from Fig. 6 (a) and (b) that, considering the track constraint, the seismic internal forces at the bottom of rigid frame pier increase with the decrease of the height of rigid frame pier, and the seismic internal forces at the bottom of transitional pier decrease with the decrease of the height of pier. The internal forces of model 1 and model 2 are equal when the rigid frame pier is 16.7 m and 16.4 m respectively. It can be seen that it is reasonable to use model 1 to study the influence of the subsequent structure. In Fig. 6(c), the internal force at the bottom of model 3 rigid frame pier is always greater than that of transition pier. Therefore, when considering the track constraint and the successor structure, when the difference between the height of rigid frame pier and the height of simple beam pier is greater than 28%, the main energy dissipation member of the structural system is the transition pier. When the difference between the height of rigid frame pier and the height of girder pier is less than 28%, the main energy dissipation component of the structure is rigid frame pier. Therefore, in the actual project, in view of the difference between the height of continuous rigid frame pier and the height of simple supported beam pier, strengthening measures are taken to strengthen the pier bottom of the transition pier and rigid frame pier respectively, or the energy dissipation capacity of the pier bottom is enhanced in the design, such as using self-resetting pier, which can effectively improve the seismic performance of the bridge.

4.3 Integrated Track Stress Analysis

Figure 7 shows the track stress distribution of model 1 and Model 2 under four working conditions.

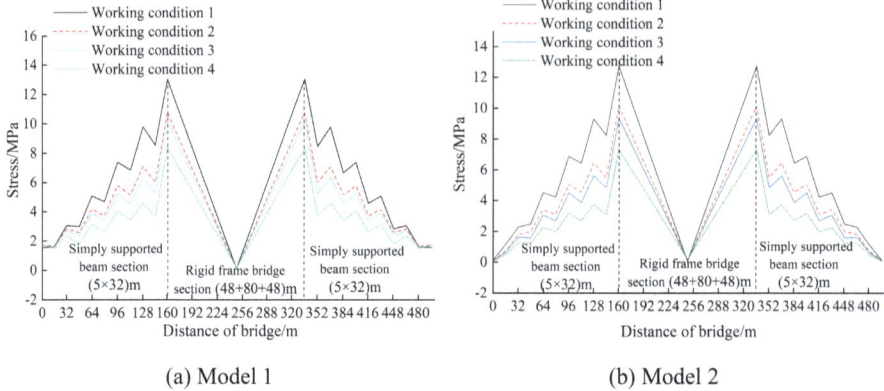

(a) Model 1   (b) Model 2

Fig. 7. Integrated track stress profile

It can be seen from Fig. 7 that the maximum stress of each span bridge occurs at the beam expansion joint, and the track stress is the largest at the beam joint at the top of the transition pier, which is the first to be damaged under the earthquake. In addition, the track stress increases with the increase of pier height. From working condition 4 to working condition 1, the maximum track stress of model 1 increases by 50%, and that of model 2 increases by 65%. It can be seen that the change of pier height has a significant effect on the seismic response of the track, and the subsequent structure has a certain inhibitory effect on the seismic response of the track, with the average inhibitory effect of 4.18% under the four working conditions.

5 Conclusion

By studying the seismic response mechanism of three calculation models of high-speed railway Bridges, the following conclusions are drawn:

(1) The track constraint reduces the natural vibration period of the structural system, and the subsequent structure has a certain inhibitory effect on the natural vibration period of the structural system and the seismic response of the track system, with a suppression effect distribution of 17.66% and 4.18%. It is more reasonable to use Model 1 when studying the seismic response of structural systems.

(2) The height difference between rigid frame pier and simply supported beam pier has a significant effect on the seismic response of the track system. The maximum rail stress increases with the increase of rigid frame pier height, and the maximum rail stress increases by 50% in model 1 and 65% in model 2. The maximum stress of the track occurs at the top beam joint of the transition pier, where the track is damaged first under the earthquake. Energy dissipation devices can be installed at the beam joint to reduce the damage of the track.

(3) When there is a 28% difference between the height of rigid frame pier and that of simple beam pier, the internal forces of the two piers are equal. When the difference is greater than 28%, the main energy-consuming member of the structural system is the transition pier, and the bottom of the pier first becomes the plastic hinge area. When it is less than 28%, the main energy-consuming member of the structural system is rigid frame pier, and the bottom of the pier first becomes the plastic hinge area. Therefore, for bridge types with different pier heights and differences, corresponding reinforcement measures should be taken at the bottom of transition pier and rigid frame pier respectively, or piers with good energy dissipation capacity, such as self-resetting pier, should be used to improve the seismic performance of the structural system.

In this paper, the response spectrum method is used to study the seismic response mechanism of each component, but the energy dissipation capacity of each component under earthquake cannot be calculated, and the energy dissipation law of each component of the structural system can be further studied by the dynamic time history method.

Acknowledgement. Fund projects: Gansu Provincial Natural Science Foundation Project (21JR1RA240).

References

1. Zhou, W., Peng, D., Jiang, L., et al.: Study on track irregularity of CRTS III ballastless track-bridge system of high-speed railway under transverse earthquake. J. Railway Sci. Eng. **20**, 2773–2784 (2023). https://doi.org/10.19713/j.cnki.43-1423/u.t20221499
2. Dehghani, E., Zadeh, M.N., Nabizadeh, A.: Evaluation of seismic behaviour of railway bridges considering track-bridge interaction. J. Roads Bridges-Drogi i Mosty **18**, 51–66 (2019). https://doi.org/10.7409/rabdim.019.004
3. Jiang, L., Zhang, Y., Feng, Y., et al.: Simplified calculation modeling method of multi-span bridges on high-speed railways under earthquake condition. J. Bull. Earthquake Eng. **18**, 2303–2328 (2020). https://doi.org/10.1007/s10518-019-00779-x
4. Livingston, E., Sasani, M., Bazan, M., et al.: Progressive collapse resistance of RC beams. J. Eng. Struct. **95**, 61–70 (2015). https://doi.org/10.1016/j.engstruct.2015.03.044
5. Jiao, X., Wang, R., Ma, H.: Influence of fault strike on the seismic response of rigid-frame bridge. J. China Earthquake Eng. J. **45**, 50–57 (2023). https://doi.org/10.20000/j.1000-0844.20210516001
6. Lai, Z., Jiang, L.: Mapping between the residual deformation in the CRTS II slab track-bridge system and the rail deformation under earthquake in the transverse direction. J. China Civil Eng. J. **56**, 87–99 (2023). https://doi.org/10.15951/j.tmgcxb.21121295
7. Yu, M., Lü, J., Jia, H., et al.: Response analysis of high-speed railway bridge-rail system subjected to near-fault pulse-type earthquake. J. Hunan Univ. (Nat. Sci.) **48**, 138–146 (2021).https://doi.org/10.16339/j.cnki.hdxbzkb.2021.09.015
8. Liang, Y., Yan, S., Zhao, B., et al.: Seismic fragility analysis of rigid frame bridge near-fault high-speed railway. J. Zhengzhou Univ. (Eng. Sci.) **43**, 80–85+91 (2022). https://doi.org/10.13705/j.issn.1671-6833.2022.04.022
9. Wei, J., Wu, X.: Influence of structural parameters on the seismic response of rigid frame continuous girder bridges with high and low piers. J. China Earthquake Eng. J. **45**, 1333–1342 (2023). https://doi.org/10.20000/j.1000-0844.20210916003

10. Zhao, J., Jia, H., Zhan, Y., et al.: Effects of near-field rotating seismic waves on seismic vulnerability of multi-span high-pier continuous rigid frame bridge. J. Vib. Shock **42**, 146–159 (2023). https://doi.org/10.13465/j.cnki.jvs.2023.01.018
11. Liu, Z.: Research on seismic performance and design method for high-speed railway bridges based on track-bridge integrated model. J. Chinese J. Rock Mech. Eng. **39**, 1080 (2020). https://doi.org/10.13722/j.cnki.jrme.2019.1000

Research on Typical Design Scheme for Mechanized Construction of 220 kV Overhead Transmission Line Foundation

Wenxiang Zhang[1(✉)] and Xiaoyu Huang[2]

[1] POWERCHINA Fujian Electric Power Engineering Co., Ltd., Fuzhou 350001, Fujian, China
495529242@qq.com
[2] State Grid Fujian Economic Research Institute, Fuzhou 350012, China

Abstract. The level of mechanization of overhead transmission lines is an important symbol of modern power construction. Good design and planning of road construction and typical planning of construction foundation can minimize the disturbance of mechanical equipment to the foundation and reduce the impact on the environment, and improving the reliability of soil and water conservation measures. Based on the sorting of commonly used mechanical construction foundation types in mountainous areas, through equipment research, this paper discuss the typical schemes for temporary road on plain area, coastal area, river area and mountain area, and use the typical design on tower site with different equipment to optimise the arrangement of construction and quantitative evaluation on excavation. The results show that: (1) the scheme of plain area are wooden boards and old tires, the scheme of coastal areas and river area are filling and temporary steel trestle, the scheme of mountain area are expand the original road and excavating new road. (2) The excavation quantity of rock anchor foundation with slope $\leq 10°$ is the less than that of pile foundation and micro pile foundation, The excavation quantity of split equipment is less than integral equipment because of the requirement of the road width. (3) the slope is an important factor to the excavation quantity, the excavation quantity of rock anchor foundation with slope $25°$ is 2.33 times than that on slope $10°$. The amount of earthwork on the tower site increases by 2.33 times with slope $25°$ of split equipment on pile and micro pile foundation.

Keywords: typical design scheme · mechanized construction · transmission line · temporary road · tower site · micro pile foundation · pile foundation · rock anchor foundation

1 Introduction

The level of mechanization of overhead transmission lines is an important symbol of modern power construction [1]. Overhead transmission line has the characteristics of long distance, crossing different region, complex and diverse terrain and harsh field construction environment, etc. All the time, affected by traditional habits and equipment investment, the mechanization degree of transmission line construction is low in the

A. Bieliatynskyi et al. (Eds.): CSTTE 2023, LNCE 603, pp. 197–215, 2024.
https://doi.org/10.1007/978-981-97-5814-2_18

past. With the innovation of the State Grid to promote the mechanized construction of transmission lines, China has also achieved more results in the field of mechanized construction. Foundation mechanized construction application rate reached more than 90% on UHV from Longdong to Shandong ±800 kV transmission line project and Hami – Chongqing ±800 kV transmission line project. At the same time, the domestic counterparts have more innovation in mechanized construction, Wang Yuan etc. [2] according to design innovation to overcome the unfavorable conditions in the river network area, improve the level of construction mechanization, and achieve a new model of mechanized construction of each process in the harsh environment.

For temporary road construction methods, Ma Li etc. [3] based on typical whole-process mechanized construction cases, proposed that the key influencing factors of temporary road construction include: topography and geomorphic conditions, road width and subgrade bearing capacity. Wang Daojing etc. [4] summarized the requirements of mechanized construction equipment on road conditions, giving the suggestion to transmission line construction. Chen Xing etc. [5] studied the mechanized construction temporary road construction scheme of paddy field terrain, and used gravel paving to build temporary roads for good geological conditions, and some towers needed to be widened. Ma Mingzhi etc. [6] proposed a comprehensive evaluation method of construction road network to realize efficient use of existing roads and rationalization of temporary road construction, while optimizing the route.

In terms of quantitative evaluation of mechanized construction, Ge Zhaojun etc. [1] proposed an evaluation method of mechanization rate applicable to overhead transmission lines, calculates the average score rate of each sub-process according to the use of mechanized equipment of each sub-process of tower construction. Zeng Shoujian etc. [7] conducted technical and economic analysis, and initially found out the cost and benefit of the whole process of mechanized construction of transmission line engineering. In terms of equipment research, Wang Xichen etc. [8] analyzed and studied the economy and necessity of using centralized mixing concrete in UHV transmission lines, and applied it in the foundation construction of 1000kV Jindongnan - Nanyang - Jingmen UHV AC transmission line project. Qin Qingzhi etc. [9] developed the special equipment of rotary digging drill for cutting and digging basic machinery of transmission tower in combination with the basic design characteristics of Xiangjiaba-Shanghai ±800 kV UHV DC transmission line project.

Although a lot of research have been carried out on mechanization, for the transmission line on mountain area is still difficult to carry out mechanization construction. The main reason is that a large area of temporary road construction and foundation excavation are required for the equipment to enter the tower site, which causes damage to the environment. The lack of planning on road construction and foundation resulted in the slope slip phenomenon along with the rainy season (Fig. 1).

To sum up, although domestic counterparts have carried out some researches on mechanization, there are few researches on typical design of temporary road construction and foundation construction. Good design and planning of road construction and typical planning of construction foundation can minimize the disturbance of mechanical equipment to the foundation and reduce the impact on the environment, reducing the cost of environmental remediation and improving the reliability of soil and water conservation measures.

Fig. 1. Road slope slip on mechanized construction

2   Introduction of Mechanized Construction Foundation

The core of influencing on temporary road and foundation construction lies in the selection of foundation type and foundation construction equipment. This section focuses on the research of foundation equipment commonly used in domestic transmission lines. Other equipment such as temporary road construction, material site transportation, concrete construction, grounding construction, tower erection, stringing and construction auxiliary can be selected on the basis of the selection of foundation excavation equipment.

2.1   Pile Foundation

The common excavation equipment for the pile foundation on mountain is the rotary drilling rig, and different bits can be selected according to different geological conditions. Short auger, suitable for clay, silt, fill, sand and weathered rock above the water table; Core auger drill, suitable for gravel soil, medium hardness rock and weathered rock; Core rotary bucket, suitable for weathered rock and cracked rock. The functional parameters of commonly equipment used in transmission line construction are shown in Tables 1 and 2.

Table 1. The functional parameter of integral rotary drilling rig

Rig type	Maximum hole diameter in strata below 30MPa (m) one-time	Drilling efficiency (penetration speed, m/h, m/h)	Maximum hole diameter in strata above 30MPa and below 60MPa (m) one-time	Width of walking state (m)	Dimension on working state Length (m)X Width (m)
KR50D	1.2	2	0.6	2.2	5.4 × 2.2
KR100D	1.5	3	0.8	2.6	6.45 × 3.6
KR110D	2.0	3.5	1.0	2.6	7.15 × 3.6
KR150D	2.0	4	1.2	3.5	7.05 × 3.6
KR125ES	1.2	4	1.2	3	7 × 3.6

Table 2. The functional parameter of split rotary drilling rig

Rig type	Width on working (m)	Length on working (m)	Self driving turning radius (m)	Self driving turning slope (m)	Length (m) Xwidth (m)	scope of application
split rotary drilling rig	2	3	1.2	25°/0.3 m	3*2	Moderately weathered rock, with a rock hardness below 40MPa

2.2 Rock Anchor Foundation

The mechanized construction of rock anchor foundation mainly focuses on the drilling of rock anchor. Air compressor + light anchor drill is the key equipment of drilling, which affects the quality of the foundation - the orientation, depth and accuracy of the drilling diameter. Through investigation, the related parameters on drilling equipment of rock anchor foundation commonly used in transmission line are shown in the Table 3.

Table 3. The functional parameter of rock anchor equipment

Type of foundation	Equipment	The character and suit range	picture
Rouk anchor foundation	QWY-30	The weight of equipment is 255kg, which can be disassembled into 5 pieces, and the maximum weight of a single piece is less than 60kg. The maximum drilling depth is 30 m, and the maximum opening diameter is 120mm	
	MGS-30	The whole machine is divided into two parts, the main engine and the hydraulic station. The maximum decomposing weight does not exceed 100kg	
	MDS-90	Split modular drill & multi-functional power platform. The maximum weight of the unit is less than 200kg,	
	JY-MD-150/20-A	Special drilling equipment suitable for the construction of rock anchor foundation of power transmission line. The weight of a single module does not exceed 200kg	

2.3 Mountain Micro Pile Foundation

Mountain micro pile foundation can reduce the amount of foundation, improve the efficiency of construction and reduce the risk of operation. Through deeply investigation of scientific research institutions and equipment manufacturers, the latest mountain micro pile foundation construction equipment and related parameters are shown in the Table 4.

Table 4. Comment micro pile equipment on transmission line

Type of foundation	Equipment	The character and suit range	picture
Mountain micro pile foundation	WZFT300A	The weight of machine is 17t. The whole machine is designed into 15 integrated modules. A single module does not exceed 2 tons. Drilling diameter range 90~400 mm;	
	Split-type micro pile foundation equipment	The strength does not exceed 40Mpa in the medium and strong weathering geology, the largest diameter is 400 mm, the deepest hole depth of 15 m, and, a single module does not exceed 400kg	

3　Typical Application of Foundation Design in Mountain Areas

Considering that the selection of foundation and construction equipment in mountain areas is a major difficulty in mechanized construction, the typical application scenarios of foundation design in mountain areas are sorted out as follows according to tower topography, geology and equipment approach conditions. Typical design parameters is shown on Table 5.

Table 5. Typical design parameters

Type	Parameter
slope of tower site	≤10 (degree)
	10~25 (degree)
Stratum	Below Class V:extremely soft rock, extremely broken rock and, frk ≤40 MPa
	Above Class V:extremely soft rock, extremely broken rock, frk ≤40 MPa
	Above Class V:extremely soft rock, extremely broken rock, frk >40 MPa
Overburden thickness	≤1.5 m
	1.5 ~3.5 m
	>3.5 m
Condition of mechanized construction	Have mechanical road construction conditions
	Do not have mechanical road construction conditions

4 Typical Design Scheme of Temporary Road Construction

4.1 Plain Area

The plain area has a flat terrain and good general traffic conditions. It is usually possible to use the existing road or earthed road, and some roads need to be widened and reinforced.

Road scheme: If heavy machinery such as tanker truck, crane or rotary drilling rig passes by, wooden boards should be laid on the cement road surface, and old tires can also be padded when the crawler equipment walks. When padding the wooden board, the board shall cover the width of the wheel or track, as shown in Fig. 2(a); When padding old tires, place two old tires per meter, as shown in Fig. 2(b).

(a)Padding wooden boards   (b) Padding old tires

Fig. 2. Diagram of cement road reinforcement

4.2 Coastal Areas

Some towers in coastal areas are located in beach areas or ponds, and general construction machinery cannot pass through, so approach roads and construction platforms need to be built. The common foundation is cast-in-place pile foundation, and the main machinery is rotary drilling rig, transport truck, crane, commercial mixing tank truck, etc.

Approach road scheme 1 Filling the construction road and construction platform to the shore elevation, with road width is 4m and slope is not more than 1:1.5. The size of the construction platform is related to the distance of the tower root, and normally can be considered by 20–30 m. Before filling, measures such as pumping, cofferdam and silting need to be carried out according to the situation. The filling the construction road is shown on Fig. 3.

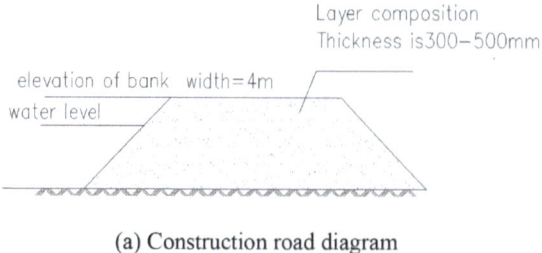

(a) Construction road diagram

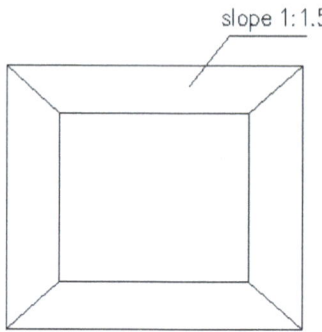

(b) Construction platform diagram

Fig. 3. Schematic diagram of construction road and construction platform constructed by filling

Approach road scheme 2: Build temporary steel trestle and steel platform. The height of trestle and platform top is the same as the top elevation of foundation cap. The length of steel trestle is determined according to the distance between each pile position and the shore. Steel trestle and steel platform are composed of steel pipe piles, bridge panels, steel beams, guardrail and so on. The steel trestle platform in the sea and the beach shall be dismantled after the construction is completed. Temporary steel trestle is shown on Fig. 4.

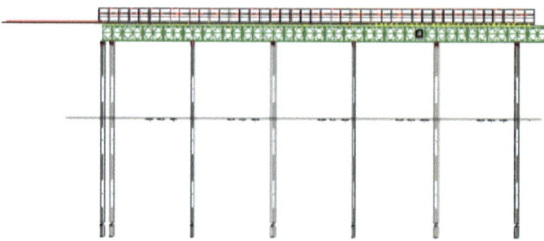

Fig. 4. Temporary steel trestle

4.3   River and Swamp Area

The tower is located in the river and swamp area, including fields where water is often accumulated, areas where mud is silted, and areas where rivers cross and affect land traffic. With Abundant surface water and soft soil, they are not conducive to mechanical passage, and general vehicle passage is prone to subsidence. The foundation types are mainly cast-in-place pile foundation and plate foundation, and the main entering machinery is rotary drilling rig, transport truck, crane, commercial mixing tank truck, etc.

Approach road scheme: When there is silt on the surface, the 0.3 m silt should be replaced and filled by 0.3 m thick pond slag and compacted. The thickness of Lay steel plate is not less than 10 mm. The steel plate is shown on Fig. 5.

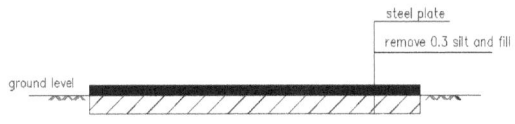

Fig. 5. The steel plate diagram

4.4   Mountainous Areas

The terrain of mountain area is complex, with a certain distance from the existing road to tower site. The overburden soil is mostly silty clay and sandy clay, and the bearing capacity is high. The temporary road construction in the mountain areas should make use of the original cultivator road, gravel road, etc., and use bulldozers, excavators, road rollers and other machinery to broaden and strengthen to meet the equipment entry requirements. Part of the weathered rock in the mountain area is exposed and broken, and the surface soil is loose and shallow. Under the action of concentrated rainwater, some surfaces soil on temporary are muddy and prone to soil erosion. The foundation type suitable for mechanized construction in mountain areas mainly include pile foundation, micro pile foundation, and rock anchor foundation; Main machinery are rotary drilling rig, micro pile drilling rig, transport vehicle, excavator etc.

Typical scheme 1: The original machine plowing road is 1-2m long, with slopes on one or both sides of the road, with slopes of less than 10°, 10–25°, 25–35°, and greater than 35°.

Approach road scheme1: Expand the original organic farming road, excavate soil and rock, and trim the slope. The new slope formed after excavation has a slope of 45–80°, a road width of 3.5 m, and a turning back site is 6m. After leveling and compacting the ground, construction equipment can pass through.

Typical scheme 2:There are no existing roads, and existing slopes are less than 10°, 10–25°, 25–35°, and greater than 35°.

Approach road scheme 2:Excavate the earth and rock and trim the slope, and the new slope formed after excavation has a slope of 45–80°. The road width is 3.5 m, and a turning back site is 6m, the ground is smooth and compact, and some muddy road need to be reinforced with steel plates. Detail is shown on Fig. 6 and Table 6.

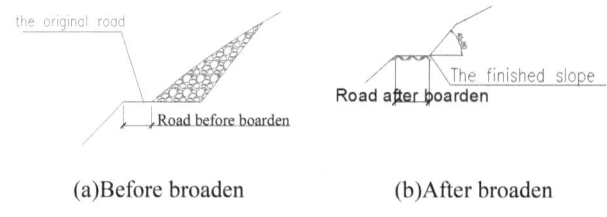

(a)Before broaden   (b)After broaden

Fig. 6. Road scheme on Mountain area

Table 6. Technical parameter of approach road scheme in mountain area (per meter)

Scheme	Slope	Item	Unit	Quantity
Broaden roads	≤10°	The amount of earth and stone	m³	0.75
	10~25°	The amount of earth and stone	m3	2.7
	25~35°	The amount of earth and stone	m³	9.7
		Retaining wall	m³	5.5
	>35°	/	/	/
	No limit	Road bed shaping	m²	1–2
New road	≤10°	The amount of earth and stone	m³	0.87
	10~25°	The amount of earth and stone	m³	2.95
	25~35°	The amount of earth and stone	m³	10
		Retaining wall	m³	5.5
	>35°	/	/	/
	No limit	Road bed shaping	m²	3.5–4
Road bed shaping + paving steel road	/	Road bed shaping	m²	3.5–4
		Steel plate	t	0.054–0.064

5   220 kV Typical Design Scheme on Tower Site

5.1   Typical Scheme 1 - Rock Anchor Foundation with Slope ≤10°

The equipment of Rock anchor foundation can be divided into self-propelled and disassembly. The climbing capacity of self-propelled equipment is 20°, and requirement of the road width is 2 m. If the disassembly type is used, it is necessary to use crawler transporters for transportation, and the road construction width is 2 m.

The self-propelled equipment of rock anchor foundation can be transport on the base surface, and the length of temporary road is 4.8 times of tower root value when transferred between tower legs (assuming the tower root value is 10 m). Since the climbing capacity of equipment is 20°, the earth excavation between the tower leg transition is not considered. Taking into account the requirements of the flatness of the construction surface, the excavation area of the working surface is 16 m². The diagram of transition between tower legs is shown on Fig. 7 and Table 7.

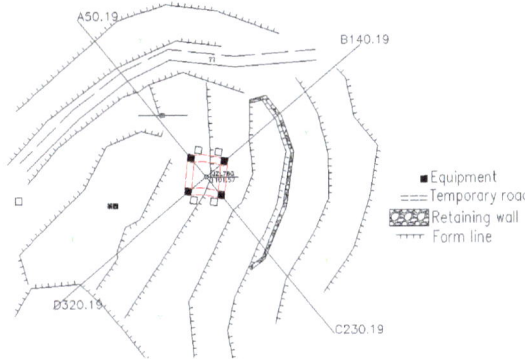

Fig. 7. The diagram of transition between tower legs on 10° slope

Table 7. The amount of excavation work of rock anchor foundation with slope ≤10°

No	Human transportation distance m	Length of Road m	Excavation quantity of earthwork for new road construction m^3	Excavation quantity of earthwork for widen road m^3	Excavation quantity of tower site surface m^3
1	100	130	64.6	55.7	12
2	200	260	129.3	111.4	12
3	300	390	193.9	167.1	12
4	400	520	258.5	222.9	12
5	500	650	323.1	278.6	12
6	600	780	387.8	334.3.1	12
7	700	910	452.4	390.0	12
8	800	1040	517.0	445.7	12

5.2 Typical Scheme 2 - Pile Foundation with Slope ≤10°

The equipment of mechanical pile foundations are mainly divided into two types: split rotary drilling equipment and integral rotary drilling equipment. Specific parameters are detailed in Tables 1 and 2. Due to different equipment, the requirements for temporary roads are also different. Therefore, the following will elaborate on the tower site scheme for different equipment.

Split Rotary Drilling Equipment
The width of the temporary road construction is 2 m, and it is transported to tower site using tracked transport vehicles. Taking into account factors such as construction efficiency, the split type rotary drilling rig adopts a self-propelled method for the tower legs transfer. When transferring between tower legs, the length of road should be 4.8 times of tower root. As the climbing capacity of the rotary drilling rig is 25°, the excavation of surface soil during tower leg transfer is not considered. With the requirement for flatness of the construction work surface, the area of the work surface is 30 m². Detail is shown on Fig. 8 and Table 8.

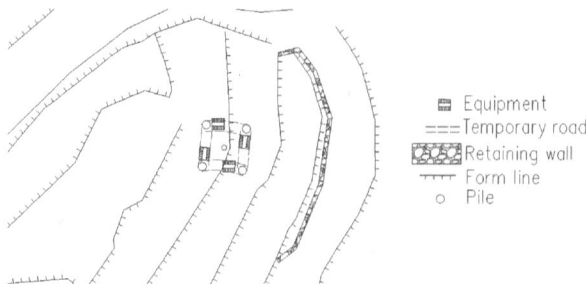

Fig. 8. The diagram of transition between tower legs on 10° slope (split rotary drilling equipment)

Table 8. The amount of excavation work of pile foundation with slope ≤ 10° (split rotary drilling equipment)

No	Human transportation distance m	Length of Road m	Excavation quantity of earthwork for new road construction m^3	Excavation quantity of earthwork for widen road m^3	Excavation quantity of tower site surface m^3
1	100	130	64.6	55.7	24
2	200	260	129.3	111.4	24
3	300	390	193.9	167.1	24
4	400	520	258.5	222.9	24
5	500	650	323.1	278.6	24
6	600	780	387.8	334.3	24
7	700	910	452.4	390.0	24
8	800	1040	517.0	445.7	24

It can be seen on Table 8 that under the condition of a 10° slope, the excavation volume of the split type on tower site is not large, accounting for 5–30% of the total volume. The proportion of earthwork excavated on tower site is higher than that of rock anchor foundation under the same conditions.

Integral Rotary Drilling Equipment

Taking KR150D rotary drilling rig as an example, the equipment needs to build a temporary road with a width of 3.5m and walk up to tower site on its own.

Considering factors such as construction efficiency, the KR150D rotary drilling rig adopts a self-propelled method for tower legs transfer. When transferring between tower legs, the base is leveled within the range of the tower legs (taking a 10 m tower as an example).

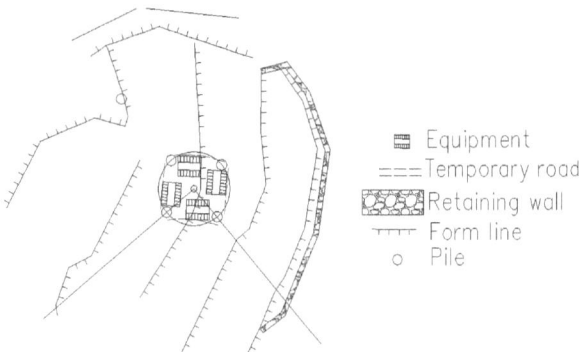

Fig. 9. The diagram of transition between tower legs on 10° slope (integral rotary drilling equipment)

It can be seen on Fig. 9 that due to the small distance of tower legs, the transfer road of equipment basically occupies the land area inside the tower legs. At this point, the corresponding earthwork volume will be slightly greater than that of split equipment.

Table 9. The amount of excavation work of pile foundation with slope ≤10° (integral rotary drilling equipment)

No	Human transportation distance m	Length of Road m	Excavation quantity of earthwork for new road construction m^3	Excavation quantity of earthwork for widen road m^3	Excavation quantity of tower site surface m^3
1	100	130	113.1	97.5	64.8
2	200	260	224.3.2	195	64.8
3	300	390	339.3	292.5	64.8
4	400	520	452.4	390	64.8
5	500	650	565.5	487.5	64.8
6	600	780	678.6	585	64.8
7	700	910	791.7	682.5	64.8
8	800	1040	904.8	780	64.8

From Table 9, it can be seen that under the condition of a 10° slope, the excavation volume of tower site surface is relatively large, accounting for about 5–50% of the total volume. Due to the high width requirements of the integrated rotary drilling rig for temporary roads, the earthwork volume of temporary roads is 1.75 times that of rock anchor temporary roads.

5.3 Typical Scheme 3 - Micro Pile Foundation with Slope ≤10°

Micro pile foundation drilling equipment are mainly divided into two types: split type and integral type, with specific parameters detailed in Table 4. Due to different equipment, the requirements for temporary roads are also different. Therefore, the following will elaborate on the schemes for different equipment.

Split Micro Pile Drilling Rig

The equipment adopts JK-MD-400/15-A split type micro pile drilling rig. As the equipment can be assembled and disassembled, the temporary road construction width is 2 m, and it is transported to tower site by tracked transport vehicles.

Considering factors such as construction efficiency, the split type micro pile drilling machine adopts a self-propelled method for the transfer of the foundation. Considering the requirement for flatness of the construction work surface, the area of the work surface is 25 m^2.

The excavation quantity of the foundation surface for the split micro pile foundation is the same as that of the split type drilled pile foundation under a 10° slope condition. The proportion of earthwork excavated on its foundation is higher than that of rock anchor foundation under the same conditions.

Integral Micro Pile Drilling Rig

Taking the WZL400/10 A micro pile drilling rig as an example, the equipment needs to build a temporary road with a width of 3.5 m.

Considering factors such as construction efficiency, the WZL400/10 A micro pile drilling rig adopts a self-propelled method for the transfer of the foundation. The excavation quantity of the foundation surface for the integral micro pile foundation is the same as that of the integral type drilled pile foundation under a 10° slope condition.

5.4 Typical Scheme 4–10-25° Rock Anchor Foundation

As the climbing capacity of the anchor rod drilling rig is 20°, it is necessary to consider increasing the length of the temporary road between tower legs to reduce the climbing slope. The length of temporary road should be 5.0 times tower legs. Due to the small size of the 220kV foundation, if a transfer is carried out within the tower leg range, it will involve excavation of a large amount of soil. Therefore, the transfer road for rock anchor rod foundation should be selected from outside the tower leg. Considering the requirement for the flatness of the construction work surface, the area of the work surface is 16m^2. Considering factors such as sloping, the excavation volume of the tower site surface is 40 cubic meters. Detail is shown on Fig. 10.

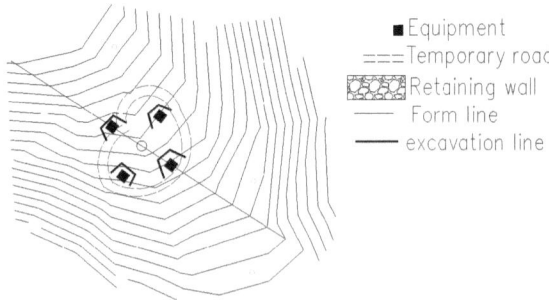

Fig. 10. The diagram of transition between tower legs on 25° slope

Table 10. The amount of excavation work of rock anchor foundation with slope 25°

No	Human transportation distance m	Length of Road m	Excavation quantity of earthwork for new road construction m³	Excavation quantity of earthwork for widen road m³	Excavation quantity of tower site surface m³
1	100	130	219.1	200.6	40
2	200	260	438.3	401.1	40
3	300	390	657.4	601.7	40
4	400	520	874.3.6	802.3	40
5	500	650	1095.7	1002.9	40
6	600	780	1314.9	1203.4	40
7	700	910	1534.0	1404.0	40
8	800	1040	1753.1	1604.6	40

From Table 10, it can be seen that under the condition of a 25° slope, the amount of earthwork for temporary road construction is about 2.39 times that of a 10° slope under the same conditions, and the amount of earthwork increases significantly. Compared to the 10° slope under the same conditions, the amount of earthwork on the tower site increases by 2.33 times. Under this typical scheme, the main amount of earthwork occurs in the construction of temporary roads.

5.5 Typical Scheme 5–10-25° Pile and Micro Pile Foundation

Split Equipment
When transferring between tower legs, the length of temporary road should be 2.5 times of tower legs distance. Due to the small 220 kV tower legs distance, the method of constructing a platform with two tower legs is considered, resulting in a larger amount of foundation soil. Due to the climbing capacity of the rotary drilling rig being 25°, earthwork excavation during tower leg transfer is not considered. Detail is shown on Fig. 11.

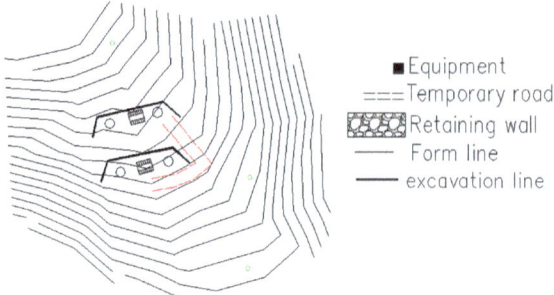

Fig. 11. The diagram of transition between tower legs on 25° slope (split equipment)

Table 11. The amount of excavation work of split equipment with slope 25°

No	Human transportation distance m	Length of Road m	Excavation quantity of earthwork for new road construction m^3	Excavation quantity of earthwork for widen road m^3	Excavation quantity of tower site surface m^3
1	100	130	219.1	200.6	84
2	200	260	438.3	401.1	84
3	300	390	657.4	601.7	84
4	400	520	874.3.6	802.3	84
5	500	650	1095.7	1002.9	84
6	600	780	1314.9	1203.4	84
7	700	910	1534.0	1404.0	84
8	800	1040	1753.1	1604.6	84

From Table 11, it can be seen that under the condition of a 25° slope, the excavation volume of the split equipment increases by 2 times compared to the rock anchor foundation under the same conditions, and the excavation volume of the temporary road increases by about 2.14 times compared to a 10° slope.

Integrated Equipment

Considering factors such as construction efficiency, integrated equipment adopts a self-propelled approach for the transfer of the foundation. Considering the small tower root opening, a work platform is set up for the two tower legs, and the length of the road construction during the transfer between the tower legs is 2.5 times tower legs distance. Due to the climbing ability of the rotary drilling rig at 25°, soil excavation during the transfer of the tower legs is not considered. Detail is shown on Fig. 12.

From Table 12, it can be seen that under the condition of a 25° slope, the excavation volume of the foundation surface increases by 94% compared to the condition of a 10°

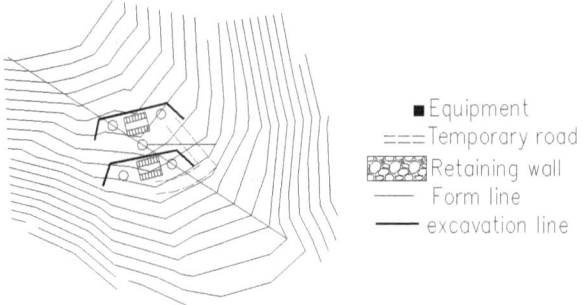

■ Equipment
=== Temporary road
Retaining wall
—— Form line
—— excavation line

Fig. 12. The diagram of transition between tower legs on 25° slope (integral equipment)

Table 12. The amount of excavation work of integral equipment with slope 25°

No	Human transportation distance m	Length of Road m	Excavation quantity of earthwork for new road construction m^3	Excavation quantity of earthwork for widen road m^3	Excavation quantity of tower site surface m^3
1	100	130	383.5	351.0	126
2	200	260	767.0	702.0	126
3	300	390	1150.5	1053.0	126
4	400	520	1534.0	1404.0	126
5	500	650	1917.5	1755.0	126
6	600	780	2301.0	2106.0	126
7	700	910	2684.5	2457.0	126
8	800	1040	3068.0	2808.0	126

slope, and the earthwork volume for road construction increases by 2.38 times compared to the condition of a 10° slope. Due to the high width requirements of the integrated rotary drilling rig for temporary roads, the earthwork volume of the temporary road is 1.75 times that of the split type temporary road.

6   Conclusions

This article is based on the sorting of commonly used mechanical construction foundation types in mountainous areas, through equipment research, typical schemes for access road and foundation design are studied, and the main conclusions are as follows:

(1) The most famous mechanized foundation types in mountain area are rock anchor foundation, pile foundation and micro pile foundation. The equipment of different foundation type can divide into split and integral type.

(2) Typical design scheme of temporary road construction include plain area, coastal areas, river and swamp area and mountain area. The scheme of plain area are wooden boards and old tires. The scheme of coastal areas and river area are filling and temporary steel trestle. The scheme of mountain area are expand the original road and excavating new road.

(3) The excavation quantity of rock anchor foundation with slope $\leq 10°$ is the less than that of pile foundation and micro pile foundation, because of the small equipment. The excavation quantity of split equipment is less than integral equipment because of the requirement of the road width.

(4) The excavation quantity of rock anchor foundation with slope $25°$ is 2.33 times than that on slope $10°$. The amount of earthwork on the tower site increases by 2.33 times with slope $25°$ of split equipment on pile and micro pile foundation. Compared to the $10°$ slope under the same conditions, the amount of earthwork of integral equipment on the tower site increases by 2.33 times.

References

1. Ge, Z., Li, X., Zhang, Q., et al.: Research on the evaluation method of construction mechanization rate for overhead transmission line engineering. Smart Grid **4**, 1252–1256 (2016)
2. Yuan, W., Xingbin, Q., Hansheng, L.: Research on mechanized design and construction of transmission lines in muddy areas of river networks. Hubei Electr. Power **40**(8), 21–26 (2016)
3. Li, M., Liping, S.: Differential study on mechanized construction measures for transmission line engineering. Autom. Instrum. **12**, 4 (2018)
4. Daojing, W., Xiaohu, Z., Min, Y.: Research on investment estimation of temporary road construction in mechanized construction engineering based on entropy weight method. Comput. Digit. Eng. **46**(12), 5 (2018)
5. Xing, C., Lin, Y., Mingwen, G., et al.: Research on optimal selection of mechanized construction foundation for transmission lines in paddy field terrain. Jiangxi Electr. Power **4**, 3 (2019)
6. Zhizhi, M., Changjie, Y.: Research on the construction of temporary roads in the whole process mechanized construction of transmission lines based on road network evaluation method. Electr. Technol. **09**, 51–53 (2022)
7. Zeng Shoujian, X., Chaochen, L.R., Ying, Z.: Technical and economic analysis of mechanized construction of transmission lines throughout the whole process. China Electr. Power Enterprise Manage. **02**, 74–78 (2016)
8. Xichen, W., Zhenyu, L., Qihai, H.: Application of concentrated mixing concrete in ultra high voltage transmission line foundation engineering. Electr. Power Construct. **12**, 52–54 (2008)
9. Qingzhi, Q., Yanjun, Z., et al.: Development and engineering application of drilling equipment for excavation foundation machinery. Electric Power Construct. **11**, 47–49 (2010)

Numerical Simulation of Ultimate Bearing Capacity of Corroded Pipeline under Coupling Load

Tong Xu[✉], Wei Sun, Wenyu Tan, and Shuai Feng

Shandong Key Laboratory of Disaster Prevention and Mitigation in Civil Engineering, Shandong University of Science and Technology, Qingdao, China
xutong1205@163.com

Abstract. Based on the ductile damage theory of metal materials, this paper systematically studies the ultimate bearing capacity of corroded pipeline subjected to axial force, bending moment and pressure by using fluid cavity simulation technology. For pipelines with different corrosion characteristics, the interaction diagram of the bearing capacity of corroded pipeline is established by considering the coupling effect of medium properties, pressure, external axial force and bending moment. The numerical simulation results show that the depth of corrosion strongly affects the bearing capacity of corroded pipeline. With the increase of defect depth, the N-M (axial force and bending moment) interaction curve of corroded pipeline gradually flattened. And the results can truly reflect the damage phenomenon.

Keywords: Corroded pipeline · Fluid cavity · Ductile damage · Ultimate bearing capacity

1 Introduction

At present, the research on the residual strength of corroded pipeline is generally based on the finite element simulation results. Irregular defect shapes have caused great distress to the research of corroded pipeline. Through a series of assumptions and simplifications of corrosion defects, some scholars [1] transformed irregular corrosion defect shapes into a regular square, circular and other shapes. Researchers [2–4] have realized the decrease in the pressurized bearing capacity of pipeline subjected to external loads by using finite element simulation technology.

Mondal [5] proposed that for corroded pipeline, it is impossible to simulate the influence of corrosion cracking and crack propagation on the pressurized bearing capacity of pipeline by using standard finite element modeling technology. It is recommended to use fracture mechanics to calculate the residual strength of corroded pipeline containing corrosion defects or crack-like defects. Lu et al. [6] studied the residual strength of pressurized equipment with structural defects and proposed to establish a finite element defect component model. The surface of the outer wall of the pipeline to the defect surface is modeled in a smooth way to prevent stress concentration. Bai et al. [7] studied the

© The Author(s) 2024
A. Bieliatynskyi et al. (Eds.): CSTTE 2023, LNCE 603, pp. 216–224, 2024.
https://doi.org/10.1007/978-981-97-5814-2_19

bearing capacity of pipeline subjected to axial force, pressure and bending moment, and proposed that different loading sequences had different effects on the bearing capacity of pipeline. The loading sequence of the first pressure and then the final axial force and bending moment of the thick-walled pipe has a greater impact than other loading sequences.

Based on the research results of Bai et al. [7], this paper will study the ultimate bearing capacity of corroded pipeline by applying pressure first and then applying axial force or bending moment to corroded pipeline.

2   The Finite Element Model

2.1   Modeling

In order to reduce the time of calculation, the shell element is used to model the finite element pipeline, and the symmetrical boundary of the finite element software is used to establish the model of a quarter of corroded pipeline. As shown in Fig. 1, the mesh size of the non-defect area is 1 mm, and the mesh size of the defect area is 3 mm.

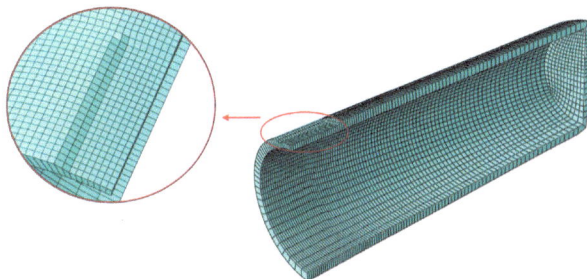

Fig. 1. The FE model of a quarter of corroded pipeline

In this paper, the simplified rectangular defect shape is used as the corrosion defect of the pipeline. and the fluid cavity simulation technology is used. The cavity is defined by specifying the surface that completely surrounds the cavity and is associated with the node called the cavity reference node.

2.2   Verify the Finite Element Model

The material parameters and size parameters of the corroded pipeline are selected from the paper of Oh et al. [8], as shown in Tables 1 and 2.

According to the numerical simulation results, the stress curve of the most unfavorable element at the defect is drawn, as shown in Fig. 2. The red dotted line is the ultimate tensile strength of the pipeline material 563.8MPa, and the black dotted line is the pressure value 24.09MPa when the ultimate tensile strength is reached by simulation. This value has an error of 0.864% (0.21MPa) with the value 24.3MPa obtained by Oh et al. [8]. Mondal et al. [2] carried out a numerical analysis of the test pipeline, and the simulation value was 24.47MPa, which had a 0.69% (0.17MPa) error with the test value in Oh et al. [8]. The simulation results obtained in this paper are more conservative than those of Mondal et al. [2] but also have high accuracy, which is conducive to the analysis of the bearing capacity of corroded pipeline during operation.

Table 1. The parameter of material [8]

Property	Value
Density, ρ (kN/m^3)	7850
Modulus of Elasticity, E (GPa)	210.7
Poisson's Ratio, ν	0.30
Yield Strength, σ_Y (MPa)	464.5
Ultimate Tensile Strength, σ_U (MPa)	563.8

Table 2. The dimensions of corroded pipe [8]

Parameter	Value
Pipe Diameter, D (mm)	762
Wall Thickness, t (mm)	17.5
Defect Depth, d (mm)	8.75
Defect Length, l (mm)	100
Defect Width, w (mm)	50

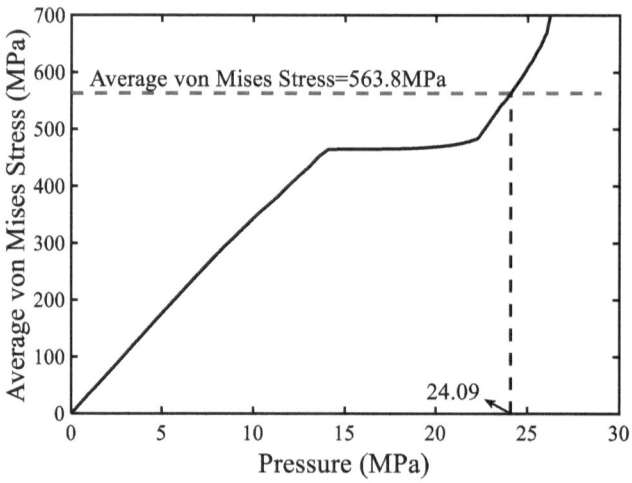

Fig. 2. Burst pressure predicted by finite element model

3   Interaction Curve of Ultimate Bearing Capacity of Corroded Pipeline

According to the defect size specification in ASME B31G [9], the shallow defect pipeline (d/t = 0.25) and deeply defect pipeline (d/t = 0.75) were designed on the basis of the intact pipeline, and the size is shown in Table 3. The material parameters in Xu [10] are selected. The stress-strain curve of the pipeline is shown in Fig. 3, and the ductile damage parameters of the material are shown in Table 4.

Table 3. The three dimensions of corroded pipeline [9]

Parameter	Intact	Shallow	Deeply
Pipe Diameter, D(mm)	60	60	60
Wall Thickness, t(mm)	4	4	4
Defect Depth, d(mm)	---	1	3
Defect Length, l(mm)	---	23	23
Defect Width, w(mm)	---	7.5	7.5

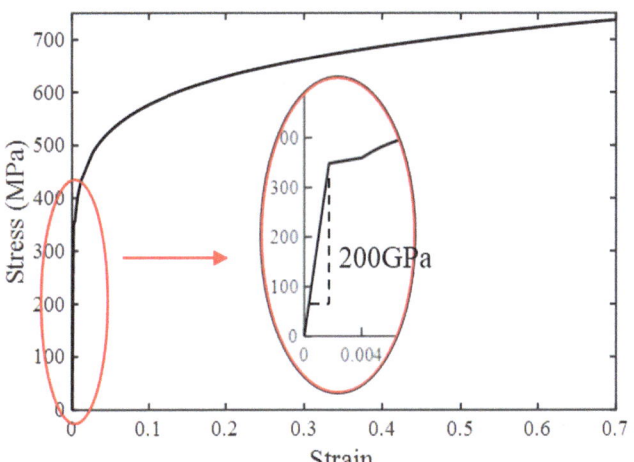

Fig. 3. The stress-strain curve [10]

As shown in Fig. 4, the axial force is positive for the axial compressive force, the bending moment is positive for the defect on the compression side. The pressure, axial force and bending moment ultimate bearing capacity of corroded pipeline are obtained by simulation.

As shown in Fig. 4 (a), the curved surface shape is like an ellipse. With the increase of pressure, the axial force and bending moment values first decrease and then increase, and finally decrease. This is because the low pressure increases the bearing capacity

Table 4. The parameter of ductile damage [10]

Damage initiation equivalent plastic strain	Stress triaxiality	Failure displacement	Index
0.708102	0.075363	0.39	0
0.627071	0.257535	0.006	0
0.272439	0.446455	0.1	−8
0.181211	0.554757	0.18	−8
0.180553	0.60314	0.17	−8
0.177627	0.647885	0.14	−8
0.221759	0.705193	0.13	−8
0.253337	0.740392	0.15	−8

of the pipeline. However, with the increase of pressure, the pipeline is easier to enter the yield stage., damage occurs in advance, and the ultimate bearing capacity of the pipeline gradually decreases. As shown in Fig. 4 (b), the axial force and bending moment ultimate bearing capacity curves of the intact pipeline are symmetrical along the M = 0 MPa line. With the increase of pressure, the ultimate bearing capacity of axial force decreases gradually, while the ultimate bearing capacity of bending moment increases first and then decreases. As shown in Fig. 4 (c), the existence of defects weakens the enhancement effect of pressure on the ultimate bearing capacity of pipeline, and also reduces the ultimate bearing capacity of pressure, axial force and bending moment of pipeline, and increases the decrease rate of the ultimate bearing capacity of axial force and bending moment of corroded pipeline under higher pressure, making pipeline more easier to damage.

As shown in Fig. 4 (e), with the increase of pressure, the ultimate bearing capacity of corroded pipeline gradually decreases, and the existence of deeply defects greatly affects the ultimate bearing capacity of corroded pipeline. As shown in Fig. 4 (d) and Fig. 4 (f), the N-M interaction curve of the corroded pipeline is not symmetrical along the M = 0 MPa line. Due to the interaction between axial force and bending moment, the ultimate bearing capacity of the pipeline subjected to axial force and bending moment will be greater than the ultimate bearing capacity of the pipeline under a single axial force or bending moment. The existence of defects reduces the ultimate bearing capacity of the pipeline under a single load. With the increase of pressure, the interaction curve of axial force and bending moment of corroded pipeline gradually flattened. As the defect depth increases, the ellipticity of the interaction curve increases, and the second-order effect of the pipeline gradually decreases.

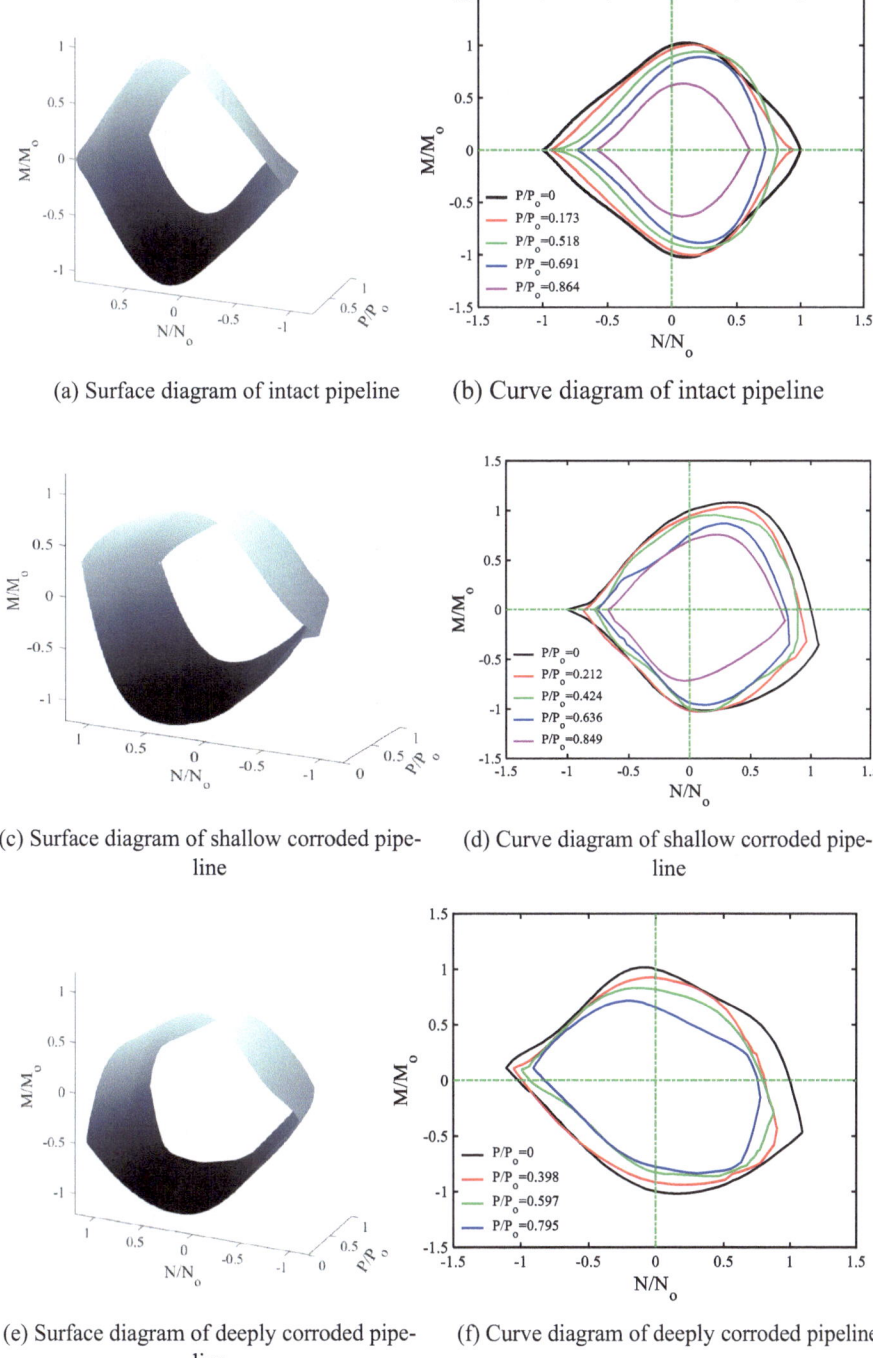

(a) Surface diagram of intact pipeline

(b) Curve diagram of intact pipeline

(c) Surface diagram of shallow corroded pipeline

(d) Curve diagram of shallow corroded pipeline

(e) Surface diagram of deeply corroded pipeline

(f) Curve diagram of deeply corroded pipeline

Fig. 4. Interaction of pressure, axial force and bending moment for corroded pipeline

4 The Failure Mode of Corroded Pipeline

Based on the ductile damage parameters of metals, the failure modes of corroded pipeline subjected to axial compressive force and bending moment are studied in this paper. Fig. 5 is the deformation diagram of the corroded pipeline subjected to axial compressive force and bending moment. The corroded pipeline is first destroyed at the defect, and the crack develops circumferentially along the pipeline.

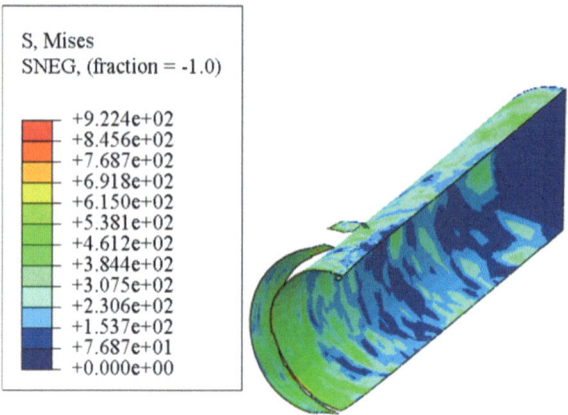

(a) The axial compressive force and closing bending moment

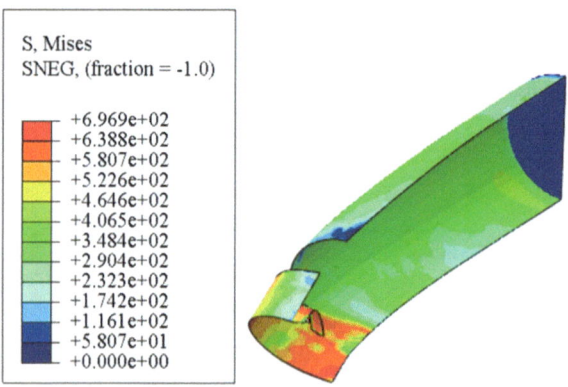

(b) The axial compressive force and closing bending moment

Fig. 5. The deformation of corroded pipeline subjected to axial force and bending moment predicted by numerical simulation

5 Conclusion

Based on the ductile damage parameters of the material, the ultimate bearing capacity of the corroded pipeline was simulated by the fluid cavity simulation technology. Finally, the interaction surface and curve of the ultimate bearing capacity of the corroded pipeline with different defect depths were obtained.

The results show that low axial force can improve the bending moment bearing capacity of shallow corroded pipeline, while for deeply corroded pipeline, it only accelerates the damage to pipeline. The ultimate bearing capacity of corroded pipeline with different defect depths is affected by pressure. Under low pressure, the axial force and bending moment bearing capacity of the intact or shallow corroded pipeline will increase, and the area contained in the corresponding N-M interaction curve will increase. The pipeline is more likely to be damaged at the location of corrosion defect, and the cracks develop along the circumferential direction.

References

1. Oh, C.K., Kim, Y.J., Park, C.Y.: Effects of local wall thinning on net-section limit loads for pipes under combined pressure and bending. Nucl. Eng. Des. **239**(2), 261–273 (2009)
2. Mondal, B.C., Dhar, A.S.: Burst pressure of corroded pipelines considering combined axial forces and bending moments. Eng. Struct. **186**, 43–51 (2019)
3. Roy, S., Grigory, S., Smith, M., Kanninen, M.F., Anderson, M.: Numerical simulations of full-scale corroded pipe tests with combined loading. J. Press. Vessel Technol. **4**, 119 (1997)
4. Taylor, N., Clubb, G., Matheson, I.: The effect of bending and axial compression on pipeline burst capacity. In: SPE Offshore Europe Conference and Exhibition (2015)
5. Mondal, B.C.: Remaining strength assessment of deteriorating energy pipelines. Memorial University of Newfoundland (2018)
6. Lu, Y.-J., Wang, C.-H.: A finite element-based analysis approach for computing the remaining strength of the pressure equipment with a local thin area defect. Eng. Fail. Anal. **131**, 105883 (2022)
7. Bai, Y., Igland, R.T., Moan, T.: Tube collapse under combined external pressure, tension and bending. Marine Struct. **10**(5), 389–410 (1997)
8. Oh, C.K., Kim, Y.J., Baek, J.H., Kim, W.S., Kim, Y.P.: Ductile failure analysis of API X65 pipes with notch-type defects using a local fracture criterion. Int. J. Press. Vessels Pip. **84**(8), 512–525 (2007)
9. American Society of Mechanical Engineers, B31G, Manual for Determining the Remaining Strength of Corroded Pipelines (2012)
10. Xu, T.: Study on the Ultimate Bearing Capacity of Defected Pressurized Pipeline under Coupling Load. Shandong University of Science and Technology (2023). unpublished.

Investigation and Analysis of Service Performance of Cable Arch Bridge Structure Under Accidental Lateral Load

Shilong Gao[✉], Zhuang Kai, Tianchong Li, and Wei Sun

Shandong Key Laboratory of Disaster Prevention and Mitigation in Civil Engineering, Shandong University of Science and Technology, Qingdao, China
gs1208023@163.com

Abstract. This article investigates the response behavior of such structures under unexpected lateral impact loads, and quantitatively studies the impact of vibration caused by impact on driving performance, pedestrian comfort, and other aspects. Based on parameter analysis methods, explore the impact of impactor quality and speed on the above behavior. The simulation results indicate that as the mass and speed of the impact body increase, the discomfort of driving increases, which is unacceptable to pedestrians.

Keywords: Unmanned aerial vehicle · Impact · Driving comfort · Pedestrian comfort

1 Introduction

In recent years, numerous scholars have conducted extensive research on driving comfort and pedestrian comfort. + quantified the concept of comfort through a calculation formula, proposed a comfort evaluation index J, and determined the standard value of comfort limits. At the same time, he also conducted research on the frequency of vibration and found that different frequencies of vibration have different factors affecting comfort. In terms of railways, German scholar Sperling first began research on train comfort, while British scholar Loach developed and improved it, proposing the Sperling comfort index. This indicator takes into account the varying sensitivity of the human body to vibrations of different frequencies. Currently, many European countries adopt this standard [2]. This article uses the allowable limit value of vertical acceleration in Yu Zhisheng [3]'s book "Automobile Theory" to determine driving comfort.

Huang Xin [4] proposed suggestions for optimizing pedestrian comfort of pedestrian bridges based on parameters that affect pedestrian comfort and proposed four measures to improve pedestrian comfort based on engineering examples of pedestrian cable-stayed bridges. The advantages and disadvantages of these four measures were comprehensively analyzed from three aspects: economy, science, and efficiency. Domestic and foreign scholars have proposed many evaluation indicators for pedestrian comfort, but there is also no unified standard. Zou R et al. [5] summarized and compared pedestrian comfort

A. Bieliatynskyi et al. (Eds.): CSTTE 2023, LNCE 603, pp. 225–234, 2024.
https://doi.org/10.1007/978-981-97-5814-2_20

standards in various countries, and the results showed that the International Organization for Standardization ISO10137 standard is the strictest; Liu Cong et al. [6] analyzed the vertical and horizontal comfort of the North Canal Bridge in Beijing based on the design specifications for pedestrian bridges in various countries; This article uses the vertical acceleration limit value proposed by the German EN03 specification [7] to determine pedestrian comfort.

2 Establishment and Verification of the Finite Element Model

2.1 Modeling

The cable arch bridge model is based on the Nannan'ao Sea Crossing Bridge, which is located in Yilan County, Taiwan, China Province, China, and was completed in 1999. As one of the iconic landmarks in the local area, it has a span length of 140 m, a bridge width of 15 m, two lanes, and a total of 13 steel cables.

The finite element model of a cable arch bridge is mainly divided into three parts: the bridge deck and arch as a whole, and the bridge piers and steel cables. The bridge deck, arch, and pier use C3D4 solid units. A total of 131851 units. The bridge deck and piers are connected as a whole using ABAQUS's built-in Tie technology. The steel cable adopts B31 beam elements, with a total of 156 elements and a circular cross-section. The impact body is defined as a rigid element attribute using the * Rigid Body command. In the performance analysis of the cable arch bridge structure under accidental lateral load, the impact body has an infinite mass and does not deform or break. The contact between the steel cable and the impact body of the cable arch bridge is analyzed using the * General contact technology provided by ABAQUS. The finite element model is shown in Fig. 1 [8]

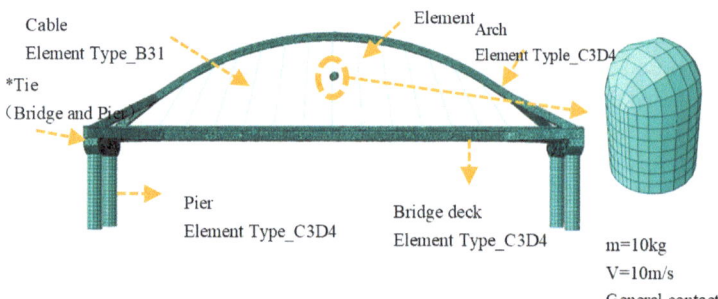

Fig. 1. Finite element model.

2.2 Material Model and Boundary Conditions

The bridge deck, arch, and pier in this article are all made of C40 concrete. According to the standard GB/T50081–2019 "Test Methods for Physical and Mechanical Properties of Concrete" [9], their mechanical properties are shown in Table 1.

Table 1. Mechanical Properties of Concrete

Title	Value
Density, ρ (kN/m^3)	2480
Modulus of Elasticity, E (GPa)	33.15
Poisson's Ratio, μ	0.19
Tensile strength Standard value (MPa)	2.39
Compressive strength Standard value (MPa)	26.8

Table 2. Concrete damage parameters [9]

Parameter	Value
expansion angle(°)	42.65
eccentricity	0.1
f_{b0}/f_{c0}	1.07
viscosity parameter	0.0005
K_{c}	0.64

The data on concrete damage parameters are shown in Table 2 [10]. The material parameters of the steel cable are shown in Table 3 [8].

Table 3. Material parameters of steel cables

Parameter	Value
Density, ρ (kN/m^3)	7850
Modulus of Elasticity, E (GPa)	201
Poisson's Ratio, μ	0.3
$\bar{\varepsilon}_0^{\mathrm{pl}}$	0.0486
$\bar{\varepsilon}_{\mathrm{f}}^{\mathrm{pl}}$	0.0493
$\bar{\mu}_{\mathrm{f}}^{\mathrm{pl}}$	0.0005
η	0.333

The impactor is a drone, with a mass generally greater than 7 kg and less than 116 kg. Its speed is less than 100 km/h, or 27.8 m/s, in full horsepower level flight [11]. In finite element simulation, the mass of the impact body is 10 kg, and the impact lifting speed is 10 m/s.

Apply a fully fixed constraint on the four bases of the bridge pier to fix the entire cable arch bridge model on the ground. The impactor adopts sliding constraints and only moves along the Z-axis direction.

3 Specification for Driving and Pedestrian Comfort

In the book "Automobile Theory", Yu Zhisheng [3] specified the allowable limit values of vertical acceleration based on the degree of human perception, as shown in Table 4.

Table 4. Relationship between vertical acceleration and driving comfort (based on automobile theory) [3]

vertical acceleration a_s (m/s^2)	Driving comfort
$a_s < 0.315$	keep comfortable
$a_s = 0.315 \sim 0.63$	Slight discomfort
$a_s = 0.63 \sim 1.25$	Quite uncomfortable
$a_s > 1.25$	Very uncomfortable

The German EN 03 specification [7] proposes a four-level comfort level based on acceleration as a pedestrian comfort index, as detailed in Table 5.

Table 5. Comfort level defined in German EN 03 [6]

Comfort level	comfort	vertical acceleration
CL1	Best	< 0.5
CL2	Medium	$0.5 \sim 1$
CL3	Minimum	$1 \sim 2.5$
CL4	Unacceptable	> 2.5

The schematic of the output acceleration position is shown in Fig. 2, and the acceleration history curve at position 5 is shown in Fig. 3.

According to Yu Zhisheng's [3] limit on vertical acceleration, the driving comfort here is slightly uncomfortable. According to German EN 03 standards, pedestrian comfort is the best.

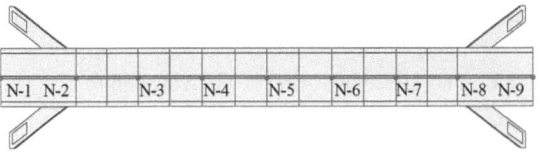

Fig. 2. Illustration of measuring the location of vertical acceleration in the bridge deck.

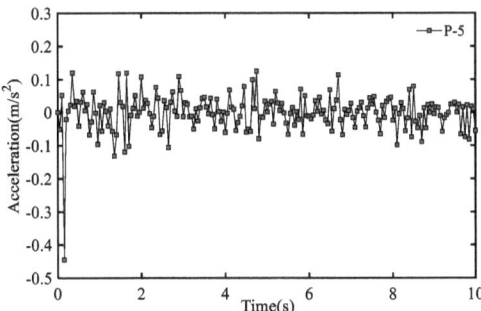

Fig. 3. Time history of vertical acceleration at No.5 positions of the bridge deck.

4   Analysis of Influencing Factors

4.1   Quality

The light unmanned aerial vehicle is used as the impactor, with a speed of 10 m/s and a mass of 10 kg, 20 kg, 30 kg, and 40 kg, as shown in Table 6.

Table 6. Impact Mass Conditions

Operating mode	Impact velocity (m/s^2)	Impact mass (kg)
Operating mode1	10	10
Operating mode2	10	20
Operating mode3	10	30
Operating mode4	10	40

The vertical acceleration history curves of different masses are shown in Fig. 4.

Table 7. Evaluation of driving and pedestrian comfort (Impact with different mass)

Operating mode	vertical acceleration a_s/(m/s^2)	Driving comfort	pedestrian comfort
Operating mode1	0.44507	Slight discomfort	Best
Operating mode2	0.51805	Slight discomfort	Medium
Operating mode3	0.69843	Quite uncomfortable	Medium
Operating mode4	0.85193	Quite uncomfortable	Medium

Table 8. Impact Velocity Conditions

Operating mode	Impact velocity (m/s^2)	Impact mass (kg)
Operating mode5	4	10
Operating mode6	10	10
Operating mode7	16	10
Operating mode8	20	10

Table 9. Evaluation of driving and pedestrian comfort (Impact with different velocities)

Operating mode	vertical acceleration a_s/(m/s^2)	Driving comfort	pedestrian comfort
Operating mode5	0.23518	keep comfortable	Best
Operating mode6	0.44507	Slight discomfort	Best
Operating mode7	0.92750	Quite uncomfortable	Medium
Operating mode8	1.27855	Very uncomfortable	Minimum

The evaluation of driving comfort and pedestrian comfort at different masses is shown in Table 7.

As the mass increases, the vertical acceleration value increases, making driving and pedestrians feel increasingly uncomfortable. This is because the vertical acceleration is related to the initial kinetic energy input, which is directly proportional to the mass. As the mass increases, the initial kinetic energy increases and the vertical acceleration value increases, making driving and pedestrians increasingly uncomfortable.

4.2 Velocity

The mass of the impactor is 10 kg, and the velocity of the impactor is taken as 4 m/s, 10 m/s, 16 m/s, and 20 m/s, as shown in Table 8.

The vertical acceleration history curves of different Velocities are shown in Fig. 5.

The evaluation of driving comfort and pedestrian comfort at different Velocities is shown in Table 9.

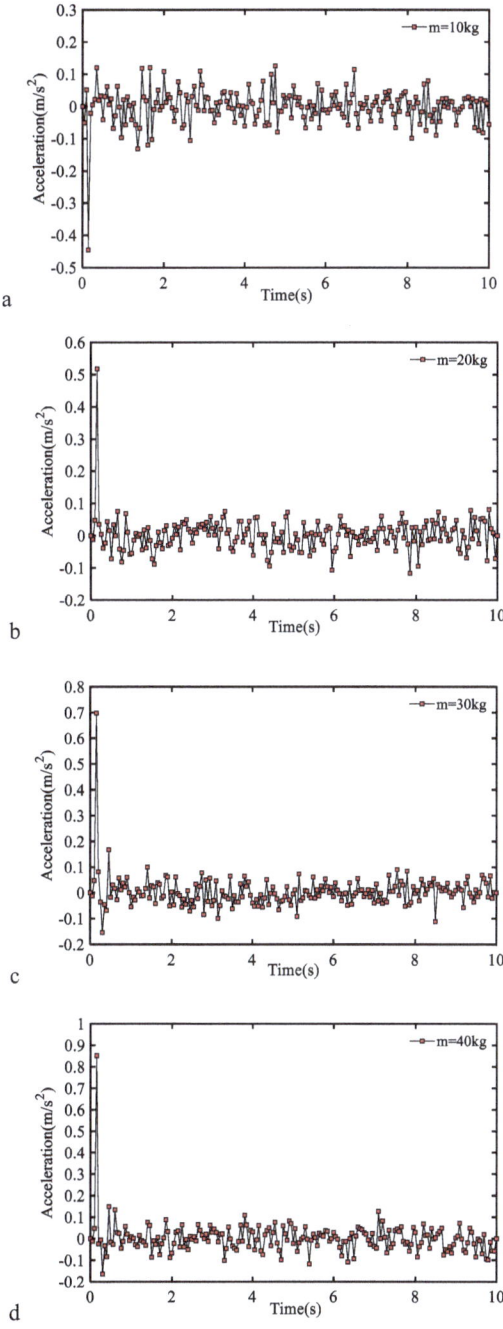

Fig. 4. Time history of vertical acceleration at No.5 positions of the bridge deck.

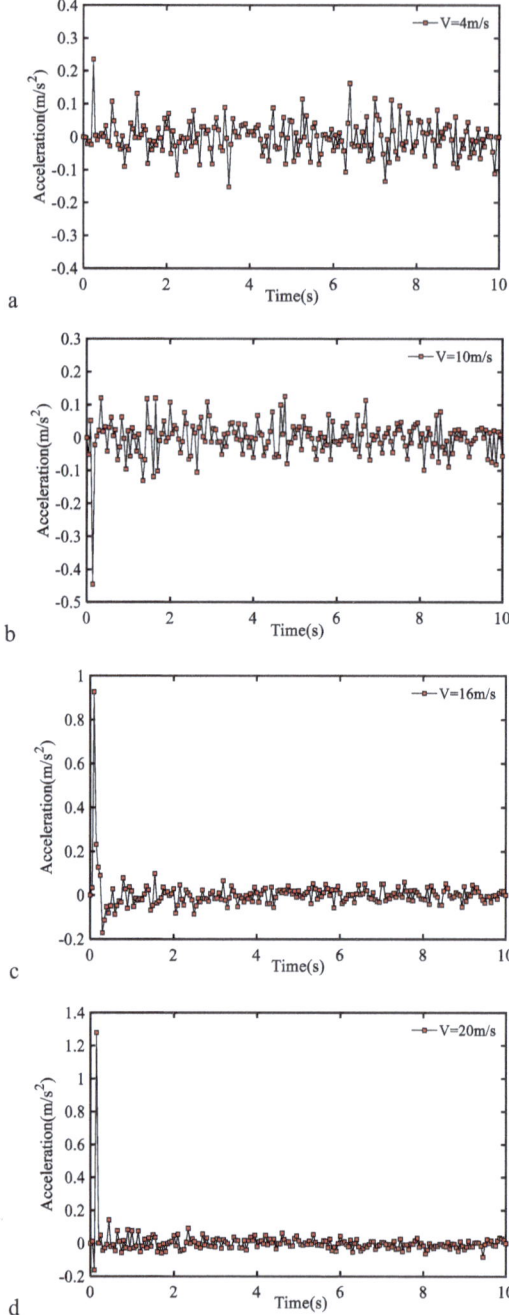

Fig. 5. Time history of vertical acceleration when a bridge is under lateral impact with different Velocity (a: Operating mode 5; b: Operating mode 6; c: Operating mode 7; d: Operating mode 8)

As the Velocity increases, the vertical acceleration value increases, making driving and pedestrians more uncomfortable. This is because the vertical acceleration is related to the initial kinetic energy input, which is proportional to the square of the velocity. As the speed increases, the initial kinetic energy increases and the vertical acceleration value increases, making driving and pedestrians increasingly uncomfortable.

5 Conclusion

After accidental load impact, the driving comfort at position 5 on the cable arch bridge showed slight discomfort, with pedestrian comfort being the best.

Quantitative study on the impact of impactor mass and speed on driving and pedestrian comfort. When the mass is less than 20kg, driving comfort is slightly uncomfortable, while pedestrian comfort is moderate. When the mass exceeds 20 kg, the driving comfort is quite uncomfortable, and the pedestrian comfort is moderate. This is because the mass is proportional to the initial kinetic energy. As the quality increases, the discomfort of driving and pedestrians increases; When the speed is less than 10m/s, pedestrian comfort is the best. As the speed increases, pedestrians become more uncomfortable, while driving comfort becomes more uncomfortable as the speed increases. This is because the initial kinetic energy is proportional to the square of the speed, and as the speed increases, driving and pedestrians become increasingly uncomfortable.

References

1. Deng, B.B.: Research on the Evaluation Method of Vehicle Vibration Ride Comfort. Hefei Polytechnic University (2005)
2. Liu, Y.: Analysis of Vehicle Bridge Coupling Vibration and Driving Comfort of Highway Cable-stayed Bridges. Central South University, Chang sha (2012)
3. Yu Z S. Automobile Theory (2000)
4. Huang, X.: Research on Comfort of Pedestrian Cable-stayed Bridges. Chongqing Jiaotong University (2020)
5. Zou, R., Zeng, D., Shen, W.A.: Comparison of National Norms for Human Induced Load Models and Comfort Evaluation J. Civil Eng. Manag (2020)
6. Liu, C., Sun, H., Ouyang, S., et al.: Research on Pedestrian Comfort Design of Long Span Pedestrian Arch Bridges Proceedings of the 2021 Industrial Architecture Academic Exchange Conference, vol. 2 (2021)
7. Design of Footbridges Guideline EN03 (2007), Germany (September 2008)
8. Gao, S.L.: Research on Collapse Behavior and Service Performance of Typical Cable Arch Bridge Structures under Dynamic Load . Shandong University of Science and Technology (2023)
9. GB/T50081–2019.Standard for Test Methods for Physical and Mechanical Properties of Concrete (2019)
10. Li, X.X.L.: Parametric study on numerical simulation of missile punching test using concrete damaged plasticity (CDP) model. Int. J. Impact Eng **144**, 103652 (2020)
11. AC-91-FS-2019–31R1-Operating Regulations for Light and Small Unmanned Aerial Vehicles. Beijing: Flight Standards Department of the Civil Aviation Administration of China (2019)

Study on the Force Transmission Mechanism of Steel-Concrete Composite Segments in Irregular Single-Tower Cable-Stayed Bridges

Kai Zhang[1], Xinqiang Wu[1], Haihui Xie[2(✉)], Musheng Ye[3], Haiyang Zhang[2], Genfa Chu[4], and Fei Xuan[4]

[1] Hefei Municipal Design Research Institute Co., Ltd., Hefei 230009, Anhui, China
[2] College of Civil Engineering, Hefei University of Technology, Hefei 230009, Anhui, China
2022170709@mail.hfut.edu.cn
[3] Anhui Feixiang Engineering Management Co., Ltd., Hefei 230009, Anhui, China
[4] Anhui Gourgen Traffic Construction Co., Ltd., Hefei 230009, Anhui, China

Abstract. In response to the complex stress distribution challenges in the steel-concrete composite segments of hybrid beam cable-stayed bridges, this study focuses on the steel-concrete composite segments of the Dianbu River Bridge in Anhui Province. A comprehensive spatial model of the entire bridge was established using ABAQUS for an overall static analysis. The analysis aimed to elucidate the stress distribution within the composite segments, including the concrete, steel plates, and rear compression plates, and to identify the specific load-bearing pathways within the structure. The results of the analysis indicate that the primary load-bearing pathways traverse the steel transition segments, the steel plates of the composite segments, the concrete of the composite segments, and the concrete transition segments, by bidirectional load transfer principles. The construction method of the steel-concrete composite segments of the bridge tower is deemed to be rational, ensuring structural safety and reliability.

Keywords: cable-stayed bridge · steel-concrete composite segment · force transmission mechanism · finite element model

1 Introduction

The main girder of a hybrid girder cable-stayed bridge consists of a steel girder and a concrete girder, which exist independently in the structural hierarchy and are combined into a single unit using connectors to jointly bear and transmit loads. The main girder consists of both steel and concrete structures along the longitudinal direction. Hybrid girders are often engineered in the form of main-span steel girders and side-span concrete girders [1]. In this system, the use of concrete forms for the side spans not only reduces the internal forces and deformations of the main span steel girders but also enhances the main span spanning and reduces the negative reaction forces at the end supports of the side spans [2]. It also allows for relatively shorter side span lengths, making the choice of bridge site more flexible.

© The Author(s) 2024
A. Bieliatynskyi et al. (Eds.): CSTTE 2023, LNCE 603, pp. 235–243, 2024.
https://doi.org/10.1007/978-981-97-5814-2_21

To ensure structural safety and smooth load transmission, a large number of load-bearing components have been incorporated in the steel-concrete composite segments of the hybrid beam, resulting in a complex structural configuration, significant cross-sectional variations, and pronounced deviation of shear flow. It is difficult to carry out the theoretical study of the structural force transfer mechanism, so the domestic and international research on the force transfer mechanism of the steel-concrete composite segments is mainly carried out through finite element simulation [3] and model test [4 ~ 5].

Zou et al. [6] proposed a simplified mechanical model for the steel-concrete composite segments of hybrid beams based on the partial combination theory. They obtained the simplified calculation methods for the interface slip distribution, maximum slip, effective utilization rate of the connector group, and the proportion of load borne by the connectors.

Zhou et al. [7] investigated the force transfer performance of the steel-concrete composite segments through model tests and finite element simulations, and their test and finite element simulation results showed that the shear nails in steel-hybrid sections had an obvious group nailing effect.

This paper carries out a study on the force transmission path of the structural steel-mixed section of the Dianbu River Bridge in Anhui Province. Duanbu River Bridge is a (84 + 152)m single tower single cable-stayed steel-mixed girder cable-stayed bridge, which is characterized by the following features compared with conventional cable-stayed bridges: The main tower has a special shape, a shaped one-tower cable-stayed bridge; The bridge is a tower-girder-pier consolidation system; it is a steel-mixed girder cable-stayed bridge. Due to the special structure, the force transmission mechanism of the steel-hybrid section of this type of bridge needs to be studied independently.

2 Project Overview

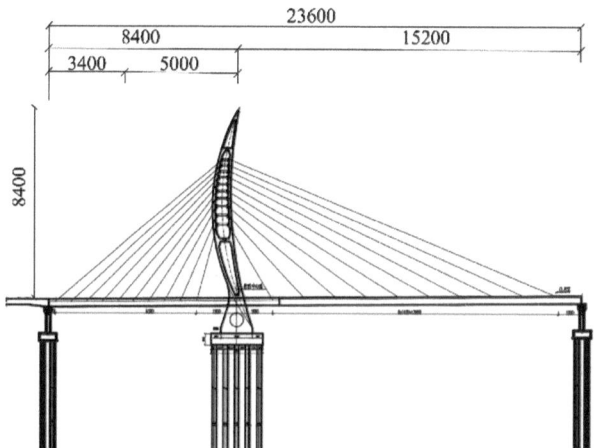

Fig. 1. Arrangement of the main bridge of the Dianbu River Special Bridge

Dianbu River Bridge adopts (84 + 152)m single tower single cable-stayed steel-mixed girder cable-stayed bridge. The main girder of side span adopts prestressed concrete box girder, the main girder of middle span adopts steel box girder, the bridge tower adopts reinforced concrete crescent tower, and the tower-pier-beam cementation system. The length of steel-mixed girder section is 6m, including 2m of steel-mixed section and 3.5m of stiffness change section, and the main part of the tower is in the shape of a moon, and the lower tower columns are set up with circular hollowing. The bridge layout is shown in Fig. 1. The units in the figures are centimeters.

3 Modelling

ABAQUS 2020 is used to establish the finite element model of the whole bridge, and the base model will be the concrete girder end and main tower section with a solid finite element model, and the steel box girder section with a shell unit model. Among them, the concrete solid finite element model, the tower-beam bonding section with 16mm thickness of covered steel plate was simulated by the C3D8R unit. Steel girders, stiffening ribs, and diaphragms are simulated using S4R units, while steel bars and diagonal cables are simulated using T3D2 units. Reinforcing steel units are built into the concrete units, and four types of steel box girders exist in the baseline model, all of which are bound and combined; Transverse bulkheads, stiffening ribs, and steel box girders are connected by binding bond; steel-hybrid bonding section will be steel plate and concrete block binding link, steel plate and concrete connecting plate is built into the concrete and bound to the steel plate.

The ABAQUS full bridge finite element model is shown in Fig. 2.

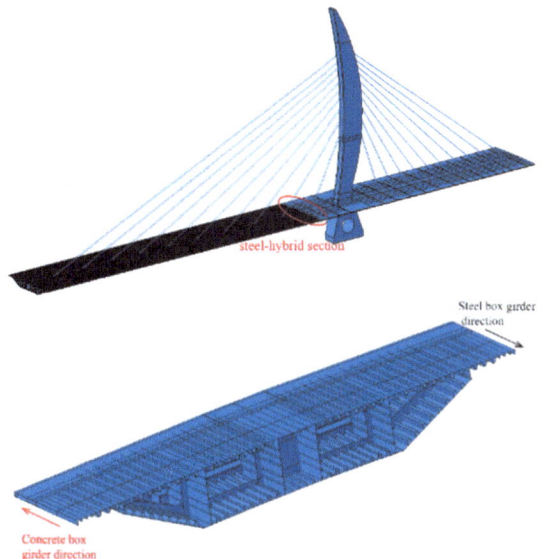

Fig. 2. Finite element model of Dianbu River Bridge and steel-hybrid section

ABAQUS was used to create a refined model of the whole bridge. For the steel-concrete composite segments, a combination of blueprints and on-site photographs were employed to ensure model fidelity with the actual construction. The design of the steel-concrete composite segments of the Dianbu River Bridge is illustrated in Fig. 3, while the detailed finite element modelling is shown in Fig. 2. To enhance the visual representation, a perspective view is presented in Fig. 4.

It is observed from Fig. 2 that both the head plate, stiffening ribs, diaphragm plate, rear pressure plate, and the steel plate of the combined section. All are represented in the model and the dimensions of each structure are consistent with the actual project.

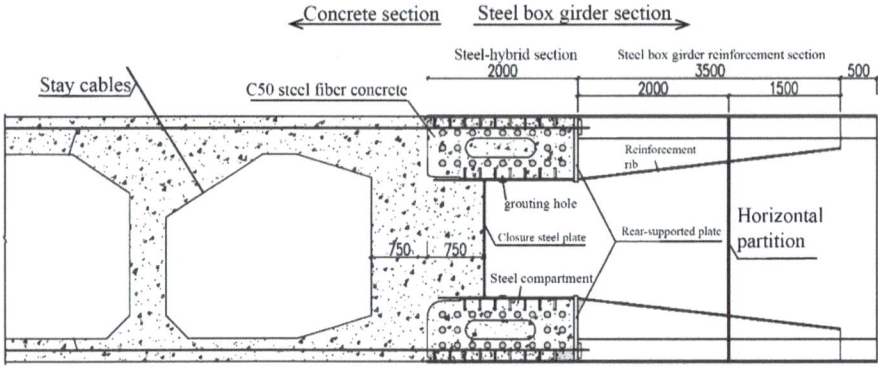

Fig. 3. Combined steel and concrete section of the Dianbu River Bridge

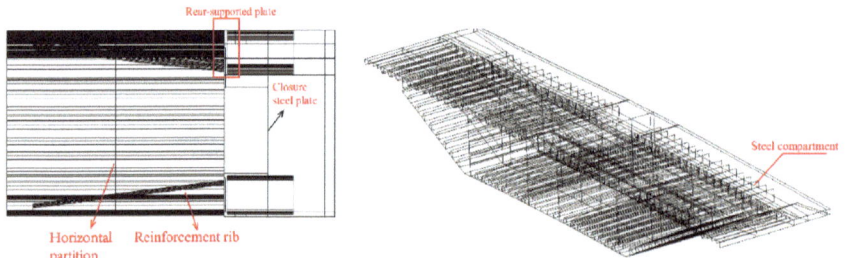

Fig. 4. Perspective view of finite element model of steel-hybrid combined section

4   Assumption of Force Transmission Path

The steel-concrete composite segments of the hybrid beam consist of the steel transition segment, the composite segment, and the concrete transition segment. Among these, the composite segment serves as the primary load-bearing component. In a macroscopic sense, the load transfer pathway within the steel-concrete composite segments of the hybrid beam can be generalized as follows: internal forces within the steel structural segment are transmitted through the steel transition segment to the composite segment and further transferred to the concrete transition segment.

The structural form classification of the steel-hybrid combined section can be divided into two categories according to the presence or absence of the lattice chamber, on this basis and then based on the form of the pressure plate and then divided in detail [8–10], its simple classification is shown in Fig. 5.

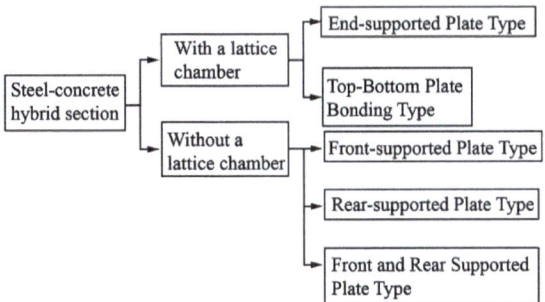

Fig. 5. Classification of Steel-Concrete Composite Segments

It can be observed from Fig. 3 that the structure has a lattice chamber and no front bearing plate, so it is a steel-hybrid section with a lattice chamber and rear bearing plate type.

It is considered that the specific force transfer paths of the steel transition section in the steel-hybrid bonded section of the hybrid beam that transfers the internal forces to the concrete transition section through the bonded section are the five shown in Fig. 6.

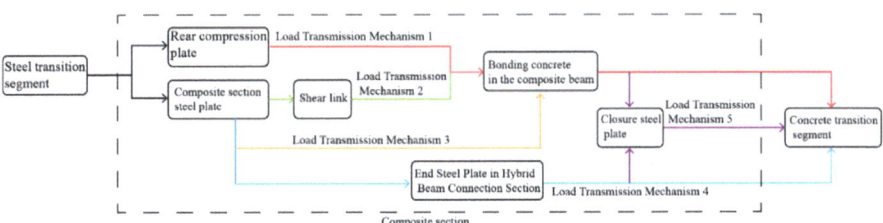

Fig. 6. Common force transfer paths in steel-cement sections with compartmentalized rear bearing plate type

5   Finite Element Result Analysis

The red and orange areas in Fig. 7 are subjected to tensile stresses, and tensile phenomena are observed in the middle of the transverse bridge direction as well as on both sides, while the rest of the area is subjected to compressive stresses. Meanwhile, the region of maximum compressive stress is located in the region of the bond section concrete near the concrete girder section, opposite to the tensile zone, indicating that the structure is stressed in both directions in the direction of the bridge. At the same time, the structure has a large volume share and is connected to a concrete box girder at one end. One end is connected to a steel-concrete section, which is a mandatory path for the longitudinal force flow.

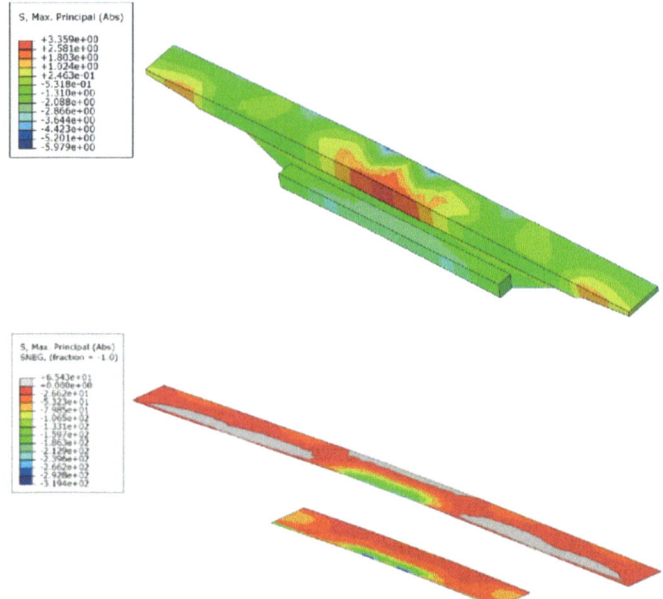

Fig. 7. Stress Contour Plot of Concrete and Steel Plate in Composite Steel-Concrete Section

The steel plate of the bond section is divided into two pieces, one covering the uppermost part of the concrete of the bond section and the other covering the lowermost part of the concrete of the bond section. In Fig. 7, the pressurized portion of the structure is marked in colour. The whole steel plate is subjected to compressive stress in the majority of the area, and the middle part of both plates is subjected to large compressive stress. It can reach 150 ~ 300 Mpa, of which the peak compressive stress is distributed in the middle of the steel plate in the lower combined section, reaching 319.4 Mpa. Slightly lower than the yield strength of the material. The structure as a whole is under high stress and is therefore considered to be a more important part of the force transfer path during the service phase.

It is observed that the full interface of the stiffening rib is under compression, for the upper stiffening rib, the middle part is subjected to higher compressive stresses, while for the lower stiffening rib, it is subjected to higher compressive stresses on both sides.

Figure 8 illustrates the stress cloud of reinforcement rib and steel lattice chamber plate. The steel latticework panels are interlocked with concrete to form several steel lattice chambers. Among them, the stress of the steel grating plate is less than 50 Mpa,

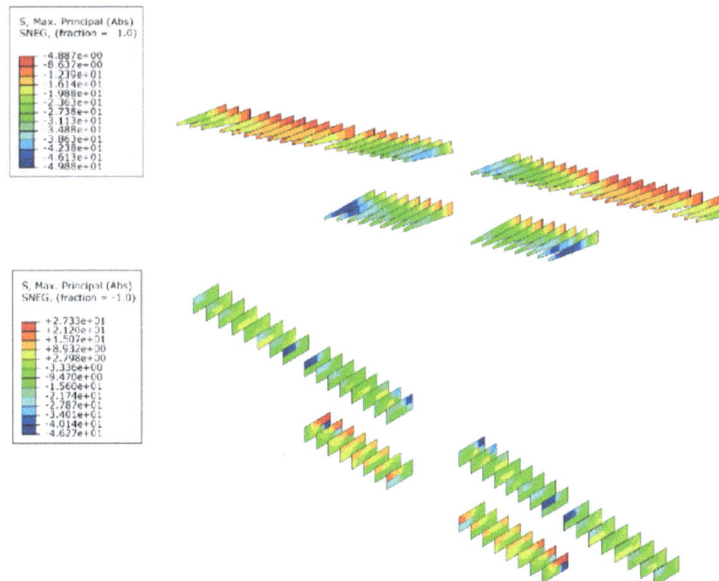

Fig. 8. Stress cloud of reinforcement rib and steel lattice chamber plate

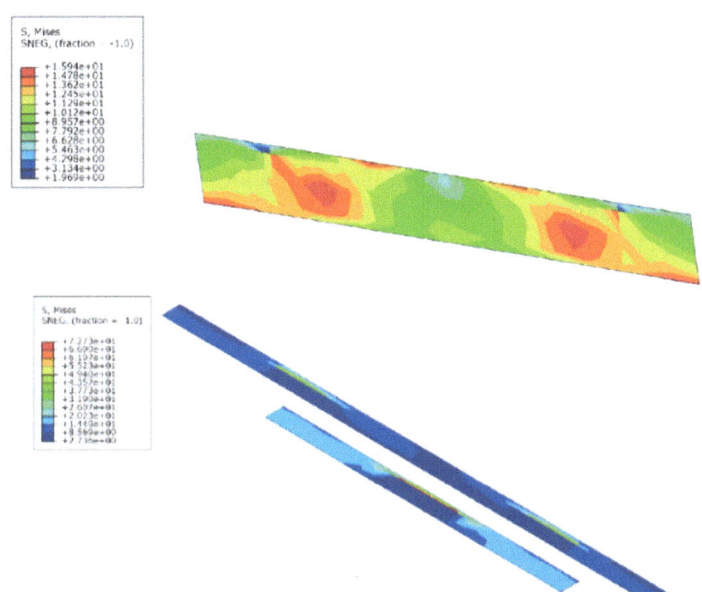

Fig. 9. Stress cloud of steel plate of end socket and steel plate at the end of the bonding section

which is much lower than the material strength, and it is considered that the structure has a role in the force transfer process, but it is not the most dominant force transfer path. At the same time, the peak stress of the steel plate of the head only reaches 16 Mpa, and it is considered that the structure does not play the role of force transmission, but the construction requirements, and plays the role of blocking. There are two steel plates at the end of the combined section, which are close to the head steel plate and are located at the upper and lower part of the head steel plate. Figure 9 illustrates the stress distribution in the steel plate of the end socket and the steel plate at the end of the bonding section. The structure has a peak stress of 73 MPa, but it is limited to a smaller area which is less than 5% for the structure, and isolating this area, the majority of the area has a stress of less than 30 MPa. It is considered that the structure, like the head plate, does not act as a force transfer but is more of a construction requirement.

Fig. 10. Stress cloud of the rear pressure plate

Figure 10 illustrates the stress cloud of the rear bearing plate. The stress distribution of the rear bearing plate is complex, but the value is low, and the whole is kept in the role of 20 Mpa, and the peak stress is 30 Mpa, which is located in the lower side rear bearing plate. Considering the low stresses, it is not considered to act as a major force transfer path. Considering that its relative position to the end concrete is similar to that of the head plate, it is considered to act more as a blocking similar to that of the head plate.

6 Conclusion

This paper utilizes a single-tower, solid-pier, single-cable-plane, irregular single-tower cable-stayed bridge as the subject. A comprehensive finite element model of the entire bridge is established for numerical simulation to analyze the load transmission pathways in the steel-concrete composite segments of such bridges. The conclusions are as follows:

1. The steel plates in the combined section are subjected to high compressive stresses and play an important role in the force transfer process. Stiffening ribs and steel lattice plates are also involved in force transfer, but not the most important path; the head steel plate, combined section end steel plate, and back pressure plate have low stress, mainly play the role of blocking and structure, not the main force transmitting parts.

2. Through finite element analysis, the primary load transmission pathways in the steel-concrete composite segments of the Dianbu River Bridge have been identified as follows: steel transition segments, composite segment steel plates, composite segment concrete, and concrete transition segments, conforming to bidirectional load transfer principles.
3. The construction method of the steel-concrete composite segments of the Dianbu River Bridge is reasonable and can safely and reliably transmit internal structural forces, meeting the design requirements.

In this paper, only the force transfer mechanism of the steel-hybrid section under static force is considered, and its effect caused by seismic loading is not taken into account, which will be further explored in the subsequent research.

References

1. Xun, G.P., Zhang, X.G., Liu, Y.Q.: Hybrid girder cable-stayed bridge. China Communications Press, Beijing (2013)
2. Kaili, C.H.E.N., Tianqing, Y.U., Gang, X.I.: Development and prospective of hybrid girder cable-stayed bridge. Bridge construction **2**, 1–4 (2005)
3. Bock, M., Gkantou, M., Theofanous, M., Afshan, S., Yuan, H.: Ultimate behaviour of hybrid stainless steel cross-sections. J. Constr. Steel Res. **210**, 108081 (2023)
4. Martins, D., Proença, M., Correia, J.R., Gonilha, J., Arruda, M., Silvestre, N.: Development of a novel beam-to-column connection system for pultruded GFRP tubular profiles. Compos. Struct. **171**, 263–276 (2017)
5. Jun, S.C., Lee, C.H., Han, K.H., Kim, J.W.: Flexural behavior of high-strength steel hybrid composite beams. J. Constr. Steel Res. **149**, 269–281 (2018)
6. Zou, Y., Zheng, K., Zhou, J., Zhang, Z., Li, X.: Mechanical behavior of perfobond connector group in steel–concrete joint of hybrid bridge. Structures **30**, 925–936 (2021)
7. Zhou, Y., Pu, Q., Shi, Z., Gou, H., Chen, X.: Experimental and numerical parametric study of the mechanical properties in a steel-concrete joint section. Inter. J. Civil Eng. **20**(12), 1431–1446 (2022)
8. Liu, Y.Q.: Combined Structures Bridges. China Communications Press, Beijing (2005)
9. Rovnak, M., Duricova, A.: Behaviour evaluation of shear connection by means of shear-connection strips. Steel Compos. Struct. Inter. J. **4**(3), 247–263 (2004)
10. Rovnak, M., Duricova, A., Ivanco, V.: Non-traditional shear connections in steel-concrete composite structures. Compos. Hybrid Struct. **1**(2), 305–312 (2000)

A Comprehensive Study on Cracks in Multi-span Simply Supported Beam Bridges Through SolidWorks Analysis

Syed Musarat Hussain[1]([✉]), Sadaqat Hussain[1], Muhammad Awais[2], and Syed Sadiq Hussain[3]

[1] School of Civil and Hydraulic Engineering,
Hefei University of Technology, Hefei, Anhui, China
`engr.smusarath@gmail.com`
[2] School of Mechanical Engineering, Hefei University of Technology, Hefei, Anhui, China
[3] School of Civil Engineering, Sir Syed University of Engineering and Technology,
Karachi, Pakistan

Abstract. This paper focuses on the significant issue of structural cracks in multi-span simply supported beam bridges. This study employs finite element analysis to assess the impact of various parameters, including load conditions, crack locations, and material qualities, on the strength of bridges. The maximum fatigue value recorded was $9.765e + 02$, while the minimum was $1.000e + 02$, suggesting potential reinforcement measures are necessary. Fatigue analysis helps identify possible structural deficiencies and assess the remaining fatigue life of the various bridge components. These results offer a thorough understanding of the behavior of bridges, hence contributing to the mitigation of risks and advancements in technology. The findings of this study can be used to improve bridge safety in the future through predictive maintenance models. Investigating novel structural materials and techniques will produce more robust, durable bridges.

Keywords: Structural Cracks · Multi-Span Simply Supported Beam Bridge · Finite Element Method · Fatigue Analysis

1 Introduction

Most infrastructure like bridges, stadiums, and skyscrapers contain mild steel beams, may crack over time owing to various stages of operation. These flaws can change the structure's elasticity and damping, increasing the risk of collapse. Vibration-based methods assess cracks' impact on aircraft, buildings, ships, and bridges, saving costs and enhancing safety by detecting structural damage [1]. The research examines the fracture parameters that affect vertical displacements in simply supported beams with cracks, emphasizing structural behavior during transient mass application. Model-based damage detection relies on modal parameter changes pre and post-faults, offering advantages over visual assessment. It aids in creating damage detection techniques by using damage-sensitive natural frequencies and mode structures [2]. Using beam-element models and

A. Bieliatynskyi et al. (Eds.): CSTTE 2023, LNCE 603, pp. 244–250, 2024.
https://doi.org/10.1007/978-981-97-5814-2_22

empirical methods, the research calculates fatigue life for key bridge components, primarily stringers, cross-girders, truss diagonals, and hangers. Focusing on fatigue-critical parts and simplified models aims to improve bridge fatigue life estimation [3].

FEM simulation using ANSYS and regression analysis assessed beam structure cracks, analyzing their impact on physical characteristics and dynamic response. Static analysis with ANSYS and Creo I-section models revealed reduced vibration with decreased fracture depth. Hu et al. demonstrated damage-adjusted bar stress in Euler-Bernoulli beams with open damage. Researchers found that vibration characteristics can detect cracks, preventing material fatigue-related failures [4]. Python, image processing, and machine learning detect concrete cracks using numerical methods and beam vibration analysis. Machine learning classifies 4-point fixed flexural tests with varying depth-to-span ratios. Joint damage causes irregular ground-concrete fractures in structure and location, which is examined by machine learning in damaged areas and hence quantified hard surface cracks. Machine learning identifies unstable ground-concrete fractures caused by joint damage. It's used for seismic damage identification, particularly in bridges, aided by YOLO's crack dimension assessment. However, installation or manufacturing issues can affect laser beam testing for crack detection in simulations and experiments [5].

This research employs SolidWorks to address crack issues in multi-span simply supported beam bridges. It involves modeling, load simulation, stress analysis, and crack detection, enhancing beam structure design and reducing cracking risks.

2   The Geometry of Beam and Deck of Bridge

In an ongoing research project, SolidWorks 2021 is employed for finite element analysis to model a simply supported steel beam bridge with eight spans, each 15 m long (total length: 120 m). The design specifications include a 0.4 m flange width, 0.03 m flange thickness, 0.01 m transverse sheet thickness, and 3 m spacing between beams. The bridge deck measures 0.15 m in height, 9.43m in width, and 120 m in length. The bridge weighs 4.6×105 kg, has a volume of 4.6×102 m^3, and a surface area of 4.9×103 m^2, as shown in Fig. 1. Specialized cutting procedures are developed for bridge deck fabrication to match strut distances and accommodate stringer ends and the HEM profile of braces.

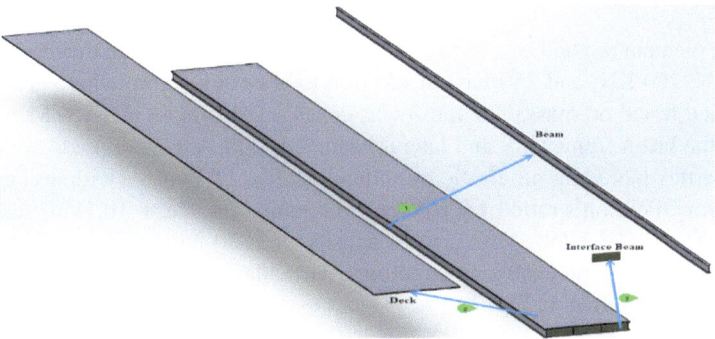

Fig. 1. T3D Drawing of Multi-Span Simply Supported Beam Bridges

3 The Fundamental Principles of SolidWorks Analysis

SolidWorks Analysis, part of Dassault Systems' software, offers engineers static, dynamic, thermal, and fatigue structural analysis tools. It supports linear and nonlinear analyses under various conditions and provides packages for skill enhancement. The software enables assessments like harmonic analysis, time history analysis, drop testing, and random vibrational analysis, utilizing parameters like eigenvectors and natural frequencies, as shown in Fig. 2.

Fig. 2. Flow Chart of Finite Element Analysis in SolidWorks

4 Static Computation and Advanced Features of SolidWorks

The finite element method was utilized for static analysis, subjecting the model to loads of 150 KN, 200 KN, and 250 KN, in addition to gravity loading, where gravity's rate was applied based on mass. The mesh was constructed, employing ASTM A36 grade steel for the beam framework and lateral beams. ASTM A36 steel possesses mechanical properties including an elastic modulus of $2e + 11$ N/m^2, yield strength of 2.5×108 N/m^2, Poisson's ratio of 0.26, shear modulus of $7.93e + 10$ N/m^2, and tensile strength of 4×108 N/m^2, with a mass density of 7850 kg/m^3. These properties are crucial in modeling and analyzing structures, particularly multi-span simply supported beam bridges, as shown in Fig. 3.

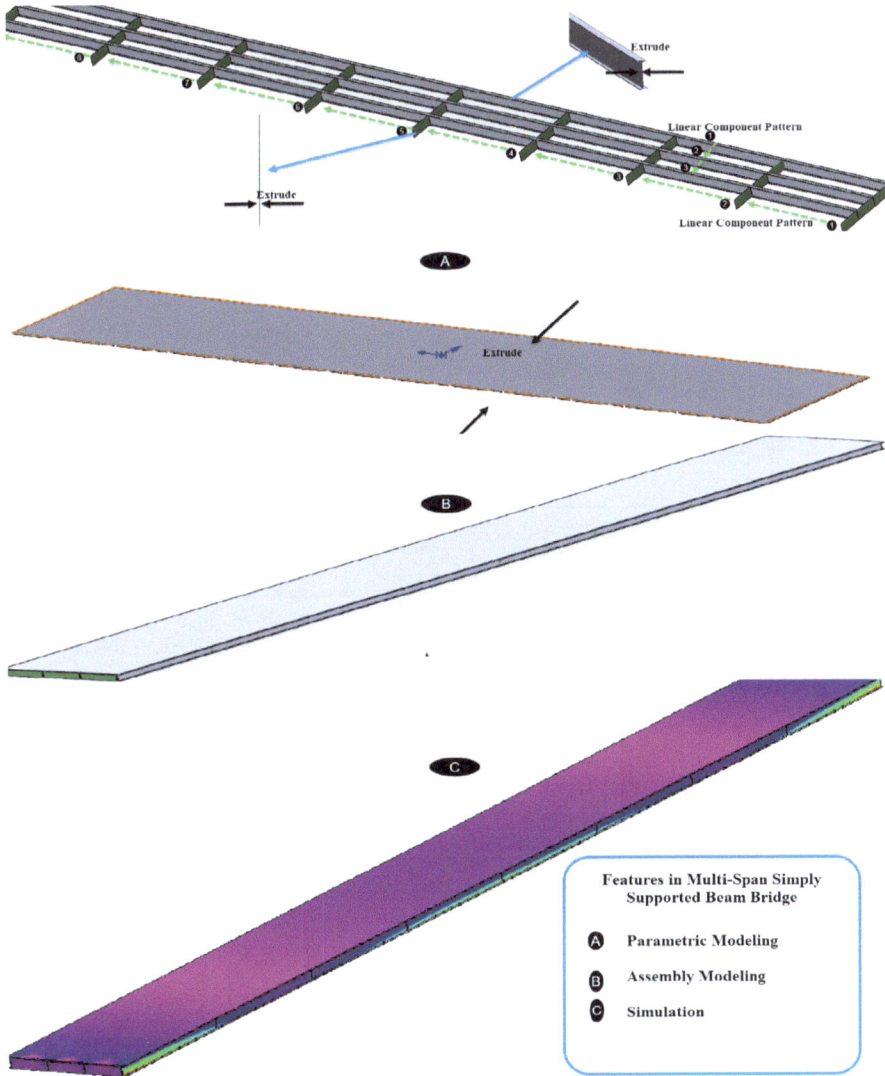

Fig. 3. Advanced Features of SolidWorks

5   Results and Discussions

5.1   Static Study of Finite Element Model

Static study assesses structural performance and safety under varying loads. Mises stress and displacement distribution was highest in steel and marginal support areas. Red indicates excessive, while blue shows moderate analysis. The informs load response graphically and statistically, which is crucial for optimizing design and evaluating load capacity.

Discussion of Stress in Bridge

The graph shows stress variations with increasing applied loads, indicating a direct relationship. Specific nodes surpass the material's yield strength, emphasizing the importance of addressing high-stress areas in structural design, as shown in Fig. 4.

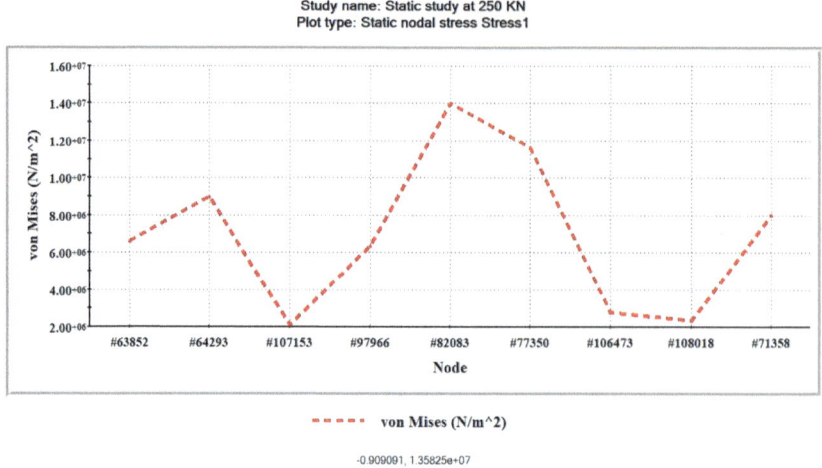

Fig. 4. Stress analysis at different loads impacting

Discussion of Displacement in a Bridge

The graph shows displacement data for nodes under varying loads, revealing a direct link between applied loads and structural deformation. Measuring displacement is vital for assessing load impact on structural integrity and performance, as shown in Fig. 5.

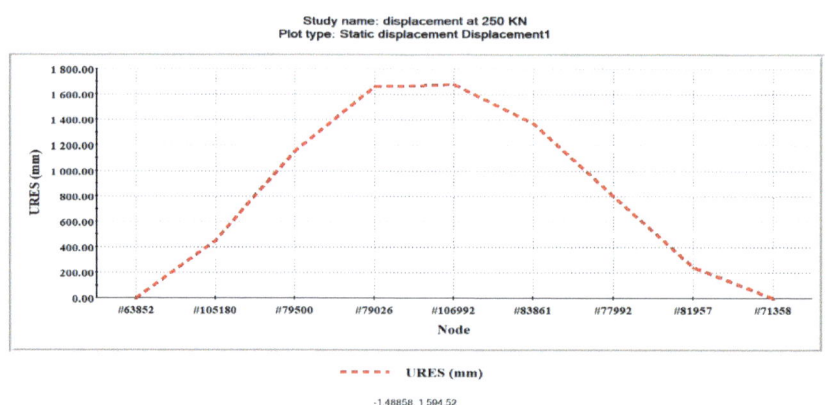

Fig. 5. Displacement analysis at different loads impacting

Meshing

This research delves into "SolidWorks" meshing options, including solid, node-based, and curve-based methods, granting precise mesh control. Fine-tuning element size, aspect ratio, and mesh density ensures high-quality meshes, vital for accurately representing structural motion, especially in complex geometries and high-stress zones. The mesh employed in this study consists of 109,632 nodes and 53,303 elements, influenced by factors like geometry complexity and element arrangement.

Fatigue

Our analysis used "SolidWorks" to assess fatigue, determining maximum and minimum damage percentages and total life cycle fatigue for a multi-span simply supported beam bridge. The maximum fatigue value, $9.765e + 02$, and minimum, $1.000e + 02$ indicate severe fatigue damage, suggesting potential strengthening needs. Estimated maximum and minimum total life cycle fatigue values provide insight into the structure's remaining lifespan, potentially requiring replacement or repairs. "SolidWorks" Simulation included a 1000000-cycle event, calculating alternating stress using von Mises equivalent stress and considering fatigue strength reduction factors, as shown in Fig. 6.

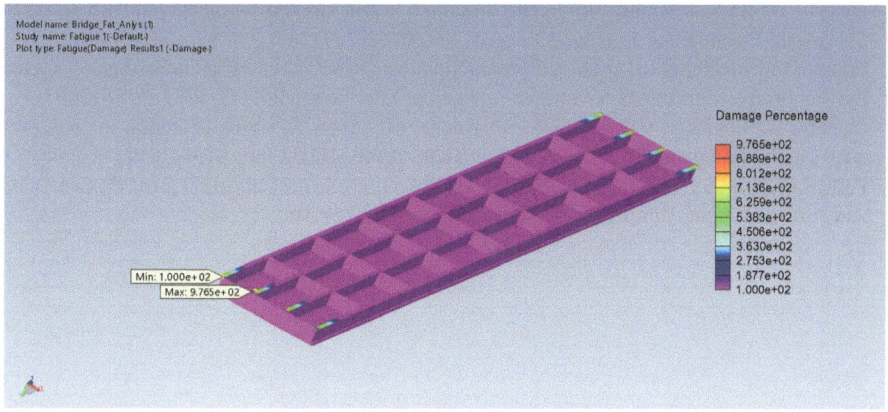

Fig. 6. Fatigue Analysis of Bridge

6   Conclusion

This article mainly studies the importance of structural cracks in multi-span simply supported beam bridges using advanced solidworks software and finite element analysis to evaluate the impact of various parameters on the bridge's strength. The paper's theme aligns with the direction of the conference submission. However, several issues need to be addressed.

The investigation revealed the influence of load magnitudes on strain, displacement, and natural frequencies. The presence and locations of cracks significantly influence a system's structural integrity, indicating the primary effect of cracks on the associated parameters.

Besides, the examination of fatigue has provided significant insights into the anticipated lifespan of the bridge, hence influencing decisions about safety protocols and maintenance strategies.

The static analysis capabilities of SolidWorks have shown to be highly beneficial in the simulation of critical processes and the design of bridge structures.

This work underscores the significance of vibration-based damage detection approaches, as demonstrated by the provided methodologies, in ensuring the durability and safety of essential infrastructure. The implementation of specific measures aids in mitigating the probability of structural cracking and improving the efficacy of bridge design.

References

1. Ahiwale, D., Madake, H., Phadtare, N., Jarande, A., Jambhale, D.: Modal analysis of cracked cantilever beam using ANSYS software. Mater. Today: Proc. **56**, 165–170 (2022)
2. Bulut, C., Jena, S., Kurt Habiboğlu, S.: Experimental and computational study on dynamic analysis of cracked simply supported structures under moving mass. Frattura Ed Integrita Strutturale-Fracture and Structural Integrity **16**(60) (2022)
3. Imam, B.M., Righiniotis, T.D., Chryssanthopoulos, M.K.: Numerical modelling of riveted railway bridge connections for fatigue evaluation. Eng. Struct. **29**(11), 3071–3081 (2007)
4. Satpute, D., Baviskar, P., Gandhi, P., Chavanke, M., Aher, T.: Crack detection in cantilever shaft beam using natural frequency. Mater. Today: Proc. **4**(2), 1366–1374 (2017)
5. Park, S.E., Eem, S.H., Jeon, H.: Concrete crack detection and quantification using deep learning and structured light. Constr. Build. Mater. **252**, 119096 (2020)

Solid Analysis of Torsional Effects on Small Radius Steel Plate Composite Beams Based on ABAQUS

Xin Liu$^{(\boxtimes)}$ and Guoxi Tang

Anhui Province Transportation Planning and Design Research Institute Co., Heifei, China
3024319405@qq.com

Abstract. In recent years, steel-mixed composite structures have been widely adopted in China, especially in projects such as urban overpasses and viaducts where steel plate composite girder bridges are widely used. However, the research on small radius steel plate composite curved girder bridges is relatively limited, and the theory of structural stress characteristics and computational analysis is lagging behind. It is found that the radius of curvature has a small effect on the torsional stress of the bridge top plate, but the torsional stress of the steel main girder is higher; the torsional stress of the steel main girder rises with the increase of the computed span diameter, and the concrete slab has a smaller effect; the number of steel girders has a small effect on the torsional stress of the bridge top plate but a significant effect on the steel main girder torsional stress. The results of the study can provide a reliable reference for future projects.

Keywords: torsional stress · radius of curvature · calculated span · number of steel beams

1 Introduction

In recent years, steel-mixed combination structure has been widely adopted in the field of bridges and building structures in China, showing excellent technical and economic benefits.

Researchers like Hu Shaowei [1] discovered through torsion tests on composite girders that the concrete wing plate significantly influences torsional load capacity, with thickness playing a crucial role. They suggested that the maximum torsional load capacity for composite girders occurs at a hoop ratio of 0.54%. Shi Qiyin [2] and colleagues identified the concrete airfoil plate and outer cladding steel as key components in the torsional load capacity of new combined beam structures. They introduced the bending-twisting ratio as a factor affecting member torsional load capacity. Nie [3] and team developed a practical longitudinal shear calculation model for combined beams based on engineering tests. Qiu Wenliang [4] proposed an analytical model for shrinkage creep in steel-concrete composite beams, accounting for concrete slab cracking effects on stiffness and strength. Given the diverse structural forms of steel plate combined girder bridges, understanding the impact of different design parameters on torsional stresses is essential for subsequent construction.

A. Bieliatynskyi et al. (Eds.): CSTTE 2023, LNCE 603, pp. 251–256, 2024.
https://doi.org/10.1007/978-981-97-5814-2_23

2 Project Overview

This project is based on a 4 × 25 m steel plate combination girder [5]. The superstructure adopts double main girders, 1, 2# middle cross girder and main girder diagonal intersection 15°; 3# middle cross girder and main girder diagonal intersection 10°, two main girder centre spacing is 5.6 m, I-beam girder height is 1.3 m; prefabricated deck slab cantilever flange thickness of 25 cm, the root thickness of 41 cm. Its deck plan is shown in Fig. 1, the fabrication arrangement diagram is shown in Fig. 2. We used ABAQUS [6–8] 2020 for modelling and the finite element model is shown in Fig. 3.

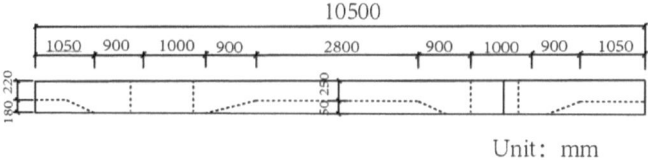

Fig. 1. Bridge cross-section

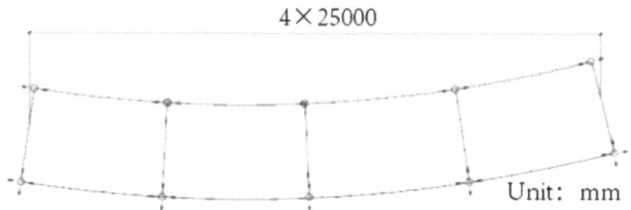

Fig. 2. Bearing arrangement plan

Fig. 3. Overall diagram of the baseline model

3 Torsional Stress Analysis of Design Parameters on the Reference Bridge Example

3.1 Influence Law of Radius of Curvature

Figure 4(a) and Fig. 4(b) depict torsional stress envelope variations in the bridge roof plate and steel main girder at pivot 1 in a small-radius steel plate combination girder under typical working conditions (constant load + out-of-lane offset arrangement). The transverse distribution of torsional stress in the roof plate remains consistent, with numerical differences. A larger roof plate radius results in a more pointed torsional stress envelope curve and uneven stress distribution. Among the outer steel, R100 to R300 exhibit main beam torsional stresses higher than the inner ones by 39.11%, 22.72%, 9.35%, 24.69%, and 11.90%, respectively, with increments less than 0.05 MPa (Fig. 4(a)). In summary, the radius of curvature has a minimal impact on the bridge roof slab's envelope torsional stress.

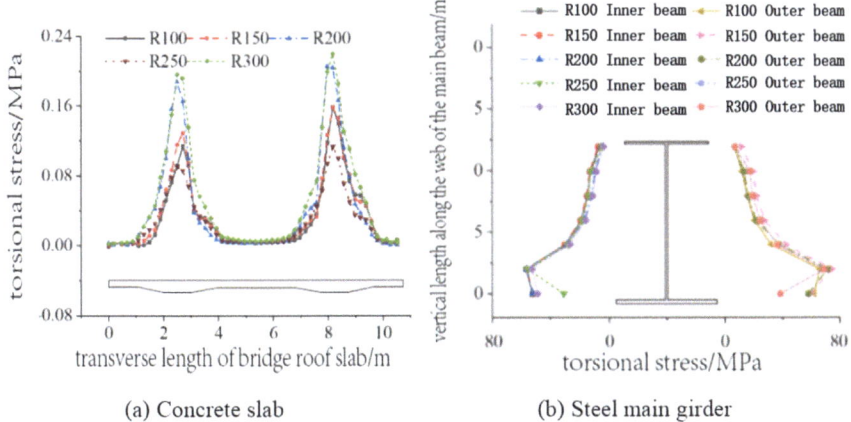

(a) Concrete slab   (b) Steel main girder

Fig. 4. Torsional stress cloud at center pivot point 1

In Fig. 4(b), torsional stress distribution along the height of inner and outer steel main girders follows a similar pattern, but with numerical differences. The torsional stress envelope of the outer steel main girder is approximately 1.2 times larger than that of the inner steel main girder. The critical torsional area in the steel main girder occurs around 1/4 of the web height from the bottom plate. For the original bridge with a radius of curvature R = 250 m, the most critical torsional stress value is 71.16 MPa. This value remains consistent at 71.16 MPa for the outer steel girder. With an increase in radius from R100 to R300, the corresponding maximum torsional stress value grows by less than 5.9%, totaling less than 8.7 MPa. Thus, the steel girder's sensitivity to curvature radius is low, but the maximum torsional stress is notably high, reaching 70 MPa, necessitating attention.

3.2 Calculation of the Law of Influence of the Span

Figure 5(a) and Fig. 5(b) show the laws of the torsional stress envelopes of the bridge roof plate and steel main girder at the mid-point 1 of the span of a small-radius steel plate composite girder bridge with the change of the computed span diameter under the typical working condition of (constant load + lane outward deviation arrangement), respectively.

Under typical working conditions, the torsional stress distribution across the bridge's transverse direction in the roof slab remains consistent, with slight numerical differences. The torsional stress envelope value for the outer steel girder is larger than that of the inner

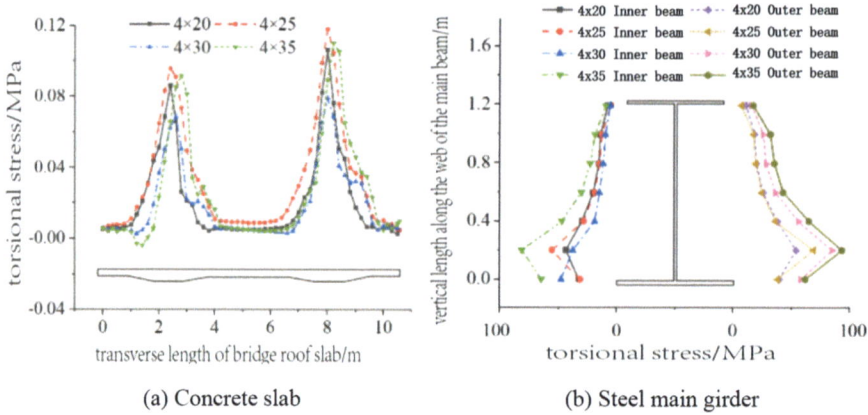

(a) Concrete slab   (b) Steel main girder

Fig. 5. Torsional stress cloud at center pivot 1

girder, indicating the outer girder experiences more unfavorable stress. The change in torsional stress envelope value with increasing computed span diameter does not follow a clear pattern (Fig. 5(a)). For instance, the maximum torsional stress of the steel main girder first increases and then decreases when the span varies from 4×20m to 4×35m. Similarly, the maximum torsional stress of the concrete slab is less than 0.12MPa in several span configurations, with changes below 0.05MPa. In summary, the influence of calculated span diameter on concrete slab torsional stress is relatively low.

The torsional stress distribution along the height of the steel main girder follows a similar pattern, with only slight numerical differences. The torsional stress envelope value for the outer steel main girder is generally larger than that of the inner girder (Fig. 5(b)). Numerically, the maximum torsional stress of the steel main girder under various computed span diameters is generally above 30 MPa, reaching 93 MPa for the outer steel main girder in the 4×35m span. With increasing span diameter, the overall torsional stress of the steel main girder shows an increasing trend; In summary, the calculated span diameter significantly influences steel main girder torsional stress and requires careful consideration.

3.3 Laws Affecting the Number of Steel Beams

Figure 6(a) and Fig. 6(b) show the laws of the torsional stress envelope of the bridge roof plate and steel main girder with the change of the number of steel girders at the midpoint 1 of the span of a small-radius steel plate composite girder bridge under the typical working condition of (constant load + out-of-lane deviation arrangement), respectively.

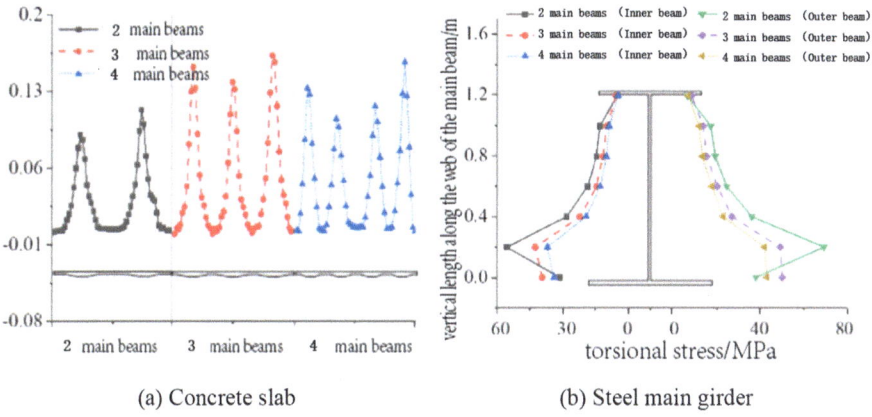

(a) Concrete slab   (b) Steel main girder

Fig. 6. Torsional stress cloud at center pivot 1

Under typical working conditions, the torsional stress distribution across the bridge roof plate's transverse direction remains consistent, differing only numerically. The outermost steel main girder experiences a larger torsional stress envelope than the inner and middle girders, making it more unfavorably loaded (Fig. 6(a)). Numerically, the maximum torsional stresses in the roof plate are 3.7% and 43.5% higher in the three-main girder configuration compared to the four-main and two-main girder configurations, respectively. Torsional stresses in the bridge roof slab are below 0.2 MPa for the three types of steel main girders, varying by less than 0.1 MPa. Hence, the bridge roof slab's torsional stress shows low sensitivity to the number of steel main girders.

The torsional stress distribution along the height of the steel main girders follows a similar pattern, with slight numerical differences. The torsional stress envelope of the outer steel main girders is generally larger than that of the inner steel main girders. Numerically, the maximum torsional stresses of the three types of steel girders generally exceed 30 MPa, indicating higher stress levels. With an increase in the number of main girders, the overall trend of the torsional stress envelope of the steel main girders decreases. In summary, the number of steel main girders significantly affects their torsional stress, with double main girders experiencing the most unfavorable conditions.

4   Conclusion

By analysing the effect of each parameter on torsional stresses in small radius combination beams, the following conclusions are drawn:

(1) The radius of curvature has a low degree of influence on the torsional stresses in the bridge roof slab, but the maximum torsional stresses in the steel main girders have higher values, especially at larger radii of curvature, which need to be taken into account by the designers. Changes in the radius of curvature will cause non-uniformity in the distribution of torsional stresses in the bridge roof plate and steel main girder, which has a potential impact on the stability of the structure.

(2) The torsional stresses of the steel main girders tend to increase with the increase of the calculated span, especially the outer steel main girders are more unfavourably stressed. However, the torsional stress of concrete slab is less affected by the calculated span diameter. It is necessary to choose the calculated span diameter carefully in the design to avoid the structure being affected by excessive torsional stress and to ensure the safety and stability of the bridge.

(3) The number of steel girders has a small effect on the torsional stress of the bridge roof slab, but it has a significant effect on the torsional stress of the steel main girder. The torsional stress of the double main girder structure is the most unfavourable, and its torsional stress is at a high stress level, which needs special attention in the design.

References

1. Hu, S., Nie, J., Xiong, H.: Torsion test and analysis of steel-concrete composite beams. J. Building Struct. **4**, 103–109 (2006)
2. Shi, Q.-Y., Cai, J.-L., Chen, Q.Q., et al.: Tests and analyses on torsion resistance of a new type of encased steel-concrete composite beam. Eng. Mech. **25**(12), 162–170 (2008)
3. Nie, J., Wang, H.: Experimental study on longitudinal shear resistance of steel-concrete composite beams. J. Building Struct. **18**(2), 13–19 (1997)
4. Qiu, W., Jiang, M., Zhan, Z.: Finite element method for shrinkage creep analysis of steel-concrete composite beams. Eng. Mech. **21**(4), 162–166 (2004)
5. Akduman S, Aktepe R, Aldemir A, et al. :Structural performance of construction and demolition waste-based geopolymer concrete columns under combined axial and lateral cyclic loading. Eng Struct. **297** (2023)
6. Qian, H., Wang, X., Li, Z., et al.: Experimental study on re-centering behavior and energy dissipation capacity of prefabricated concrete frame joints with shape memory alloy bars and engineered cementitious composites. Eng Struct. **277** (2023)
7. Liu, Y., Ma, H., Li, Z., et al.: Seismic behaviour of full-scale prefabricated RC beam -CFST column joints connected by reinforcement coupling sleeves. Structures **28**, 2760–2771 (2020)
8. Jingke, Z., Peng, L., Chang, H., et al.: Torsional behavior of I-steel–concrete composite beam considering the composite effects. Struct. Concr. **23**(2), 1151–1175 (2022)

Research on the Selection Model of Apron In-Ground Pit Based on Utility Tunnel Technology

Jie Ouyang[✉], Zhiying Song, Xueqing Qiao, and Xiaowei Li

School of Transportation Science and Engineering, Civil Aviation University of China, Tianjin 300300, China
ou_yangjie@163.com

Abstract. This paper studies the selection of apron in-ground pits locations, using the Analytic Hierarchy Process to identify key factors influencing the selection of apron in-ground pits locations. It establishes a multi-objective optimization model for apron in-ground pits locations based on utility tunnel technology, considering the construction distance cost, the human resource cost, and the in-ground pits installation cost. The model is validated through simulation using particle swarm optimization algorithm. The results show that the proposed model can achieve the integration of apron in-ground pits. Compared with the existing pit system at Zhuhai Airport, the total cost has reduced by approximately 50.4%, 52.97%, and 34.01% for the three different aircraft parking stand layouts, respectively. This can provide guidance for the application of utility tunnel in airport construction, promote the improvement of apron operation efficiency and unmanned apron realization, and contribute positively to the development of green airports.

Keywords: Utility tunnel · Selection of apron in-ground pits locations · Multi-objective optimization · Analytic Hierarchy Process · Particle Swarm Optimization · Green airport

1 Introduction

The utility tunnel is a widely used technology in urban construction, which has the advantages of intensive land use and avoiding repeated excavation of road surfaces. In the construction of new large airports such as Qingdao Jiaodong International Airport and Beijing Daxing International Airport, the construction of utility tunnel is also being considered. However, in the process of apron ground service support, the location and quantity of the apron in-ground pits have a significant impact on the safety and efficiency of apron operations. Currently, research on apron utility tunnel mainly focuses on alignment planning [1, 2], cross-sectional design [35], fueling spigot well optional domains [6], and drainage design [7]. There is relatively little research on the selection of apron in-ground pits locations based on utility tunnel technology from multiple perspectives.

Funding: National Natural Science Foundation of China Civil Aviation Joint Research Fund (Project Approval Number: U2333204): Research on key technologies for capacity improvement of airport terminal bay area based on mutual feedback between design and operation.

Therefore, this paper focuses on the selection of apron in-ground pits locations. Unlike traditional shaft selection, the selection of apron in-ground pits locations needs to consider factors such as the diameter and quantity of pipelines, as well as the service range after extraction. Based on the determination of the target area, cost function, and constraints, the design principles of in-ground pit selection are used to minimize total cost, optimize overall service level, or maximize social benefits, thus determining a rational and efficient in-ground pit network structure. Based on the demand for service support interfaces from aircraft, the paper discusses the establishment of a multi-objective optimization model from the perspectives of economic affordability, operational efficiency, and labor costs. The goal is to obtain the optimal aggregation, location, and quantity of apron in-ground pits, in order to reduce the use of ground support vehicles and improve apron operational efficiency.

2 Multi-objective Optimal Site Selection Model for Apron In-Ground Pits

2.1 Analysis of Factors Affecting the Selection of Apron In-Ground Pits Locations

This paper analyzes the influencing factors of in-ground pits site selection and establishes the cost index system of in-ground pits site selection model consisting of 3 first-level evaluation indexes, including the safety maintenance cost B1, the labor cost B2, and the facility construction cost B3, and 9 s-level indexes, as shown in Fig. 1.

Aircraft in the process of sliding into, launching and safeguarding operations will produce a variety of unsafe factors, the need to optimize the traffic flow of the apron support area, reduce the use of special support vehicles and provide sufficient parking space for other vehicles, effectively reduce environmental pollution at the same time to optimize the allocation of ground resources, therefore the introduction of the safety operation cost C1, the environmental protection cost C2, and the construction distance cost C3 as an evaluation index; The consideration of labor cost mainly focuses on reducing the labor intensity of ground staff and simplifying the service guarantee process, so as to improve the work efficiency. Especially for airports with high labor cost, reducing the number of ground staff will effectively save human resource cost, so the human resource cost C4, the service time cost C5, and the service complexity cost C6 are selected as evaluation indicators. The cost of facility construction takes into account the economic cost in the whole cycle of in-ground pits construction, such as the development of infrastructure, equipment configuration, and maintenance of in-ground pits, can effectively reduce energy demand and operating costs in the long run. This helps in saving airport operational expenses and material consumption. Therefore, the evaluation criteria include the land development cost C7, the pipeline laying cost C8, and the in-ground pits installation cost C9.

The weights and rankings of the first-level and second-level indicators are obtained through expert scoring and consistency tests, as shown in Table 1.

From the Table 1, it can be seen that the in-ground pits device cost C9, the human resource cost C4 and the construction distance cost C3 have the greatest influence on the in-ground pits site selection, and the optimization calculation of these three objectives can get the better result of the site selection of in-ground pits.

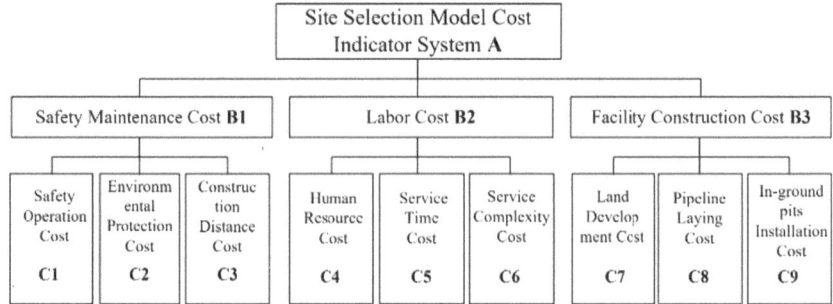

Fig. 1. Site Selection Model Cost Indicator System

Table 1. Ranking of Criteria Weights

Indicators		Weights	Arrange in Order
First-Level Indicators	B1	0.37	2
	B2	0.15	3
	B3	0.92	1
Second-Level Indicators	C1	0.47	4
	C2	0.26	7
	C3	0.85	3
	C4	0.93	2
	C5	0.33	5
	C6	0.17	8
	C7	0.28	6
	C8	0.16	9
	C9	0.95	1

2.2 Model Building

Most of the siting problems through the alternative set of points for discrete siting, but in practice the siting problem is often carried out in continuous space, that is, the demand point is in continuous space and the facility can be located anywhere in the space. Continuous siting problem is usually difficult to calculate and model. Some scholars have proposed methods to solve the siting problem, which can be summarized as follows: 1) Abstract the continuous space demand object into points, lines or polygons; 2) Adopt some rules to transform the infinite facility candidate points into a finite set of points; 3)

Adopt a discrete siting model for siting [8]; This paper adopts a method of transforming a continuous plane into a finite set of points, discretizing the continuous plane of the machine position to obtain a finite set of candidate facility points. Based on this, the article combines the maximum set cover model and the P-median model to establish a multi-objective site selection model for apron well. The decision variables are as follows:

$$X_j = \begin{cases} 1, \text{ set in } - \text{ ground pits head at j} \\ 0, \text{ else} \end{cases} \tag{1}$$

$$Y_{ij} = \begin{cases} 1, \text{ j serves docking port i} \\ 0, \text{ else} \end{cases} \tag{2}$$

The site selection of the apron in-ground pits has special characteristics. It needs to achieve full coverage of aircraft service interfaces with the minimum number of in-ground pits, while also considering optimization issues in practical application scenarios. In multi-objective optimization, it is necessary to consider the trade-off relationship between multiple objectives. This relationship is usually controlled using weights (λ) to transform the multi-objective problem into a single-objective optimization problem. However, this problem has a characteristic that differs from traditional optimization methods, that is, once the location of the in-ground pits is determined, it is difficult to change. Therefore, in the selection process, it is necessary to determine the weight relationship between several objectives, meaning that for a specific implementation plan, the weights between several objectives are always static. This problem focuses on three types of cost: the construction distance cost, the human resource cost and the in-ground pits installation cost.

The In-Ground Pits Installation Cost
In-ground pits device is the integration of various types of pipeline equipment, which can be equipped with a single pipeline or multiple pipelines in one in-ground pit, the higher the integration, the fewer the construction quantity and the lower the cost. But the work site is more complex, the integration and construction quantity of the in-ground pits need to be considered comprehensively. According to the different integration levels of the pipelines, the construction cost of the in-ground pits varies as shown in (3).

$$M_j = 2.5 * \sum_{i \in I} Y_{ij} + 2.5 \tag{3}$$

The Construction Distance Cost
The construction distance is the distance from in-ground pits to the aircraft service interface. Due to the different pipe diameters, the service distance for larger diameter pipelines should be shorter during construction, while the service distance for smaller diameter pipelines can be appropriately extended. The combination of pipelines with different diameters has a significant impact on the construction distance of in-ground pits facilities. In summary, the larger the service distance, the lower the work efficiency. The Euclidean distance is used in the model.

$$d_{ij} = \sqrt{(x_i - x_j)^2 + (y_i - y_j)^2} \tag{4}$$

The Human Resource Cost

In practical operations, a certain number of staff members are needed to handle the connection and conveyance of various interfaces. The specific cost of human labor services is difficult to calculate accurately. This paper uses the number of staff as a measure, with one worker assigned to each apron in-ground pit. As the number of staff members increases, so does the associated cost.

Combining the above 3 costs yields the model as follows:

$$\text{MinZ} = \alpha * \sum_{j \in J} M_j * X_j + \beta * f * \sum_{i \in I} \sum_{j \in J} d_{ij} * Y_{ij} + \gamma * k * X_j \tag{5}$$

$$N(i) = \left\{ j \in J | d_{ij} \leq D_i \right\} \tag{6}$$

$$Y_{ij} \leq X_j, \forall i \in I, \forall j \in J \tag{7}$$

$$\sum_{i \in I} Y_{ij} \leq X_j, \forall j \in J \tag{8}$$

$$\sum_{j \in N(i)} Y_{ij} = 1, \forall i \in I \tag{9}$$

In (5): I is the set of aircraft service interfaces, that is, demand points i, and J is the set of in-ground pits facility candidate points j; d_{ij} is the distance from in-ground pit j to aircraft service interface i; D_i denotes the acceptable service radius of interface i; and α, β, and γ are the weights of each cost, the values of the three are determined by the weights of the corresponding indicators calculated in Table 1, and $\alpha + \beta + \gamma = 1$. k stands for manual salary, the specific value of which is determined on a case-by-case basis.

In this model, if an aircraft service interface is located within the service range of one or more in-ground pits facilities, it is considered to be covered. (7) indicates that interface can be provided only when in-ground pit is established, and (8) indicates that at least one aircraft interface is serviced by any in-ground pits. In a single aircraft position, only one in-ground pit with the same function needs to be established for simplification. (9) ensures that aircraft service interface is serviced by only one in-ground pit. Due to the different magnitudes of construction cost and distance cost in the model, a cost coefficient f is introduced for normalization: $f = 1$.

2.3 Algorithmic Solution

In order to solve the above modeling problem, this paper uses the particle swarm algorithm, which is designed by simulating the feeding behavior of bird flocks, starting from a set of random particle positions, and seeking the optimal position of particles through continuous iteration [9]. In the site selection problem, the position of each particle is regarded as the position of the candidate point, and the advantages and disadvantages of the point can be evaluated by calculating the fitness function value of each particle, and the point with the highest fitness can be obtained as the final site selection point after continuous iteration, particle swarm algorithm flow as Fig. 2.

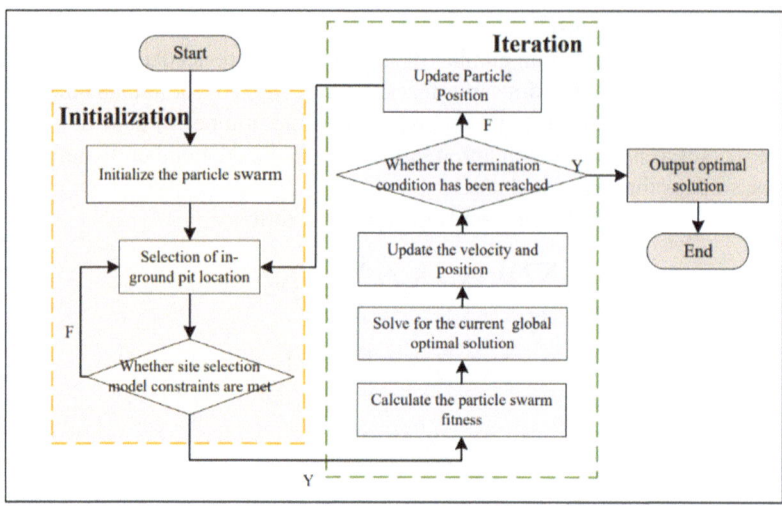

Fig. 2. Particle Swarm Algorithm Solution Process for In-ground Pits Siting Models

3 Example Analysis

3.1 Data Processing

In this paper, the selection of class C aircraft as the design aircraft, according to the *Civil Airport Flight Area Technical Standards* [10] provisions, class C aircraft on the apron net distance should not be less than 4.5 m, the in-ground pits in the service need to be raised, so the distance from the aircraft should not be less than the maximum value specified, the in-ground pits of the feasible domain should be outside the airfield security line.

The service radius of the pipelines in the in-ground pits is also specified. The diameter and service radius of the pipelines may vary according to different aircraft types, system designs and operational requirements, and may be adjusted as needed during actual work. This paper uses empirical data as a reference.

Referring to the provisions in *MHT6031–2018* [11] and *Civil Aircraft Oxygenation* [12]: the length of the oxygenation hose should be not less than 10 m, and the distance of the oxygenation equipment from the aircraft should not be less than 2 m; *MHT6014–2018* [13] stipulates that the inner diameter of the potable water hose should be 25 mm, and the length should be not less than 5 m; *MHT6015–2014* [14] stipulates that the inner diameter of the sewage truck receiving hose should be 100 mm and the length should be not less than 5 m; according to *Aircraft Ground Equipment Operator*, the cable length of the power supply unit to the external power connector of the aircraft should be not less than 10 m. The diameter of air conditioning ducts ranges from 100 to 914 mm, and the effective working distance is 15 m [15]; the diameter of the hoses for compressed air sources ranges from 26 to 76 mm, and their extraction length is about 40 m. Taking the B737-NG as an example and referring to the aircraft type manual, the interfaces required by the aircraft are shown in Fig. 3, and specific data are located in Table 2.

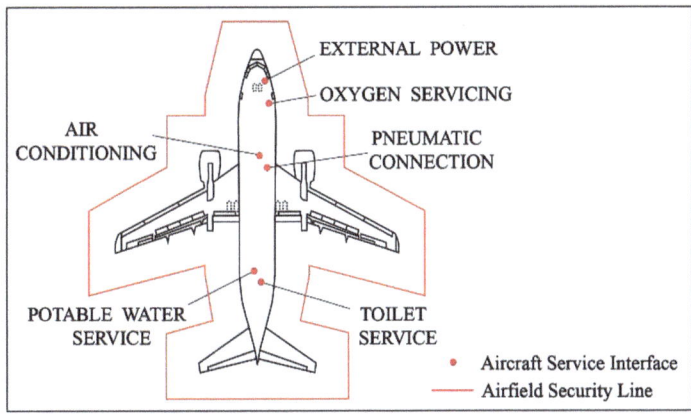

Fig. 3. Particle Swarm Algorithm Solution Process for In-ground Pits Siting Models

Table 2. Service Radius of Each Pipeline

Service Pipeline Type	Hose Diameter (millimeter)	Reasonable Service Radius (meter)
Oxygen Servicing	4 –20	30
External Power	11.5–23	30
Potable Water Service	25	30
Pneumatic Connection	26–76	20
Toilet Service	100	20
Air Conditioning	100–914	20

3.2 Aircraft Parking Stand Layout Mode

The aircraft positions are arranged continuously along the terminal shoreline, and the following situations often occur:

- For the distal end of the finger corridor terminal building may appear to arrange only a single-airplane position.
- For the forefront-type terminal building, the aircraft parallel to each other.
- For the finger corridor terminal building, the bottom of the finger often forms a harbor area, and the head of the aircraft points to the two finger corridors respectively to form a tail concentration.

The three aircraft layout patterns are shown in Fig. 4, and the above three types of aircraft combinations are modeled and solved.

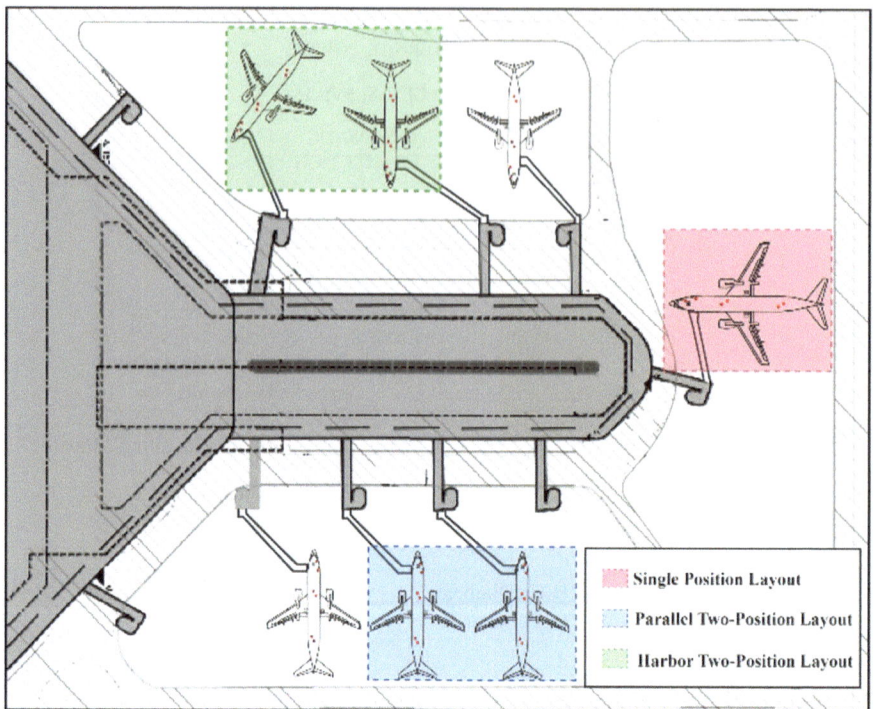

Fig. 4. Coordinate System and Corresponding Service Interface Position of Class C aircraft

3.3 Calculation Result

The results of the calculations for three different aircraft parking stand layout modes are as follows: As Fig. 5(b), the fitness evolution curve reaches the optimal solution within 20 iterations, and as Fig. 5(a), for a single aircraft parking position, the integrated in-ground pit No.1 is located on the right side of the aircraft, providing services to aircraft service interfaces A and B. Integrated in-ground pit No.2 integrates the air conditioning and compressed air pipelines, providing services to aircraft service interfaces C and D. Similarly, integrated in-ground pit No.3 provides services to aircraft service interfaces E and F, achieving the integration of the single aircraft parking position in-ground pit.

When the model converges, the number of iterations is less than 40, which proves that the algorithm is very efficient, as Fig. 6(b). In the case of parallel aircraft parking positions, as Fig. 6(a), the service interfaces A and B of the aircraft on both sides share integrated in-ground pit No.1. The service interfaces C and D of the aircraft are limited by the service radius and cannot be shared between the two sides, therefore, they are provided by integrated in-ground pits No.2 and No.3 located on both sides. Integrated in-ground pit No.4 integrates sewage and clean water services, supplying the left side service interfaces E and F, while integrated in-ground pit No.5 serves the right aircraft service interfaces E* and F*.

In the case of harbor aircraft parking positions, the model iterates less than 20 times to reach the optimal solution. As Fig. 7(a), the service interfaces A and B of the aircraft on

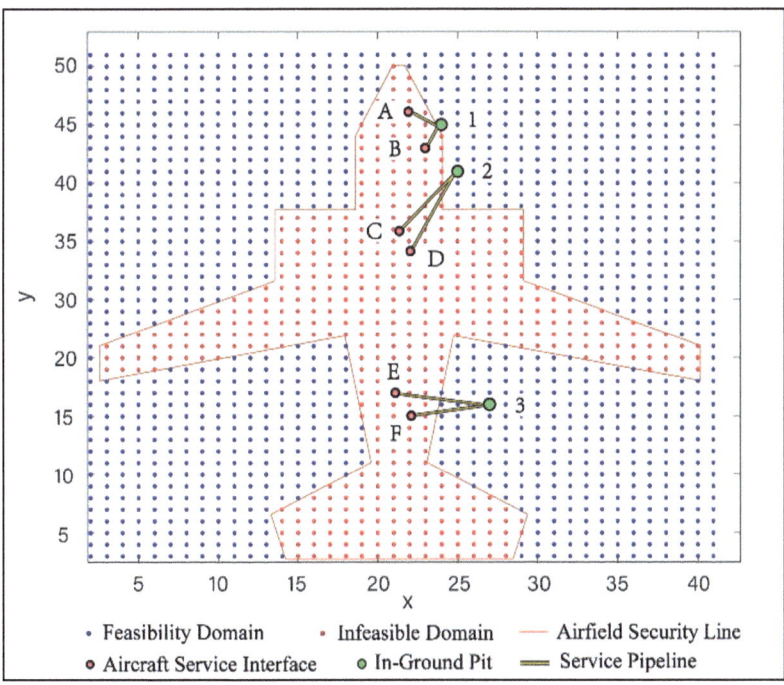

(a) Location Result

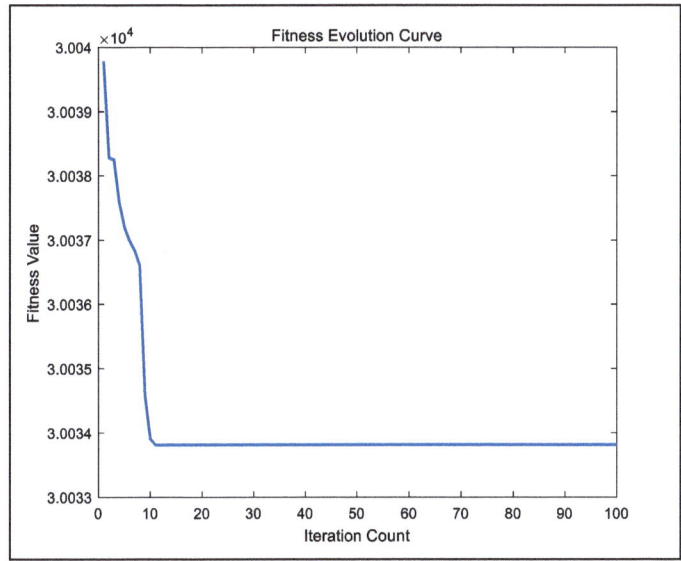

(b)Iteration curve

Fig. 5. Calculation Results for Single-Airport Integrated In-ground Pits

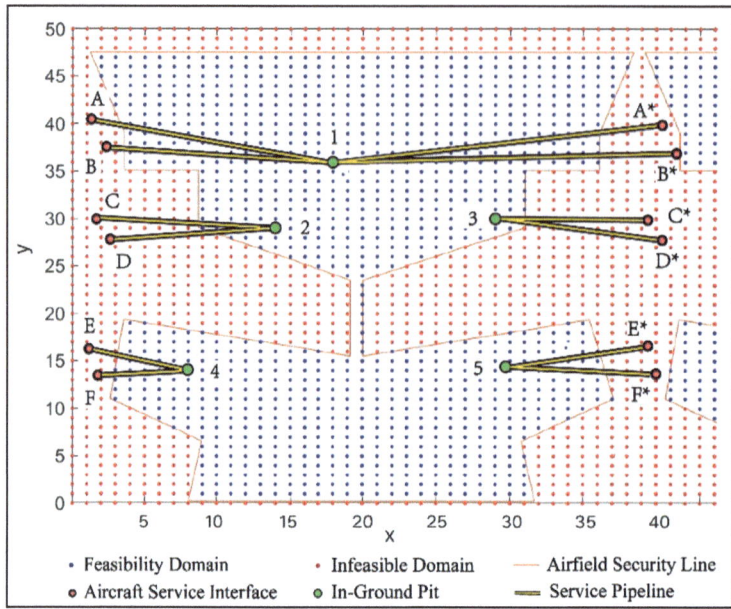

(a) Location Result

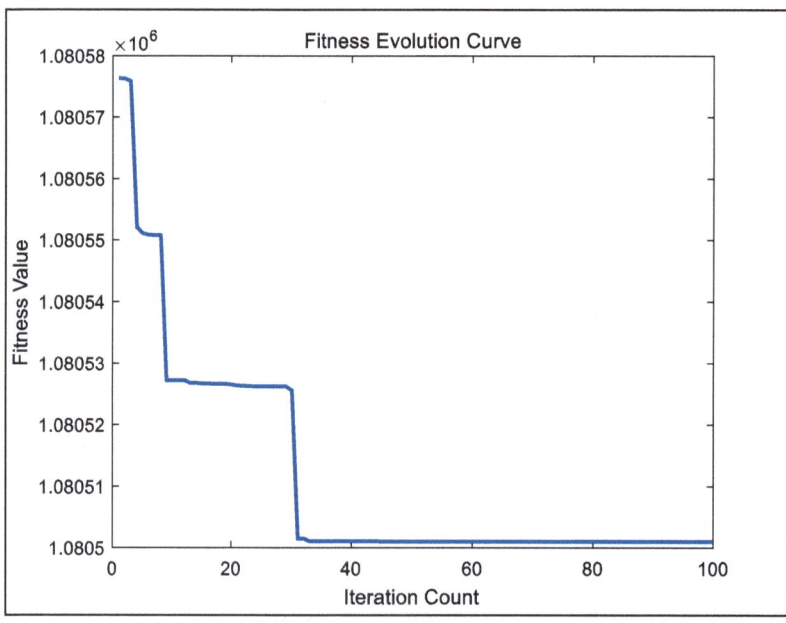

(b) Iteration curve

Fig. 6. Calculation Results for Parallel-Airports Position Integrated In-ground Pits

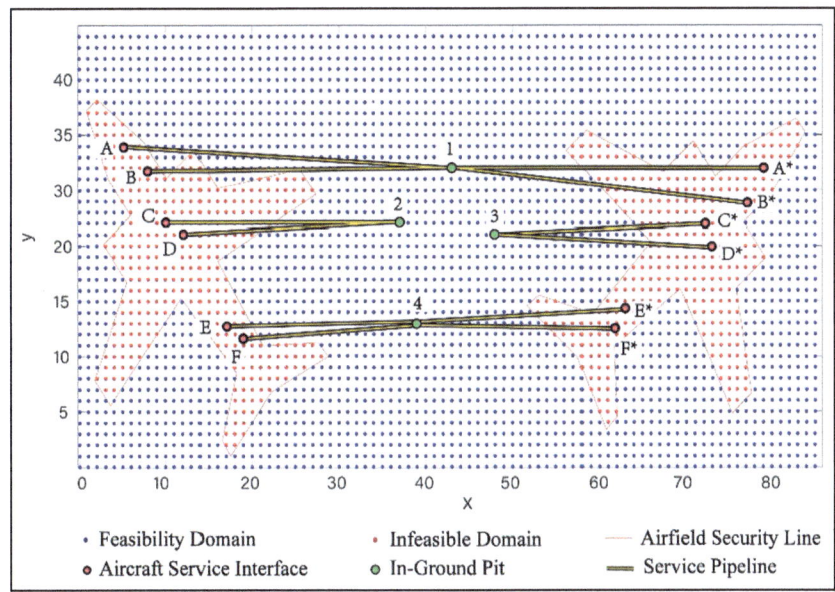

(a) Location Result

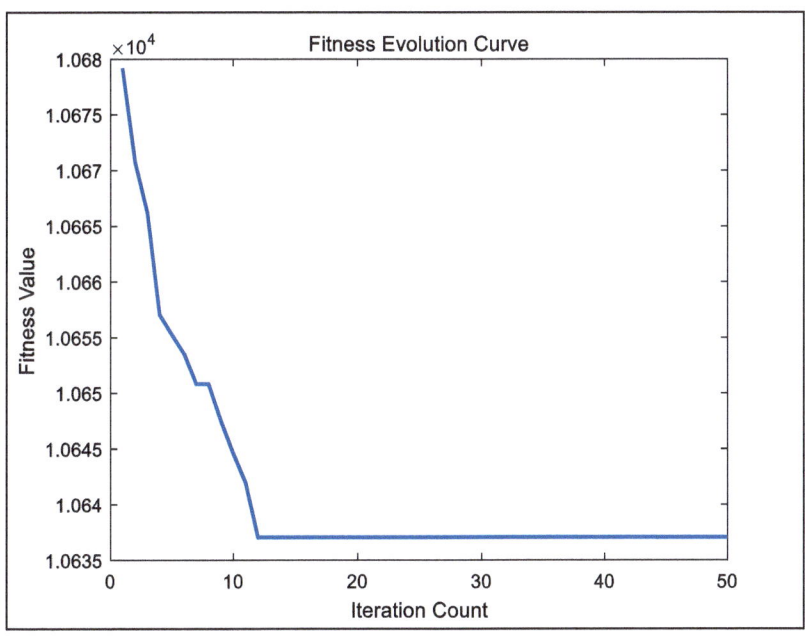

(b) Iteration curve

Fig. 7. Calculation Results for Habor-Airports Position Integrated In-ground Pits

both sides share integrated in-ground pit No.1. The service interfaces C and D are similar to the parallel aircraft parking position, provided by integrated in-ground pits No.2 and No.3 located on both sides. Integrated in-ground pit No.4 simultaneously supplies the service interfaces E and F for aircraft on both sides.

From the Fig. 8, it can be seen that regardless of the layout mode, the total cost of the site selection model is always less than the total cost of the traditional site selection for Zhuhai Airport. After calculation, it is found that compared to the traditional site selection for Zhuhai Airport, the site selection model has reduced costs by approximately 50.47%, 52.97%, and 34.01% in different layout modes, effectively achieving the cost-saving goal. At the same time, the comparison of costs among the three groups of the same layout mode shows that in the total cost composition, this study pays more attention to the proportion of the in-ground pits installation cost and the human resource cost, and the final result also proves that the site selection model effectively reduces these costs.

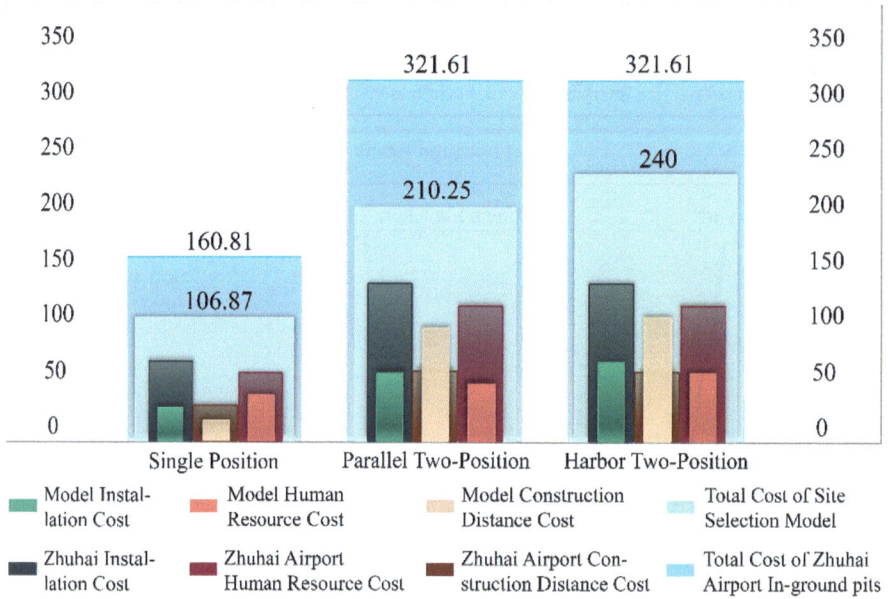

Fig. 8. Comparison of Site Selection Model and In-Ground Pits Costs at Zhuhai Airport under Three Aircraft Layout Models

4   Conclusions

This paper uses the Analytic Hierarchy Process to identify the key factors that have the greatest impact on the selection of apron in-ground pit location. These factors include construction distance cost, human resource cost, and in-ground pits installation cost. By combining the Maximum Coverage Set model and the P-median model, a multi-objective optimization site selection model is established. The model is then computed using the Particle Swarm Optimization algorithm. The results indicate that in the three layout modes of single aircraft position, parallel aircrafts position, and harbor aircrafts position, the total cost of apron in-ground pits can be effectively reduced. This allows for the consolidation of various service pipelines and the innovative sharing of apron in-ground pits between adjacent stand, further reducing the use of ground support vehicles and personnel. This has practical significance for unmanned apron construction and low-carbon airport development.

There are still some shortcomings in this study: Firstly, AHP is subjective and its consideration of index judgment matrix is limited. Secondly, this paper uses Euclidean linear distance to indicate the path of service pipeline, but there may be some deviation in the actual situation. Therefore, in the future research, we can further explore the problem of in-ground pits based on utility tunnel technology.

Acknowledgement. Civil Aviation University of China Graduate Innovation and Entrepreneurship Project Funding. *(2022YJS075)* Based on the utility tunnel technology, a modeling and simulation study of the time-sequential guarantee service for the apron multi-pipe gallery.

Funding. National Natural Science Foundation of China Civil Aviation Joint Research Fund (Project Approval Number: U2333204): Research on key technologies for capacity improvement of airport terminal bay area based on mutual feedback between design and operation.

References

1. Li, F., Wu, J.: A preliminary study on the planning and construction of comprehensive pipeline corridors in large airports–taking Chongqing airport comprehensive pipeline corridor as an example. Comprehensive Trans. **39**(1), 86–89 (2017)
2. wang, Y.: Discussion on the comprehensive arrangement of pipelines in the south working area of Kunming Changshui International Airport. China Sci. Technol. Inform. **18**, 66–68 (2013)
3. Lu, G., Jiang, N., Fang, B., et al.; A few considerations on sewage pipes into corridors. Water Supply Drainage **53**(11), 99–101 (2017)
4. Chen, M.: Research and analysis of gas pipeline entry mode in urban comprehensive pipeline corridor–Taking Xiamen Airport North Road comprehensive pipeline corridor as an example. Fujian Construction Sci. Technol. **03**, 94–98 (2020)
5. Yuan, M., Zhang, Z.: Construction of comprehensive pipeline corridor at Jiaodong Airport takes a new step forward. China Construction Inform.(09), 52–54 (2022)
6. Jiang, G., Liu, F.: Determination of optional domains of refueling spigot wells in apron refueling pipeline network design. Oil Gas Storage Trans. **35**(10), 1106–1111 (2016)
7. Wang, Y., Deng, D., Ren, Q., et al.: Exploration on the design of underground comprehensive pipeline corridors--taking an airport in China as an example. J. Hebei Univ. Architec. Eng. **37**(2), 89–94 (2019)
8. Wu, Q., Wang, K., Fan, W., et al.: A site selection method for maximum coverage of public libraries based on continuous spatial demand--a case study of the main urban area of Wuhan. Geography Geographic Inform. Sci. **36**(1), 27–34+99 (2020)
9. Kennedy, J., Eberhart, R.: Particle swarm Optimization. In: IEEE International Conference on Neural Networks, Proceedings. IEEE (1995)

10. Civil Aviation Administration of China. Technical standards for civil airport flight areas: MH5001–2021. Beijing: Civil Aviation Administration of China (2021)
11. Civil Aviation Administration of China. Aircraft oxygenation equipment: MH6031–2018. Beijing: Civil Aviation Administration of China (2018)
12. Civil Aviation Administration of China. Maintenance for civil aircraft- Ground safety- Prat 25: Oxygen charging for civil aircraft: MH/T3011.25–2006. Beijing: Civil Aviation Administration of China (2006)
13. Civil Aviation Administration of China. Aircraft clearwater vehicle: MH6014–2018. Beijing: Civil Aviation Administration of China (2018)
14. Civil Aviation Administration of China. Aircraft sewage truck: MH6015–2014. Beijing: Civil Aviation Administration of China (2014)
15. Zewei, M.A.: Design points of airport aerodrome shaft lifting device. Airport Construction **4**, 68–70 (2022)

Study on the Combined Bearing Characteristics of Cutoff Wall-Double Row Piles Composite Foundation

Libo Wang[1], Wu Liu[2(✉)], Hao Jang[1], Jinhang Shang[2], Huayan Yao[2], and Renjie Li[1]

[1] Shandong Electric Power Engineering Consulting Institute Co. Ltd., Jinan 250013, China
{wanglibo,jianghao,lirenjie}@sdepci.com
[2] College of Civil Engineering, Hefei University of Technology, Hefei 230009, China
{liuwu168,yaohuayan}@hfut.edu.cn

Abstract. The combination system of cutoff wall and double row piles, as a new type of composite reinforcement structure, can take into account the foundation reinforcement needs of the upper structure's bearing capacity and the water-rich strata seepage interception. Based on a case of the cutoff wall and double-row piles combination reinforcement foundation at a large ship lock project in Lu'an City, Anhui Province, China, the combined bearing characteristics of the composite reinforcement structure are systematically studied in this study through three-dimensional refined modeling of the geological materials and the reinforcement structural systems. The numerical modeling results indicate that the joint-bearing characteristics of the composite foundation are significantly influenced by the elastic modulus of the material in the linking region between the cutoff wall and the upper structure, as well as the size of the upper load. As the elastic modulus of the linking region increases, the stress, displacement, and load-sharing ratio of the cutoff wall increases. The load-sharing ratio contributed by the cutoff wall decreases significantly from 22% to 3.8% when the elastic modulus of the linking region decreases from 20 GPa to 0.2 GPa. The smaller the upper load, the smaller the load-sharing ratio contributed by the cutoff wall. To protect the cutoff wall and better utilize its seepage-proofing performance, a flexible connection is suggested to be used in the linking region between the cutoff wall and the upper structure, and the upper load level should be controlled. The research has reference significance for better studying the bearing characteristics of composite foundations.

Keywords: cutoff wall-piles composite foundation · 3D numerical simulation · joint-bearing characteristic · Load sharing of cutoff wall

1 Introduction

In recent years, with the increasingly complex construction environment of large infrastructure projects, such as water conservancy ship locks and urban subways, the difficulty of foundation reinforcement during the construction process has gradually increased. The foundation reinforcement technology has achieved rapid development and breakthroughs, and the selection of reinforcement structures pays more attention to matching

A. Bieliatynskyi et al. (Eds.): CSTTE 2023, LNCE 603, pp. 271–283, 2024.
https://doi.org/10.1007/978-981-97-5814-2_25

the actual characteristics of the projects [1]. The double-row piles supporting structure formed by bored piles and high-pressure rotary jet grouting piles has the advantages of small space occupation, short construction period, and large bearing capacity, and is widely used in foundation reinforcement engineering [2]. Considering that the contact area between the bored pile and the adjacent high-pressure rotary jet grouting pile is a weak area, which is prone to cracking under complex upper loads and causing groundwater leakage, this article proposes adding a cutoff wall with reliable anti-seepage performance between the double-row piles to form a composite structure for foundation reinforcement.

Driven by large-scale foundation engineering constructions, extensive studies on the bearing characteristics of double-row piles have been conducted by many researchers through the methods of model tests, theoretical analyses, and numerical simulations. Peng et al. [3] studied the influence of row spacing and excavation depth on the bearing behaviors of four groups of double-row pile supporting systems by carrying out large-scale indoor model tests and obtained the optimal value of row spacing of the double-row pile support method. Zhou et al. [4] analyzed the pile deformation during the construction of double-row-pile support engineering in a foundation pit by conducting a large-scale model test based on the similarity principle. Cao et al. [5] had proposed a new calculation model for double-row-pile structure in foundation pit by improving the equivalent model of soil between double-row piles, and studied the stress and deformation characteristics of the double-row piles retaining structure in a practical case. Xu et al. [6] used the finite element numerical software ABAQUS to analyze the spatial effect of the bearing deformation of the special double-row pile support system, and proposed the optimization design measure for the support system near the corner of the foundation pit. Zhang et al. [7] studied the load transfer characteristics of pile-group foundations in collapsible loess stratum by conducting small pile-group model tests. Wu et al. [8] used the model test method to compare the bearing characteristic of the pile-group and diaphragm wall with similar material dosages and found that the ultimate bearing capacity of the diaphragm wall foundation was about 1.1 times of the pile-group foundation. Shi et al. [9] proposed a new numerical calculation model for double-row sheet piles considering the actual conditions of weak strata, and analyzed the horizontal displacement of double-row sheet pile cofferdams arranged in weak strata. Wang et al. [10] analyzed the bearing characteristics of double-row piles by building an indoor test model for deep foundation pit engineering, and found that soil reinforcement between piles, passive area on the side of piles, and soil reinforcement at the end of piles could effectively improve the bearing performance of double-row piles. Hassen et al. [11] proposed to transform the pile reinforcement area into equivalent homogeneous anisotropic continuous material, and adopted the Moore-Coulomb strength criterion to estimate the upper limit of the ultimate bearing capacity of the pile foundation. Shivashankar et al. [12] proposed that in a soft foundation, the properties and layer thickness of soft soil in the middle and upper strata played a key role in the strengthening effect of piles. Wang et al. [13] used finite element software PLAXIS 3D FOUNDATION to numerical study the bearing characteristics of double-row piles, considering the influences of the excavation depth, the embedded depth of piles, the distance between rows, and the diameter of piles. Wang et al. [14] studied the surface settlement law of a new prefabricated circular double-row pile support system suitable for silty clay foundations upon excavation through indoor model tests and numerical simulations.

Although the above studies could well reveal the bearing characteristics of the double-row pile structure under different circumstances, there is still a lack of research on the bearing characteristic of the new type of composite foundation formed by adding a cutoff wall between the double-row piles. Therefore, this article systematically studied the combined bearing characteristics of the cutoff wall-double row piles composite foundation, by taking a case of the composite foundation structure, formed by the double-row of bored piles and high-pressure jet grouting piles and the intermediate concrete cutoff wall, in a large ship lock project as the research background. Based on three-dimensional finite element refined modeling, the influence of the elastic modulus of materials in the linking region between the cutoff wall and the upper structure, as well as the size of the upper load on the bearing characteristics of the composite foundation are simulated and analyzed.

2 Cutoff Wall-Piles Composite Foundation

The new type of composite foundation formed by the cutoff wall and the double-row piles in this study is derived from a large double-track ship lock project under construction, located at the junction of Huoqiu County and Yingshang County in the middle reaches of the Huaihe River in Lu'an City, Anhui Province, China. The downstream longitudinal cofferdam of this project is closely adjacent to the first-line ship lock channel to the north, as a result, the construction site is narrow. During the construction process, the downstream longitudinal cofferdam also needs to be in place of the original traffic roads for the passage of numerous vehicles. The strata in the engineering area are the Quaternary Holocene artificial accumulation layer (Q_4^{ml}) and the Quaternary Upper Pleistocene alluvium layer (Q_3^{al}). The specific stratigraphic information is shown in Table 1. Although using double-row piles formed by the combination of bored piles and high-pressure jet grouting piles for cofferdam foundation reinforcement can solve site limitations and meet bearing requirements, cracks may occur at the contact area between the bored pile and the high-pressure jet grouting pile during traffic load, leading to groundwater leakage problems. To improve the seepage-proofing reliability of the foundation reinforcement structure, a reinforcement system combining the cutoff wall and the double-row piles is proposed for the downstream longitudinal cofferdam foundation of the project, as shown in Fig. 1. The upper part of the foundation is a reinforced concrete box cofferdam, and the double-row piles are rigidly connected to the upper box structure. The cutoff wall located in the middle of the double-row piles is connected with the upper box structure by setting an additional linking region for the purpose of preventing damage to the cutoff wall that affects the seepage-proofing effect under the action of upper loading. Based on the regional geological characteristics and practical engineering experience, the dimensions of the composite foundation reinforcement structure are as follows: The diameter of the bored pile in the composite foundation is 1.6 m and the depth is 40 m, the spacing between neighboring bored piles in the same side row is 2 m; The high-pressure jet grouting piles are arranged in the middle of neighboring bored piles, with a diameter of 0.8 m and a depth of 22.9 m; The cutoff wall arranged in the middle has a width of 0.4 m and a depth of 22.9 m; The distance between the rows of bored piles is 8 m. The double-row pile structure composed of bored piles and high-pressure jet grouting piles can also exert a certain seepage-proofing effect, which is conducive to improving the anti-seepage reliability of the whole reinforcement structure.

Table 1. Stratigraphic information description

Units	Average thickness (m)	Descriptions
Layer I Qml 4	6.5	Mainly composed of silty clay. The color is brown yellow or grayish brown
Layer II Qal 3–1	2.3	Mainly composed of silty clay and clay. The color is brownish yellow, with partial brownish gray
Layer III Qal 3–2	7.5	Mainly composed of silty clay mixed with silt. The color is brownish yellow, with partial brownish gray
Layer IV Qal 3–3	2.2	Mainly composed of silty clay mixed with silt The color is brown yellow and gray-green
Layer V Qal 3–4	14.6	Mainly composed of silty clay and clay. The color is mainly brown yellow and grayish yellow
Layer VI Qal 3–5	6.9	Mainly composed of silty clay and clay The color is mainly gray-green and brown-yellow
Layer VII Qal 3–6	40	Mainly composed of silty clay and clay The color is mainly brown and yellow

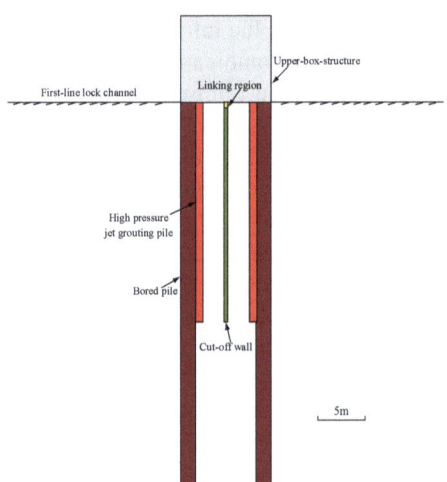

Fig. 1. Schematic diagram of the combined foundation

3 Modeling Analysis of Composite Foundation's Joint-Bearing

3.1 Numerical Calculation Model

To reveal the joint-bearing characteristics of the combined foundation formed by the cutoff wall and the double-row pile, a three-dimensional finite element numerical model is established according to the surrounding strata information and the geometric size of the combined foundation structure, as shown in Fig. 2. The Bored piles, high-pressure jet grouting piles, cutoff wall, and linking region are finely modelled in the numerical model. The boundary size of the calculation model is 90 m (length) × 4 m (width) × 80 m (height), with a total of 40032 grid elements and 45862 nodes. The FLAC3D software is adopted to simulate the bearing characteristics of composite foundations, in which the soil mass and high-pressure jet grouting piles are simulated by the Mohr-Coulomb model. The elastic constitutive model is utilized for the composite reinforcement structure due to its high stiffness. Considering the significant difference in stiffness between the composite reinforcement structure and the surrounding soils, the thickness-free contact surface elements are set up in their contact area. Note that the joint-bearing of the composite foundation is mainly studied by the simulation method in this paper, the results are affected by the reliability of the simulation model, and more efforts will be made with the model test method to further verify the simulated results in our future study.

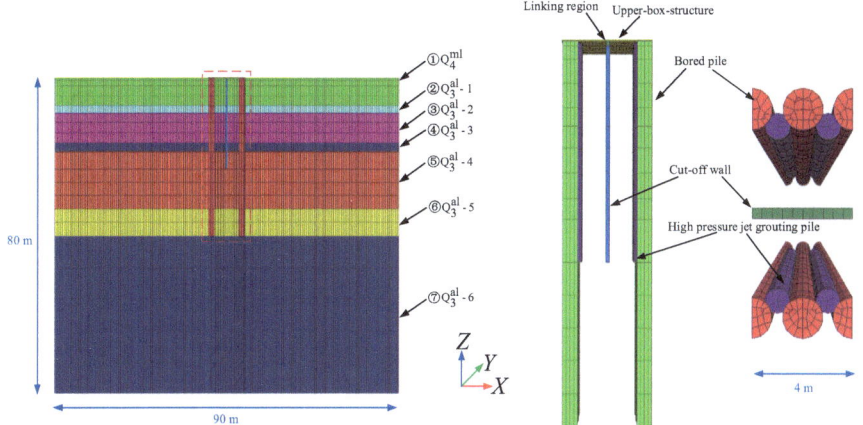

Fig. 2. Numerical calculation model for the cutoff wall-double row piles composite foundation

3.2 Calculation Parameters and Conditions

According to the engineering geological survey report, the calculation of each material in the numerical calculation model is determined as shown in Table 2. To ensure the accurate application of load, a solid element simulation is conducted on the bottom plate of the pile-top concrete box structure, and the upper load acts on the bottom plate of the box structure during the loading process. The simulation study on the bearing characteristics of the composite foundation is conducted using a graded loading method with a load stage difference of 500 kN until the ultimate load state is reached. The boundary conditions of the model are set as below: zero normal displacement constraints are set at the bottom and lateral boundaries; At the top boundary, a uniformly distributed stress boundary is applied at the box structure area based on the upper loading, and the other regions are free. The normal stiffness and tangential stiffness of the contact surface elements between the composite reinforcement structure and the surrounding soils are set as 1×10^8 Pa according to engineering experience.

Table 2. Physical and mechanical parameters of the strata and structural materials

Category	Elastic modulus E/MPa	Poisson ration v	specific weight γ/kN•m^{-3}	Cohesion c/kPa	Internal friction angle φ/°
Qml 4	48	0.30	19.6	19	23
Qal 3–1	35	0.38	19.7	23	23
Qal 3–2	40	0.38	19.5	20	22
Qal 3–3	47	0.36	19.8	12	25
Qal 3–4	44	0.35	20.1	19	23
Qal 3–5	47	0.35	20	25	21
Qal 3–6	43	0.35	20.4	29	20
Bored pile	25000	0.20	25.0	/	/
High-pressure jet grouting pile	600	0.30	23.0	120	32
Cut-off wall	2000	0.20	22.0	/	/
Upper-box-structure	30000	0.15	25.0	/	/
Linking region	2000	0.20	22.0	/	/

4   Result Analysis

4.1   Simulation Verification of Single Pile Test Results

To verify the reliability of the calculation parameters and simulation process selected in this article, simulation validation is first conducted on the single pile test results at the engineering site. The comparison between the numerical simulated vertical displacements of the pile top center under different vertical loads and the test results is shown in Fig. 3. One can find from Fig. 3 that with the increase of pile top vertical loading, the vertical displacement of the pile top center increases linearly at first and then increases sharply after exceeding about 7500 kN. The numerical simulation results are in good agreement with the test pile results, indicating the effectiveness of the numerical research method used in this study.

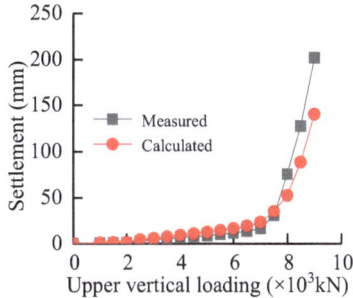

Fig. 3. Calculated and measured displacement-load curves of single pile test

4.2 Analysis of Combined Bearing Characteristics for Composite Foundation

Figure 4 shows the variation relationship between the displacement at the center of the upper structure plate and the upper vertical load for the cutoff wall-double row piles composite foundation. The loading-displacement curve of the composite foundation overall shows a linear variation and belongs to a slowly increasing type. According to the standard code JGJ 94–2008 [15], for the slowly increasing type loading-displacement curve, the settlement corresponding to 0.04 times the diameter of the bored piles can be selected to determine the vertical ultimate bearing capacity. Therefore, the vertical ultimate bearing capacity of the composite foundation corresponding to the settlement of 64mm is determined as 15000 kN.

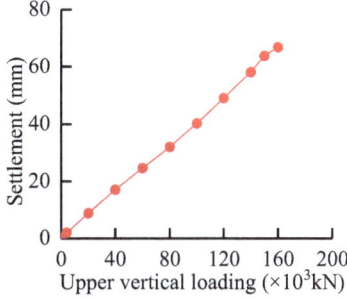

Fig. 4. Simulated displacement-load curve of composite foundation

Under the action of ultimate loading, the distributions of the vertical stress, the maximum shear stress, and the plastic zone for the composite foundation are respectively shown in Figs. 5, 6 and 7. It can be seen that the vertical stress and maximum shear stress distributed in the reinforced structure are significantly greater than those in the surrounding soil strata. The vertical compressive stress distributed in the bored piles is obviously higher than that in the cutoff wall, indicating the bored piles play a major bearing role in the composite foundation. Under ultimate vertical loading, the plastic zone in the soil stratum concentrates occurring at the bottom area of the bored piles, and the shear failure region below the two rows of piles is connected.

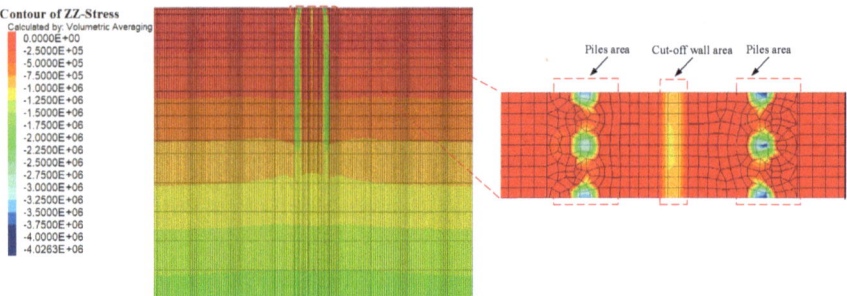

Fig. 5. Vertical stress distribution of the composite foundation under ultimate loading (Pa)

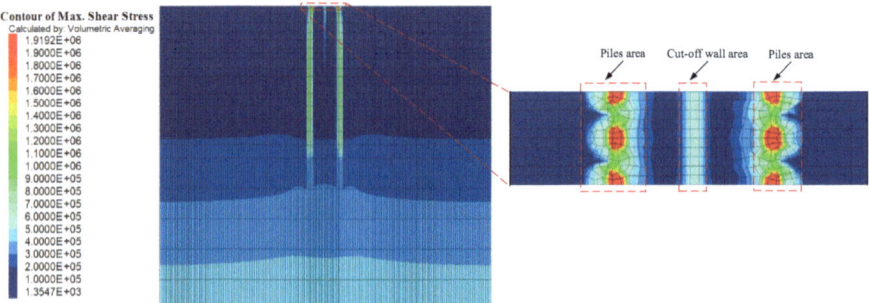

Fig. 6. Maximum shear stress distribution of the composite foundation under ultimate loading (Pa)

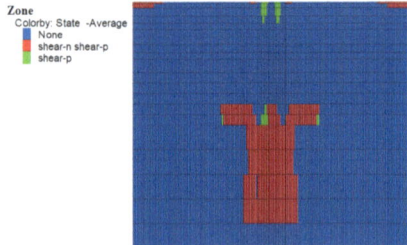

Fig. 7. Plastic zone of the composite foundation under ultimate loading

4.3 Influence of Linking Region's Elastic Modulus on the Joint-Bearing Behavior

Considering that the cutoff wall may cause damage after sharing a large vertical load, which greatly reduces the seepage-proofing effect of the composite foundation, the ways to reduce the load sharing of the cutoff wall should be studied. In this section, the impact of different connection methods between the concrete cutoff wall and the upper box structure on the bearing characteristics of the composite foundation is analyzed. It is considered that different connection methods will result in different elastic moduli of the linking region. Five elastic moduli, 0.02 GPa, 0.2 GPa, 2 GPa, 10 GPa, and 20 GPa, are selected to study to influence of the linking region's elastic modulus on the joint-bearing behavior of the composite foundation under the vertical loading of 15000 kN.

The vertical stress and displacement distributions in the cutoff wall under the linking region's elastic moduli of 0.2 GPa and 2 GPa are shown in Figs. 8 and 9. With the increase of the linking region's elastic modulus, the stress and displacement of the cut-off wall increase. As the linking region's elastic modulus increases from 0.2 GPa to 2 GPa, the maximum compressive stress distributed in the cut-off wall increases from 0.90 MPa to 1.03 MPa, the maximum vertical displacement of the cut-off wall increases from 7.16 cm to 7.26 cm. It can be foreseen that the damage risk of the cutoff wall will increase with the increase of the linking region's elastic modulus, using a low elastic modulus connection form in the linking region is beneficial for reducing the compressive stress acting on the top of the cut-off wall.

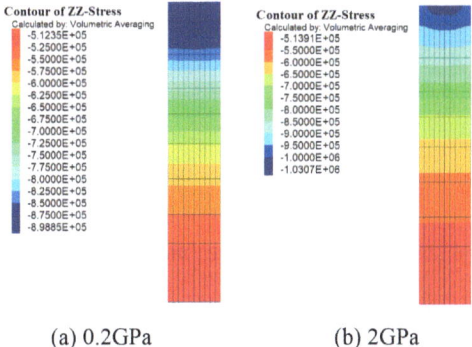

(a) 0.2GPa   (b) 2GPa

Fig. 8. Vertical stress distribution of cutoff wall under different linking region's elastic moduli (Pa)

The influence of the elastic modulus of the linking region on the load sharing of the cutoff wall is shown in Fig. 10. It can be seen from Fig. 10 that the load-sharing ratio of the cutoff wall for the whole upper loading increases obviously with the increase of the elastic modulus of the material at the connection area between the cutoff wall and the upper structure. When the elastic modulus of the linking region increases from 0.2 GPa to 20 GPa, the load-sharing ratio contributed by the cutoff wall increases significantly from 3.8% to 22%. The upper load is mainly born by the double-row piles, and its load-sharing ratio maintains at about 70% under the vertical loading of 15000 kN. The

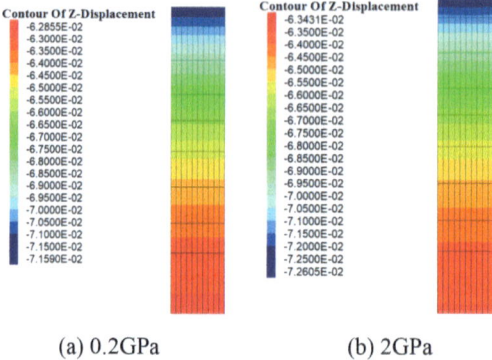

(a) 0.2GPa   (b) 2GPa

Fig. 9. Vertical displacement distribution of cutoff wall under different linking region's elastic moduli (m)

elastic modulus of the linking region between the cutoff wall and the upper structure has little influence on the load sharing of the double-row piles. It can be seen that the degree of softness and hardness of the connecting area between the cutoff wall and the upper structure mainly influences the load sharing of the cutoff wall.

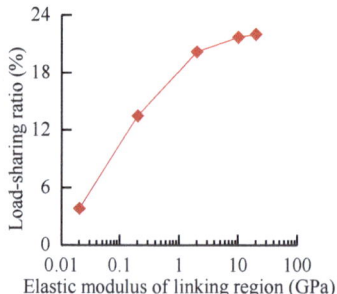

Fig. 10. Variation of the load-sharing ratio for the cutoff wall with the linking region's elastic modulus

4.4 Influence of the Upper Loading Level on the Joint-Bearing Behavior

The influence of the upper loading level on the load sharing of the composite reinforcement structure is studied in this section. Let the elastic modulus of the linking region between the cutoff wall and the upper structure be 2 GPa, the load-sharing ratio of each structure in the composite foundation under different load levels is calculated, as shown in Fig. 11. With the increase of loading level imposed on the bottom plate of the upper box structure, the load-sharing ratio of the cutoff wall and the double-row piles increases gradually. The influence of load level change on the load sharing of the cutoff wall and double-row piles in the composite foundation is obvious when the load is small. The load sharing of the cutoff wall is affected greater by the change in the upper loading level than the double-row piles. For the cutoff wall, its load-sharing ratio increases rapidly with the loading level when the upper load is less than 6000 kN, and after that, the load-sharing ratio basically remains unchanged. Therefore, controlling the level of upper loading can reduce the load sharing of the cutoff wall, which is beneficial for protecting the cutoff wall.

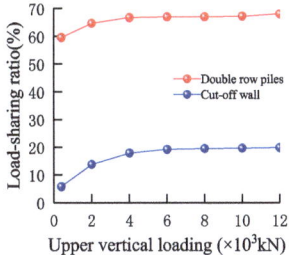

Fig. 11. Variation of load-sharing ratio for the composite reinforcement structure with the upper loading level

5 Conclusions

This article uses a three-dimensional finite element numerical simulation method to study the joint-bearing characteristics of a new type of reinforced foundation with a combination of cutoff wall and double-row piles. The main research conclusions are as follows:

(1) Under the action of the upper load, the load-bearing characteristics of the combined foundation of the cutoff wall and double-row pile are significantly affected by the combined structure system. The vertical stress in the area of the double-row piles and cutoff wall is greater than that in the surrounding soil, and the largest value is in the bored pile.

(2) The combined load-bearing characteristics of the composite reinforced foundation are greatly affected by the elastic modulus of the connection area above the cut-off wall. The smaller the linking region's elastic modulus, the smaller the vertical stress, settlement, and load-sharing ratio of the cutoff wall, when the elastic modulus decreases from 20GPa to 0.02GPa, the load-sharing ratio contributed by the cutoff wall decreases from 22% to 3.8%. Adopting a flexible connection between the cutoff wall and the upper structure can effectively reduce the damage of the cutoff wall under the action of upper loadings.

(3) The upper loading level has a significant impact on the joint-bearing characteristics of the composite foundation reinforcement system. With the increase of the upper load, the load-sharing ratio of the cutoff wall increases significantly, and the load-sharing ratio of the double-row piles increases slightly. Controlling the upper load grade is also conducive to protecting the cutoff wall and better utilizing its seepage-proofing performance.

It is worth noting that this article has not fully considered the impact of factors such as groundwater seepage, foundation pit excavation and upper horizontal load, on the combined bearing characteristics of the composite foundation, and only utilizes the simulation method the explore the law, in our future work, relevant research will be improved by combining model tests or other methods.

Acknowledgments. This work was funded by the Anhui Provincial Natural Science Foundation (2208085ME153), the University Synergy Innovation Program of Anhui Province (GXXT2022–020), and the National Natural Science Foundation of China (51709072).

References

1. He, Q.: Analysis of structural characteristics in construction of deep water double-row steel sheet pile cofferdam. Jinan University (2020)
2. Yan, B., Hu, K., Cao, M.: Analysis of the influences of reinforced soil parameters between piles on double row pile supporting structure. Chinese J. Underground Space Eng. **18**(201), 226–232 (2022)
3. Peng, W., Liu, B.: Indoor model test on row spacing of double-row piles supporting deep excavation pit. J. Hunan Univ. (Natural Sciences) **45**(1), 121–127 (2018)
4. Zhou, Y., Yao, A., Li, H., Zheng, X.: Correction of earth pressure and analysis of deformation for double-row piles in foundation excavation in Changchun of China. Adv. Mater. Sci. Eng. **2016**, 1–10 (2016). https://doi.org/10.1155/2016/9818160
5. Cao, J., Qian, G., Gao, Y., Zuo, H.: Study on equivalent calculation model of soil between piles in double-row piles supported by foundation pit. Chinese J. Underground Space Eng. **16**(3), 749–757 (2020)
6. Xu, S., Fan, Q., Cui, F.: 3D finite element simulation for special type of double-row piles supporting structure relying on ABAQUS. Chinese J. Underground Space Eng. **11**(6), 1514–1521 (2015)
7. Zhang, Y., Wang, X., Liang, Q., Jiang, D., Li, J.: Model tests on bearing behavior of pile groups in collapsible loess ground under water immersion. Chinese J. Geotech. Eng. **43**(S1), 219–223 (2021)
8. Wu, J., Cheng, Q., Wen, H., Cao, J.: Vertical bearing behaviors of lattice-shaped diaphragm walls and group piles as bridge foundations in soft soils. Chinese J. Geotech. Eng. **36**(9), 1733–1744 (2014)
9. Shi, Y., Zhang, T., Yuan, C., Huang, Z., Hou, S.: Simplified calculation method of horizontal displacement of double-row sheet pile cofferdam in soft strata. J. Build. Struct. **44**(9), 246–254 (2022)
10. Wang, X., Liao, Z., Zheng, T., Zhu, D.: Model test study on bearing characteristics of double-row piles based on reinforcement effect of soil around piles. J. Environ. Eng. **43**(2), 19–25 (2021)
11. Hassen, G., Gueguin, M., de Buhan, P.: A homogenization approach for assessing the yield strength properties of stone column reinforced soils. Europ. J. Mech. - A/Solids **37**, 266–280 (2013). https://doi.org/10.1016/j.euromechsol.2012.07.003
12. Shivashankar, R., Dheerendra Babu, M.R., Sitaram Nayak, V., Rajathkumar: Experimental Studies on behaviour of stone columns in layered soils. Geotech. Geol. Eng. **29**(5), 749–757 (2011). https://doi.org/10.1007/s10706-011-9414-0
13. Wang, Z., Zhou, J.: Three-dimensional numerical simulation and earth pressure analysis on double-row piles with consideration of spatial effects. J. Zhejiang Univ.-Sci. A **12**(10), 758–770 (2011)

14. Wang, R., et al.: Model test and numerical simulation of a new prefabricated double-row piles retaining system in silty clay ground. Underground Space **13**, 262–280 (2023)
15. JGJ 94–2008: Technical code for building pile foundation (in Chinese). China Architecture & Building Press (2008)

Stability Analysis of Surrounding Rock in Portal Section of Plateau Loose Rock Pile Tunnel Under Different Construction Methods

Zhimin Chen[1], Mingyang Yi[1(✉)], Gengwang Zhang[2], and Zheng Zhang[1]

[1] Lanzhou Jiaotong University, School of Civil Engineering, Lanzhou 730070, China
czm@mail.lzjtu.cn, 973086752@qq.com
[2] Zhongjiao Second Highway Survey and Design Institute Co., Ltd, Wuhan 430056, China

Abstract. Rock pile is a rock block formed by weathering of bedrock, and its stability is lower than that of other rock and soil. Therefore, there are numerous difficulties in the construction of tunnels through rock pile. In this paper, TBM method, three step method and CD method are chosen as the basis of tunnel stability analysis, and the numerical model of surrounding rock excavation stability of tunnel passing through rock pile is established. In this paper, the tunnel excavation is simulated by numerical simulation software, and the stability analysis of tunnel under different construction methods is performed. Combined with the engineering background and calculation analysis, compared with the three construction methods, it is found that the CD method has the strongest comprehensive protection for tunnel construction and is a more suitable construction method for crossing the rock pile tunnel.

Keywords: loose rock pile · surrounding rock stability · mechanical property · construction methods · numerical simulation

1 Introduction

Due to the high altitude and significant temperature differences in the border area of Sichuan and Tibet, the bedrock undergoes further erosion, resulting in the formation of loose rock pile [1, 2, 3]. The unstable tunnel through the rock pile is prone to instability and failure during excavation. In order to address these challenges, it is important to select a suitable construction excavation method that can effectively adapt to the geological conditions of the surrounding rock, minimize construction disturbances, and ensure the safety and progress of tunnel construction.

In order to solve the construction problems of the tunnel, scholars at home and abroad have studied the stability of the tunnel in different aspects. Deng Yongjie [4] summed up the deformation law of surrounding rock in the excavation process. Zhou Xueqing [5] studied the plastic zone of surrounding rock of loose rock pile tunnel under three construction methods, and optimized the construction method. Li Wei [6, 7] analyzed the stability of the tunnel face when the shield tunnel is tunelling along the inclined

© The Author(s) 2024
A. Bieliatynskyi et al. (Eds.): CSTTE 2023, LNCE 603, pp. 284–300, 2024.
https://doi.org/10.1007/978-981-97-5814-2_26

stratum. Anagnostou G [8] studied the face stability of slurry-shield-driven tunnels. Lin Qingtao [9, 10] constructs the stability model of shield tunnel in pebble stratum.

This study aims to analyze the stability of tunnel surrounding rock in rock mass under different construction methods, considering the high stability requirements for tunnel crossing rock mass in southwest China. Through point measurements and numerical simulations, the stability of surrounding rock is investigated.

2 Engineering Overview and Construction Method Analysis

Rock piles, especially loose ones, are susceptible to instability. The varying composition of sand and rock in rock piles can lead to changes in their mechanical properties [11, 12, 13]. Improper excavation methods may result in mountain collapse, posing difficulties in tunnel construction, maintenance, and use. Therefore, it is crucial to formulate the tunnel construction method based on specific engineering conditions. Given the higher mountain height and larger top load in the southwest region, this study focuses on selecting an appropriate tunnel construction method for further investigation.

2.1 Analysis of Engineering Overview

The background project of this paper is C.D. Tunnel, which is located in the east of Ya 'an to Linzhi section. The tunnel was built through loose rock piles.

This section mainly analyzes the general situation of rock piles at the entrance and exit of the tunnel. The rock pile at the exit section of the C.D. Tunnel is located at the foot of the slope and extends to the lower part of the slope. It is continuously accumulating along the right bank of the Frozen Cuoqu, with a zonal distribution. The rock pile is approximately 100–200 m long, 10–50 m thick, and over 1.5 km wide. The average volume per meter is about 5.3×103 m^3, making it a massive accumulation. The rock mass consists mostly of gravel, which is prone to instability when exposed to vibrations.

2.2 Analysis of Construction Method

The safety of tunnel excavation depends on the construction method used, as different methods result in different stress redistribution. Therefore, it is important to choose a suitable construction method based on the specific construction environment. Currently, there are several methods commonly used in tunnel construction, including the Tunnel Boring Machine (TBM) method, the three-step method (including the multi-step method), the CD method (Center Diaphragm method), the core soil construction method, the CRD method (tunnel cross middle wall method), and the tunnel double side wall heading excavation method. Through the analysis of these construction methods, this study focuses on the TBM method, the three-step method, and the CD method, as they are more suitable considering the construction background. A comparative research will be conducted using these three methods in this article.

The TBM method, also known as the full-face excavation method, involves excavating the entire section once and then lining the construction [14]. This method offers the advantages of a large excavation section and high efficiency. However, it is prone to causing deformation in the surrounding rock. In practice, it also has drawbacks such as a high tool wear rate [15, 16], high driving cost and easy to cause shield jamming [17, 18].

The three-step method is the specific construction method involves dividing the working face into three parts for cyclic excavation. Each time, one step is excavated and after each excavation, the initial support is implemented. The secondary lining is carried out after excavating the lower step for 16 m, with a length of 4 m in a single construction [19]. The three-step method offers advantages such as a smaller operation space and multiple processes. However, it may cause disturbances to the surrounding rock due to repeated excavation using the step method. Additionally, the multiple lining constructions may lead to stress concentration [20].

The CD method, which stands for tunnel middle wall excavation method, involves excavating one side first, supporting the middle wall, and then excavating the other side during the excavation process. This method has some disadvantages, such as difficult construction, complex process, small construction space, and the need to remove the middle partition wall. However, it offers the advantage of timely applying initial support to each part during construction [21, 22], allowing for the quick formation of a ring. This reduces disturbance to the surrounding rock and effectively controls settlement deformation. The CD method is widely used in engineering practice, especially when dealing with broken surrounding rock [23, 24, 25].

3 Surrounding Rock Stability Model of Rock Pile Tunnel

3.1 Computation Module Establishment

In this section, numerical simulation is used to simulate the construction process of different excavation techniques in the construction process of rock pile tunnel, and the variation law of displacement, stress and strain of lining and surrounding rock in tunnel excavation process is analyzed and compared. The tunnel excavation section is 11.6 m high (z direction) and 13.6 m wide (x direction). Because in the process of tunnel excavation, the release of load will change the stress field and strain field of surrounding rock. The stress field and strain field are about less than 5% outside 3 times the diameter of the hole, and less than 1% outside 5 times the diameter of the hole. Therefore, the excavation diameter of the soil on both sides is 4 times the excavation diameter. The bottom of the model is taken to 35 m below the tunnel excavation contour line. The longitudinal (y direction) length of the model is 100 m, of which the tunnel excavation length is about 80 m. The model is shown in Fig. 1.

The surrounding rock is assigned parameters according to the Mohr-Coulomb constitutive model. The required materials are preferentially obtained from the engineering geological survey data. For the lack of material properties, the research literature and related specifications are selected [26, 27].

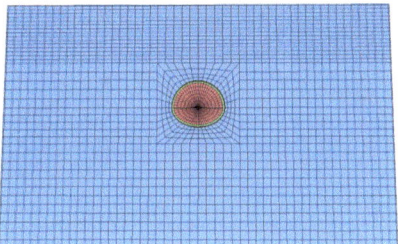

Fig. 1. Computation module

3.2 Measuring Point Arrangement

The data observed by the measuring point has the characteristics of high efficiency and reliability, so the layout of the measuring point needs to clarify the purpose of the layout [28]. The stability of the tunnel at the entrance section mainly includes the stability of the tunnel body, the stability of the slope and the stability of the lining structure. In order to solve these problems, the vault settlement, surrounding displacement, bottom uplift, surface subsidence and internal force of the supporting structure are mainly monitored in the numerical calculation process. The surface subsidence provides data support for slope safety analysis, and the monitoring data in the cave provides the basis for structural safety analysis. With the advance of the construction section, the buried depth of the tunnel is also increasing. Therefore, in order to facilitate the analysis, six sections of the tunnel buried depth of 5 m, 10 m, 15 m, 20 m, 25 m and 30 m were selected for monitoring.

4 Numerical Simulation Analysis of Tunnel Under Three Kinds of Tunnel Construction Methods

To analyze the stability of the tunnel, we should start from the displacement and stress of the tunnel, the change of displacement and stress of tunnel surrounding rock is closely related to the construction method. After analyzing the change trend of displacement and stress of tunnel surrounding rock, this section further links it with the construction method, and gradually analyzes the relationship between the change trend of displacement and stress of tunnel surrounding rock and the construction method.

4.1 Tunnel Settlement Analysis

Analysis of Ground Settlement

The numerical simulation of the comparison of the construction methods requires the settlement of the tunnel as the basic data. The data of the detection results of the surface settlement in the numerical simulation calculation are extracted and plotted as follows.

According to Fig. 2, the maximum surface settlement among the three construction methods occurs directly above the tunnel vault. Therefore, the analysis focuses on the surface settlement directly above the vault. The figure shows that the TBM method results in the largest surface settlement, while the three-step method has the smallest settlement, and the CD method falls in between. This observation differs from our previous

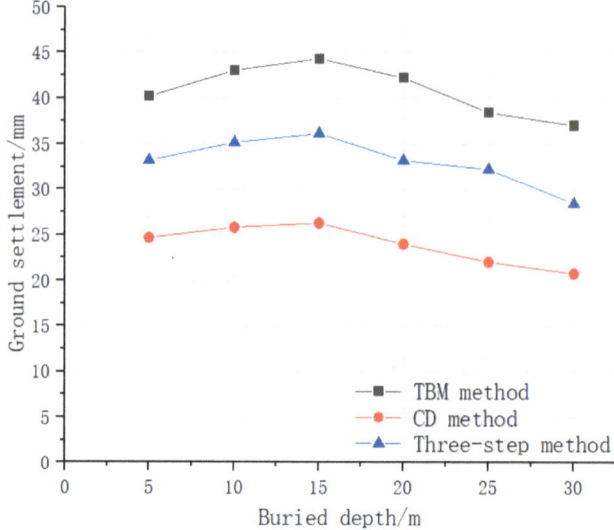

Fig. 2. Curve of ground settlement with buried depth

engineering experience, possibly due to the poor self-stability of the surrounding rock at the entrance of the tunnel. The direct excavation by TBM creates a large free surface, which hampers the stability of the surrounding rock after excavation. Consequently, significant displacement occurs rapidly, leading to large settlement. Although the excavation section of the CD method is smaller, the numerous construction procedures and the long closed-loop time of the lining structure result in long-term deformation of the surrounding rock, causing significant displacement as well. On the other hand, the three-step method, with a smaller excavation section than TBM and an earlier lining forming time compared to the CD method, exhibits the smallest displacement.

The maximum surface settlement occurs at a buried depth of 15 m, possibly due to the transition of the tunnel into the deep buried stage beyond this depth.

Analysis of Vault Settlement

During the study, it was observed that the maximum vertical displacement of the tunnel excavated using these three methods occurred at the vault. This highlights the significance of the vault in the actual project. The numerical simulation yielded the following final settlement values for the vault after the excavation using the three construction methods:

The settlement of the cross-section vault at different buried depths was monitored. The results are shown in Fig. 3 above. The smallest settlement was observed in the vault excavated by the CD method, indicating the best control of the surrounding rock. On the other hand, the three-step vault exhibited the largest settlement. As the excavation progressed, the vault settlement initially increased rapidly and then gradually stabilized. The curve reached an inflection point at approximately 15 m, suggesting a change in settlement behavior between 15–20 m. The transition from shallow to deep burial resulted in the formation of a stress arch in the tunnel, leading to a less pronounced change in settlement in the second half of the curve. Overall, the excavation methods of TBM

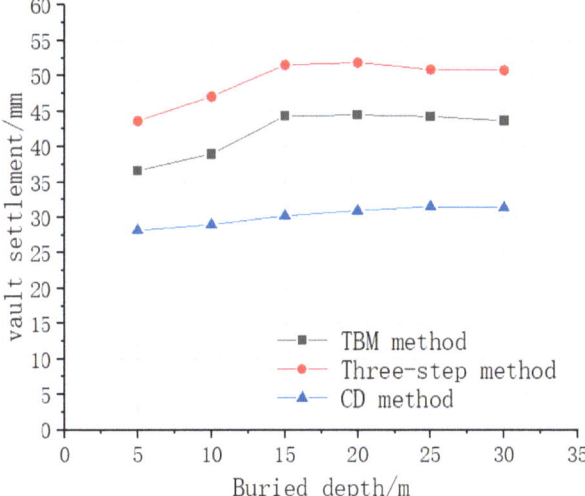

Fig. 3. Vault settlement curve

and three-step showed clear trends and inflection points. The poor geological conditions of the surrounding rock mass in shallow buried section, when subjected to the larger section of TBM excavation and the three-step method, caused significant disturbance to the original state of the surrounding rock, resulting in a sharp displacement change. In contrast, the CD method exhibited a slower rate of change, emphasizing the importance of initial support ringing. Therefore, it is crucial to pay attention to support ringing in each section during actual construction to reduce deformation rates and enhance the stability of the surrounding rock.

Through the above analysis, for the control effect of vault settlement in the construction process, the three-step method is not ideal, and the CD method has the best effect, followed by the TBM method. Therefore, as long as the excavation surface can be closed in time and quickly, the deformation control effect can be better.

Analysis of Bottom uplift of Tunnel
Under the three construction methods, the bottom uplift of the tunnel at different buried depths is analyzed, and Fig. 4 is drawn.

The figure above illustrates that the vertical deformation of the arch bottom varies across the three construction schemes as the buried depth increases. The TBM method exhibits the largest change trend. The bottom uplift curve initially increases rapidly and then gradually stabilizes with the increase in buried depth. In general, the excavation of the TBM method causes the largest uplift value of the arch bottom, reaching a maximum of 1.81 mm. Comparatively, the three-step method and the CD method show relatively smooth change trends, with a value that increases initially and then gradually stabilizes. The vault uplift generated by the three-step method is greater than that of the CD method, with maximum values of 0.46 mm and 0.19 mm, respectively. The maximum difference between them is only about 0.25 mm, which is relatively small overall.

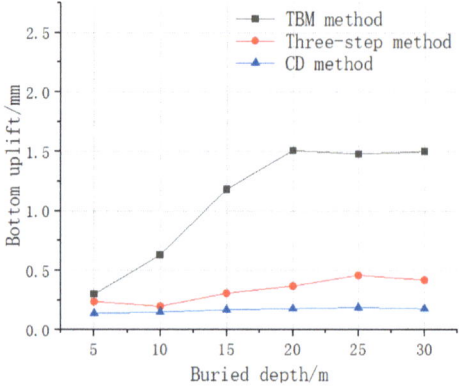

Fig. 4. Bottom uplift curve of tunnel

Horizontal Displacement Analysis of Surrounding Rock Around Tunnel

The horizontal displacement of surrounding rock in each part of the three construction methods was monitored and recorded. The final horizontal displacement deformation of each measuring point is shown in Fig. 5:

The Figure above shows that the three excavation methods result in varying degrees of horizontal displacement. The largest displacement occurs at the arch waist, attended by the side wall, while the vault experiences the least horizontal displacement, almost negligible.

When the buried depth is shallow, the horizontal displacement caused by the three construction methods is analogous. However, as the buried depth increases, they start to vary. The three-step method exhibits the largest horizontal displacement at the arch waist, followed by the TBM method, while the displacement caused by the CD method is the smallest. Notably, all three construction methods demonstrate a distinct change tendency around 20 m. Prior to this depth, the curve change slope increases with the burial depth, but after 20 m, the curve change stabilizes. The horizontal displacement of the side wall is described in the above diagram. Parallel to the trend at the arch waist, the three construction methods exhibit a similar change pattern. The curve slope decreases as the buried depth increases and generally stabilizes after 20 m. The largest horizontal displacement is observed with the three-step method, followed by the CD method, while the TBM method falls in between.

In order to explore the change of horizontal displacement during tunnel excavation at the same buried depth, the tunnel section when the buried depth is 20 m is selected, and the arch waist part is taken as an example for analysis, as shown in Fig. 6 below.

For the TBM method, the horizontal displacement generated by the excavation of the upper steps is large, which is due to the stress release caused by the one-time large-area excavation. Therefore, it is extremely important to control the horizontal displacement in time after the excavation of the section. If the support is not closed in time during the construction process, it will not only increase the horizontal displacement of the surrounding rock of the tunnel, but also increase the uplift of the bottom of the tunnel. For the CD method, the horizontal displacement of the arch waist caused by each step of

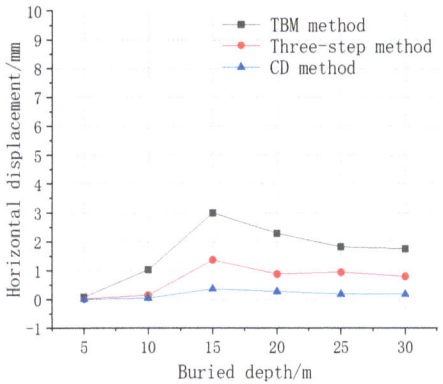

(a) Horizontal displacement of vault

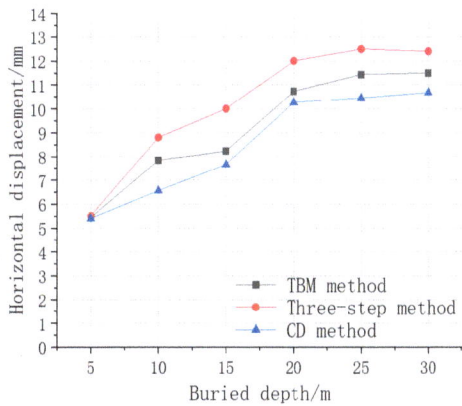

(b) Horizontal displacement of arch waist

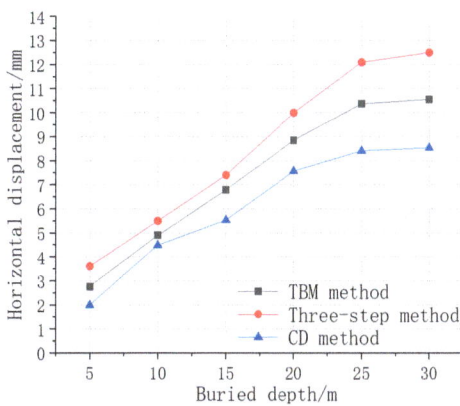

(c) Horizontal displacement of side wall

Fig. 5. Horizontal displacement of each part of the tunnel

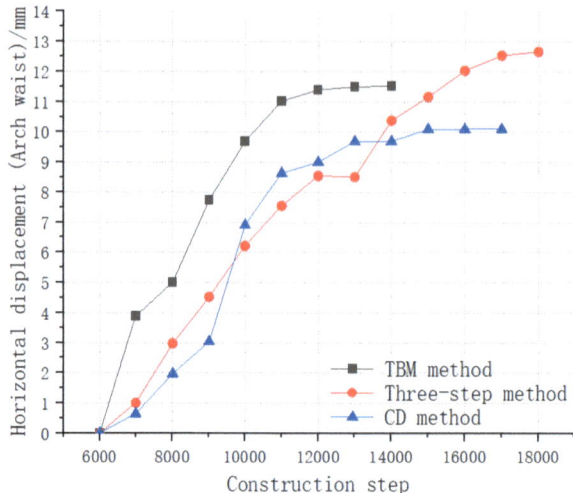

Fig. 6. Horizontal displacement of arch waist with the construction step change curve

the excavation of the left pilot tunnel is small. This is because the excavation area of the left pilot tunnel is small and the temporary middle wall will be set up in time after the excavation is completed, which effectively controls the deformation of the surrounding rock of the arch waist. However, it should be pointed out that when the excavation of the upper bench of the right pilot tunnel is carried out, the horizontal displacement tends to increase sharply. This may be because the excavation of the right pilot tunnel increases the area and changes the momentary state of the surrounding rock and the middle wall, which causes the release of the surrounding rock stress and increases the horizontal displacement. For the three-step method, the horizontal displacement change of the arch waist in each step is basically stable, and there is no dramatic displacement change, which has a good control effect on the deformation of the surrounding rock. This is explained by the fact that the three-step method divides the section into three parts, and excavates separately. The three steps are staggered between each other, and each construction step is excavated and supported separately and the excavation soil is less. Therefore, the displacement caused by a single construction step is small, so the three steps have a beneficial effect on the control of the deformation of the surrounding rock of the arch waist. However, because there are many processes and the initial support is closed into a ring time, the overall cumulative displacement is the largest of the three construction methods.

In summary, it can be seen that in the construction of the tunnel through the rock pile, the displacement change of each construction step during the excavation of the three-step method is relatively stable, but the cumulative horizontal displacement is the largest after the construction is completed. The cumulative horizontal displacement caused by CD method is the smallest, but the sudden increase of horizontal displacement should be given attention to during excavation. When TBM method is used, the initial horizontal displacement of each construction step is the largest, but the ultimate cumulative displacement is between the other two methods. During construction, attention should be paid to timely support after section excavation to avoid excessive horizontal displacement deformation.

4.2 Stress Analysis of Surrounding Rock of Tunnel

Horizontal and Vertical Stress Analysis of Tunnel

The displacement and deformation in the process of tunnel excavation are analyzed, but the stress state of surrounding rock in the process of excavation also has a great influence on the stability of surrounding rock. Therefore, the horizontal stress and vertical stress near the tunnel during the excavation of the three construction methods are monitored. The three parts of the vault, arch waist and side wall are focused on, and the horizontal stress and vertical stress are obtained respectively. The results are as follows Tables 1 and 2 show:

Table 1. Horizontal stress

Construction method Buried depth	TBM method			Three-step method			CD method		
	Vault	Hance	Sidewall	Vault	Hance	Sidewall	Vault	Hance	Sidewall
5	0.1	0.12	0.16	0.05	0.08	0.07	0.01	0.04	0.01
10	0.2	0.2	0.27	0.16	0.15	0.1	0.1	0.1	0.05
15	0.36	0.29	0.36	0.23	0.19	0.15	0.15	0.15	0.11
20	0.42	0.32	0.42	0.245	0.2	0.21	0.16	0.15	0.12
25	0.48	0.31	0.44	0.27	0.21	0.22	0.17	0.17	0.13
30	0.47	0.31	0.44	0.28	0.2	0.21	0.18	0.17	0.13

Table 2. Vertical stress

Construction method Buried depth	TBM method			Three-step method			CD method		
	Vault	Hance	Sidewall	Vault	Hance	Sidewall	Vault	Hance	Sidewall
5	0.07	0.15	0.16	0.04	0.08	0.11	0.01	0.03	0.04
10	0.21	0.28	0.21	0.06	0.13	0.18	0.02	0.04	0.09
15	0.23	0.41	0.33	0.1	0.14	0.27	0.07	0.05	0.19
20	0.26	0.64	0.53	0.12	0.12	0.29	0.071	0.04	0.2
25	0.26	0.68	0.59	0.14	0.15	0.28	0.078	0.05	0.21
30	0.28	0.7	0.58	0.14	0.15	0.31	0.075	0.03	0.2

The simulation results of the vertical stress and horizontal stress of the surrounding rock of the three construction methods are indicated above. The vertical surrounding rock stress and horizontal surrounding rock stress of the vault, hance and side wall of the CD method are the smallest. The table show that: The maximum horizontal stress of the vault of the CD method is 62% lower than that of the TBM method, and the side wall is reduced by 75%, and the stress of the surrounding rock at the excavation section of the tunnel is under pressure. Therefore, in the process of excavation, attention should be given to the phenomenon of horizontal stress concentration at the vault, and the maximum vertical stress appears at the side wall, which should also be paid attention to.

In general, the horizontal stress and vertical stress of TBM construction are greater than those of the other two schemes, which may be caused by the large excavation section.

Tunnel Surrounding Rock Shear Stress

The buried depth of the tunnel has a certain influence on the shear stress state of the tunnel section. Taking the TBM method as an example, the variation law of the shear stress of the surrounding rock under different buried depths is analyzed. Through the observation of the results, the shear stress in the XZ plane can more effectively reflect the stress concentration and stress. The final results are shown in Fig. 7.

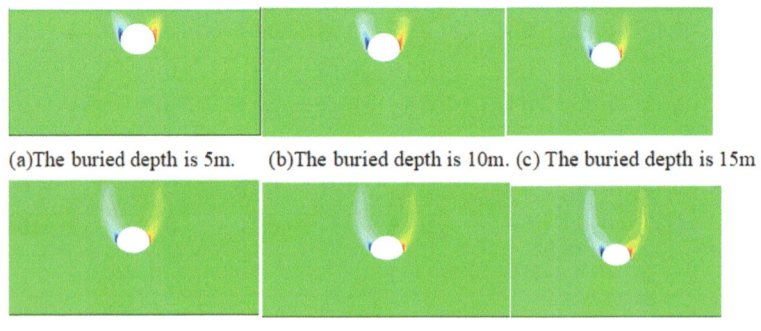

(a)The buried depth is 5m.   (b)The buried depth is 10m. (c) The buried depth is 15m

(e)The buried depth is 25m.   (f)The buried depth is 30m (d) The buried depth is 20m

Fig. 7. Shear stress diagram under different buried depth

From the above shear stress diagram, it can be observed that the shear stress of the tunnel under different buried depths is mainly concentrated near the entrance of the tunnel, mainly concentrated in the arch waist and side wall of the excavation section. Comparing the shear stress concentration area, it is found that the maximum shear stress may appear at the arch waist. For the maximum shear stress under different buried depths, it is found that with the increase of buried depth, the maximum shear stress increases first, then increases first and then decreases. The maximum value of appears at a depth of 15 m, and the maximum value is approximately 90 kPa. It shows that when the buried depth is more important than a certain range, the shear stress will decrease. Therefore, attention should be paid to the concentration of shear stress near the haunch and side wall during excavation, especially near the buried depth of 15 m.

Stress Analysis of Excavation Section.

When evaluating the stability of surrounding rock, the stress state of the rock is a crucial factor. The stress cloud diagram below illustrates the maximum and minimum principal stresses under various excavation methods. Based on the previous analysis, it was observed that the maximum displacement of the surrounding rock occurred within the range of 15–20 m. Therefore, a tunnel section with a burial depth of 15 m was chosen for further analysis. The obtained results are presented in Fig. 8 below.

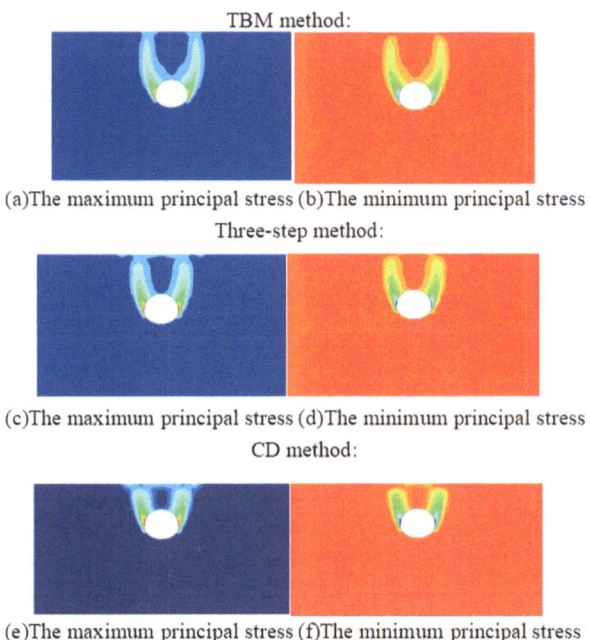

TBM method:

(a)The maximum principal stress (b)The minimum principal stress

Three-step method:

(c)The maximum principal stress (d)The minimum principal stress

CD method:

(e)The maximum principal stress (f)The minimum principal stress

Fig. 8. The maximum principal stress and minimum principal stress under different construction methods

This paper presents an analysis of the stress characteristics of the surrounding rock around the tunnel arch ring. The figure above illustrates that the maximum principal stress is predominantly compressive stress. During tunnel excavation, the unloading effect causes changes in the stress field of the surrounding rock, leading to a decrease in compressive stress around the tunnel. Consequently, it is crucial to pay attention to stress concentration in the range of the side wall during each step of excavation when employing the CD method or three-step excavation. This is particularly important in the tunnel passing through the southwest alpine environment, as it may create weak areas and require special attention to ensure construction safety.

According to the upper minimum principal stress diagram, the stress in the surrounding rock is predominantly compressive, with no tensile stress. Although the three-bench and CD methods have better control over surface settlement compared to the TBM method, each section excavation still leads to stress concentration. When the lining structure is not fully closed into a ring, force transmission becomes uneven, potentially causing damage to local parts of the structure. Therefore, it is important to address this issue during construction and consider reinforcing the corresponding areas if necessary.

In order to analyze the variation law of the maximum principal stress with the buried depth of the tunnel, the TBM method is selected to study the maximum principal stress when the buried depth is 5, 10, 15, 20, 25 and 30 m, as shown in Fig. 9.

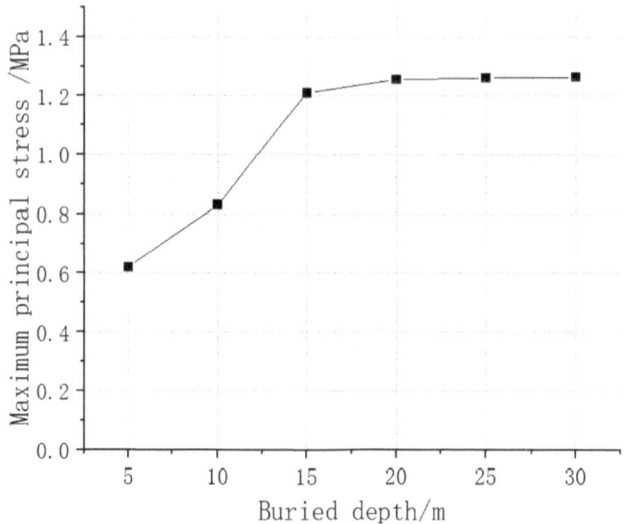

Fig. 9. Curve of maximum principal stress changing with buried depth

It can be seen from the above figure that the maximum principal stress of the surrounding rock is affected by the gravity field. During the excavation construction, with the continuous circulation of the excavation, the buried depth increases, and the corresponding maximum principal stress also gradually increases. With the increase of buried depth, the maximum principal stress curve increases rapidly and then tends to be gentle, and the maximum value is 1.067 MPa. The inflection point of the curve transformation slope is near the buried depth of 15 m. When the buried depth is less than 15 m, the maximum principal stress increases approximately linearly, and the rate is faster. When the buried depth is greater than 15 m, the curve area is smooth, and the variation of the maximum principal stress changes little with the increase of the buried depth. This may be that after 15 m, the tunnel enters the deep buried state from the shallow buried state, forming the stress arch effect, so the variation is very small.

4.3 Comprehensive Comparative Analysis of Tunnel Construction Methods Combined with Engineering Background

The safety of the tunnel vault is of utmost importance in the southwest region due to its high mountains and susceptibility to landslides. Additionally, the accumulation of rock piles can exert a significant load on the slope's base, necessitating control over the stability of other sections. Among the three construction methods, the CD method demonstrates the smallest settlement, maximum displacement, and surrounding rock stress, indicating its superior protection for the vault. In terms of secondary factors, the CD method exhibits the smallest displacement of the haunch side wall, as well as the lowest vertical and horizontal surrounding rock stress during construction. Conversely, the TBM method suffers from drawbacks such as a large excavation section, resulting in substantial surface settlement, tunnel bottom uplift, and increased horizontal and vertical stress. The three-step method causes more soil disturbance, leading to significant vault settlement that fails to meet the project's stringent stability requirements.

Overall, the CD method should be prioritized when constructing tunnels traversing rock piles in the southwest region.

5   Conclusion

In this paper, the numerical simulation analysis of tunnel crossing rock mass is performed by numerical simulation. The TBM method, three-bench method and CD method are selected and set up, and the displacement and stress of surrounding rock are monitored. The stress variation law and displacement variation law of surrounding rock during tunnel excavation are analyzed, and the reasons for the different changes of stress and displacement of tunnel surrounding rock under different construction methods are analyzed, which provides a reference for the construction technology of tunnel crossing rock mass. The main contents are as follows:

(1) During the construction of TBM method, the surface settlement is the most important thing, the surface settlement of the three-bench method is the smallest, and the CD method is in the middle. The uplift value of arch bottom caused by TBM excavation is the most important thing. Relatively speaking, the change trend of three-bench method and CD method is quite smooth, but there is still a value that increases first and then gradually stabilizes. The vault uplift produced by the three-step method is larger than that of the CD method, but it is generally smaller; the settlement of the vault excavated by the CD method is the smallest, the control of the surrounding rock is the best, and the settlement of the three-bench vault is the largest, while the TBM method is between the two.

(2) The horizontal displacement of the arch waist obtained by using these three methods is the largest, followed by the side wall, and the horizontal displacement deformation of the vault is the smallest, and far less than the other two parts, which can be approximately regarded as no displacement. The maximum displacement of the vault of these three methods occurs at a buried depth of 15 m, and the maximum displacement of the haunch and the side wall occurs at a buried depth of 30 m. Among the three methods, the maximum displacement of the vault is the largest when the TBM method is constructed, followed by the three-bench method, and the CD method is the smallest. The displacement of the haunch and side wall is the most important thing when the three-bench method is constructed, while the displacement of the haunch side wall is the smallest when the CD method is constructed.

(3) The vertical surrounding rock stress and horizontal surrounding rock stress of vault, haunch and side wall of CD method are the smallest. The change trend is curvilinear growth, and the slope changes from large to small. The maximum horizontal stress

of the three methods occurs at the vault, and the maximum vertical stress occurs at the side wall. With the change of buried depth, the distribution law and shape of shear stress are relatively similar, and the maximum value of shear stress appears at the arch waist. For the maximum shear stress under different buried depths, it is found that with the increase of buried depth, the maximum shear stress increases first, then increases first and then decreases. During TBM excavation, the stress concentration occurs near the side wall. When the CD method is excavated, stress concentration occurs in each part of the excavation, and the range of stress concentration is larger than that of the TBM method, and the stress concentration range has a tendency to diffuse from the side wall to the arch waist.

(4) Coupled with the numerical simulation data and engineering background, the CD method has the strongest protection for the vault, and has better safety guarantee for other parts. It is the best thing construction method for the tunnel crossing the rock pile region.

Acknowledgments. This research was funded by National Natural Science Foundation of China (12262018); Special Funds for Guiding Local Scientific and Technological Development by The Central Government (22ZY1QA005).

References

1. Xiangyang, J.: Characteristics of rock pile and its influence on tunnel engineering. Sichuan Build. Materials **41**, 171–172 (2015). (in Chinese)
2. Shi, X., Dai, Y., Guo, J.: Calculation and analysis of excavation footage in shallow buried section of rock pile tunnel portal : Taking Zhaojiawu Tunnel of Mazhao Expressway in Yunnan as an Example. Tunnel Construct. 2015, **35**(8), 787 (2015) (in Chinese)
3. Liu, C., Yan, D., Xie, W.: Cause analysis and discussion of rock pile. Highway **4**, 81–83 (2019). (in Chinese)
4. Deng, Y.: Research on tunneling technology of shallow buried partial pressure long span tunnel entrance section. Southwest Jiaotong University, Chengdu (2013)
5. Zhou, X., Zhangxi, Zhang, J., et al. Partial pressure tunnel construction method of thick layer loose accumulation body optimization study. Sichuan Build. Mater. **40**(12), 4925–4934 (2019)
6. Li, W., Zhang, C., Shiqin, Tu., Chen, W., Ma, M.: Face stability analysis of a shield tunnel excavated along inclined strata. Underground Space **13**, 183–204 (2023). https://doi.org/10.1016/j.undsp.2023.03.007
7. Zhang, C., Li, W., Zhu, W., et al.: Face stability analysis of a shallow horseshoe-shaped shield tunnel in clay with a linearly increasing shear strength with depth. Tunnell. Underground Space Technol. **97**(C) (2020)
8. Anagnostou, G., Kovári, K.: The face stability of slurry-shield-driven tunnels. Tunnell. Underground Space Technol. **9**(2), 165–174 (1994). https://doi.org/10.1016/0886-7798(94)900 28-0
9. Lin, Q., Dechun, L., Chunming Lei, Y., Tian, Q.G., Xiuli, D.: Model test study on the stability of cobble strata during shield under-crossing. Tunnell. Underground Space Technol. **110**, 103807 (2021). https://doi.org/10.1016/j.tust.2020.103807

10. Lin, Q., Dechun, Lu., Chunming Lei, Yu., Tian, F.K., Xiuli, Du.: Mechanical response of existing tunnels for shield under-crossing in cobble strata based on the model test. Tunnell. Underground Space Technol. **125**, 104505 (2022)
11. Rapp, A., Fairbridge, R.W.: Talus fan or cone; scree and cliff debris, pp. 1106–1109. Springer, Berlin Heidelberg (1968)
12. Vallejo, L., Mawby, R.: Porosity in lluenee on the shear strength of granular material–clay mixtures. Eng. Geol. **58**(2), 125–136 (2000)
13. Innacchione, A.T., Vallejo, L.E.: Shear strength evaluation of clay-rock mixtures. Slope Stability **2000**, 209–223 (2000)
14. Wang, R., Zhang, L.: A theoretical method and model for TBM tunnelling trajectory adjustment. Appl. Sci. **13**(19), 10876 (2023). https://doi.org/10.3390/app131910876
15. Sun, R., Mo, J., Zhang, M., Yemao, Su., Zhou, Z.: Cutting performance and contact behavior of partial-wear TBM disc cutters: A laboratory scale investigation. Eng. Failure Anal. **137**, 106253 (2022). https://doi.org/10.1016/j.engfailanal.2022.106253
16. Xu, H., Gong, Q., Zhou, X., Yang, F., Han, B.: Influence of the assisted kerf depth on cracks pattern and cutting performance of TBM cutter. Int. J. Rock Mech. Mining Sci. **170**, 105516 (2023). https://doi.org/10.1016/j.ijrmms.2023.105516
17. Hai, Z.: Study on the correlation between TBM tunnel construction efficiency and rock mechanics parameters. Eng. Construct. **55**(1), 25–30 (2023). (in Chinese)
18. Wang, L.: Stability analysis of TBM tunnel under unfavorable geology of large dip angle fault. Coal J. (2023) (in Chinese)
19. Changming, L.: Comparative study on construction methods of shallow buried large section highway tunnel. J. Guangdong Tech. Coll. Water Resources Electric Power **20**(3), 4–8 (2022). (in Chinese)
20. Cai, L.L., Yang, X.Y., Guo, N.: Numerical simulation study on three-step method construction for neighborhood tunnel. Appl. Mech. Materials **580–583**, 1327–1330 (2014). https://doi.org/10.4028/www.scientific.net/AMM.580-583.1327
21. Rui, Z.: Analysis for bearing performance and construction mechanical behavior of supporting structure for the large cross-section tunnel by half bench CD method. PLoS ONE **16**(8), e0255511–e0255511 (2021)
22. Honglong, M.: Study on the influence of construction method of shallow buried bias small spacing tunnel. Scientific and technological innovation **02**, 149–154 (2023). (in Chinese)
23. Lin, Q., Peng, L., Zhang, Z., Zhang, X., Zhao, C.: Research on the construction scheme of CD method to step method for shallow buried large section tunnel. In: Proceedings of the 2022 National Civil Engineering Construction Technology Exchange Conference (Volume I), Construction Technology Editorial Department: 153–156 (2022) (in Chinese)
24. Qian, H.: Study on the safe dismantlement length of CD method excavation in shallow buried large section tunnel. Chongqing Jiaotong University (in Chinese)
25. Jie, Z.: Numerical simulation analysis of CD method construction of shallow buried tunnel in soft and rich water stratum. Fujian Architect. **07**, 135–139 (2018). (in Chinese)
26. National Standard Writing Group of the People 's Republic of China. GB50218—94 Standard for classification of engineering rock masses.Beijing : China Planning Press,1995 (in Chinese)
27. Engineering geological handbook writing committee: Manual of Engineering Geology, (4th Edition) China Building Industry Press, Beijing (2007). (in Chinese)
28. Wang, G., Xiao, X.: Discussion on highway tunnel monitoring project selection and measuring point arrangement. Eng. Invest. **10**, 59–62 (2008). (in Chinese)

Research on Intelligent Detection Technology for Highway and Bridge Engineering

Xiangyu Meng[✉]

Architecture and Civil Engineering, Faculty of Infrastructure Engineering,
Dalian University of Technology, Dalian 116024, China
mengxiangyu941226@163.com

Abstract. Bridges are a crucial component of infrastructure, and the detection of their structure and health status has always been a focus of attention in the engineering field. However, there are a series of challenges in the application of traditional bridge detection methods in complex environments, especially in fine detection of special areas, which usually only rely on manual visual inspection, which limits the ability to comprehensively understand the health status of bridges. In response to this issue, research and summary on intelligent detection technology and equipment for highway bridges indicate that timeliness and accuracy are key factors in the process of bridge health diagnosis, evaluation, and maintenance. In modern engineering practice, intelligent detection technology and equipment have been widely applied, and rapid development has made machine vision based detection technology the main means. This trend has brought significant improvements to bridge inspection, and the introduction of various non-destructive testing technologies has provided diversified data support for the inspection process, making the inspection results more comprehensive and reliable. High performance intelligent detection equipment has emerged in this context, with its ability to carry various advanced sensors gradually covering various bridge detection needs.

Keywords: Bridge engineering · Intelligent detection · Highway bridges

1 Introduction

With the vigorous development of the global economy, countries have invested in promoting infrastructure construction, with the construction of highway bridges becoming a significant trend. This development is not only a reflection of economic strength, but also an inevitable requirement for the modernization of the transportation system. While increasing support for transportation infrastructure, the country has also accelerated the pace of bridge construction to improve the density and efficiency of the transportation network [1].

However, with the rapid increase in the number of motor vehicles, road traffic is facing a series of serious problems. Overloading, overloading, and large traffic flow have become the main bottlenecks that constrain the efficiency of highway transportation [2]. This not only puts an excessive burden on roads, but also gradually exposes the limitations

© The Author(s) 2024
A. Bieliatynskyi et al. (Eds.): CSTTE 2023, LNCE 603, pp. 301–311, 2024.
https://doi.org/10.1007/978-981-97-5814-2_27

of design, construction, and maintenance levels for bridges that have been used for a long time. When facing these pressures, bridges may cause various diseases, thereby threatening the smooth operation of traffic.

In some countries, the impact of natural disasters further exacerbates the vulnerability of highway bridges. Natural disasters such as earthquakes, mudslides, and strong winds may cause catastrophic damage to bridges, placing a heavy burden on transportation infrastructure. Therefore, the maintenance and management of highway bridges has become crucial, and countries are gradually shifting the focus of maintenance to highway transportation infrastructure, focusing on strengthening specialization and informatization, and improving the technical system of data collection, detection and diagnosis, and maintenance and treatment.

Bridge damage not only brings direct social impacts, but also comes with huge economic cost losses [3]. The collapse events of the I-10 highway bridge in the United States in July 2015 and the Morandi bridge in Italy in August 2018 were typical examples, leading to a sharp decline in traffic efficiency, irreversible loss of life, and significant economic losses. Poor bridge maintenance may also lead to corrosion and other issues, and the corrosion problem of the Morandi Bridge ultimately led to the collapse of the bridge.

Bridge collapse not only directly limits the availability of infrastructure, but may also cause strong traffic interference and negative impacts on the surrounding road network, including indirect costs such as loss of life, user delays, and alternative route planning. Therefore, the detection and maintenance of highway bridges are particularly important.

Although natural disasters are difficult to avoid, by improving inspection standards, developing non-destructive testing technology, and applying intelligent testing equipment, bridges can be evaluated, tested, and maintained in a timely manner, effectively reducing the direct and potential indirect costs of repairing bridges [4]. In this context, this study focuses on the current status and problems of intelligent detection technology and equipment for highway bridges in the form of literature review, aiming to propose a series of ideas for optimizing the process of highway bridge detection.

Overall, the construction and management of highway bridges have become a global focus of attention. In the era of high connectivity, ensuring the safety and stable operation of highway bridges is not only related to national economic development, but also to the convenience of people's lives. Therefore, we need to delve deeper into and understand the construction and maintenance management of highway bridges in order to cope with more complex and ever-changing challenges in the future.

2   Research Status

In order to promote the intelligent development of bridge detection technology and combine the spatial layout characteristics of bridge structural components, the research and development of intelligent detection equipment for bridges has become a hot topic in this research field. Among them, in order to achieve the detection of bridge surface diseases, mechanical equipment equipped with image acquisition equipment has made significant progress in image acquisition and analysis in recent years.

Liu et al. conducted crack and surface damage detection on bridge piers based on traditional commercial drone equipped with image acquisition equipment (Fig. 1), and reconstructed three-dimensional images based on the detection results. The results showed that drone image acquisition can achieve free perspective switching, adapt to a wide range of bridge types, and have good detection flexibility [5–8]. However, this method usually requires manual control of drones, and the detection results are greatly affected by the drone's flight attitude and path.

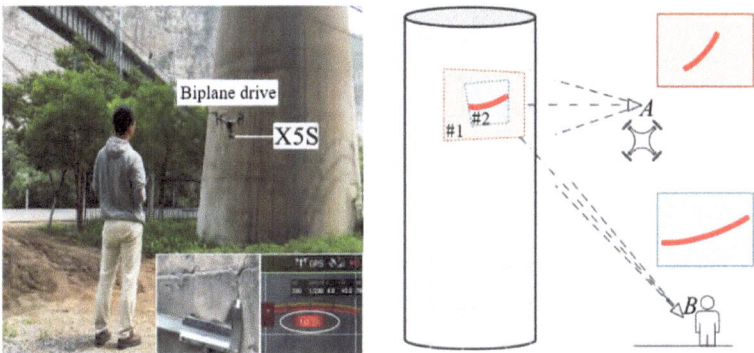

Fig. 1. Application of drone technology in bridge surface disease detection.

To overcome the problems of blurred image acquisition and difficult positioning caused by unstable drone platforms in drone image acquisition, Cuevas et al. developed an attachable drone monitoring device, which can achieve deformation monitoring and detection of high-altitude bridge beams on the basis of bridge surface disease inspection, as shown in Fig. 2. The detection device consists of four parts: ① unmanned aerial vehicle carrying platform; ② Attached device; ③ Deformation testing device; ④ Computer control system.

Fig. 2. Attachable drone monitoring device.

Phung et al. proposed using point cloud scanning technology to plan the detection path of unmanned aerial vehicles (UAVs) in response to the difficulty of locating under the bridge and the possibility of collision between the UAVs and the bridge when using UAV detection, and conducted practical engineering verification.

Due to the difficulty in image positioning caused by the installation of image acquisition devices on drones, Diaz et al. developed a mobile device for synchronous image acquisition and ranging, and conducted experimental tests.

Xie et al. installed a multi-point 3D camera and laser irradiation system on the robotic arm of the inspection vehicle (Fig. 3) and designed an intelligent detection device based on 2D point images and line positioning. The device has good applicability in rapid detection of beam bridge diseases.

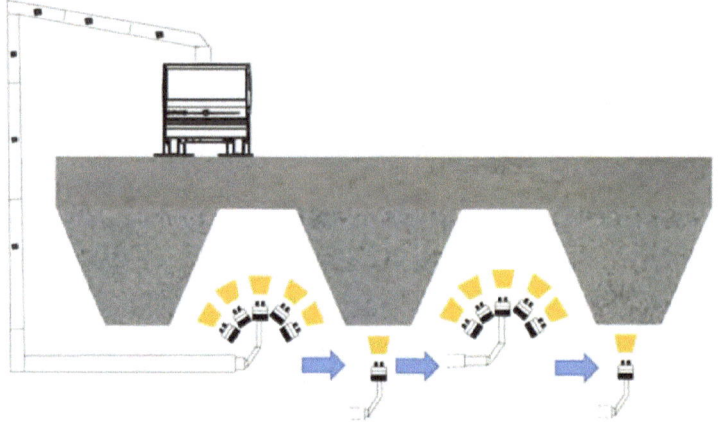

Fig. 3. Multi point 3D camera and laser irradiation intelligent detection robot

In addition, Wang et al. developed an adsorption type steel bridge detection device based on the magnetic adsorption characteristics of steel in steel bridge structures [9]. The device consists of a magnetic shaft and sensors. In response to the electromagnetic characteristics changes caused by steel bridge defects, a strong magnetoresistance array is used to detect steel structure bridge damage based on the magnetic leakage characteristics of the steel bridge.

To achieve performance detection of bridge bearings, Peel et al. developed a detection tool that combines mobile robots and image acquisition technology. This device can achieve aging and cracking detection of bridge bearings.

3 Highway Bridge Disease Detection Technology Under the Background of Artificial Intelligence

3.1 Deep Learning

Deep learning is an important branch of machine learning and artificial intelligence. Its core idea is to train a large number of data samples layer by layer and construct multi-layer neural networks for sample detection. This method has shown enormous advantages and potential in fields such as computer vision, speech recognition, and natural language processing [10]. Especially in the application of semantic segmentation models, a deep learning branch, it can be successfully applied to the detection of highway bridge diseases. The publicly available dataset provides good data support for this detection task, providing a solid foundation for model training and performance improvement. Against the backdrop of the rapid development of artificial intelligence, deep learning technology has played an important role in constructing efficient semantic segmentation models through meticulous preprocessing and effective feature extraction of disease images [11, 12]. This model has significant application significance in highway bridge disease detection, and provides a powerful tool for improving the accuracy and efficiency of detection. This highlights the widespread applicability of deep learning technology in solving practical problems, providing people with more reliable and intelligent solutions.

3.2 PSPNet Semantic Segmentation Network

The semantic segmentation of images plays a crucial role in deep learning research, especially in scene understanding. In the field of driverless driving, achieving high-quality semantic segmentation of road scenes is crucial to ensure the safe driving of autonomous vehicle. This technology is not only limited to unmanned driving, but also plays an important role in areas such as highway bridge disease detection. By quickly and accurately segmenting bridges, the overall safety of the road can be effectively improved. In deep learning, especially in the improvement of semantic segmentation networks, PSPNet (Pyramid Scene Segmentation Network) is a noteworthy model. PSPNet is an improvement on the basis of fully convolutional network (FCN), introducing PSP module and pyramid pooling module to improve the ability to obtain global information. The key to the PSP structure is to divide the feature layer into grids of different sizes, and average pooling is performed within each grid to effectively aggregate contextual information from different regions. Specifically, in the PSPNet model, the input feature layers are cleverly divided into grids of different sizes (6×6. 3×3. 2×2. 1×1) Subsequently, perform an average pooling operation within each grid to obtain the final output result. The design of this structure enables the network to better capture semantic information at different scales in images, thereby improving the accuracy and effectiveness of semantic segmentation. Overall, the introduction of PSPNet has injected new vitality into the field of image semantic segmentation, providing powerful tools and methods for solving complex problems in practical scenarios. The steps are shown in Fig. 4.

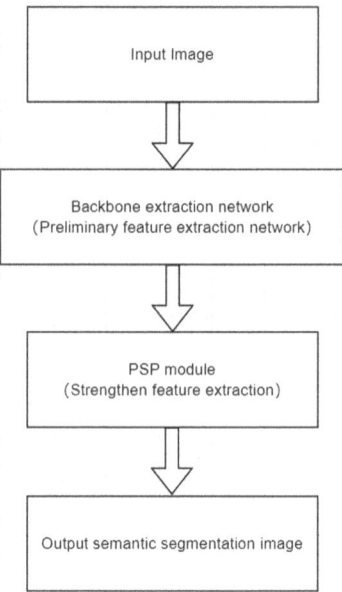

Fig. 4. Basic structure of PSPNet

4 Highway Bridge Disease Detection Based on PSPNet Model

4.1 Data Production

Personnel using artificial intelligence technology to replace highway bridge disease detection need sufficient image data support. In this process, the size of the semantic segmentation label should be consistent with the image size to ensure that each pixel value corresponds to a different target area. However, there are relatively few label sets for highway bridge disease images in the current public dataset, which requires us to actively collect existing public data and use Labelme tools for point labeling. The scope of point marking covers multiple aspects of highway bridge diseases, including exposed reinforcement, cracks, honeycomb and pitted surfaces, etc. In order to meet the requirements of semantic segmentation, this process is not only a simple labeling of the image, but also precise target classification for each pixel. In order to improve efficiency, we can also adopt methods such as cutting to reduce resource occupancy. At the same time, data augmentation through geometric transformations or color adjustments can help improve the robustness and generalization ability of the model. The organic integration of these steps will provide a more reliable data foundation for the application of artificial intelligence in the field of highway bridge disease detection.

4.2 Network Training

In the process of deep learning network training for highway bridge disease detection, we adopted the Keras framework, which allows us to flexibly adjust network parameters to optimize performance. In this process, key adjustments mainly include network classification settings and learning rate settings. Firstly, we divide the network into two categories to address the dual nature of highway bridge image data: one is diseased image data, and the other is disease-free image data. This grouping strategy helps the network better learn and distinguish between two situations, improving the accuracy of disease detection.

During the parameter adjustment process, we set the initial learning rate to 0.0001. This setting has been carefully considered and aims to make the model converge more stably in the early stages of training. By selecting an appropriate learning rate, we can accelerate the training process and improve the model's adaptability to data.

In order to complete the entire network training process, we conducted 45 iterations. The selection of this iteration number is based on balancing the training effect and computational cost, ensuring that the model can achieve sufficient convergence in a limited time and achieve satisfactory detection performance. Overall, by adjusting the parameters of the Keras deep learning framework reasonably, we have successfully established a deep learning model that can effectively detect highway bridge diseases.

4.3 Concrete Step

We first successfully created an effective dataset and conducted comprehensive training. Next, we used drones to collect video images of highway bridges in the field, and these images were transmitted to a computer for processing. On the computer side, we adopted the PSPNet semantic segmentation model for further training and calculation, in order to accurately identify and classify various diseases that may exist on highway bridges. The goal of the entire process is to quickly and accurately detect potential problems. Once our system detects any diseases on the bridge, we will immediately record these issues and take necessary measures to eliminate or repair them. This comprehensive method not only improves the efficiency of highway bridge inspection, but also ensures timely measures to maintain and strengthen the structural integrity of the bridge. By integrating advanced technology and data processing methods, we provide an efficient and reliable solution for bridge management and maintenance. The specific steps are shown in Fig. 5.

5 Experiment and Analysis

5.1 Dataset Collection

The core objective of this paper is to achieve intelligent detection methods by studying the use of PSPNet for highway bridge disease detection. To achieve this goal, a dataset containing highway bridge diseases was first constructed. These diseases mainly include various types such as cracks, exposed reinforcement, honeycomb and pitted surfaces. In order to obtain sufficient and diverse data, this article collected a total of 1000 images of highway bridge diseases through the internet and on-site photography. This dataset is divided into two parts, with 900 sheets used as the training set and the remaining 100 sheets forming the test set.

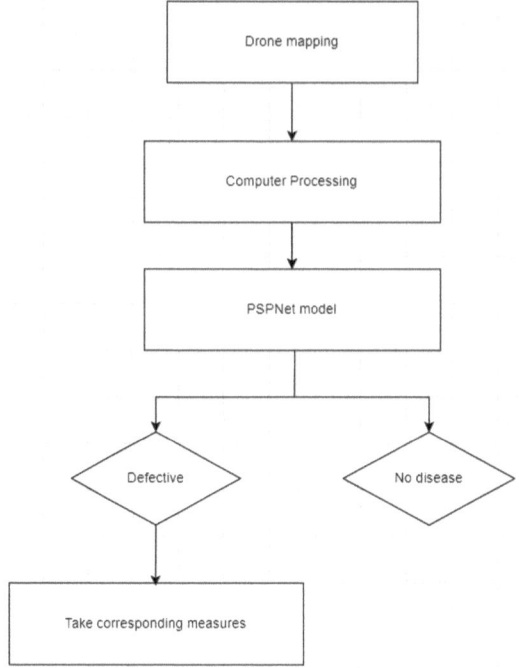

Fig. 5. Steps for Highway Bridge Disease Detection Based on PSPNet Model

In order to generate meaningful results for subsequent model training and testing, this article used the labelme tool to annotate the images in detail. The process of annotation is complex and detailed, and the specific steps and annotation methods can be referred to Fig. 6 in the paper.

In summary, through this study, we not only established a large-scale dataset of highway bridge diseases, but also laid a solid foundation for the training and testing of subsequent PSPNet models. The collection and annotation of this dataset provides rich experimental materials for research, and provides strong support for the development of intelligent highway bridge disease detection methods.

5.2 Experimental Process and Analysis

This study aims to achieve effective crack detection by training on the highway bridge crack dataset using the PSPNet semantic segmentation platform. During the training process, we adopted a learning rate of 0.0001, batch_ Set the size to 4 and epoch to 50, and successfully generate the corresponding weight file.

Through the analysis of the results, we conclude that this method performs well in crack detection and successfully identifies cracks in highway bridges. This achievement not only proves the effectiveness of the PSPNet semantic segmentation platform in this task, but also emphasizes the rationality of the training settings used. Overall, this method

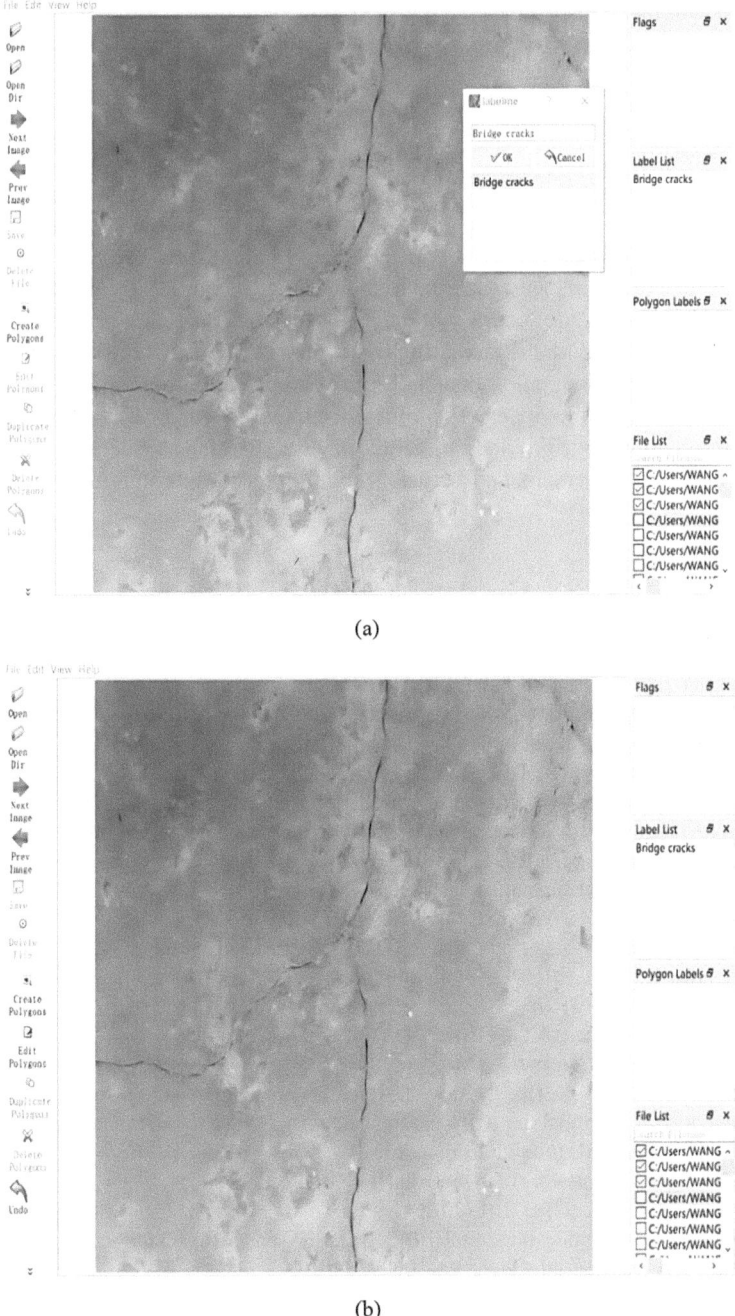

(a)

(b)

Fig. 6. Dataset production

not only accurately detects cracks, but also significantly improves the convenience and efficiency of detection, providing strong technical support for personnel in related fields.

6 Conclusion

The detection of highway bridge diseases plays a crucial role in management, but traditional manual detection has problems such as high cost and danger. In the current context of artificial intelligence, it is particularly important to study the detection of highway bridge diseases. This paper takes the PSPNet semantic segmentation platform as an example to verify the accuracy of the model in crack detection of highway bridges through experiments. With the increasing usage time of highway bridges, adopting deep learning semantic segmentation models for disease detection not only helps to reduce maintenance costs, improve detection efficiency, but also ensures the safety of operators. Therefore, this study provides an advanced, effective, and safe detection method for highway bridge management, and a feasible solution for long-term bridge maintenance.

References

1. Liu, Y.F., Nie, X., Fan, J.S., Liu, X.G.: Image-based crack assessment of bridge piers using unmanned aerial vehicles and three-dimensional scene reconstruction. Comput.-Aided Civ. Infrastruct. Eng. **35**(5), 511–529 (2020)
2. Sanchez-Cuevas, P.J., Ramon-Soria, P., Arrue, B., Ollero, A., Heredia, G.: Robotic system for inspection by contact of bridge beams using UAVs. Sensors **19**(2), 305 (2019)
3. Phung, M.D., Quach, C.H., Dinh, T.H., Ha, Q.: Enhanced discrete particle swarm optimization path planning for UAV vision-based surface inspection. Autom. Constr. **81**, 25–33 (2017)
4. Martínez-Díaz, S.: 3D distance measurement from a camera to a mobile vehicle, using monocular vision. J. Sens. **2021**, 1–8 (2021)
5. Xie, R., et al.: Automatic multi-image stitching for concrete bridge inspection by combining point and line features. Autom. Construct. **90**, 265–280 (2018)
6. Wang, R., Kawamura, Y.: An automated sensing system for steel bridge inspection using GMR sensor array and magnetic wheels of climbing robot. J. Sens. **2016**, 1–15 (2016). https://doi.org/10.1155/2016/8121678
7. Peel, H., Luo, S., Cohn, A.G., Fuentes, R.: Localisation of a mobile robot for bridge bearing inspection. Autom. Constr. **94**, 244–256 (2018)
8. He, H., Hu, G., Ren, J., Hu, J., Zhao, H., Nie, T.: Structure design of cable-stayed bridge intelligent detection robot. In: IOP Conference Series: Earth and Environmental Science, vol. 632, No. 2, p. 022061. IOP Publishing (2021)
9. Zhou, P., et al.: A framework of myocardial bridge detection with x-ray angiography sequence. Biomed. Eng. Online **22**(1), 101 (2023)
10. Yang, M., Xu, H.: Application of fiber Bragg grating sensing technology and physical model in bridge detection. Results Phys. **54**, 107058 (2023)
11. Giglioni, V., Poole, J., Venanzi, I., Ubertini, F., Worden, K.: On the use of domain adaptation techniques for bridge damage detection in a changing environment. ce/papers **6**(5), 975–980 (2023). https://doi.org/10.1002/cepa.2143
12. Rapaport, G.: Applications of State of the Art NDT Techniques in Bridge Inspections. ce/papers **6**(5), 155–162 (2023). https://doi.org/10.1002/cepa.2083

Study on Improving Crack Resistance of Concrete by Waste Rubber Powder and Its Mechanism

Yunfang Meng[1(✉)], Xiaoping Ma[2], Li Han[2], Yanzhu Huang[3], and Zhiqing Wan[3]

[1] College of Energy Resources of Yinchuan, Yinchuan 750000, China
2979526409@qq.com
[2] Research and Innovation Team of College of Energy Resources of Yinchuan,
Yinchuan 750000, China
[3] Team of Civil Engineering Teachers of Huang Danian of Ningxia University, Lanzhou, China

Abstract. In this study, we explore the use of an innovative mixture incorporating compound ecological fiber, waste rubber powder, fly ash, and single ore to enhance the cracking resistance of concrete. Through experimental analysis, we investigated the mechanical properties and underlying mechanisms associated with this composite material. The findings indicate that while the inclusion of mixed fiber and waste rubber powder does not significantly impact the compressive strength of concrete, these components notably enhance its toughness and early crack resistance. Specifically, the addition of these materials, along with single ore and fly ash, effectively reduces the brittleness coefficient of concrete, thereby markedly improving its resistance to cracking. This synergy between the materials demonstrates that the crack resistance of concrete can be significantly optimized through their combination.

Keywords: Waste rubber powder · Composite ecological fiber · Ash-fly · Crack resistance property

1 Introduction

Concrete, due to its high compressive strength, economic benefits, and the wide availability of raw materials, has become an indispensable structural material in various construction projects. Despite its many advantages [1], concrete is a non-homogeneous, long-range disordered, brittle composite material with poor tensile strength, low toughness, and a propensity to crack. This can lead to structural damage or even failure, severely impacting the normal use and service life of the project. Therefore, exploring ways to improve concrete's crack resistance is an urgent issue for engineering technicians. In response to the heterogeneous brittle characteristics of concrete [2], this paper attempts to enhance its crack resistance by utilizing waste rubber powder, fly ash, and ecological fibers [3].

A massive amount of waste rubber originates from the rapid development of the automotive industry. According to relevant data, there are currently 4 billion waste tires

A. Bieliatynskyi et al. (Eds.): CSTTE 2023, LNCE 603, pp. 312–325, 2024.
https://doi.org/10.1007/978-981-97-5814-2_28

globally, with an annual increase of 1.1 billion. As a significant consumer of rubber, China generated 400 million waste tires in 2023 alone. Simultaneously [5], the amount of waste rubber products and scraps from other rubber industries has been steadily increasing [4]. These industrial wastes not only occupy a considerable amount of limited land but also pose pollution problems related to natural resources, energy, and the environment. Therefore, finding effective methods for the treatment and recycling of waste rubber and fly ash has become a focus in the engineering field, holding significant importance in today's era of increasingly scarce energy and resources [7].

The waste rubber powder selected in this study mainly comes from scrapped automobile tires and waste rubber products produced by the rubber industry, as well as large quantities of leftover scraps from mass production. These waste rubbers can be mechanically crushed or ground into fine particles, becoming granules and powder [8].

2 Experimental Materials and Methods

2.1 Raw Materials

(1) Cement: Saima Sign Cement Co., Ltd. in Ningxia produces P.052.5 MPa ordinary Portland cement. The cement mortar strength is tested according to the *Test Method of Cement Mortar Strength (ISO method)* (GB/T17671-2020). Its physical and mechanical properties, chemical composition, mineral composition, and single mineral chemical composition, as well as hydration heat detection indicators, are shown in Tables 1, 2, 3, and 4.

Table 1. Physical property test index of ordinary Portland cement by Ningxia Saima Sign P. 052.5 MPa

Test items	Fineness 0.08 μm (residue on sieve/%) ≤ 10.0	Standard consistency water requirement (%)	Density (kg/cm^3)	Setting time (h: Min)		Stability	Compressive strength (Mpa)		Flexural strength (Mpa)	
				Initial set ≥ 0: 45	Final set ≤ 10: 00		3d	28d	3d	28d
Actual results	1.8	27.82	3.16	2: 26	3: 46	Qualified	32.6	53.8	6.2	8.6

Table 2. Chemical composition (by mass) of cement by Ningxia Sign P.052.5 MPa

Name	SiO_2	Al_2O_3	Fe_2O_3	CaO	MgO	K_2O	Na_2O	TiO_2	MnO	SO_3	IL
Result	24.36	8.76	2.56	55.76	2.49	0.66	0.19	0.45	0.17	3.86	4.68

(2) Fly ash: The fly ash is derived from the Grade I fly ash produced by the Lingwu Power Plant in Ningxia. Its chemical composition is shown in Table 5, and particle size distribution indicators are tested as shown in Table 6, meeting national standards.

Table 3. Mineral composition and hydration heat test results of cement by Ningxia Saima Sign P.052.5 MPa

Name	C$_3$S	C$_3$A	C$_2$S	C$_4$AF	f-CaO	Heat of hydration (j/g)			
						1d	3d	5d	7d
Result	52.36	3.86	22.96	18.72	0.89	92.81	198.17	238.96	266.48

Table 4. Chemical composition and mineral composition of single ore of cement by Ningxia Saima Sign P.052.5 MPa

Mineral name	Analysis results (%)									Residue on 80 μm sieve (%)
	SiO$_2$	Al$_2$O$_3$	Fe$_2$O$_3$	f-CaO	C$_4$AF	B$_2$O$_3$	ZnO	β-C$_2$S	Total	
β-C$_2$S	0.18	/	/	0.27	/	1.07	1.09	98	99.32	7.48
C$_4$AF	/	0.46	0.62	0.87	98	/	/	/	99.95	5.72

Table 5. Chemical composition (by mass) of Fly ash

Composition	SiO^2	Al^{2}O^3	Fe^{2}O^3	CaO	MgO	K^{2}O	Na^{2}O	TiO^2	MnO	SO^3	IL
Fly ash	48.72	29.69	3.18	7.77	0.78	1.19	0.27	1.29	0.08	1.31	3.69

Table 6. Particle size distribution of fly-ash

Material name	Specific surface area (cm^2/g)	Particle size distribution (volume) /%			
		< 10 μm	10~37 μm	37~60 μm	> 60 μm
Fly ash	1474	43.28	18.16	35.83	4.83

(3) Fine and coarse aggregates: Aggregates account for 70%-80% of the total volume used in concrete, with 5–150 mm as coarse aggregates and 0.16–5 mm as fine aggregates.

1) Physical property testing and grading analysis of fine aggregates:

Fine aggregates are natural sand from Zhenbeibao in Ningxia. The apparent density, bulk density, and dust content are evaluated, as shown in Table 7. The results of the sieve analysis of fine aggregates are shown in Table 8, and the grading curve is illustrated in Fig. 1.

Table 7. Physical properties test results of fine aggregate

Name	According to standard	Apparent density (kg/m^3)	Bulk density (g/cm^3)	Mud content (%)	Fineness modulus	Evaluation result
Fine aggregate (natural sand)	*Sand for Construction* (GB/T14684–2011)	2695	1.67	0.54	2.68	Medium sand

The fineness modulus of sand is calculated using the following formula:
$$\mu_1 = \frac{(A_2+A_3+A_4+A_5+A_6)-5A_1}{100-A_1} = 2.88 \in (3.0 \sim 2.3) \text{ is medium sand.}$$

Table 8. Test results of fine aggregate sieve analysis

Sieve aperture size (mm)	Mass of residue on each sieve (g)			Graded residue (%)	Cumulative residue (%)	Specified by standard (%)
	1	2	Mean			
4.75	20.00	21.00	20.50	4.10	4.10	10~0
2.36	70.60	69.40	70.00	14.00	18.10	25~0
1.18	120.00	110.00	115.00	23.00	41.10	50~10
0.60	90.00	105.00	97.50	19.50	60.60	70~41
0.30	110.00	95.00	102.50	20.50	80.10	92~70
0.16	75.0	85.00	80.00	16.00	97.10	100~90
Bottom of sieve (g)	14.40	14.60	14.50	2.90	100.00	

2) Physical property testing and results of coarse aggregates:

Coarse aggregates play a vital role in enhancing concrete strength and preventing shrinkage. The aggregates used are artificially crushed stones with rough surfaces and good bonding properties with cement, rich in angular shapes, with particle sizes of 525 mm and 4.7516 mm. These are tested according to the standards for *Pebble and Crushed Stone for Construction* (GB/T14685-2001), with results shown in Table 9. Upon testing, both fine and coarse aggregates meet the national standard specifications.

(1) Mixing and curing water: Tap water is used, and its quality meets the standards required for concrete mixing water.
(2) Additives: FDN naphthalene-based water reducer produced by Beijing Muhu Additives Co., Ltd. is used, achieving a water reduction rate of 25%-30%, with a dosage of 0.3%-0.5% of the cementitious material.

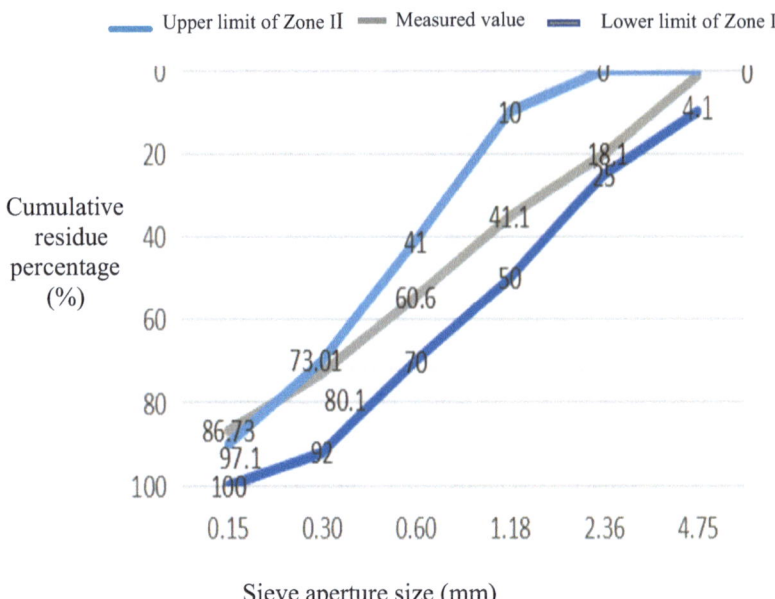

Fig. 1. Gradation Curve of Fine Aggregate

Table 9. Physical properties test results of coarse aggregate

Item	Maximum particle size	Apparent density	Bulk density	Porosity	Mud and dust content	Crushing index
Unit	(mm)	(kg/m^3)	(kg/m^3)	(%)	(%)	(%)
Test value	34	2940	1520	47	0.31	9.7

(3) Ecological fiber: UltraFiber 500, a product from Shanghai Luoyang New Material Technology Co., Ltd., is used with a dosage of 0.6–1.2%. Performance indicators are shown in Table 10, and the actual product is illustrated in Fig. 3.

(4) Rubber powder: Waste rubber powder is produced from waste rubber products through mechanical processes and then processed into powders of various fineness according to different uses. Particles smaller than 1.5 mm are rubber powder, and 8–20 mesh are granules, used for underlays, lawns, road base layers, tracks, elastic layers for roads, and sports field paving; 30–40 mesh is coarse rubber powder, used for producing activated rubber powder, reclaimed rubber, paving, and rubber boards; 40–60 mesh is fine rubber powder, used for plastic modification, rubber product production; 60–80 mesh is fine rubber powder, applied in car tires, rubber products, and building materials; 80–120 mesh is microfine rubber powder, used in rubber products, military products; 200–500 mesh is ultrafine rubber powder, used for high polymer roll material modification. In this experiment, 60 mesh rubber powder produced by Henan Jiaozuo Hongrui Rubber Co., Ltd. is used, with a bulk

density of 0.375 g/cm^3 and a density of 1.22 g/cm^3. The actual product is shown in Fig. 2.

Table 10. Performance index (UltraFiber 500) of Bokai ecological fiber

Material indices	Length (mm)	Denier (g/9000 m)	Fiber diameter (μm)	Density (g/cm^3)	Specific surface area (cm^2/g)	Elastic modulus (N/mm^2)	Tensile strength (N/mm^2)
Index value	1.9~2.3	2.0~3.0	14~17	1.1	25000	8500	600~900

Fig. 2. 60-mesh waste rubber powder

Fig. 3. BoKai 500 ecological fiber produced in Shanghai

2.2 Experimental Methods and Mix Design

The experimental design utilizes an L18 (3^5) orthogonal array, with the concrete designed for a strength class of C30. The strength and forming tests are conducted according to the *Standard for Test Method of Concrete Physical and Mechanical Properties* (GB/T 50081–2002) GB/T1767. A 10 × 10 × 10 cm^3 model is chosen. Five factors at three levels, namely water-cement ratio, fly ash, waste rubber powder, sand ratio, and ecological fibers, are identified as the influential factors in the experimental design, with the orthogonal experiment factor level table shown in Table 11.

Following the specifications of the *Standard for Test Method of Concrete Physical and Mechanical Properties* (GB/T 50081–2002) GB/T1767, the orthogonal experiment plan is outlined in Table 12. The total volume of binder (C + F) in the concrete is set at 450 kg/m^3, with fly ash replacing an equivalent amount of cement. Composite ecological fibers are mixed in as a percentage of the volume of the binder materials. The designed slump for the concrete is 30–50 mm. Specimens are molded and maintained in the initial setting with the mold for 24 h, demolded, and then cured under standard conditions until the age of testing [9]. The test results and comparisons are presented in Table 13.

Table 11. Orthogonal test table of factor levels L_{18} (3^5)

Level	Factor				
	Water-cement ratio A	Fly ash (%) B	Waste rubber powder (%) C	Sand ratio (%) C	Fiber content kg/m^3 E
1	0.33	10	1.5	35	0.6
2	0.35	15	3.0	40	0.9
3	0.37	20	4.5	45	1.2

Table 12. Design table of orthogonal scheme

Test number	Water-cement ratio A	Fly ash/% B	Rubber powder (%) C	Sand ratio (%) D	Ecological fiber/kg/m^3 E	Blank column	Blank column
1	1 (0.30)	1 (10)	1 (1.5)	1 (35)	1 (0.6)	1	1
2	1 (0.30)	2 (15)	2 (3.0)	2 (40)	2 (0.9)	2	2
3	1 (0.30)	3 (20)	3 (4.5)	3 (45)	3 (1.2)	3	3
4	2 (0.33)	1 (10)	1 (1.5)	2 (35)	2 (0.9)	3	3
5	2 (0.33)	2 (15)	2 (3.0)	3 (40)	3 (1.2)	1	1
6	2 (0.33)	3 (20)	3 (4.5)	1 (45)	1 (0.6)	2	2
7	3 (0.37)	1 (10)	1 (1.5)	1 (35)	3 (1.2)	2	3
8	3 (0.37)	2 (15)	2 (3.0)	2 (40)	1 (0.6)	3	1
9	3 (0.37)	3 (20)	3 (4.5)	3 (45)	2 (0.9)	1	2
10	1 (0.30)	1 (10)	1 (1.5)	3 (45)	2 (0.9)	2	1
11	1 (0.30)	2 (15)	2 (3.0)	1 (35)	3 (1.2)	3	2
12	1 (0.30)	3 (20)	3 (4.5)	2 (40)	1 (0.6)	1	3
13	2 (0.33)	1 (10)	1 (1.5)	3 (45)	1 (0.6)	3	2
14	2 (0.33)	2 (15)	2 (3.0)	1 (35)	2 (0.9)	1	3
15	2 (0.33)	3 (20)	3 (4.5)	2 (40)	3 (1.2)	2	1
16	3 (0.37)	1 (10)	1 (1.5)	2 (40)	3 (1.2)	1	2
17	3 (0.37)	2 (15)	2 (3.0)	3 (45)	1 (0.6)	2	3
18	3 (0.37)	3 (20)	3 (4.5)	1 (35)	2 (0.9)	3	1

3 Experimental Results and Analysis

(1) Orthogonal experiment results and brittle coefficient analysis

Comparing results from Table 13, the baseline concrete has the highest brittle coefficient of 9.85, which is 37.8% higher than the average brittle coefficient of 7.16 in the

Table 13. Analysis of orthogonal test results and comparison of brittleness coefficient

Group	Water (kg)	Cementitious materials (kg)	Aggregate (kg)	Rubber powder (kg)	Composite ecological fiber (kg)	28-day compressive strength (Mpa)	28-day split tensile strength (Mpa)	Brittleness coefficient
JZH	135	450	1420	0	0	68.4	6.94	9.85
1	135	450	1420	0.675	0.27	62.4	7.93	7.87
2	135	450	1420	0.675	0.41	60.08	7.87	7.63
3	135	450	1420	0.675	0.54	51.84	6.96	7.44
4	149	450	1380	1.35	0.27	55.48	6.46	8.58
5	149	450	1380	1.35	0.41	54.58	6.85	8.45
6	149	450	1380	1.35	0.54	53.36	6.95	7.67
7	167	450	1370	2.03	0.27	53.76	7.64	7.04
8	167	450	1370	2.03	0.41	51.65	7.45	6.93
9	167	450	1370	2.03	0.54	49.87	6.25	7.98
10	135	450	1420	0.675	0.27	52.38	6.78	7.23
11	135	450	1420	0.675	0.41	51.75	46.66	7.77
12	135	450	1420	0.675	0.54	51.35	6.78	7.57
13	149	450	1380	1.35	0.27	50.97	6.56	7.76
14	149	450	1380	1.35	0.41	49.83	6.67	7.47
15	149	450	1380	1.35	0.54	48.96	6.58	7.44
16	167	450	1350	2.03	0.27	46.83	6.54	7.16
17	167	450	1350	2.03	0.41	45.97	6.53	7.03
18	167	450	1350	2.03	0.54	44.94	6.39	7.26

modified mixtures. The inclusion of rubber powder, fly ash, and fibers does not significantly affect the compressive strength or may slightly decrease it. However, the tensile strength is significantly increased, and the brittle coefficient is considerably reduced, as shown in tests 3, 7, 8, 10, 15, 16, 17, and 18, where the optimal crack resistance is achieved with 1.5% - 4.5% rubber powder and 0.9% -1.2% fiber content. This indicates that rubber powder, fly ash, and fibers can enhance the toughness of concrete and play a vital role in crack resistance.

(2) Analysis of the effect of mineral components on the brittle coefficient

The mineral components of the binder materials significantly impact the toughness of the concrete. In cement, the mineral components C_2S and C_4AF reduce the heat of hydration and drying shrinkage. C_4AF also functions to reduce brittleness and improve toughness. In Table 14, when C_2S at 14% and C_4AF at 2%, and C_2S at 17% and C_4AF at 2% are added (groups JC11 and JC12), the 7-day brittle coefficients are respectively 32.5% and 15.3% lower than the baseline concrete, and 28-day brittle coefficients are 10% and 8% lower. A reduction in the brittle coefficient indicates an increase in toughness.

(3) Analysis of performance test results from baseline and optimized composites

Results from Tables 15 and 16 indicate that the 7-day and 28-day brittle coefficients of baseline concrete (JZH) are higher than those of the composite groups JC5, JC9, JC10, and JC11. With the addition of single minerals, rubber powder, fly ash, and fibers, the brittle coefficient shows a decreasing trend. Performance data from Table 16 reveals that splitting tensile strength and columnar tensile strength are on the rise, while the modulus of elasticity gradually decreases, and the tensile modulus shows an increase. The mechanical performance tests indicate that rubber powder, fly ash, and fibers can enhance the crack resistance of concrete, playing a significant role in extending the service life of engineering projects.

Table 14. Effect of mineral composition on brittleness coefficient

Test number	Binder and supplementary mineral components (%)			Water-cement ratio (%)	Mortar (compressive/flexural) strength (Mpa)			Brittleness coefficient		
	C	C_2S	C_4AF		3d	7d	28d	3d	7d	28d
JZH	100	0	0	0.37	26.2/6.8	40.2/7.3	63.1/9.4	3.85	5.51	6.71
JC1	96	4	0	0.37	24.3/5.2	37.1/6.8	62.1/9.6	4.67	5.45	6.47
JC2	91	9	0	0.37	21.5/4.9	34.1/6.5	57.0/9.0	4.39	5.25	6.33
JC3	86	14	0	0.37	22.9/4.9	35.1/7.2	63.6/10.1	4.67	4.88	6.30
JC4	83	17	0	0.37	23.4/6.4	36.7/7.7	60.6/9.4	3.65	4.77	6.38
JC5	95	4	1	0.37	26.2/6.4	39.7/8.1	63.4/10.1	4.09	4.90	6.27
JC6	91	9	1	0.37	24.6/6.5	36.2/7.7	62.7/10.0	3.78	4.70	6.27
JC7	86	14	1	0.37	22.2/5.3	36,3/7.0	68.9/10.8	4.19	5.19	6.40
JC8	83	17	1	0.37	20.8/5.0	32.7/6.5	59.8/9.8	4.16	5.03	6.10
JC9	96	4	2	0.37	26.1/5.2	32.3/6.4	61.6/10.2	5.01	5.04	6.04
JC10	91	9	2	0.37	22.8/5.1	38.2/6.7	62.7/10.4	4.47	5.70	6.03
JC11	86	14	2	0.37	20.8/5.0	32.0/7.7	60.4/9.9	4.08	4.16	6.10
JC12	83	17	2	0.37	20.9/4.9	34.4/7.2	60.9/9.8	4.27	4.78	6.21

4 Microscopic Testing and Analysis

(1) Comparative analysis of Figs. 4 and 5 shows that, after failure under compression, the baseline concrete specimens exhibit significant damage (as shown in Fig. 4). In contrast, concrete specimens with a blend of a single mineral, rubber powder,

Table 15. Strength test results of optimized composite mortar

Test number	Supplementary materials (%)				Mortar (compressive/flexural) strength (Mpa)			Brittleness coefficient		
	XJ	C_2S	C_4AF	SXW	3d	7d	28d	3d	7d	28d
JZH	0	0	0	0	26.2/6.8	40.2/7.3	63.1/9.4	3.85	5.51	6.71
JC5	3.0	4.0	1.0	1.2	26.2/6.4	39.7/8.1	63.4/10.1	4.09	4.90	6.27
JC9	4.5	4.0	2.0	0.9	26.1/5.2	32.3/6.4	61.6/10.2	5.01	5.04	6.04
JC10	1.5	.0	2.0	0.9	22.8/5.1	38.2/6.7	62.7/10.4	4.47	5.70	6.03
JC11	3.0	14.0	2.0	1.2	20.8/5.0	32.0/7.7	60.4/9.9	4.08	4.16	6.10

Table 16. Test results of partial concrete performance after composite of reference and optimization

Test number	R_{28} (MPa)	$R_{Column28}$ (MPa)	E_{28} (GPa)	R_{L28} (MPa)	E_{L28} (GPa)	ε_{28} ($\times 10^{-6}$)
JZH	35.8	23.8	33.8	2.33	36.88	87
JC5	40.0	24.3	29.5	2.97	38.88	105
JC9	43.5	25.3	31.4	2.80	37.97	118
JC10	36.9	28.2	30.4	2.92	37.70	109
JC11	38.8	26.8	28.8	2.75	38.62	102

and fibers maintain their integrity and do not disintegrate upon failure (as shown in Fig. 5), demonstrating entirely different failure morphologies.

(2) The concrete with composite binder materials, after 28 days of curing, shows continuous strength growth. Dense crystals and binder reaction products gradually form around the fly ash spheres, as illustrated in Figs. 6(a), (b), (c), and (d). Some fly ash spheres remain exposed, and as curing time progresses, reactions on the surfaces of the particles continue, producing hydration products of varying sizes and increasingly complete polymer encapsulation layers. This demonstrates the micro-level grading and microfiller effects of polymers. Therefore, the inclusion of multiple components is beneficial for both the strength and crack resistance of concrete, consistent with macroscopic test results.

(3) Figure 7 presents concrete with a mix of single minerals, rubber powder, fibers, and fly ash. The concrete shows low early strength but significant gains in the later stages, particularly when the hydration products of fibers, rubber powder, and single mineral combine to form a dense structure, significantly reducing porosity. This indicates that the combined addition of these materials has a synergistic effect. The chaotic nature of the fibers helps prevent the propagation of cracks, and the elastic properties of rubber powder, the micropowder effect of fly ash spheres, and the toughness of single mineral C_4AF all contribute to the composite effect. Coupled with the testing

and calculations of the mechanical properties of the concrete, these findings confirm that the combined performance meets the expected targets for crack resistance.

Fig. 4. Baseline concrete failure mode

Fig. 5. Single mineral rubber powder fiber concrete failure mode

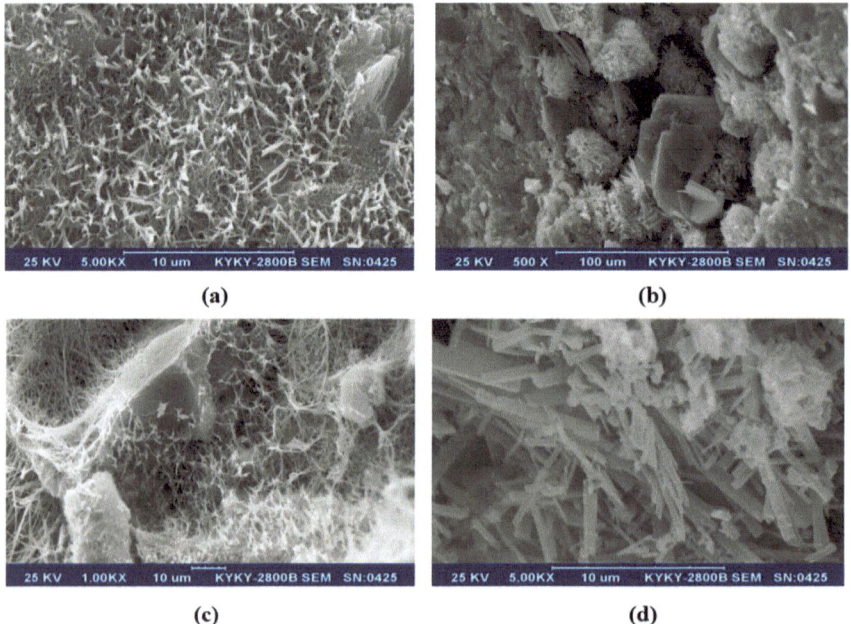

Fig. 6. SEM characteristics of standard curing for 28 days Baseline Concrete

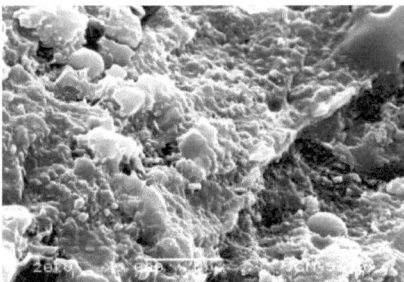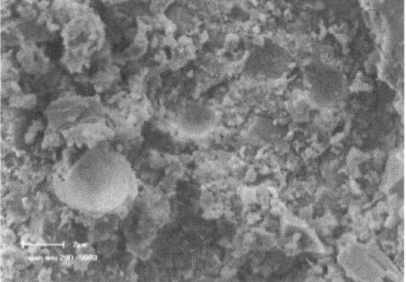

Fig. 7. SEM condition of optimized composite concrete after 28 days

5 Mechanism Analysis

(1) Benefits of fly ash micropowder micro-grading morphological characteristics

Incorporating an appropriate amount of fly ash in concrete can effectively enhance the workability of the concrete mix. Replacing cement with fly ash not only reduces cement usage and cost but also decreases the heat of hydration in concrete, which is beneficial for large-volume concrete structures. Fly ash particles are molten aluminosilicate spherical vitreous bodies that are dense and elastic with a smooth surface, which favorably impacts concrete's crack resistance. Fly ash contains large amounts of SiO_2 and Al_2O_3, which react with the hydration product $Ca(OH)_2$ of cement, forming C-S-H and C-A-H gel materials. These materials gradually transform into fibrous crystals over time, increasing in number and interlocking to form a chained structure. This fills the voids in the mixture, playing the role of microfiller grading and micro-framework, positively influencing the strength enhancement and significantly contributing to the later strength of concrete, making the structure of hardened concrete denser [10].

(2) Elastic effects of micro-spring bodies in rubber powder.

1) Experimental results indicate that the strength of rubber powder concrete is slightly reduced compared to the baseline concrete. Rubber powder, being an elastic organic polymer material, has relatively weak bonding with cement paste. This reduces the interfacial bond strength, increasing the number of weak points within the concrete and thus reducing its compressive strength. However, the high elasticity of rubber complements the brittleness of concrete, enhancing its ductility and effectively reducing brittleness and the onset of cracking.

2) Rubber powder acts as micro-spring bodies distributed within the concrete. External forces cause the cement matrix around the rubber particles to crack due to stress concentration. Rubber itself has excellent tensile properties that hinder the propagation of cracks, thereby maintaining the integrity of the specimens by preventing cracks caused by compression from becoming continuous.

3) There is an optimal content of rubber powder for enhancing crack resistance, as shown in the orthogonal test results, where the best crack resistance effects are evident at rubber powder content ranges between 1.5% and 4%. In comparative experiments, the strength of concrete with optimal rubber content decreases less compared to the

baseline concrete. As part of the aggregate mix, rubber powder optimizes aggregate grading and fills voids within the concrete.

4) The combined effect of rubber powder and fly ash, along with coarse and fine aggregates and cement, where each component's average particle size exists on different scales, is advantageous. This composite arrangement allows for complementarity: voids in coarse aggregates are filled by fine aggregates and rubber powder, voids in fine aggregates by cement particles, and cement particle voids by the effect of fly ash, optimizing micro-aggregate grading, hindering crack propagation, enhancing concrete's toughness, reducing brittleness, and improving the density and crack resistance of the structure, with a combined effect superior to single additions [6].

(3) Crack resistance effect of composite ecological fibers' random distribution.

From Fig. 7, it can be observed that the hydration particles are abundant, and the structure is relatively dense. Due to the reaction of SiO_2 in fly ash with $Ca(OH)_2$, the edges become irregular, and the formation of CSH not only fills the internal voids of the concrete but also intersects with the boundary fibers. In the concrete interface transition zone, the $Ca(OH)_2$ crystals are smaller, porosity is reduced, and the CSH gel forms a dense network structure. The small diameter of fibers and their random orientation within the concrete allow them to bear some of the tensile stress caused by shrinkage in the cement stone, alleviating stress concentration at the tips of micro-cracks and preventing the formation and propagation of micro-cracks. They cross each other, forming a more complex multidimensional random distribution structure within the concrete, making the microstructure of the concrete dense, with few pores and micro-cracks, aligning with macroscopic test reslust [11]. When the fiber content reaches the optimal range of 0.6–0.9%, its resistance to columnar indices RL282.97 MPa and split tensile modulus E reach maximum values of 38.88 GPa.

6 Conclusion

(1) The combined use of multiple additives exhibits various synergistic effects. Rubber powder, single minerals, fibers, fly ash, coarse and fine aggregates, and cement have average particle sizes at several different scales. Their combined use facilitates complementary benefits, optimizing the functions and effects of each material. This arrangement restricts crack propagation, enhances the toughness of the concrete, reduces brittleness, and increases the density of the structure, which is beneficial for improving crack resistance. The overall effect is superior to using single additives.

(2) Concrete with only rubber powder as an additive has a lower compressive strength compared to the baseline concrete, indicating that rubber powder has an adverse effect on the growth of concrete's compressive strength. However, the rubber itself possesses excellent tensile properties that can hinder the progression of cracks, preventing the cracks caused by compression from penetrating through, thus maintaining the integrity of the concrete specimens.

(3) Concrete with a composite mixture of rubber powder, fibers, single minerals, and fly ash demonstrates superior overall performance. Experimental results and calculated data on various crack resistance indices show that the performance of the composite mixture surpasses that of single additives and achieves the anticipated expectations.

Acknowledgement. Projects: Team of Huang Danian-style Civil Engineering Teachers at Ningxia Universities; Research and Innovation Team of Yinchuan College of Energy Resources; Innovation and Entrepreneurship Team of Yinchuan College of Energy Resources.

References

1. Ge, X.Y., Wu, Z.G., Peng, B., et al.: Study on the performance of polymer and fly ash modified mortar. Concrete **6**(236), 110–119 (2009)
2. Guo, J.S., Yun, J.Q.: Research progress on basalt-polypropylene hybrid fiber concrete. Concrete World **5**, 88–90 (2017)
3. Huang, C.K.: Fiber Concrete Structures. China Machine Press, Beijing (2004)
4. Aslani, F., Khan, M.: Properties of high-performance self-compacting rubberized concrete exposed to high temperatures. J. Mater Civileng. **31**(040190405) (2019)
5. Guo, J., Huang, M., Huang, S., Wang, S.: An experimental study on mechanical and thermal insulation properties of rubberized concrete including its microstructure. Appl. Sci. **9**(14), 2943 (2019)
6. Li, S.L.: Study on the crack resistance test methods and evaluation system of cement-based materials. [Master's Thesis], Fuzhou: Fuzhou University, pp. 63–65, April 2020
7. Zheng, Q., Wang, T.Q.: Study on the effects of coarse aggregate and basalt fiber on the performance of ultra-high performance concrete. Concrete World **42**(11), 68–71 (2021)
8. Chen, B.Y., Liu, X., Li, Z.S., et al.: Impact of fibers and types of sand on the compressive strength of ultra-high performance concrete. Jiangsu Build. Mater. **41**(4), 27–29 (2021)
9. Wang, D.M., Zhang, S.Y., Yao, S.W., et al.: The effect of modified silica fume and fly ash on the performance of ultra-high performance concrete (UHPC). Concrete Cement Products **48**(11), 1–5 (2021)
10. Tawich, K., Nhat Ho, T.T., Sambath, M., Chai, J., Weerachart, T.: Strength, chloride resistance, and water permeability of high volume sugarcane bagasse ash high strength concrete incorporating limestone powder. Constr. Build. Mater. **311** (2021)
11. Kim, W., Lee, T.: A Study to Improve the Reliability of High-Strength Concrete Strength Evaluation Using an Ultrasonic Velocity Method. Materials **16**(20) (2023)

Key Construction Techniques for Cable Hoisting of Long-Span Steel Pipe Concrete Arch Bridge

Yan Yang[✉]

School of Intelligent Manufacturing, PanZhiHua University, Panzhihua 617000, Sichuan, China
37362770@qq.com

Abstract. Large span steel tube concrete arch bridges with the characteristics of high construction risk and high technical difficulty are usually constructed by cable hoisting technology. In this parper, the self-balance of the load-bearing main cable and optimized scheme for the main tower and anchor were proposed by conducting an iterative theoretical design calculation on the main cable. The design and key technology of the cable lifting system for large-span concretes filled steel tube arch bridge were investigated. Moreover, the "inverted loading method" was used for simulation analysis, which provides a scientific basis for the construction stage. The engineering practice results show that the optimization and innovation of the cable hoisting system can greatly improve design efficiency, enhance the safety of the cable system, and obtain good economic value.

Keywords: Bridge Engineering · Steel Pipe Concrete Arch Bridge · Cable Hoisting

1 Introduction

Concrete-filled steel tubular arch bridges, known for their large spans and excellent structural performance, are widely used in mountainous road bridges. Cable hoisting technology plays a pivotal role in addressing the construction challenges of these bridges, characterized by high technological requirements, strong specialization, and elevated construction difficulty. Taking the Xianfeng Chaoyang Bridge as a case study, this paper utilizes Visual Basic (VB) visual programming for the force analysis of the main cable under different working conditions. Finite element simulations are employed for validation. The design schemes for the main cable, main tower, and main anchor of the cable hoisting system are optimized, using the Xianfeng Chaoyang Bridge as a case study, with valuable insights for reference in similar projects.

2 Bridge Overview

The Xianfeng Chaoyang Bridge has a width of 12 m, with a bridge span arrangement of 30 m (prestressed box girder) +195 m (superstructure steel pipe concrete arch) + 25 m (prestressed box girder). The total length of the bridge is 271 m, and the width is

A. Bieliatynskyi et al. (Eds.): CSTTE 2023, LNCE 603, pp. 326–335, 2024.
https://doi.org/10.1007/978-981-97-5814-2_29

12 m. It is a second-level highway with two lanes in each direction. The main bridge adopts a net span of 195 m steel pipe concrete truss superstructure arch bridge, with a net span-to-rise ratio of 1/5. The arch axis is aligned with the catenary line, and the arch axis coefficient is m = 2.2. The arch ribs are of equal-section steel pipe concrete truss structure. The construction of the main bridge adopts the cable hoisting construction process. The entire bridge consists of two truss arch ribs, and each main arch rib is hoisted in 13 segments on each side, with a total of 26 arch rib hoisting segments for the entire bridge. The arch foot section has the heaviest arch ribs, with a single segment weight of 75.032 tons. The columns on the arch are made of steel pipe concrete, with a lighter weight and not used as hoisting control. Each cap beam weighs 65.52 tons. The 16-m small box beam on the arch edge is relatively heavy, with a maximum weight of 42.38 tons. The approach bridge 30-m box beam has a maximum weight of 104.52 tons, and the 25-m box beam has a maximum weight of 79.56 tons. The overall layout is shown in Fig. 1.

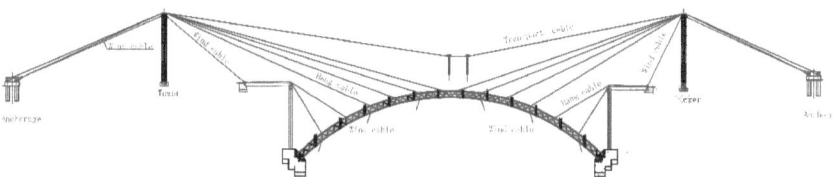

Fig. 1. Cable Hoisting System Overall Layout Diagram.

3 Optimization Research on Cable Hoisting Technology

In traditional cable hoisting systems, the calculation of the load-bearing cable is challenging, and obtaining accurate values through tension equations is difficult. The calculation workload is also significant, especially when considering various conditions and temperature changes, making the calculation process complex. Load-bearing cables often span large distances and bear significant weights, falling into the category of long-span cables. With numerous cable lines in each group, the installation of cables in an unloaded state results in inconsistent forces and shapes due to various factors. Additionally, under heavy-load conditions during the operation of the trolley, forces continuously change, leading to uneven tension in each load-bearing cable and a potential inclination of the trolley, making adjustments difficult [1, 2].

Due to limitations in conventional universal joint materials, the traditional assembled main tower cannot meet the load requirements of large-span, heavy arch rib main coupling systems, resulting in poor structural safety. Conventional main anchors using gravity-type anchors have large construction volumes, long construction cycles, and do not fully exploit the mechanical properties of the rock and soil, resulting in poor economic efficiency.

Given the shortcomings in traditional cable hoisting system design outlined above, optimization and technological innovations are proposed.

(1) Optimization of Load-Bearing Cable Calculation. In traditional calculations, the process is cumbersome and lacks precision. Applying the principles of static equilibrium, a software programming calculation was employed to optimize the calculation of the load-bearing cable length. Two approaches were taken: first, assuming the sag of the load-bearing cable and calculating the length S based on geometric relationships; second, assuming the sag of the load-bearing cable and obtaining the elastic elongation ΔS by calculating the tension within the main cable, resulting in the adjusted length $S' = S0 + \Delta S$. When $S \approx S'$ (within the required precision), the assumed sag of the load-bearing cable is considered solved. Iterative calculations were performed after software programming, ensuring high speed and precision, meeting practical requirements, and significantly improving work efficiency.

(2) Optimization of Load-Bearing Cable Tension Balance. In conventional cable hoisting systems without a balancing pulley, the trolley often tilts during operation, posing safety risks. To ensure even force distribution in the suspension cables, a series of large-tonnage pulleys totalling 150 tons were connected in tandem to automatically adjust the tension uniformly. The arrangement of the balancing pulleys is shown in Fig. 2.

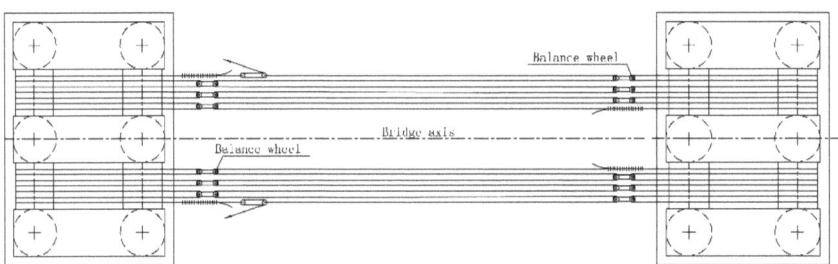

Fig. 2. Balancing Pulley Tandem Arrangement Diagram.

(3) Optimization of Main Tower Structure. The traditional universal joint tower frame fails to meet the requirements of large-span, heavy cable hoisting due to issues with tower deformation and material strength. To address this, the tower frame was modified based on the assembly materials of the universal joint tower frame. The columns were constructed using φ630 × 14 steel pipes, connected longitudinally through flanges, with four on each side, totalling eight steel pipe columns for the entire tower. The remaining components were assembled using universal joint materials, significantly improving both safety and cost-effectiveness.

(4) Optimization of Main Anchor Structure. The traditional gravity-type anchor has a large construction volume. To address this, a combination anchor consisting of "pile + baffle" was introduced, fully exploiting the mechanical properties of the rock and soil. This greatly enhanced the anchor's resistance to pulling and tilting. The main anchor structure is depicted in Fig. 3.

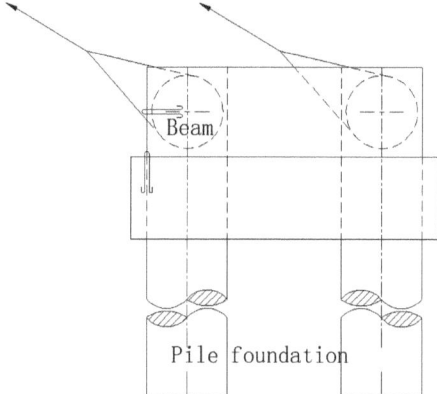

Fig. 3. Main Anchor Structure Diagram.

4 Overall Design Scheme

The overall arrangement of the load-bearing cables for the cable hoisting machine spans (100 + 341 + 86) m, with one set of load-bearing cables installed on the upstream and downstream sides, respectively. When considering the in-place positioning of the arch ribs, a single set of main cables supports the load. The 16-m small box beam is hoisted by a single set of main cables, while the 25-m and 30-m box beams are lifted and installed by two sets of main cables. Therefore, the entire bridge is designed with two sets of transport cables, each with a rated load capacity of 76 tons. Two sets of lifting pulley assemblies, each with a rated net lifting weight of 38 tons, are installed on each transport cable. When both sets of main cables evenly bear the load, the maximum lifting weight can reach 152 tons. The cable hoisting system consists of main cables, lifting cables, traction cables, front and rear wind ropes, trolleys, lifting pulley assemblies, cable saddles, towers, anchors, and winches.

(1) Two sets of 7φ56 mm (6 × 37S + FC) synthetic fiber core steel cables are arranged at the top of the tower as main cables, aligned with the installed rib axis. Both sets of main cables can be laterally moved in an unloaded state. Considering the requirements for hoisting attachment cables, equipment maintenance, transporting small machinery, and construction assistance, one set of 1φ56 mm (6 × 37S + FC) synthetic fiber core steel cables is arranged on the inner side of each set of main cables at the top of the tower (1.2 m from the center of the main cables) as working cables.

(2) The structures of the tower frames on both sides are essentially the same, with a total height of 47.255 m, and the tower tops on both sides are at the same elevation. The tower frame columns are constructed using φ630 × 14 steel pipes, connected longitudinally through flanges, with four on each side, totalling eight steel pipe columns for the entire tower. Horizontal and diagonal belly rods within the column frame use universal joint components 2N4 and 2N5, respectively. The horizontal diagonal belly rods use 2N5 components. The 2N4 and 2N5 belly rods are connected to the column steel pipes by bolts and welded splice plates. The tower frame is 2.4 m

wide longitudinally (center-to-center distance between front and rear columns), with a width of 2.4 m on each side of the column horizontally, an 8.4 m width at the central bracket, a full width of 17.6 m at the top of the tower, and a full width of 13.2 m at the tower base. Universal joint components are used for the lateral connection systems between upper and lower river columns at the tower top and central portions. For the central lateral connection systems, 2N4 and 2N5 components are used. For the upper connection system, 4N3(2N3) components are used for lateral diagonal belly rods based on the force requirements, 4N4 components are used for upper and lower lateral horizontal rods, and 4N1(2N1) components are used for vertical rods, with the remaining using 2N4 and 2N5 components.

(3) For each installation segment of the steel pipe arch rib upstream and downstream, one set of fastening cables is installed, totaling six sets on each side of the rib. Cables 1 to 4 in each set use $2\varphi40$ mm steel cables ($6 \times 37S$ + FC 1670 Mpa), and cables 5 and 6 use $2\varphi48$ mm steel cables ($6 \times 37S$ + FC 1670 Mpa). Each set consists of one cable on each side of the rib. Fastening cables for Sects. 1 and 2 on both sides are guided over the bearing pulley at the top of the abutment and anchored on the pre-embedded tension plates inside the bridge tower. Fastening cables for Sects. 3 to 6 are guided over the seat pulley at the top of the tower and anchored on the front anchor beam in front of the main anchor after passing through the pulley [3].

(4) Each shore is equipped with a structurally identical main anchor, used for anchoring the main cable, working cable, fastening cable, tower rear wind cable, etc., during the installation of the arch rib on that shore. The main anchor is symmetrically arranged relative to the axis of the cable hoisting machine. The main anchor is designed as a pile foundation support structure. The design includes six reinforced concrete piles for each anchor, with a pile diameter of 2.5 m and a pile anchoring depth of 8.5 m. The support platform has a plan dimension of 13.5 m × 9.5 m, a height of 2.5 m, and the front and rear rows of piles protrude from the support platform. Horizontal horizontal anchor beams are set for anchoring steel cables, and the anchor beam has a diameter of 2.2 m. On both the left and right sides, the rear anchor beams anchor a set of main cables and corresponding lifting traction connection cables. The front anchor beams on the left and right sides anchor the working cables, fastening cables, and tower rear wind cables on their respective sides. The anchor is constructed with C40 reinforced concrete, with the requirement that the pile foundation and support platform are embedded in the rock layer, and the foundation bearing capacity is not less than 0.5 Mpa.

(5) The wind cable ropes for the arch ribs use $2\varphi20$ mm (6×37 + FC) synthetic fiber core steel cables. The angle between the wind cable and the ground should not exceed 30°, and the horizontal projection angle of the wind cable with the bridge axis should not be less than 50°. The actual placement is adjusted based on on-site topography measurements, aiming to meet the aforementioned requirements for the wind cable angles. The initial tension of the wind cable should be adjusted according to the actual wind cable angles. To reduce the non-elastic impact of the wind cable sag, the initial tension on one side of the wind cable is controlled at 60 KN, and the tension on the other side is calculated based on the principle of equal lateral horizontal force. Since the lateral stability during the installation of the double ribs is relatively good, wind cables are only installed on the 1st, 3rd, and 5th segments

of the arch ribs. For both ribs, a total of $4 \times 6 = 24$ wind cable ropes are required. In addition to maintaining lateral stability during the arch rib installation process, the arch rib wind cables are mainly used to control and adjust the transverse axis of the arch rib during the installation process.

5 Installation of Steel Pipe Arch Ribs

Each arch rib is divided into 13 hoisting segments, with a total of 16 hoisting segments for the entire arch rib. There are 12 K-braces between ribs, and a horizontal brace in the shape of the letter "M" is installed at the top of the arch. The steel pipe arch ribs are installed symmetrically from the arch foot to the arch top, meeting the symmetry requirements on both sides and upstream and downstream. Each arch rib segment should be installed after the installation of the horizontal brace to ensure the lateral stability of the arch rib. The specific sequence is shown in Fig. 4, and the installation is carried out in numerical order from small to large according to the numbering.

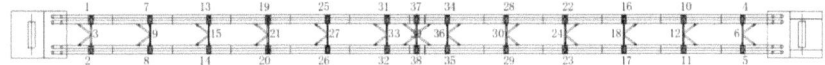

Fig. 4. Arch Rib Installation Sequence Diagram.

Installation of the Arch Foot Hoisting Segment: For each shore's arch foot hoisting segment, the arch rib is divided into left and right sides, creating two hoisting segments. Initially, the downstream side of the arch rib truss piece from the Chaoyang Temple shore is hoisted to the side of the arch seat using cable hoisting. Gradually, the arch rib segment's foot end is placed above the arch seat. With the help of pre-embedded components on the arch seat and the use of chain pulleys and lifting pulley assemblies, the foot end of the arch rib segment is gradually adjusted to the position along the arch rib axis. The temporary hinge seats on both sides slowly insert into the embedded seat plates on the arch seat. Adjust one side's hinge seat hole alignment, fix it with M150 hinge seat bolts, then adjust the alignment of the other side's hinge seat hole, and fix it with M150 hinge seat bolts. At this point, the foot end of the first segment is in place. Adjust the arch rib axis using lateral cable wind at one end of the span, simultaneously install the fastening cables, adjust the installation elevation according to the design benchmark height, and gradually loosen the front hoisting cables. When all the force is transferred to the fastening cables and the arch rib elevation and axis meet the design and code requirements, release the hoisting hooks. Then, using the same method, install the downstream side of the arch rib truss piece for the Chaoyang Temple shore's arch foot segment. Once the arch foot segments on both upstream and downstream sides of the Chaoyang Temple shore are installed, begin installing the wind braces for the arch foot segment. Lift the entire assembly using two sets of main cables. After the wind braces are in place, temporarily secure them with planks and bolts, and then proceed with welding the joints. Use the same method to install the arch foot segment for the Jimingba shore [4].

Installation of General Hoisting Segments (Segments 2 to 6): The construction of general hoisting segments follows the hoisting procedure and construction method of the arch foot hoisting segment. Starting from the first segment of the arch seat, the construction proceeds symmetrically from both shores towards the mid-span, assembling up to the sixth segment. After the installation of symmetrical segments upstream and downstream is complete, the corresponding wind braces should be installed before proceeding to install the subsequent segments.

Transport the arch ribs using the four hoisting points of the two sets of main cables. Adjust the spatial orientation of the arch ribs by slightly lifting or lowering each hoisting point. After adjusting the arch rib truss pieces in place, temporarily connect the lower end joint with the adjacent upper end joint using flanges and bolts. Initially, connect 1 to 2 bolts for each joint, leaving the bolts slightly loose. Once all the bolts for the lower end four-legged steel pipe joints are fastened, tighten each joint bolt in a loop. Then release the lower end hoisting hook, hang the fastening cables, and adjust the lateral adjusting wind cables. Finally, tension the fastening cables. After adjusting the elevation and the axis of the arch rib to meet the design requirements, release the upper end hoisting hook and weld the circumferential weld seams between segments. The elevation of the arch rib is adjusted using the fastening cable pulley assembly, and the lateral axis is adjusted using the arch rib wind cables.

Following the hoisting procedure, after each set of fastening cables is in place, a thorough inspection of the previously fastened cables is required to determine if any cable adjustment operation is needed. The cable adjustment operation, based on the adjustment forces and arch rib elevations jointly issued by the design and monitoring parties on-site, involves using corresponding fastening cables, pulley assemblies, and hoists for each fastening cable. The operation is synchronized, carried out in symmetrical stages, and uses spectrum analyzers to test the cable forces to ensure a smooth cable adjustment process. This ensures the safety of the weld seams between sections, the weld seams of the lateral and transverse connections, and the safety of the bolted connections of the structural components. For each fastening segment, the arch rib axis and elevation must be checked to avoid cumulative errors in the arch rib's alignment and elevation, which may lead to difficulties in adjustment. This ensures effective control of the installation accuracy.

Stability measures during the arch rib hoisting process: For every other section of the arch rib hoisted in place (set at 1st, 3rd, and 5th sections), a pair of $2\varphi20$ mm wind cables is set up upstream and downstream. There are a total of 24 arch rib wind cables throughout the bridge. This is done to adjust the arch axis and ensure its safe stability during the cantilever construction phase. The wind cable forces are calculated to ensure a safety factor of no less than 3, ensuring a certain margin in case the actual cable forces during construction differ from the calculated values.

Installation of the Closure Section: After the installation of the arch ribs and transverse braces is completed, preparations for the closure section are initiated. Before the closure of the arch ribs, the structure is observed in its maximum cantilever state for at least 24 h. Temperature impact observations are carried out within the design temperature range, and precise measurements of the closure length are taken. Based on the measurement results, precise cutting is carried out. The closure section is assembled according to the design requirements, with adjustments and positioning structures set as per the design specifications. Joint plates are installed at the closure, and the closure section is positioned accordingly. Following the closure requirements provided by the monitoring unit, the closure section is welded in place within the specified temperature range [5].

The closure section hoisting is typically scheduled during lower temperatures. As the temperature rises, in accordance with the principle of thermal expansion of steel structures, the spacing between the closure section and the two sides of the arch ribs gradually decreases. When it reaches the design dimensions, positioning horseboards are installed, welded, and locked, completing the closure construction.

6   Analysis of Cable Hoisting Calculation

Utilizing SAP2000 for the calculation of a spatial truss structure, the connections between the chord and the web members within each segment are considered as fully restrained. During the installation process of the steel tube arch ribs, the arch foot is treated as a hinged structure, taking into account the hinged support points. The calculation of the tie rods is incorporated using two-end hinged bar elements, and the connections between the tie rods and the anchoring towers or anchorages are considered as fixed hinges. The tie rods and wind cables are simulated using only tension-only truss elements [6]. The finite element simulation model is illustrated in Fig. 5.

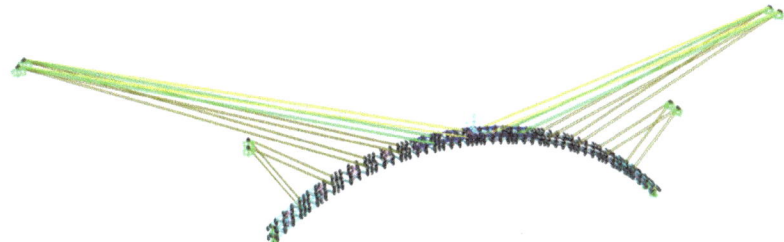

Fig. 5. Finite Element Simulation Model.

First, calculate the tension forces of each cable using the 'zero bending moment method' (considering the hinge joints between the arch feet and each section). Then, use the calculated tension forces along with the self-weight of the arch to perform calculations for the pre-closure state (at this stage, the arch feet are hinged, and sections are considered fixed between them). Adjust the tension forces slightly to ensure that the vertical displacements of the arch nodes meet the specifications (height difference within L/3000 and ±50 mm). At the same time, check that the bending moments in the arch sections are small, and the stresses are within the allowable range according to the specifications. The determined tension forces at this point are considered the tension forces for each cable in the pre-closure state. Next, use the 'inversion method' to calculate the tension forces, arch stresses, and node displacements for each stage. This is done to control the tension forces and deformations of the arch during the dismantling process. If significant deformation occurs during the dismantling process, make slight adjustments to the tension forces. Ensure that the stresses in the arch sections are within the allowable range for each stage, and design the tension forces with a safety factor not less than 3.0 [7]. The calculated maximum tension forces for a single cable are shown in Table 1.

Table 1. Table of Results for Maximum Cable Tension due to Self-Weight of Arch Rib (KN)

Number	T1	T2	T3	T4	T5	T6
Chaoyang Bucking force T_{max} (KN)	165.466	187.690	193.630	228.114	327.500	346.288
Safety factor K	5.33	4.7	4.56	3.87	3.88	3.67
Number	T1	T2	T3	T4	T5	T6
Jiming Bucking force T_{max} (KN)	167.567	188.877	180.602	219.001	301.139	316.459
Safety factor K	5.26	4.67	4.88	4.03	4.22	4.01

In the actual process of arch rib installation, the construction control principle of 'mainly focusing on shape control, with cable tension as a supplement' is followed. As the arch rib segments are installed, the cable tension shows a uniformly increasing trend, and the uniformity is good both upstream and downstream. The measured tension is controlled within 10% of the theoretical value, indicating high accuracy in cable tension calculation. The shape of the arch rib after cable loosening approaches the theoretical control target, with a shape deviation controlled within 20 mm, meeting the requirements for standard errors and demonstrating good shape control [8, 9]. The stress analysis results of the arch rib are shown in Fig. 6.

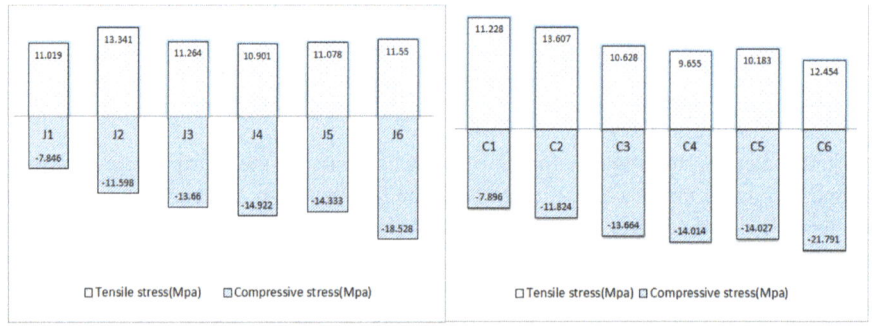

Fig. 6. Arch Rib Stress Analysis Results Chart.

7 Conclusion

Through the study of the key construction technology of cable hoisting for the large-span steel-concrete arch bridge of Xianfeng Chaoyang Bridge, it has been successfully applied in engineering practice, achieving excellent results.

(1) The optimization and innovation of the cable hoisting system have achieved remarkable results in the installation and construction control of the arch ribs, especially the development of software for the calculation of the suspension cables, greatly improving the design efficiency.
(2) For the tension self-balancing of the supporting cables and the optimization of the main tower and main anchor structure, it not only ensures the safety of the cable system but also achieves significant economic value, saving 35% compared to traditional structural construction measures.

References

1. Zhang, Y.: Key Steel Box Girder Hoisting Techniques for Second Changshou changjiang River Bridge in Chongqing. World Bridges **51**(02), 34–38 (2023)
2. Qing, Z.: Innovative technology for cable hoisting of arch bridges in complex geological environments. Highway **60**(08), 120–124 (2015)
3. Chen, Y.: Research on lifting construction technology of large span steel pipe concrete arch. Highway **66**(07), 165–169 (2021)
4. Peng, Z.: Key technologies for lifting reinforced beams in the cable beam intersection area of Wenzhou Oujiang Beikou BRIDGE. Bridge Constr. **52**(06), 140–146 (2022)
5. Liang, Y.: Application of unsupported cable hoisting construction technology in large-span steel pipe truss basket arch bridge. Highway **63**(12), 151–154 (2018)
6. Krahl, A.P., Martins, O.D.D., Carrazedo, R., et al.: Experimental and analytical studies on the lateral instability of UHPFRC beams lifted by cables. Compos. Struct. **209**, 652–667 (2019)
7. Size and post-tensioning cable force optimization of cable-stayed footbridge. Atmaca Barbaros. Structures
8. Yi, J., Wei, C., Huang, Z., Li, Q., Liao, P., Lin, W.: Cable force calculation of cable hoisting of CFST arch bridge research. Buildings **13**(9) (2023)
9. Junjia, Y., Alias, H.A., Haron, A.N., et al.: A bibliometrics-based systematic review of safety risk assessment for IBS hoisting construction. Buildings **13**(7) (2023)

Key Construction Technologies for Steel Truss Bridges in High Water Level and High Flow Velocity Conditions: A Case Study for Shoupanyan Bridge

Haiyong Liu[1], Ming Li[2], Yuelong Dai[3], and Mintao Ou[4(✉)]

[1] Guangzhou CCCC Construction Engineering Co., Ltd., Guangzhou 511466, Guangdong, China
[2] CCCC Second Harbor Engineering Co., Ltd., Wuhan 430040, Hubei, China
[3] CCCC Second Harbor Engineering Sixth Engineering Branch Co., Wuhan 430040, Hubei, China
[4] School of Civil and Hydraulic Engineering, Huazhong University of Science and Technology, Wuhan 430074, Hubei, China
577828404@qq.com

Abstract. Temporary steel truss bridges are increasingly used in engineering projects due to the large-scale construction and development of sea-crossing bridges. However, the construction of these bridges is greatly affected by environmental factors and can face significant challenges when dealing with complex conditions such as fast currents, water depth, and large fluctuations in water level. Therefore, advanced construction techniques are necessary to overcome these challenges. This paper discusses the design, analysis, and construction of a steel truss bridge over water under high water level and rapid flow conditions, using the Shupan Yan Bridge construction project as an example. The paper proposes novel techniques for drilled piles, lateral support, and bridge deck structure. Additionally, a detailed finite element model is established. The results of the finite element analysis demonstrate the suitability and practicality of the novel technology design proposed in this study for constructing steel truss bridges in complex water environments. The technology improves safety during construction. The research findings can provide guidance for similar projects.

Keywords: Bridge construction · construction technology · steel truss bridges · finite element analysis

1 Introduction

In recent years, the rise of marine resource development has provided immense opportunities for the global construction of cross-sea bridges [1]. In order to better utilize China's abundant marine resources, the country has constructed numerous marine platforms and cross-sea channels in coastal areas [2], achieving significant advancements in the theory and technology of bridge engineering. Chen et al. [3] took the Qingdao Bay Bridge as an

A. Bieliatynskyi et al. (Eds.): CSTTE 2023, LNCE 603, pp. 336–345, 2024.
https://doi.org/10.1007/978-981-97-5814-2_30

example to predict the service life of reinforced concrete structures in harsh marine environments. Zhao et al. [4] studied the interaction mechanism between waves, currents, and bridge piers under extreme marine conditions caused by hurricanes or tsunamis, revealing the interplay of waves and currents. Simultaneously, with the development of new materials and equipment, an increasing number of bridge construction projects, both domestically and internationally, will face complex hydrogeological conditions. The research on key technologies for the construction of long-span continuous beam bridges over rivers under complex hydrogeological conditions is expected to see further development and widespread application.

Nevertheless, in these complex environments, the construction of waterborne bridges faces a series of unpredictable challenges. For instance, local erosion caused by ocean currents, waves, heavy rainfall, and wind can impact the stability and fatigue life of sea-crossing bridge foundations[5]. The construction process is also influenced by adverse factors such as swift water flow, deep water, significant water level variations, high waves, and the difficulty of water transport during construction. In such situations, the erection of temporary truss bridges can effectively address these issues. Temporary truss bridges can serve as transportation pathways during construction and as construction platforms for the substructure, transforming waterborne construction into land-based construction. This ensures the normal progress of construction in adverse conditions, significantly shortens construction periods, and has therefore found widespread application in bridge construction projects both domestically and internationally.

However, in general, the construction environment for temporary steel truss bridges is often challenging, and the construction of these bridges is significantly influenced by the surrounding conditions. Particularly for temporary steel truss bridges spanning bodies of water, the impact of complex environmental loads such as wind, waves, and currents is substantial. With numerous calculation parameters involved, comprehensive considerations are essential in the design and construction of these truss bridges [6].

Currently, various forms of temporary steel truss bridges have been constructed in China. With the increasing construction of more sea-crossing and river-crossing bridges, as well as the growth in bridge spans, greater water depths for foundation submersion, and the complexity of hydrological and geological conditions, future bridge construction is expected to rely more on temporary construction steel truss bridges. This further highlights the crucial technical requirements for bridge construction under specific natural conditions. On the other hand, while there is extensive research on the interaction between waterborne bridges and waves domestically and internationally [7–9], there is limited systematic research on trestle bridge projects over water. Existing references often focus on the assessment of losses and damages for steel trestle bridges [10–14]. Therefore, it is essential to conduct a systematic study on the design and construction of temporary trestle bridges over water.

This paper, based on the Panzhihua Shoupanyan Bridge project, utilizes Midas software to conduct finite element analysis on the structural design of steel truss bridges in high water levels and fast-flowing environments. Additionally, it explores the key construction technologies for safely withstanding floods in the context of high water levels and fast-flowing conditions. The aim is to promote the development of temporary construction truss bridge projects, making the design and construction of truss bridges more scientifically rational. Through an in-depth study of the Panzhihua Shoupanyan Bridge project, we hope to provide valuable experience and technical support for addressing construction challenges in complex conditions such as deep water, high flow rates, and shallow cover layers where navigation is not possible.

2 Project Overview

2.1 Overview of the Shoupanyan Bridge

The Shoupanyan Bridge is situated in Panzhihua City. Spanning the Yalong River, the bridge has a total length of 258.0m and a width of 22.0 m. The bridge features a variable-section prestressed concrete continuous box girder layout. As shown in Fig. 1, Piers 1 and 2 serve as the main piers of the Shoupanyan Bridge, located within the main riverbed, with a height of 26.0 m. The main piers adopt a circular end solid structure, with a pier section dimension of 14.0 m × 3.0 m and rounded chamfers at both ends with a radius of 1.806 m. The abutment dimensions are 16.0 m × 9.2 m × 4.0 m, supported by six bored cast-in-place piles with a diameter of 2.5 m each beneath the abutment.

Fig. 1. The layout of the Shoupanyan Bridge span.

The geological conditions at the site of the Shoupanyan Bridge are primarily composed of an 8.6 m layer of conglomerate/gravel sandstone (including 25 cm floating stones) beneath Pier 1, and a shallow covering of 1.2 m conglomerate/gravel sandstone and 9.9 m of silty clay beneath Pier 2.

2.2 Hydrological Conditions

The Shoupanyan Bridge project is located downstream of the Yalong River, with complex hydrological conditions. The Tongzilin Hydrological Station is situated 2.5 km upstream from the bridge site, followed by the Tongzilin Hydroelectric Station at 5.5 km upstream. Due to the control of the hydroelectric station's power generation and the impact of flood discharge during the flood season, the designed bridge site experiences significant daily fluctuations in flow rates. According to hydrological station survey data, during the dry season (November to May), the water flow velocity can reach 3.87 m/s, with a maximum daily flow rate variation of 2500 m³/s. This corresponds to a fluctuation in water level of around 4 m. The flood season occurs from June to October, with the peak flow velocity reaching 5.69 m/s during the main flood period from July to October. The flow rate typically ranges from 3700 to 6500 m³/s. Based on data from the past six years, flow rates exceeding 7000 m³/s occur for a maximum of 7 days during the flood season. Therefore, the design of the truss bridge assumes of normal usage for flow rates below 7000 m³/s.

Table 1. Statistical table of main hydrological parameters for steel truss bridge design.

Serial Number	Flow Rate (m³/s)	Water Level (m)	Remarks
1	3770	3.87	Maximum during Nov-May
2	7000	4.83	Maximum normal usage for truss bridge
3	10000	5.69	Design maximum flow rate (mid-Sept)

3   Structural Design and Analysis of Steel Truss Bridges

3.1   Steel Truss Bridge Structural Design

To facilitate rapid construction and convenient dismantling, the upper structure of the truss bridge adopts an assembled configuration of I-beams and Bailey beams, while the lower structure is designed as a steel pipe pile with steel beam load-bearing structure. The foundation incorporates a combination of bored cast-in-place piles and steel pipe piles. The Bailey steel truss bridge employs a continuous beam structure with a width of 6.5 m. The steel pipe piles for the truss bridge are arranged in pairs, consisting of 630 × 10 mm steel pipes, spaced at intervals of 6.0 m. The top distribution beams of the Bailey beams are two HM488 steel crossbeams, spaced at 0.75 m intervals. Six transverse Bailey beams are arranged, with each beam spaced at 0.9 m intervals. Bridge abutments are set at both ends, with concrete construction walkways cast behind the abutments. Protective railings are installed on both sides of the truss bridge. The specific details of the plan layout for the steel truss bridge are illustrated in Fig. 2.

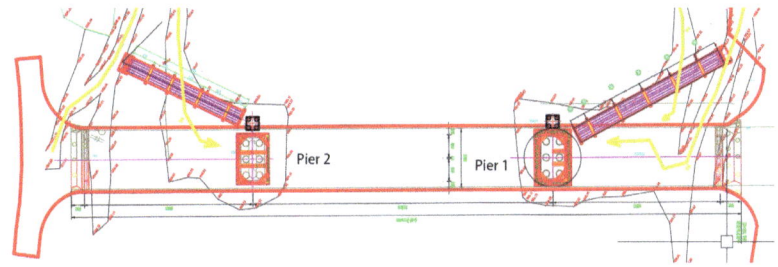

Fig. 2. Layout plan of the steel truss bridge of the Shoupanyan Bridge across the river.

3.2 Hydraulic Flood Assessment for the Steel Truss Bridge

Following the structural design outlined above, a comprehensive modeling and analysis were conducted using Midas numerical simulation software. The resulting model is depicted in Fig. 3.

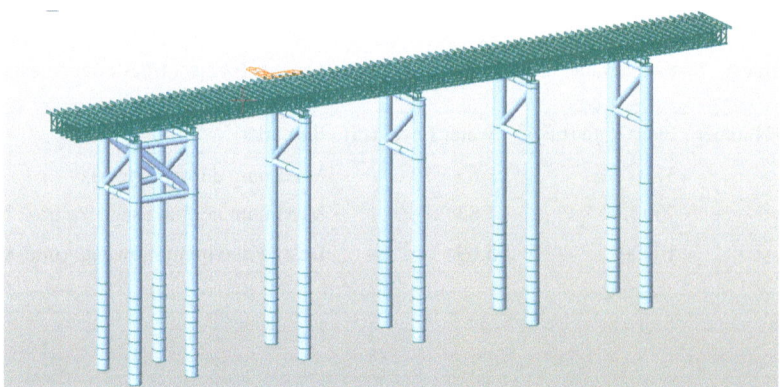

Fig. 3. Model of the steel truss bridge in Midas.

When analyzing the forces on the pile cap distribution beams and steel pipe piles, the interaction between the steel pipe piles and the soil is simulated using the m-method with a specified parameter of m = 5000 kN/m^4. The connections between the upper and lower distribution beams are assumed to be fully pinned, meaning that the bending moment from the upper distribution beams does not transfer to the lower distribution beams. The analysis considers the loads at the design maximum water level and flow velocity. At this point, there are no additional construction loads on the truss bridge or platform. The calculated values for the water flow forces on the steel truss bridge are illustrated in the Fig. 4.

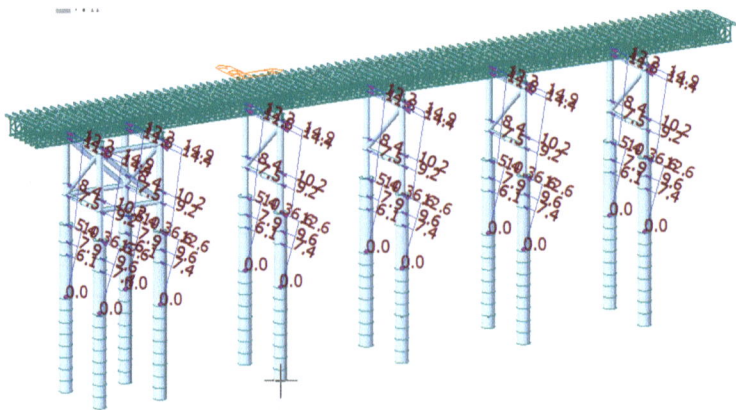

Fig. 4. Results of water flow force calculation on the steel truss bridge.

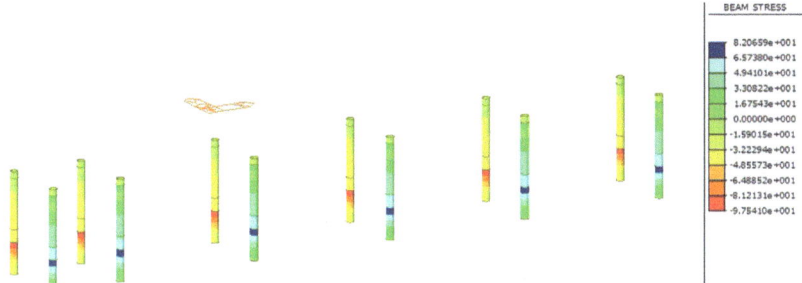

Fig. 5. Steel pipe-pile combined stress (MPa).

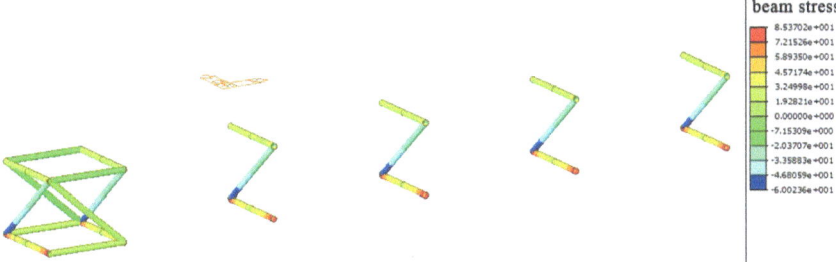

Fig. 6. Combined stress for the connection system (MPa).

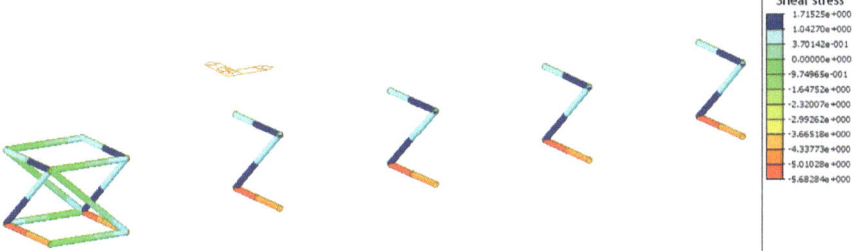

Fig. 7. Shear stress for the connection system (MPa).

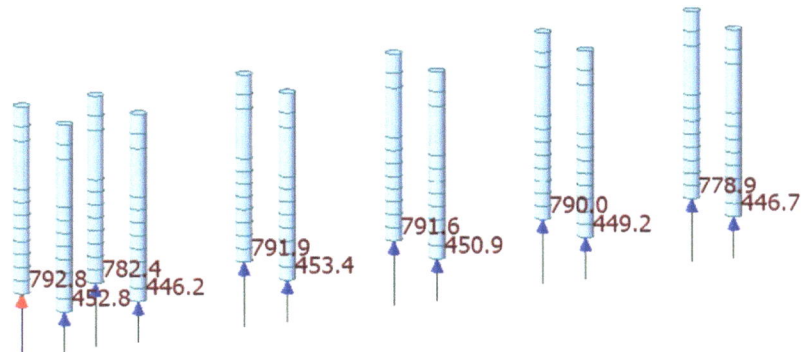

Fig. 8. Drilled pile reaction force (kN).

As shown in Figs. 5, 6, 7 and 8, it can be observed that the maximum combined stress in the steel pipe and connection system is 97.5 MPa, with a maximum shear stress of 5.8 MPa, meeting the design requirements. The maximum vertical reaction force on the drilled pile is 793 kN with no tensile force generated. The design of the steel truss bridge meets the specified requirements.

4 Key Construction Technologies for Steel Truss Bridge

4.1 Pile Foundation Construction

The pile foundation construction involves the use of drilled cast-in-place piles along with steel pipe piles. The pile diameter is 1.2 m, and each section of the steel pipe pile is 10m in length. Given that the truss bridge foundation incorporates embedded rock piles, the construction employs an impact drilling rig. The key to the construction of the pipe pile foundation is to ensure that the steel pipe piles, and the cast-in-place piles are in the same plane and concentric, with their own verticality aligned in a straight line. To achieve this, the construction adopts positioning assistance measures, a level, and a cross-guidance tool.

The specific construction steps are as follows: once the drilled cast-in-place piles are completed and concrete is poured, reinforcement and positioning assistance measures are added around the steel casing. A cross-guidance tool is installed at the top of the casing to ensure concentricity when inserting the steel pipe piles into the cast-in-place piles. The steel pipe piles are then lifted and lowered using a tracked crane, with a level controlling their verticality during the piling process. Once the required alignment is achieved, the insertion of the pipe piles is halted after reaching the designated position before the concrete solidifies.

4.2 Lateral Bracing Construction

After the installation of a row of steel pipe piles, it is crucial to promptly initiate the assembly of lateral bracing between the piles to link the fully driven steel pipe piles into a unified structure, preventing any deviation in their positioning.

The key to successful lateral bracing lies in accurately determining the installation positions. In the lateral bracing installation phase, a precise control method is employed using positioning steel plates and a chain hoist with guidance. The construction process is as follows: initially, measurements are taken to determine the lateral bracing positions, marked with chalk for reference. Subsequently, a positioning steel plate is welded beneath the lateral bracing steel pipe, lifted into the designated position using a chain hoist, and then secured with fixed welds. Full-section welding is then carried out. During the adjustment of lateral bracing positions, a chain hoist is used for additional assistance in positioning.

4.3 Bridge Deck Structure Construction

The bridge deck structure consists of transverse distribution beams and specialized deck panels. The installation of these distribution beams utilizes custom-designed locks, as illustrated in Fig. 9. These custom locks facilitate a more convenient and secure connection between the distribution beams and the Bailey beams. Before laying the distribution beams, the locks are pre-installed on the upper chords of the Bailey beams. During the installation of distribution beams, the fixture's two lips clamp onto the sides of the distribution beam flanges, and securing bolts fasten the fixture, ensuring a tight connection with the flange plate. The connection between the distribution beam and Bailey beam is illustrated in Fig. 10.

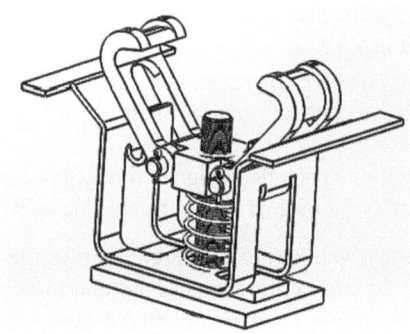

Fig. 9. Custom lock for distribution beam-bailey beam connection.

Fig. 10. The distribution beam-Bailey beam connection

Once the transverse distribution beams are laid, the installation of deck panels commences. The specialized deck panels have segmented dimensions of 3000 × 10 × 5500 mm, and they are interconnected using custom fixtures. The prefabricated steel panels have slots and bolt holes at the corners and the central length, facilitating connection through custom locks and reserved slots. At the center of the deck panel, a custom fixture is employed to connect with the transverse distribution beams. The application of a series of custom fixtures enhances the stability of the connection between the bridge deck panels and the beams.

5   Conclusion

This study examines the construction of a sea-crossing bridge under complex hydrological conditions, using the construction of the Shoupanyan Bridge in China as an example. The aim is to construct a steel truss girder bridge over water, despite the challenges posed by high water levels and fast currents. The study proposes the following novel techniques to achieve this goal:

(1) This task involves achieving precise positioning and vertical control of steel pipes that are inserted into cast-in-place piles during the pile foundation construction stage. This can be achieved using positioning aids, levelling tape, and cross guiding jigs.
(2) For the lateral support construction stage, it is recommended to use positioning steel plates and chain hoists to ensure precise control of the lateral support installation position.
(3) During deck construction, specially designed fixing devices can be used to connect distributor girders and berth girders, as well as bridge deck slabs.

These technologies improve the safety and reliability of steel truss bridge construction. The deformation of the truss bridge under maximum water flow, as obtained from construction monitoring, is consistent with theoretical expectations. This study presents technical insights and experiences that can be applied to the construction of steel truss bridges under complex hydrological conditions.

References

1. Tao, W., Guang-shun, H., Li-jing, D., Rui, Z., Lu, Y., Yue, Y.: The framework design and empirical study of China's marine ecological-economic accounting. Ecological Indicators **132** (2021). https://doi.org/10.1016/j.ecolind.2021.108325
2. Liu, P., Zhu, B., Yang, M.: Has marine technology innovation promoted the high-quality development of the marine economy? ——Evidence from coastal regions in China. Ocean Coastal Manage. **209** (2021). https://doi.org/10.1016/j.ocecoaman.2021.105695
3. Chen, D., Guo, W., Wu, B., Shi, J.: Service life prediction and time-variant reliability of reinforced concrete structures in harsh marine environment considering multiple factors: a case study for Qingdao Bay Bridge. Eng. Failure Anal. **154** (2023). https://doi.org/10.1016/j.engfailanal.2023.107671
4. Zhao, E., Xia, X., Gao, J., Jiang, F., Chen, X., Liu, R.: Performance of coastal circular bridge pier under joint action of solitary wave and sea current. Ocean Eng. **250** (2022). https://doi.org/10.1016/j.oceaneng.2022.111033
5. Chen, J., Qu, Y., Sun, Z.: Protection mechanisms, countermeasures, assessments and prospects of local scour for cross-sea bridge foundation: a review. Ocean Eng. **288** (2023). https://doi.org/10.1016/j.oceaneng.2023.116145
6. Trueheart, M.E., Dewoolkar, M.M., Rizzo, D.M., Huston, D., Bomblies, A.: Simulating hydraulic interdependence between bridges along a river corridor under transient flood conditions. Sci. Total. Environ. **699**, 134046 (2020). https://doi.org/10.1016/j.scitotenv.2019.134046
7. Ti, Z., Zhou, Y.: Frequency domain modeling of long-span sea-crossing bridge under stochastic wind and waves. Ocean Eng. **255** (2022). https://doi.org/10.1016/j.oceaneng.2022.111425

8. Yang, R., Li, Y., Xu, C., Yang, Y., Fang, C.: Directional effects of correlated wind and waves on the dynamic response of long-span sea-crossing bridges. Appl. Ocean Res. **132** (2023). https://doi.org/10.1016/j.apor.2023.103483

9. Li, C., Wu, G.-Y., Li, L.-X., Liu, C.-G., Li, H.-N., Han, Q.: A comprehensive performance evaluation methodology for sea-crossing cable-stayed bridges under wind and wave loads. Ocean Eng. **280** (2023). https://doi.org/10.1016/j.oceaneng.2023.114816

10. Pu, B., Zhou, X., Liu, Y., Liu, B., Jiang, L.: Mechanical behavior of concrete-filled rectangular steel tubular composite truss bridge in the negative moment region. J. Traffic Transp. Eng. (English Edition) **8**, 795–814 (2021). https://doi.org/10.1016/j.jtte.2021.09.002

11. Patil, V., Ahiwale, D.: Damage detection of warren truss bridge using frequency change correlation. Mater. Today Proc. **56**, 18–28 (2022). https://doi.org/10.1016/j.matpr.2021.11.483

12. Lima, J.M., Bezerra, L.M., Bonilla, J., Barbosa, W.C.S.: Study of the behavior and resistance of right-angle truss shear connector for composite steel concrete beams. Eng. Struct. **253** (2022). https://doi.org/10.1016/j.engstruct.2021.113778

13. Zhou, X., Kim, C.-W., Zhang, F.-L., Chang, K.-C.: Vibration-based Bayesian model updating of an actual steel truss bridge subjected to incremental damage. Eng. Struct. **260** (2022). https://doi.org/10.1016/j.engstruct.2022.114226

14. Torres, B., Poveda, P., Ivorra, S., Estevan, L.: Long-term static and dynamic monitoring to failure scenarios assessment in steel truss railway bridges: a case study. Eng. Failure Anal. **152** (2023). https://doi.org/10.1016/j.engfailanal.2023.107435

Analysis of Reinforcement Effect of High Pressure Jet Grouting Pile During Water Inflow of Loess Tunnel Basement During Operation Period

Zhanhu Yang[1], Jinpeng Dai[1(⊠)], Qicai Wang[1], Xincheng Tian[2], and Hao Liu[2]

[1] College of Civil Engineering, Lanzhou Jiaotong University, Lanzhou, China
{daijp,wangqc}@mail.lzjtu.cn
[2] China Second Metallurgy Group Corporation Limited, Baotou, China

Abstract. In the built loess tunnel, a large number of diseases have been found during the operation period, especially when the tunnel basement is after water inflow. At present, a large number of high pressure jet grouting pile are used to reinforce the tunnel basement, but there are few studies on the reinforcement effect of high pressure jet grouting pile when the tunnel basement is after water inflow during the operation period. Due to the operation control of the expressway, it is difficult to carry out field research during the operation period. In this paper, by changing the physical parameters of the water inflow layer of the surrounding rock of the tunnel basement, numerical simulation is used to analyze the reinforcement effect of high pressure jet grouting pile on the tunnel basement under the loads, and field monitoring means are used to confirm the correctness of the three-dimensional model. The results show that the reinforcement of high pressure jet grouting pile can prevent the uplift caused by the softening of tunnel basement surrounding rock. The error of monitoring data and simulation results is also within a reasonable range.

Keywords: Reinforcement of Loess Tunnel Basement · High Pressure Jet Grouting Pile · Operation Period · Numerical Simulation · Tunnel Monitoring

1 Introduction

A large number of diseases have been found in the completed loess tunnels, such as inverted arch cracking, excessive surface deformation, and lining failure [1]. Considering the characteristics of loess itself, the strength loss of the surrounding rock of the basement is large in the case of water ingress during the operation period, so the reinforcement of the loess tunnel foundation has always been a key research direction.

At present, high pressure jet grouting pile [2] are well used in loess tunnels due to their advantages of high strength, simple construction equipment, small floor area, fast construction speed, and good reinforcement effect on high compressibility and strong collapsible loess.

Some scholars at home and abroad have conducted a lot of researches on the high pressure jet grouting pile method, and Li Xiaojie et al. [3] compared and analyzed the measured and calculated values of the bearing capacity and settlement of the composite foundation of the high pressure jet grouting pile with the background of engineering practice. Lai Jingxing et al. [4] used numerical simulation combined with field measurements to analyze the consolidation of the foundation of the loess tunnel reinforced by high pressure jet grouting pile. Xuan Loi Nguyen et al. [5] took the Zhonghua Tunnel project in Vietnam as an example to study the principle and application effect of high pressure jet grouting pile in underground engineering reinforcement in Vietnam by comparing two methods: theoretical calculation formula and on-site monitoring and measurement.

To sum up, although a lot of research has been done by scholars at home and abroad, it is impossible to carry out research on tunnel water intake during the operation period due to the operation control of expressways, and the current research in this area is also very lacking. In this paper, the research is carried out by numerical simulation, and the correctness of the model is demonstrated by the monitoring data during the construction period, so as to ensure the reliability of the research.

2   Project Overview

The relying project is located in Ning Jiaping Village, Qing Shuiyi Township, Yuzhong County, Lanzhou City, with a total length of 674 m, a maximum buried depth of 97.0 m, an excavation width of 12 m, and a tunnel longitudinal section as shown in Fig. 1. Tunnel profile (Source: Rely on engineering design documents). The strata in the area where the tunnel is located can be divided into three layers, from top to bottom, the Quaternary Upper Pleistocene aeolian slightly dense loess (Q_3^{eol}), the Quaternary Upper Pleistocene aeolian meso-dense loess (Q_3^{eol}) and the Cretaceous Lower Cretaceous estuarine group 1 (K1hk1) sandstone. In addition, the collapsibility grade of the loess is grade 4, which has serious collapsibility, and the natural bearing capacity of the surrounding rock at the bottom of the tunnel cannot meet the design requirements, so the tunnel bottom must be reinforced to ensure the stability of excavation and the safety of the operation period.

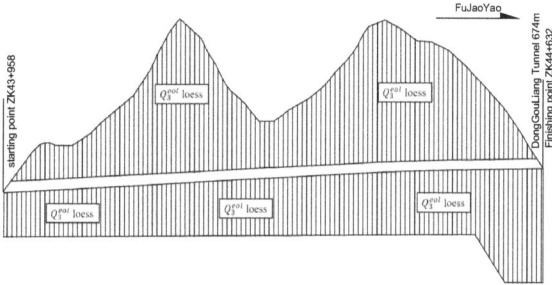

Fig. 1. Tunnel profile (Source: Rely on engineering design documents)

3 Substrate Reinforcement Scheme and Tunnel Monitoring Scheme

3.1 Reinforcement Schemes

The tunnel reinforcement scheme and the parameters of the high pressure jet grouting pile are shown in Fig. 2

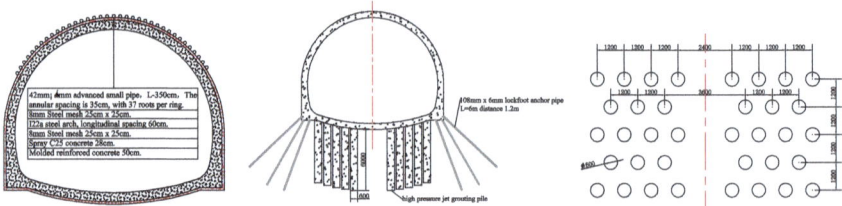

Fig. 2. Tunnel reinforcement scheme (Source: Rely on engineering design documents)

3.2 Tunnel Monitoring Program

In this paper, a total of 8 earth pressure cells are buried at the monitoring section, which are respectively buried at the vault, spandrel, side wall, arch foot and inverted arch center, and the supporting automatic acquisition system is used to collect data at 1 h intervals, please see Fig. 3 for details.

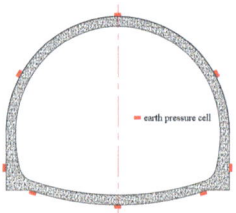

Fig. 3. Pictures of the tunnel monitoring site (Source: Design documents and picture taken by the author)

4 Numerical Simulation Under Water Ingress Conditions During the Operation Period of the Tunnel Bottom

4.1 Selection of Vehicle Loads and Pavement Loads

According to the design documents, the pavement load acting on the inverted arch is calculated by gravity of the surface course. In the design of underground structures, the vehicle load is often treated according to the uniform load, and the road design code is treated according to the concentrated force, and these two treatment methods can not well reflect the vibration effect of the vehicle load. In this paper, the dynamic load coefficient is introduced to convert the dynamic load into a quasi-static load [6], and it is found that the dynamic load coefficient of the vehicle reaches 0.1–0.4 in the many researches [7].

In the conventional vehicle moving dead load method, considering the high-frequency characteristics of the vehicle load, the vehicle load is simplified into a uniform load distributed along the road surface in a strip, the width is taken as the width of the standard contact surface b of the front wheel of the vehicle, and the value of b is not changed by the vehicle overload, and the length of action is taken as the length d of the vehicle, and the moving load is distributed on the two contact surfaces, and the calculation diagram is shown in Fig. 4. The vehicle moves the dead load $P = (1 + \mu)P_0$, P_0 is the load of the standard vehicle. The backing project is a highway-I. class. The vehicle gravity standard value is 550 kN, the left and right wheel tracks of the vehicle are 1.8 m, the wheelbase is $3 + 1.4 + 7 + 1.4$ m, the front wheel landing length and width is 0.3 m × 0.2 m, the middle and rear wheel landing length and width are 0.6 m × 0.2 m, the vehicle dimensions are 15 m × 2.5 m, and the vehicle dynamic load coefficient is 0.4. The calculated vehicle load is 128 Kpa, taking into account the most unfavorable scenario, according to the two-lane arrangement.

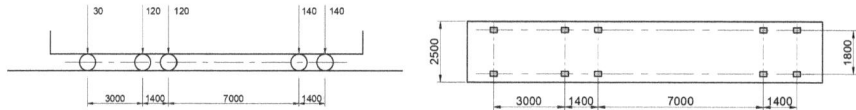

Fig. 4. Vehicle load layout (axle load unit: kN; Size unit: m) (Source: Technical standards for highway engineering)

4.2   Parameter Selection of Inlet Layer at the Bottom of the Tunnel

Jian Tao et al. [8] analyzed the shear modulus of undisturbed loess with different moisture content under different confining pressures, and found that when the confining pressure is constant, the higher the moisture content, the smaller the shear modulus, and when the moisture content increases from 3% to 18%, the shear modulus decreases between 13%–60%. There are many scholars at home and abroad who have found the same law. In addition, a large number of experiments show that when the elastic modulus of loess is determined, its physical parameters have a clear value range, and the current physical parameters of loess have been obtained through experiments, and the attenuation degree of elastic modulus of 15%, 30%, 45% and 60% is used to reflect the degree of water inflow at the bottom of the tunnel.

4.3 Establishment of Finite Element Modeling

Basic Assumptions

The rock mass material is assumed to be homogeneous and isotropic, and the more applicable Drucker-Prager criterion is used for the constitutive relationship of rock and soil, which overcomes the Mohr-Coulomb criterion by considering the influence of the medium principal stress compared with the Moore-Coulomb criterion The main drawback of the Coulomb criterion. The shotcrete and micro anchor pipes in the supporting structure are regarded as linear elastic materials, and nonlinearity is not considered. The initial stress field of the rock mass only considers the self-weight stress, the effect of the grouting of the advanced small conduit is simulated by improving the physical parameters of the reinforcement layer, and the supporting effect of the steel arch and the reinforcement mesh is converted to the shotcrete, and the elastic modulus and gravity of the second-lined reinforcement mesh are also converted to the second-lined concrete.

Although the model based on the above assumptions is different from the actual situation, these assumptions have been verified by time, and it can effectively reflect the law of force and deformation, and it is convenient to operate, which is a powerful tool for scientific researchers.

Model Parameters

The parameters of the surrounding rock material and high pressure jet grouting pile are shown in Table 1.

Table 1. Material parameters (Source: Specifications for Design of Highway Tunnels Sect. 1 Civil Engineering)

Name	Weight $\gamma/(kN/m^3)$	Elasticity modulus $E/(MPa)$	Poisson's ratio μ	Internal friction angle $\varphi/(°)$	Cohesive force $C/(kPa)$
Loess (slightly less dense)	14	25	0.35	21.1	13
Loess (medium density)	15	70	0.35	26	22
C25 primary lined concrete	23.552	31536	0.2	/	/
C35 secondary lined concrete	25	32756	0.2	/	/
high pressure jet grouting pile	21.5	1200	0.25	/	/
Lock foot anchor pipe	78.5	206000	0.25	/	/
Pipe reinforcement	15.2	604	0.3	23	120
C15 invert backfill concrete	23	22000	0.2	/	/

Model Building

The three-dimensional model is established in strict accordance with the relying project, and the stratum is divided into two layers, the upper layer is a slightly dense loess layer with a depth of 0–20 m, and the lower layer is a medium-dense loess layer with a depth of more than 20 m, see Fig. 5. The normal stiffness modulus (K_n) and shear stiffness modulus (K_t) of the pile interface unit are automatically calculated by the software interface assistant, and the final shear force and pile end bearing capacity are related to the ultimate side friction resistance of the pile and the ultimate end resistance of the pile, according to the "Technical Code for Building Pile Foundation" JGJ94-2008 and considering the effect of the rotary grouting pile gourd-like pile body on the improvement of bearing capacity, the pile end spring stiffness can be taken according to the previous experience [9], and the specific parameters are shown in Table 2.

Table 2. 1D pile element parameters (Source: Software automatically generates and references 9)

Name	Finally Shearing force /(kN/m^2)	Shear stiffness modulus /(kN/m^3)	Normal stiffness modulus /(kN/m^3)	Pile tip bearing capacity /kN	Pile tip spring stiffness /(kN/m)
Pile interface	250	244855.967	2693415.64	/	/
Tip of pile	/	/	/	400	80000

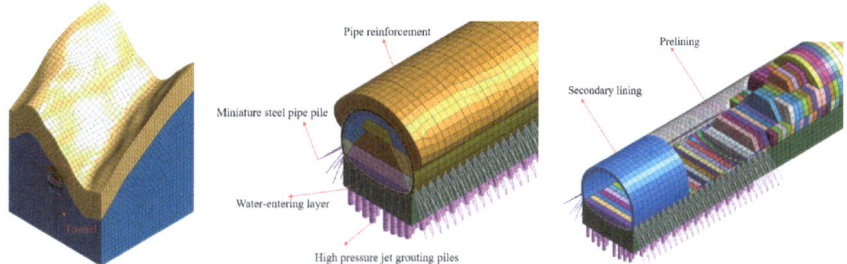

Fig. 5. 3D FEM model (Source: The author built the model by himself)

5 Calculation Results and Analysis

In order to analyze the variation law of tunnel bottom stress and displacement with the construction stage, five transverse analysis points were selected at the monitoring section, which were located on the surrounding rock at the tunnel bottom below the "inverted arch initial lining-40" (longitudinal center section of the tunnel), and the analysis points were numbered from left to right as point C' at the foot of the left wall, point B' in the middle of the left half of the inverted arch, point A in the center of the tunnel, Point B in the middle of the right half of the inverted arch and point C at the foot of the right wall, see Fig. 6. Analysis points position.

Fig. 6. Analysis points position (Source: Author's drawing)

5.1 Analysis of Soil Displacement Between Piles

As can be seen from Fig. 7, the vertical settlement at the analysis point becomes larger and larger with the decrease of the elastic modulus of high pressure jet grouting pile. The vertical settlement of the analysis points on the left and right arch feet is close to linear growth, and the growth rate of the vertical settlement of the analysis points in the middle of the inverted arch and the left and right halves of the inverted arch (the slope of the vertical displacement connection line of the adjacent two working conditions) gradually decreases with the decrease of the elastic modulus of the inlet layer, because the vertical displacement at the analysis point of the surrounding rock of the inverted arch is an absolute displacement, which is the difference between the overall settlement of the structure at the inverted arch and the uplift of the inverted arch under the action of load.

When there is no high pressure jet grouting pile at the bottom of the tunnel, the vertical displacement at the analysis point when the elastic modulus is not attenuated under the load is larger than that when there is a jet grouting pile at the bottom of the tunnel. Until "E decreases by 15%", the displacement growth rate at the analysis point is positive. After the "E decreases by 15%", except for the left arch foot, the displacement growth rate of the rest of the positions turns negative, that is, the uplift begins to rise, and with the continuous decrease of the elastic modulus, the uplift growth rate decreases continuously (the absolute value gradually increases), because the elastic modulus of the inlet layer continues to decrease without the reinforcement of the jet grouting pile, and the ability to resist the uplift of the surrounding rock at the bottom of the tunnel is getting worse and worse. And because the buried depth at the left arch foot is much greater than that of other positions, the phenomenon of bulge only occurs after "E is reduced by 45%".

It can be seen that the high pressure jet grouting pile is very important for the reinforcement of the tunnel base, which can effectively increase the ability to resist deformation after the water ingress at the tunnel bottom, and prevent the uplift phenomenon (relative displacement) caused by the softening of the surrounding rock at the tunnel bottom, and its most unfavorable position is at the center of the inverted arch.

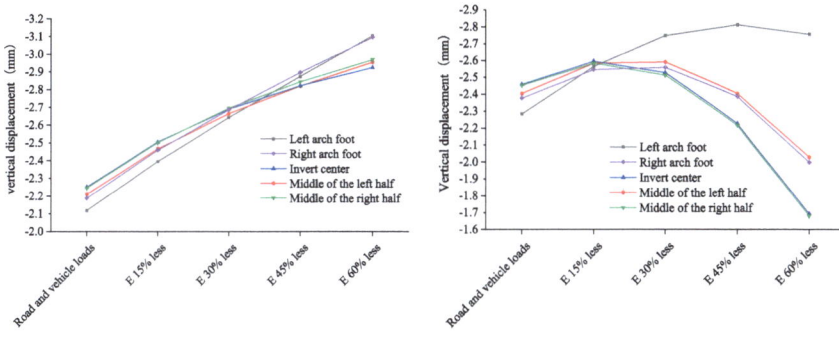

(a) High pressure jet grouting pile   (b) Without high pressure grouting jet pile

Fig. 7. Displacement of tunnel basement with attenuation of elastic modulus of submerged layer (Source: The author draws according to the calculation results of the model)

5.2 On-Site Measurement and Analysis

The number and position of the earth pressure cell at the monitoring section are shown in Fig. 8 (Source: Author's drawing). Figure 9 (Source: The author draws according to monitoring results) shows the variation of earth pressure over time, due to the impact of the epidemic, it was not possible to inspect the equipment on site in time. After the earth pressure cell is buried for 4 h, the data acquisition equipment is powered off, and the data acquisition equipment is re-powered at 1080 h, but does not affect the overall distribution law of the radial earth pressure of the surrounding rock, in the 8 earth pressure cells, the data is not collected for 523955 damage, and 523960 is damaged at the 3283 h, but its data has been stabilized. The rest of the earth pressure cell data is normal, and its pattern is gradually increasing with time until it is stable. As can be seen from Table 3, the surrounding rock is mainly subjected to compressive stress, with the pressure at the arch foot being the largest, followed by the spandrel and the lowest pressure at the center of the inverted arch. Although there is a certain deviation between the measured values and the numerical simulation results, they are within a reasonable range, and the basic law of the distribution of surrounding rock and soil pressure is consistent, which shows that the calculation results are credible.

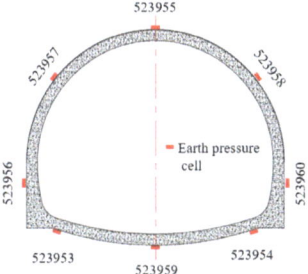

Fig. 8. Earth pressure cells arrangement

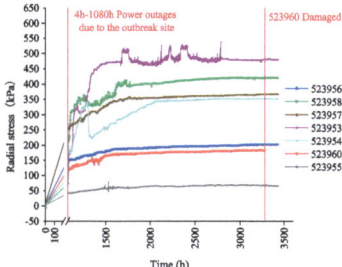

Fig. 9. Earth pressure monitoring data

Table 3. Radial pressure of the surrounding rock after the tunnel is completed (Source: The author draws according results)

Location	Numerical simulation value (kPa)	Measured value (kPa)	Error (Column 2- Column 3)/ Column 3
523955	−204.85	0	/
523957	−399.15	−367.22	8.7%
523958	−450.89	−419.3	7.5%
523956	−231.11	−201.82	14.5%
523960	−198.65	−174.81	13.6%
523953	−527.54	−480.06	9.9%
523954	−394.53	−351.13	12.4%
523959	−69.37	−65.4	6.1%

6 Conclusion

After the high pressure jet grouting pile reinforces the tunnel bottom, the uplift phenomenon caused by the softening of the surrounding rock at the tunnel bottom can be effectively inhibited when the water enters the tunnel bottom during the operation period, so that the initial lining of the inverted arch is in a state of compression, and the monitoring results also prove that the model calculation results are reasonable.

It can be seen that the bottom of the tunnel reinforced by the high pressure jet grouting pile is different from the natural foundation, which not only has to resist the settlement of the inverted arch surrounding rock under the load, but also resists the uplift of the inverted arch surrounding rock when the tunnel bottom enters the water, and the pile body stress is very complex in this process, and it should be considered in the design, in addition, the drainage must be done in the tunnel maintenance.

References

1. Youyun, L., Yawei, Z., Jianguo, Y., et al.: Numerical analysis of stress and deformation of jet grouting pile foundation under creep condition of large section loess tunnel. J. Chin. Rail. Soc. **43**(03), 166–174 (2021)
2. Cheng, S.-H., Chao, K.-C., Wong, R.K.N., et al.: (2023) Control of jet grouting process induced ground displacement in clayey soil. Transp. Geotechn. **40**, 100983 (2023). https://doi.org/10.1016/j.trgeo.100983
3. Li, X.: Analysis of calculation methods for bearing capacity and settlement of high pressure chemical churning pile composite foundation. Rock Soil Mech. **25**(9), 1499–1502 (2004). 94.1000–7598.2004.09.032
4. Lai, J., Fan, H., Xie, Y., et al.: Consolidation analysis of jet grouting pile reinforcement in Joess Tunnel. J. Chang'an Univ. (Nat. Sci. Ed.) **36**(2), 74–79 (2016). CNKI:SUN:XAGL.0.2016-02-011

5. Nguyen, X.L., Li, W., Nguyen, K.T., et al.: Application research of high pressure jet grouting pile in an underground engineering in Vietnam. Arch. Civ. Eng. **66**(3), 575–593 (2020)
6. Quan, Y.: Calculating method of vehicle dynamic load caused by uneven pavement. Xi'an University of Science and Technology (2018)
7. Buhari, R., Rohani, M.M., Abdullah, M.E.: Dynamic load coefficient of tyre forces from truck axles. Appl. Mech. Mater. **2685**(405–408), 1900–1911 (2013)
8. Tao, J., Lingwei, K., Wei, B., et al.: Experimental study on effect of water content on small strain shear modulus of undisturbed loess. Chin. J. Geotech. Eng. **44**(S1), 160–165 (2022)
9. Shan, C., Gan, L., Wang, Y., et al.: Analysis of mechanical behaviors and deformation characteristics of the loess tunnel reinforced by lime-soil compaction piles. Modern Tunneling Technol. **57**(2), 86–95 (2020). https://doi.org/10.13807/j.cnki.mtt.2020.02.013

Analysis of Tunnel Excavation Deformation Based on FLAC3D

Wensheng Ge[1,2(✉)]

[1] Maanshan Engineering Technology Research Center for Water Resources Efficient Utilization in Hilly Region, Maanshan 24303, Anhui, China
38445589@qq.com
[2] Wanjiang University of Technology, Maanshan 24303, Anhui, China

Abstract. Deformation analysis of tunnel excavation is one of the core problems. FLAC3D is used to simulate tunnel excavation, including tunnel modeling, excavation and support, and analyzes the force and deformation characteristics in the construction process. On the basis of collecting tunnel data, made reasonable monitoring measurement scheme, and measurement of the real-time monitoring data for the initial analysis, analysis the excavation time and excavation length regarding the deformation of the surrounding rock, predicted the deformation trend, determine the total displacement of the surrounding rock deformation, and map the relationship to time to provide a reasonable time for the secondary lining, using FLAC3D software for numerical simulation of tunnel excavation, including tunnel modeling, excavation, support, and analyzed the force and deformation characteristics in the construction process. Different combined values of K and E are designed and applying the model built by the software as the normal calculation program of numerical simulation to obtain different displacement values of arch subsidence and hole circumference convergence.

Keywords: tunnel · FLAC3D · numerical analysis

1 Introduction

With the application and popularization of computer technology in rock mass mechanics and geotechnical engineering, the application of numerical calculation in geotechnical engineering has developed rapidly [1]. However, because of the heterogeneity and discontinuity of rock materials, the rock mass parameters are generally difficult to determine, even if indoor tests and large field tests are often used in engineering, can not fully reflect the nature of the rock mass within the scope of the project, which causes the theoretical calculation value and the actual value is very different [2], based on the actual displacement can effectively determine the parameters of numerical calculation method. The establishment of a systematic and practical rock mass parameter inversion method has a good guiding significance for the tunnel design and construction [3], but also will produce significant economic benefits.

The elastic plasticity model was implemented using the FLAC3D program, which simulated the deformation and stress of mud shale during the construction process.

A. Bieliatynskyi et al. (Eds.): CSTTE 2023, LNCE 603, pp. 356–366, 2024.
https://doi.org/10.1007/978-981-97-5814-2_32

(Huang et.al., 2009) the modeling and monitoring of a clay shale tunnel is studied [4]. This method can be used to study the effect of construction on the performance of mud shale tunnel. Combined with the actual engineering of Changling Tunnel of Guangdong Highway (Dengetal., 2011) [5], the influence of different sizes and positions of karst caves on the deformation of double-track tunnel was analyzed by FLAC3D. Since the cave is located above or below the left tunnel, the simulation indicates that the deformation characteristics of the surrounding rock are similar [6]. Based on the numerical simulation of the elastic response and EDZ behavior around the tunnel openings (in clay caused by excavation and ventilation), (Li et.al., 2013) found a larger radial elastic deformation tunnel of the roof and floor, and a significantly different stress distribution than SPHM [7]. (Li et al., 2013) focused on the numerical study of the mechanical response of excavation and ventilation around tunnels in clay rock. Throughout the tunnel, the patterns and expansion of the TPHM calculated for SPHM were clearly different from those calculated for SPHM. The entire failure process of the model was tracked by the 800 mm 800 mm 200 mm physical model (PM) (Zhao et al., 2015). Under high stress conditions, the deformation localization phenomenon in SR becomes severe [8], and different spatial positions exhibit different deformation features. A multiple grouting method is proposed to analyze the mechanical deformation behavior of surrounding rock using FLAC 3 D fast Lagrangian analysis) software (Gong et al., 2018).In the test site work, the grouting scheme reduced the maximum displacement from 300 t/h to 40 t/h, with no obvious deformation and abnormal stress in the tunnel.(Wang et al., 2019) Large deformation of long wall coal mine tunnel is studied. Stress-related elastic properties were not considered. The model provides a useful tool for assessing the safety edges of the underground tunnel. The numerical simulation and parameters of shield tunnel are introduced in detail, and the mechanism and characteristics of the site deformation caused by shield tunnel are systematically discussed (Li et al., 2020).The influence of construction parameters on the site deformation, including construction sequence, tunnel spacing and chamber pressure [9]. (Kabwe et al., 2020) present, a constitutive model of viscoelastic viscoplasticity (FDVP) with fractional energy estimation of compression delay deformation. Then, the constitutive equation is implemented in FLAC3D and applied to simulate the constitutive deformation of compression within the tunnel for validation using the built-in constitutive model. (Li et al., 2020) Objective To discuss the stress transfer mechanism of pile foundation and the influence of shield tunnel construction on pile stability. The structural deformation of shield tunnel lining mainly occurs near the pile foundation support area, mainly from the settlement deformation with small horizontal displacement. (Xu et al., 2021) a numerical method is proposed to study the damage evolution of secondary tunnel lining under the combination of preliminary supporting corrosion and surrounding rock creep deformation. The improved numerical method includes three main models: (1) tensile failure of corroded rock bolts; (2) yield of corroded supporting steel arch; and (3) elastic-plastic failure of secondary tunnel lining [10].

2 The Numerical Simulation

2.1 The Constitutive Model Employed in the FLAC3D

The constitutive model is actually a mathematical expression of the stress-strain relationship, which in a sense represents the general relationship between stress and strain. FLAC3D In order to meet the needs of all kinds of engineering research, a good simulation of the damage limit of 12 types of geotechnical constitutive model, including a zero model nulll, three elastic model (isotropic, isotropic and orthogonal anisotropic) and eight plastic models (Drucker-Prague (Drucker-Prager), Moore-coulen (Morh-Coulomb), strain hardening/softening, throughout joint, bilinear strain hardening/softening throughout joint, modified Cambridge and Hookbrown). Due to the use of finite difference method display scheme to handle field control differential equation, combined with the use of mixed discrete method, therefore, it can well simulate the material yield, plastic flow, softening large deformation, especially in the material elastic plasticity analysis, large deformation analysis and construction process simulation has unique advantages.

2.2 FLAC3D Calculation Process

The required calculation object is discretized into a finite difference grid with hexahedral cells, and each hexahedral cell can be further divided into several constant strain tetrahedral cells. Now to present the calculation process with a tetrahedron. The tetraheon shown in Fig. 1, the number of nodes is 1 to 4, and the n th surface is the face of node n, shown in Fig. 2.

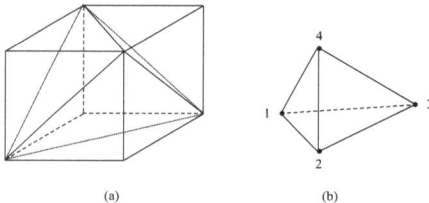

(a)   (b)

Fig. 1. The cube units are divided into five constant strain delta cone units

Suppose that the velocity component at a point within this tetrahedron is v_i By applying the Gaussian divergence theorem to the triangular cone shape unit, we can derive:

$$\int_V v_{i,j} dV = \int_S v_i n_j dS \tag{1}$$

The integral in the formula represents the integral of the volume V and area S of the tetraheon cell, and njIs represents the outer normal vector of the tetrahedral surface.

For the tetrahedral units of a constant strain, v_iIs a linear distribution, and n_jBe unchanged on the same face. Then it can be obtained by summing over the upper formula:

$$v_{i,j} = -\frac{1}{3V} \sum_{l=1}^{4} v_i^l n_j^{(l)} S^{(l)} \tag{2}$$

FLAC3D With the node as the calculation object, both the force and the mass are concentrated on the node, and then solved in the time domain through the equation of motion. The motion equation of the node is expressed as follows:

$$v_i^l\left(t + \frac{\Delta t}{2}\right) = v_i^l\left(t - \frac{\Delta t}{2}\right) + \frac{F_i^l(t)}{m}\Delta t \tag{3}$$

$F_i^l(t)$ Where: It is the unbalanced force component of time t l node in the i direction, and m is the concentrated mass of node i.

The unit strain increment form of a certain step is expressed as:

$$\Delta e_{ij} = \frac{1}{2}\left|\frac{\partial v_i^l}{\partial x_j} + \frac{\partial v_j^l}{\partial x_i}\right|\Delta t \tag{4}$$

Δe_{ij} Where: it is the strain increment of the unit.

As the strain increment increases, the stress increment is obtained by the constitutive equation, and then the total stress is obtained by the superposition of the stress increment at each step. Then, the node imbalance force of the next step can be found by the virtual work principle, and the calculation of the next step can be entered.

This paper uses the most widely used Moore-Coulomb strength criterion in the rock mass calculation, and its yield function is:

$$|\tau| = c + \sigma_n \tan\varphi \tag{5}$$

where τ is the shear stress on the yield surface; σ_n is the normal stress on the shear surface; the c is the impedance per material area; the φ is the internal friction angle of the rock mass, depending on the roughness of the shear surface. From Fig. 2 the ultimate stress circle is:

$$f(\sigma_1, \sigma_2, \sigma_3) = \frac{1}{2}(\sigma_1 - \sigma_3) - \frac{1}{2}(\sigma_1 + \sigma_3)\sin\phi - c\cos\phi = 0 \tag{6}$$

In e.g. $\sigma_1, \sigma_2, \sigma_3$ Main stress.

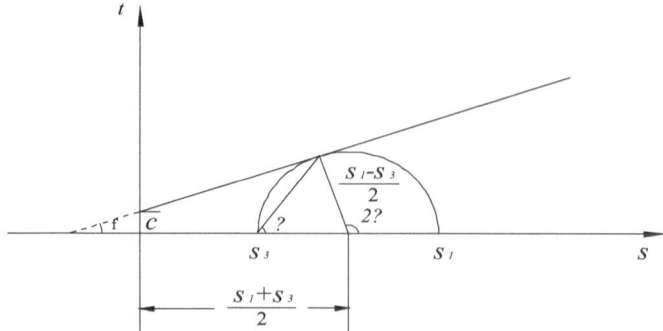

Fig. 2. Ultimate stress circle

3   Using the Template

This calculation only simulated the excavation and the lower steps, because the actual arch excavation lag many, displacement can be stable, in the process of parameter inversion is not simulation, at the same time the right hole excavation behind the left hole 50 m, also is not simulation, and because is V surrounding rock, using the core soil annular excavation.

Modeling of the model was carried out in ANSYS and then imported into FLAC3D. Other modeling, calculation, and analysis were all carried out in FLAC3D. The built model is shown in Fig. 3.

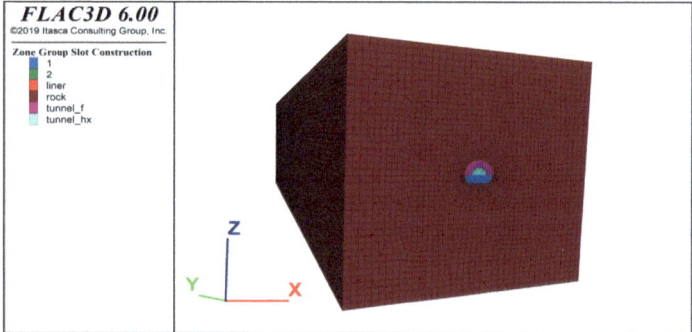

Fig. 3. FLAC3D Numerical model and mesh division

4   Numerical Calculation

List a set of values (E = 0.3 GPa, K = 1.2) to calculate, the initial ground stress of the shallow buried section is mainly generated by the stress field, according to formula $\sigma_{xx} = \sigma_{yy} = K\sigma_{zz}$. Due to rock mass fragmentation and more cracks, the Moore-Coulomb constitutive model is adopted.

The tunnel excavation is divided into three parts: the non-core soil part of the upper steps, the core soil part of the upper steps and the lower steps, and the whole process is written by FLAC3D built-in FISH. There are 3 cycle steps, 1 large cycle and two nested small cycles. In the simulation process, immediately after the tunnel excavation, the steel arch, grouting and the anchor are installed, without considering the construction gap. The built model is shown in Fig. 4.

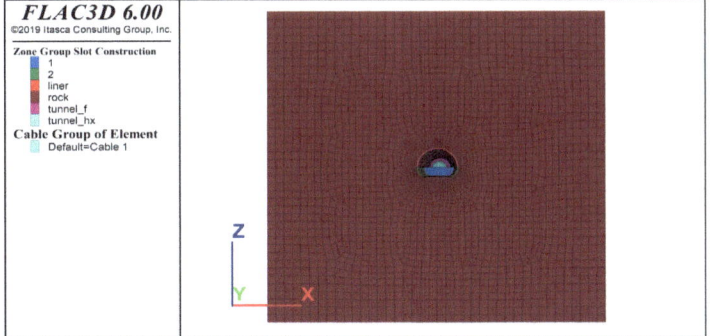

Fig. 4. Form and support after 10 m of tunnel excavation

Section surface subsidence, vault subsidence and surrounding convergence at 10 m were always monitored throughout the simulation, and these values were recorded in the storage documents.

4.1 Stress-Field Analysis

After tunnel excavation, due to the unloading effect of excavation, the stress redistribution occurs, the rock stress state around the tunnel changes, and the surrounding rock in the unfavorable stress state will be damaged. The built model is shown in Fig. 5.

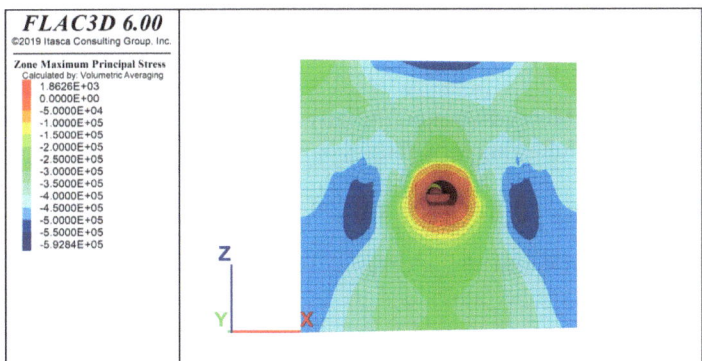

Fig. 5. Maximum main stress of 10 M in tunnel excavation

In the secondary stress field formed after excavation, the main stress direction of the rock around the hole is obviously deflected. In general, the maximum main stress direction is parallel to the excavation wind surface, and the minimum main stress direction is perpendicular to the air surface. As shown in Fig. 4, the closer the surface, the greater the deflection from the surface. The stress of the lining is significantly higher than that of the surrounding rock, as shown in Fig. 5, indicating the majority of the lining support, the maximum main stress is the compression stress along the cutting direction of the tunnel excavation surface, the hole axis of the minimum main stress, the minimum main stress of the arch part ① the compression stress at the excavation stage, As the palm surface forward the compressive stress gradually into tensile stress; ② The minimum main stress after the excavation is the tensile stress, And gradually decreases as the palm surface moves forward; ③ The minimum main stress of the lower step is always the compressive stress, And gradually increased with the advance of the palm surface (the same trend as ②); This is because the arch lining gradually bears the main load after excavation, And manifested as longitudinal pressure and lateral tension. The built model is shown in Fig. 6.

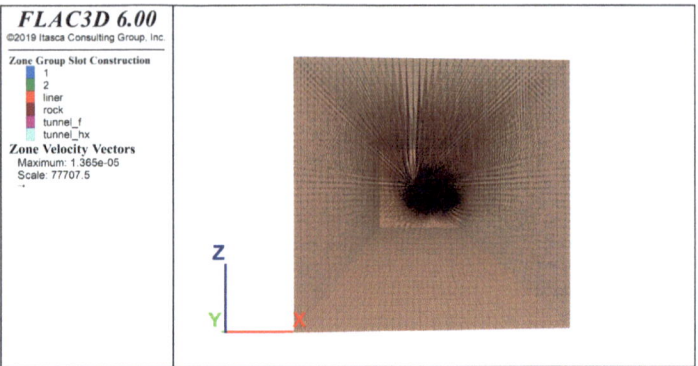

Fig. 6. Vertical head view of the stress vector tunnel

4.2 Analysis of the Plastic Areas

When calculating the Moore-Kulun constitutive model in FLAC3D, the yield of rock mass is judged by the following formula:

$$f_x = \sigma_1 - \sigma_3 N_\varphi + 2c\sqrt{N_\varphi} \qquad (7)$$

In formula: $N_\varphi = (1 + \sin\varphi)/(1 - \sin\varphi)$, σ_1 Is the maximum principal stress, With σ_3 Is as the minimum principal stress, φ as the internal friction angle and c as cohesion.

Like $f_x < 0$ means the rock mass; $f_x 0$, indicating that no shear yield has occurred in the rock mass.

When the normal stress is tension stress, it is beyond the mechanical validity range of the Moore-Kulun criterion, and the minimum principal stress should not exceed the tensile mild σ of the rock mass$_t$ Otherwise, the rock mass will appear in tension and yield, and the discrimination formula is:

$$f_t = \sigma_3 - \sigma_t \tag{8}$$

like $f_t > 0$, indicating the tensile yield of the rock mass; $f_t 0$, indicating that no tension yield occurs in the rock mass.

At the same time, because FLAC3D uses all the dynamic motion balance equations to solve the stress and strain problems, the output destruction area distribution data has the concept of relative time, which is divided into present (n) and past (p). Combined with the above two destruction modes, they are divided into the following five situations:

Class I (hear-n): indicates that a region is in a state of shear failure;
Class II (hear-p): means once went into shear yield state and now quit;
Class III (tension-n): indicates that a region is in a state of tensile failure;
Class IV (tension-p): means it has entered the tensile yield state and has now withdrawn;
Class V (none): indicates no damage.

Figure 7 shows the plastic zone marking when the tunnel excavation depth is 10 m. As can be seen from the figure, the plastic area occurs in the side wall and lower steps and excavation palm surface after the range of 3 m, and after the side wall surrounding rock with shear damage, and the bottom with stretching damage, in the arch near the plastic area, this is because of the timely excavation and anchor support, so that the stability of surrounding rock does not affect the development of deep rock strata. In the surrounding rock, the anchor rod strengthens some surrounding rock around the tunnel, increasing the strength of the rock mass, thus maintaining the stability of the surrounding rock.

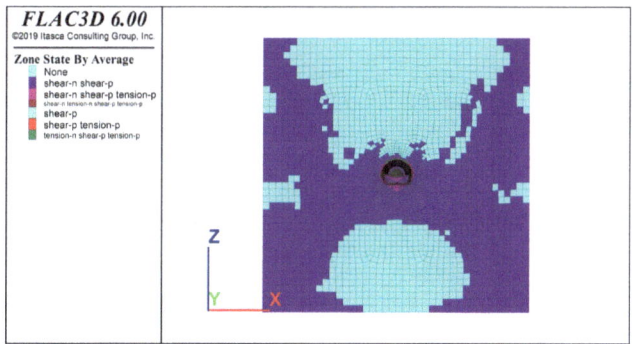

Fig. 7. Plastic area after 10 m of tunnel excavation

4.3 As for the Shift Field Analysis

After the tunnel excavation, the rock displacement, the overall displacement trend is to the hole, arch surrounding rock moved lower, the bottom around the rock to the above, Fig. 8 shows the vertical displacement cloud tunnel 10 m, Fig. 9 shows the 10 m tunnel 16 m profile vertical displacement cloud, from the Fig. 10 of surrounding rock because no support measures after excavation for upward uplift state, to about 15 mm..5.6

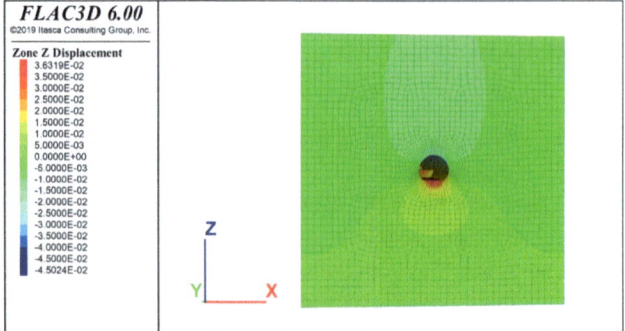

Fig. 8. Vertical displacement cloud map at 10 m of tunnel excavation (m)

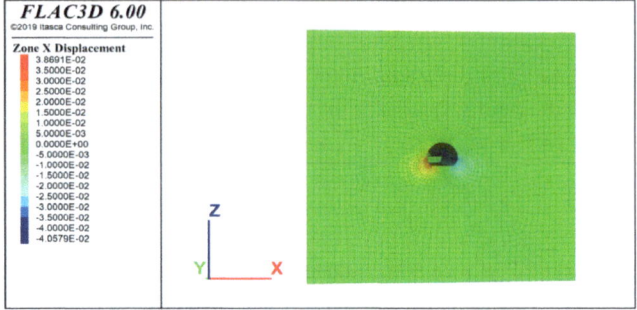

Fig. 9. Horizontal displacement cloud map at 10 m of tunnel excavation (m)

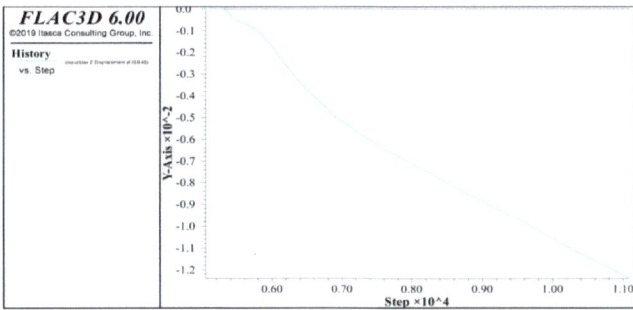

Fig. 10. Top plate displacement curve during 10 M tunnel excavation

5 Conclusion

For the study of tunnel surrounding rock stability, scholars and geotechnical engineers from various countries have done a lot of research, no matter which method, it aims to use the most economical and reliable scheme in the best time to ensure the safety of the construction and operation of the project. Currently only support with surrounding rock coupling is not reach quantitative level, not to mention other including stress field, displacement field and so on multiple field coupling, but the numerical calculation method and computer intelligence is increasing, the surrounding rock parameters can establish diversity, large parameter matrix is possible, therefore, can weaken the influence of some secondary factors, so as to reach the level of a half definite quantitative, using a variety of factors coupling mechanism between the accuracy of quantitative analysis, will provide a new analysis mode for rock mechanics and engineering research.

Funding. This study was funded by the Open Fund of the Engineering Technology Research Center for Efficient Utilization of Water Resources in Maanshan Hills Region under grant number WREU202002.

References

1. Huang, L., Xu, Z., Zhou, C.: Modeling and monitoring in a soft argillaceous shale tunnel. Acta Geotechnica (2020)
2. Deng, Q.H., Ma, F., Guo, J., Zhang, L.P., Yu, R.: The effect of karst cave space distribution on deformation law of double track tunnel surrounding rock. Adv. Mater. Resl. (2021)
3. Li, L.C., Liu, H.-H.: A numerical study of the mechanical response to excavation and ventilation around tunnels in clay rocks. Int. J. ROCK Mech. Mining Sci. (2023)
4. Zhao, Y., Zhang, Z.: Mechanical response features and failure process of soft surrounding rock around deeply buried three-centered arch tunnel. J. Cent. South Univ. **22**(10), 4064–4073 (2015)
5. Gong, B., Jiang, Y., Okatsu, K., Wu, X., Teduka, J., Aoki, K.: The seepage control of the tunnel excavated in high-pressure water condition using multiple times grouting method. Processes **6**(9), 159 (2018)
6. Wang, M., Zheng, D., Niu, S., Li, W.: Large deformation of tunnels in longwall coal mines. Environ Earth Sci **78**(2), 45 (2019)
7. Li, L., Du, X., Zhou, J.: Numerical simulation of site deformation induced by shield tunneling in typical upper-soft-lower-hard soil-rock composite stratum site of Changchun. KSCE J. Civ. Eng. **24**(10), 3156–3168 (2020)
8. Kabwe, E., Karakus, M., Chanda, E.K.: Creep constitutive model considering the overstress theory with an associative viscoplastic flow rule. Comput. Geotech. **124**, 103629 (2020)
9. Li, Z., Chen, Z., Wang, L., Zeng, Z., Gu, D.: Numerical simulation and analysis of the pile underpinning technology used in shield tunnel crossings on bridge pile foundations. Underground Space **6**(4), 396–408 (2021)
10. Xu, G., Gutierrez, M.: Study on the damage evolution in secondary tunnel lining under the combined actions of corrosion degradation of preliminary support and creep deformation of surrounding rock. Transport. Geotech. **27**, 100501 (2021)

Analysis of Collapse Mechanism and Seismic Measures of RC Frame Structure with Side Corridor

Huan Jin[✉], Shengqiang Li, and Yaokai Tan

Guangdong University of Petrochemical Technology, Maoming 525000, China
103579101@qq.com

Abstract. RC frame structure with side corridor is widely used in the construction of school buildings in southern China, but its seismic performance under strong earthquakes is much lower than the design expectations, and the seismic damage is very serious and even collapsed. Based on the seismic damage of the school building of in Wenchuan Xuankou Middle School, from experimental research to theoretical analysis, and from the RC frame unit model to the spatial frame model, the seismic collapse mechanism and reasons of RC frame structure with side corridor is deeply explored. The study shows that the complex interaction between the masonry infilled wall and the frame cannot be ignored. When infill panels is not arranged reasonably, it is easy to cause the stiffness imbalance of the structure and the relevant failure of the frame column, thereby causing the collapse of the structure, and some seismic measures are proposed.

Keywords: frame structure with side corridor · school building · masonry infilled wall · collapse mechanism · seismic measure

1 Introduction

The reinforced concrete frame structure with side corridor has good ventilation and lighting effects, and is widely used in densely populated building structures such as teaching and dormitory buildings in the south, once an earthquake disaster occurs, the collapse of such buildings will cause serious casualties and property damage. However, previous earthquake disasters have shown [1–3] that the seismic performance of the frame structure with side corridor under strong earthquakes is much lower than the design expectations, and the earthquake damage is very serious and even collapsed, such as the typical Xuankou Middle School teaching building that collapsed in the 5.12 Wenchuan earthquake, which caused over 50 teachers and students to lose their lives. Xuankou Middle School is located in Yingxiu Town, Wenchuan County, and the earthquake damage was very serious. As shown in Fig. 1, the main school buildings of the side corridor style RC frame structure on both north and south sides have completely collapsed, while the inner corridor style RC frame structure and masonry structure buildings located on the same site have achieved "extreme earthquake resistance". From this, it can be known

that the collapse reasons of side corridor frame structure system with typical and double unequal span in the school building are worthy of further studies and reflections,
and the seismic performance of the corridor style frame structure system has obvious
shortcomings.

The corridor style RC frame structure includes two types, single span corridor type
and double span corridor type. The single span corridor type is clearly not recommended
in the specifications, and the double span corridor type is still widely used. The biggest
characteristic of the double span external corridor frame structure is the large size difference between the two spans and the uneven layout of the plan. Some studies [4–6]
have shown that the seismic performance of the external corridor frame structure is far
inferior to that of the internal corridor frame structure, and the layout of the structure
plan and facade has a significant impact on the seismic performance of the external
corridor frame structure, especially the arrangement of the infill wall cannot be ignored.
At present, the impact of infill walls on the collapse performance of external corridor
frame structures is mainly studied through vibration table tests and numerical simulation methods for the macroscopic performance of the structure, and there has not been
in-depth and systematic research on collapse mechanism from plane to space, and from
experiments to theories.

This article takes the side corridor RC frame structure as research object, the school
building of Xuankou Middle School in Wenchuan as example. Through systemic seismic
damage analysis, experiments and theoretical research, the earthquake collapse mechanism and reasons of the side corridor RC frame structure are explored. The influence of
infilled walls on the seismic performance and collapse mechanism of the side corridor
RC frame structure is summarized, and optimization design suggestions and seismic
measures are proposed.

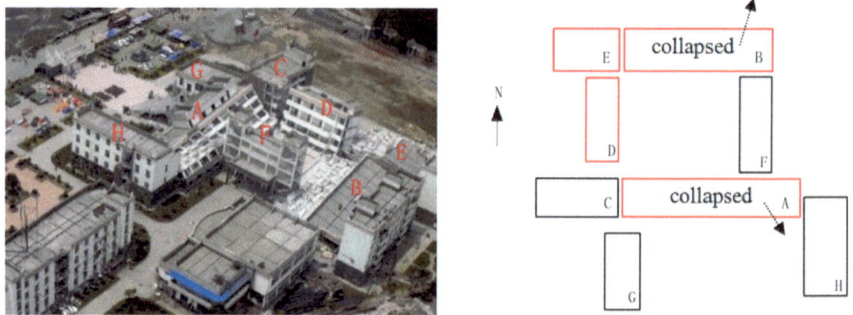

Fig. 1. Seismic damage to teaching buildings of Xuankou Middle School

2   Earthquake Damage Analysis

Wenchuan Xuankou Middle School is located in Yingxiu Town, the epicenter of the
earthquake. It consists of 14 buildings, including teaching buildings, faculty dormitories,
student dormitories, and canteens. The school building is composed of eight independent

frame structures separated by expansion joints, as shown in Fig. 1, forming a circular teaching building group. Buildings A, B, D, and E all collapsed in the earthquake. After investigation and analysis, it was found that the four completely collapsed teaching buildings were all side corridor style RC frame structure. The architectural plan of teaching building A is shown in Fig. 2, the transverse direction of the building is basically two spans. The classroom span is much larger than the corridor span. The collapse forms of the four teaching buildings are basically the same. The damage to the frame columns is significantly more serious than that to the frame beams, and the damage to the top of the bottom column is more serious than that to the bottom of the column. The failure of the first floor frame columns resulted in the overall collapse of the second and above floors towards the classroom side. Therefore, why did the corridor style frame structure suffer such serious seismic damage?

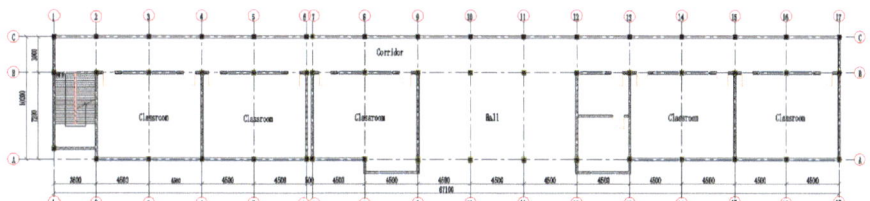

Fig. 2. Layout Plan of teaching building

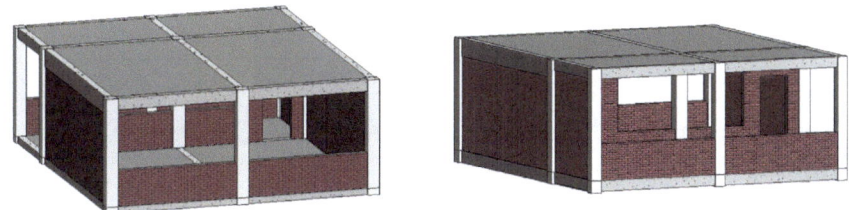

Fig. 3. Layout plan of frame units

The RC frame structure includes three typical types of frame units in transverse axis, the pure frame, fully infilled wall frame in large-span, and fully infilled wall frame in two span. The frame structure has large door and window openings on the three longitudinal axes. The low infilled wall is arranged on the side corridor of the A-axis, the infilled wall with larger door and window openings is arranged on the B-axis, and the partial window wall is arranged on the classroom side of the C-axis, as shown in the Fig. 3. Through seismic damage analysis, it is found that due to the failure of the A-axis frame column on the bottom classroom side, the whole structure collapsed to the classroom side from the second and above floors in a cascading collapse mode. The frame column on the side corridor did not show significant damage, but the top of the bottom corridor column was broken due to the overall collapse towards the classroom side, as shown in Fig. 4.

At present, the seismic performance and the collapse reasons of Xuankou Middle Teaching building have been analyzed by many scholars [6, 7], including the structure's low bearing capacity reserve, poor deformation capacity, poor energy dissipation capacity, low redundancy and poor integrity, etc. These are analyzed only from a macroscopic seismic performance perspective without detailed analysis and suggestions. In summary, all the reasons of the collapse are mainly attributed to the failure mechanism of "strong beam and weak column" in the structure, which is contrary to the design expectation. The root cause is related to the neglect of "super" influence of cast-in-place floor and the associated failure mechanism caused by the setting of filled wall [8–11]. In the following, through the further analysis of pseudo-static test and shaking table test results, from the two aspects of the impact of floor and filled wall on structure, the serious earthquake damage of the corridor frame structure is traced.

(a) South side   (b) North side

Fig. 4. Collapse of teaching building [13]

3   Research on the Plane Frame Units of Side Corridor Frame Infilled Walls

As we all know, there is a complex interaction between masonry filled walls and RC frame columns, and the unreasonable arrangement of the infilled walls can easily cause the "column hinge" failure mechanism of the frame structure that is beyond expectation [12–14]. In order to further our Understanding of the effect of masonry infills on side corridor RC frames, based on Xuankou middle school building, three representative frame units are selected, and quasi-static testson three 1/2-scale masonry-infilled RC frame specimens have been conducted under in-plane reversal cyclic load [15, 16]. The strength, stiffness, energy dissipation capacity and failure mechanisms of one story, two-bay RC frames with aerated block infills were investigated. The parameters of the test specimens are summarized in Table 1, Speimen 1 is the reference bare frame, specimen 2 is fully infilled with aerated block walls in the larger span, and Specimen 3 was infilled with half height aerated block walls. Geometrical dimensions and reinforcements of the two specimens were selected to be the same. The experimental setup consisted of rigid floor, reaction wall, loading equipment, instrumentation, lateral bracing and data acquisition system, as is shown in Fig. 5. All of specimens were tested under reversed cyclic lateral loading simulating seismic action, the specimens were loaded in-plane displacement control, and the loading history is shown in Fig. 6.

Table 1. Parameters of test specimens

Spec. No.	Infill arrangement	Floor width (mm)	Masonry Type	Masonry strength (mm)	Infill thickness (mm)	Brick(block) standard (mm)
1		1300	——	——	——	——
2		1300	Aerated block	A3.5 B06	120	600×120×240
3		1300	Aerated block	A3.5 B06	120	600×120×240

Fig. 5. Test setup   **Fig. 6.** Loading history

Through analysis of experimental phenomenon, the following conclusions can be drawn. Specimen 1 demonstrates a typical "strong-beam and weak-column" flexure failure mechanism, the top of the column yields earlier than the bottom of the column, and plastic hinges appeared first in columns compared to beams. The whole failure mode of specimen 1 is shown in Fig. 7. The failure mode of specimen 2 is completely different from specimen 1, due to the restraint effect on the frames of masonry infill panels, the diagonal shear failure was extremely serious in the RC frame fully in-filled panels. The whole failure mode of specimen 2 is shown in Fig. 8. Specimen 3 did not demonstrate a typical the "short column" shear failure mechanism under the effect of aerated block walls, flexural/shear crack induced the structure destruction, as shown in Fig. 9. Thus, the failure mechanism of test models were consistent with the actual damage of prototype structure in Wenchuan earthquake.

Through comparative analysis of experimental results, the initial stiffness of the large-span filled wall frame is 2.43 times that of the pure frame, its bearing capacity is

1.77 times that of the pure frame, and its energy dissipation capacity is 1.5–3 times that of the pure frame. However, the initial stiffness of partially-filled wall frame is 1.56 times that of the pure frame, and its bearing capacity is 1.26 times that of the pure frame, but its energy dissipation capacity is lower than that of the pure frame and the fully-filled wall frame, which indicates that the contribution of the partial-filled wall to the earthquake resistance of the structure is not obvious. When the strength of the filled wall material is too high, it is easy to cause the short column effect, and then the energy dissipation capacity of the frame structure is reduced [17–19]. Therefore opening windows through the walls is unfavorable to the seismic resistance of the frame structure.

Fig. 7. Ultimate failure modes of specimens 1 [13]

Fig. 8. Ultimate failure modes of specimens 2 [13]

In addition, based on specimen 3, the influence of the height H_w of the infilled wall and the masonry material on the failure mechanism of the frame structure was studied through nonlinear finite element numerical analysis [20]. The infilled wall is equivalent to the diagonal support of the frame column. As the height of the infilled wall arrangement increases, the height of the upper free section column decreases, and the bending point of the frame column moves upward, as shown in Fig. 10, the internal force is redistributed. Through analysis, it can be seen that when the height of the infilled wall is big, the strength of the masonry material is high, and the shear span ratio of the upper

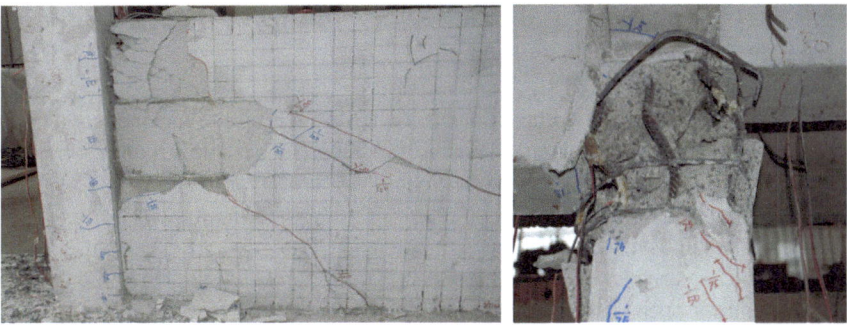

Fig. 9. Ultimate failure modes of specimens 3 [13]

free section column is reduced, the upper column is prone to short column shear failure. When the wall is fully infilled, due to cracking and damage at the corners of the infilled wall, short columns are formed in the upper section of the frame column. This results in shear failure at the top of the column, as shown in specimen 2. When the height of the infilled wall arrangement is relatively small, due to the upward displacement of the yield position and reverse bend point of the longitudinal reinforcement of the constrained column, it is easy to cause bending failure of the short column in the frame structure. For example, in specimen 3, although the bending failure of the short column is not as harmful as the shear failure of the short column, it will also reduce the seismic performance of the frame structure. From this, it can be known that the collapse of the side corridor style school building in Xuankou Middle School is closely related to the layout of the infilled wall. It is obviously unreasonable to ignore the effect of the infilled wall in the structural design process.

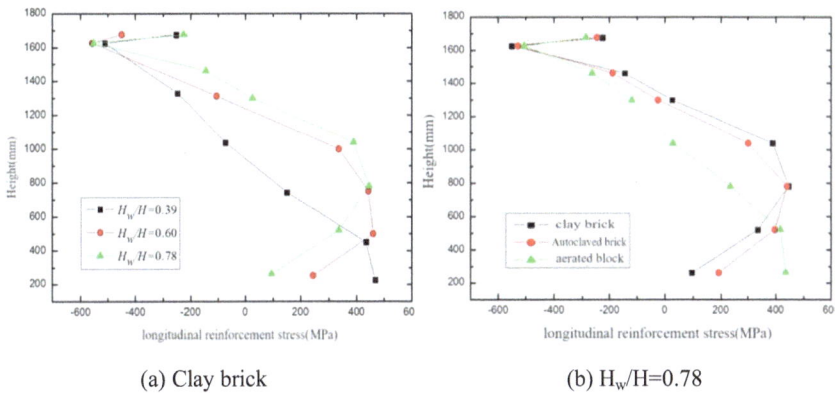

(a) Clay brick   (b) $H_w/H=0.78$

Fig. 10. Longitudinal reinforcement stress curves along the column height

4   Research on the Spatial Framework Model of Side Corridor Frame Infilled Walls

In order to further explore the impact of infill walls on the failure mechanism of frame structures, analyze the collapse mechanism of side corridor frame structures, based on the teaching building of Xuankou Middle School as a prototype, a comparable 1:5 scale side corridor infilled wall frame test models were designed, as shown in Fig. 11. The experimental model is made of particle concrete and Q235 galvanized iron wire, and the infill walls are constructed with MU3.5 concrete hollow blocks and Mb5 mixed mortar, and the similarity relationship between the scaled model and the prototype is derived according to the consistent similarity rate of seismic simulation experiments [15]. The layout of the infilled panels is designed according to the prototype of the collapsed teaching building of Xuankou Middle School, and the seismic motion includes 15 input conditions of El Centro wave and Wolong wave.

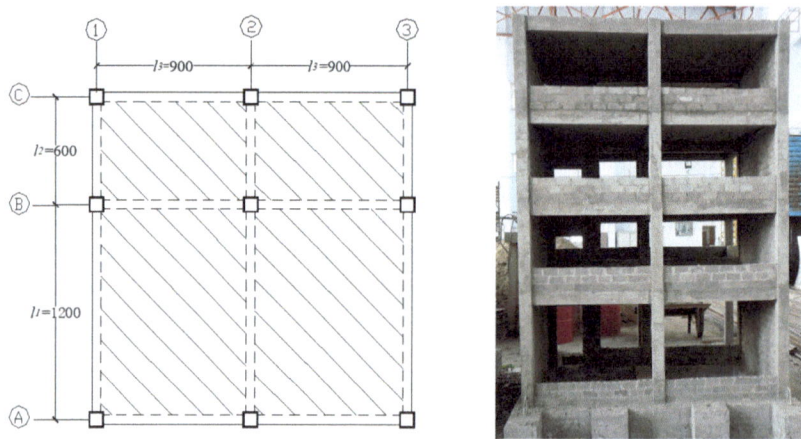

Fig. 11. Basic overview of experimental model

Through the analysis of the results of the model vibration table test, it was found that the infilled wall cracked firstly as the first seismic fortification, the top columns of frame yielded earlier than the bottom columns, and the frame column yielded earlier than the frame beam. The fully infilled walls caused oblique shear failure at the top of the column, and the partially infilled walls caused short column bending failure at both ends of the free section column. In conclusion, the arrangement of infilled walls has changed the internal force redistribution and failure mode of the frame structure, and the damage of columns is very serious, which is consistent with the results of the pseudo static test of the frame elements.

In addition, the experiment shows that there is a significant difference in the failure of the three frame units in transverse direction, as shown in Fig. 12. The side frame of axis ① and ③ is placed with infilled walls in large spans, and there is obvious cracking and damage at the top of the column. However, the middle frame of axis ② is not equipped with infilled walls, and the cracking and damage are not significant. From this, it can be seen that after being fully infilled with masonry walls, the stiffness of the two frames on both sides significantly increases, and the allocated seismic shear force also increases, resulting in severe shear failure at the top of columns in the axes ① and ③. The failure phenomenon is consistent with the pseudo static frame element test. The damage to the frame columns in the middle axis ② is not significant, especially in the early stage of loading when cracking is not obvious. With the failure of the infill wall, the overall stiffness of the structure decreases. The damage to the A-column in the second axis is more severe in the later stage.

Fig. 12. Destructive behavior of model A [13]

Through comparative analysis of the results of longitudinal frame unit tests, it can be seen that both the A-axis and C-axis are arranged with partial infilled walls, resulting in a short column constraint effect. There are obvious horizontal cracks at the top of the frame columns and walls, as shown in Fig. 12. The transverse spacing between AB columns is relatively large and fully infilled walls, while the longitudinal A-axis is arranged with partial infilled walls, resulting in significant damage of A-axis frame columns comparing to other axis columns. The spacing between BC columns is relatively small, and although the longitudinal C-axis frame columns are arranged with partial infilled walls, the damage is not significant.

The model has two spans in both transverse and longitudinal directions, and it has fewer arrangements of infilled wall and weaker stiffness in the longitudinal directions than the transverse directions. Therefore, before collapse, the A-axis frame column on the first floor failed and tilted along the longitudinal direction. Finally, due to the large self-weight on the side of the large-span classroom, the structure collapsed layeredly towards the classroom side, as shown in Fig. 13. In conclusion, the collapse mode of the test model is basically consistent with the collapse mode of the school building at Xuankou Middle School. Therefore, opening larger doors and windows in the infilled wall causes uneven stiffness of the structure and column hinge failure, and reduces the energy dissipation performance and anti-collapse ability of the structure, which is the main reason for the collapse of the Xuankou Middle School teaching building [21].

Fig. 13. Final failure modes of model A and model B [13]

5 Conclusions

Through experimental research and related numerical analysis of RC frame units and spatial frame models with masonry infilled walls, the collapse mechanism and reasons of the corridor style frame structure school building were deeply explored, and the following conclusions were drawn:

Firstly, based on the characteristics of the teaching building of Xuankou Middle School, which is a typical side corridor frame structure, the difference of transverse span is large. When the masonry infilled walls are not arranged reasonably, it is easy to cause uneven stiffness of the whole structure, and local frame columns fail, thus cause the "column hinge" failure mechanism. From the analysis, it can be seen that the frame columns on the long span classroom side are severely damaged, while the damage on the short span corridor side is not severe. Therefore, it is recommended to increase the bearing capacity of the large span frame columns on the classroom side, it is necessary to increase the cross-sectional size or reinforcement during design.

Secondly, masonry infilled walls can have stiffness and constraint effects on the frame structure, with both advantages and disadvantages. However, unreasonable layout of infill panels can turn the advantages into disadvantages. The layout of the infilled walls in the side corridor style school building of Xuankou Middle School is obviously unreasonable. The infilled walls frame in the longitudinal direction has larger door and window openings, while the transverse frame has a large span and is infilled with infilled walls. The stiffness ratio of the longitudinal and transverse frames is imbalanced. In addition, the partial infilled walls in the longitudinal frame did not achieve the purpose of the first seismic fortification, and at the same time it had a negative impact on the energy consumption of the frame structure. Therefore, it is recommended that the longitudinal infilled wall frame at the side of classroom should not arrange thorough windows. To improve the stiffness and energy dissipation capacity of the structure, and avoid the "short column" effect, the measures of setting up a wall between windows or replacing masonry walls with lightweight infill wall panels are recommended.

Finally, the arrangement of infilled walls increases the stiffness of the frame structure and enhances its load-bearing capacity. However, the increase of the load-bearing capacity of the frame columns cannot offset the increased seismic force caused by the increase in stiffness and the failure of column end correlation caused by constraints. From this case, it can be known that the influence of stiffness on the anti-collapse ability of building structures is crucial. It's not the fact that the higher the stiffness, the better the seismic performance of the structure. It is necessary to achieve the best balance between stiffness and ductility, which requires further quantitative exploration.

Acknowledgements. The study described in this paper was supported by Projects of Talents Recruitment of GDUPT (2019rc084). This support is gratefully acknowledged.

References

1. Ye, L., Li, Y., Pan, P.: Investigation of the seismic damages of building structures of Xuankou Middle School in Yinxiu Town. Building Strut. **39**(11), 54–57+29 (2009). https://doi.org/10.19701/j.jzjg.2009.11.015
2. Xu, Y.: Lessons from Wenchuan seismic damage: introspection on the collapse of classroom buildings. Build. Struct. **39**(11), 50–53 (2009). https://doi.org/10.19701/j.jzjg.2009.11.014
3. Tang, Z.: Research on optimal design and seismic capacity of RC frame structures with exterior verandah. Master Dissertation, Institute of Engineering Mechanics, CEA (2023). https://doi.org/10.27490/d.cnki.ggjgy.2022.000073
4. Li, J., Li, F., et al.: Seismic performance analysis of exterior-corridor type and middle-corridor type architecture. J. Phys: Conf. Ser. **1578**, 012242 (2020). https://doi.org/10.1088/1742-6596/1578/1/012242
5. Huang, Q.: Study on improvement of seismic performance of side corridor frame with infill wall. Master Dissertation, Hunan University (2017)
6. Huang, S.: Mechanism of seismic damage and collapse of RC frame structure with external corridor. Doctoral dissertation, Institute of Engineering Mechanics, CEA (2012)
7. Ye, L., Lu, X., Zhao, S., Li, Y.: Seismic collapse resistance of RC frame structures-case studies on seismic damages of several RC frame structures under extreme ground motion in Wenchuan Earthquake. J. Build. Struct. **30**(06), 67–76 (2009). https://doi.org/10.14006/j.jzjgxb.2009.06.009
8. Demir, A., Cengiz, M.M.: Effect of infill wall properties on seismic response of RC structures. Comput. Concr. **6**, 27 (2021)
9. Dong, X., Guo, X., Luo, R., et al.: Seismic damage investigation and structural force analysis of Luding Ms 6. 8. J. Build. Struct. **44**(S2), 11–19 (2023). https://doi.org/10.14006/j.jzjgxb.2023.S2.0002
10. Haoran, W., Shuang, L., Changhai, Z.: Research progress on influence of infill wall on progressive collapse of structure. Eng. Mech. **39**(10), 1–16 (2022)
11. Sharma, V., Madan, S.K.: Seismic response of reinforced concrete frames with masonry infills. (2022).https://doi.org/10.1007/978-981-16-6557-8_16
12. Wararuksajja, W., Srechai, J., et al.: Seismic design of RC moment-resisting frame with concrete block infill walls considering local infill-frame interactions. Bull. Earthq. Eng. **18**(14), 6445–6647 (2020). https://doi.org/10.1007/s10518-020-00942-9

13. Duran, B., Tunaboyu, O., et al.: Structural failure evaluation of a substandard RC building due to basement story short-column damage. J. Perform. Constr. Facil. **4**, 34 (2020). https://doi.org/10.1061/(ASCE)CF.1943-5509.0001455

14. Khelfi, M., Bourahla, N., Remki, A.M., et al.: Performance evaluation of masonry Infilled RC frame structures under lateral loads. Gradevinar **73**(3), 219–234 (2021). https://doi.org/10.14256/JCE.2647.2019

15. Jin, H.: Research on seismic failure mechanisms and key aseismic measures for masonry-infilled RC frame structures. Doctoral Dissertation, Institute of Engineering Mechanics, CEA (2014)

16. Xu, D., Dai, J., et al.: Study on numerical simulation of seismic collapse of RC frame structure with infilled wall. Adv. Civil Eng. **4**, 1–24 (2022). https://doi.org/10.1155/2022/1890091

17. Duo, Z., Huang, Q., Wei, R.: Experimental study of the seismic behavior of irregularly brick infilled RC frames. Chin. Civil Eng. J. **43**(11), 46–54 (2010). https://doi.org/10.15951/j.tmgcxb.2010.11.011

18. Akid, A.S.M., Rashid, M.H., Sobuz, M.H.R.: Effect of masonry infill wall with opening on reinforced concrete frame due to seismic loading: parametric study. Int. J. Struct. Eng. **11**(1), 84–105 (2021). https://doi.org/10.1504/IJSTRUCTE.2021.112103

19. Xue, L., Qi, T., Xiaolan, Y., et al.: Study on progressive collapse performance of infilledRC frame with opening. Build. Sci. **39**(03), 135–145 (2023). https://doi.org/10.13614/j.cnki.11-1962/tu.2023.03.018

20. Jin, H., Junwu, D., et al.: Effect and anti-seismic measures of masonry-infilled walls on the failure mechanism of short column in RC frame structure. Earthquake Resistant Eng. Retrofitting **43**(04), 1–7 (2021). https://doi.org/10.16226/j.issn.1002-8412.2021.04.001

21. Zhu, Y., Yang, W., Zhang, J., et al.: Study on the seismic collapse resistance of RC frame structure considering infill wall plane stiffness distribution. J. Qingdao Technol. Univ. **45**(01), 10–18+35 (2024)

The Effect of Heavy Loads and Cable Breakage on the Line Shape of Main Girders of One-Tower Cable-Stayed Bridges

Chaoying Chen[✉] and Yinjun Shen

School of Civil Engineering, Changsha University of Science and Technology, Hunan, China
570181183@qq.com

Abstract. To study the adverse vibrations of overloading and cable breakage, a single tower cable-stayed bridge model was established using finite element software ANSYS. The fracture effects of cables at different positions were simulated, and the overall linear changes of the bridge before and after fracture were analyzed. Equivalent vehicle loads were applied to obtain the influence of the most unfavorable fracture on the main beam line shape when the load was distributed at the most unfavorable position. The analysis results show that the impact of cable breakage varies at different positions. Cable breakage at locations with higher cable forces and bending moments will cause more significant vibration in the main beam, resulting in more significant changes in the main beam line shape. It is only when the three cables break that the displacement exceeds the allowable value in the specifications.

Keywords: cable-stayed bridge · finite element analysis · broken cables · vibration · amplification factor

1 Introduction

As the main stress components of cable-stayed bridges, cable-stayed cables transfer the loads from the main girders to the main towers and play a role in supporting and dispersing the loads, so their integrity plays a crucial role in the safety of the whole cable-stayed bridge. However, in the actual operation of cable-stayed bridges, there are corrosion [1], fatigue [2], fire [3], explosion [4], impact [5] and other factors, and usually a variety of diseases together [6], cable-stayed cables are easy to be damaged, reducing the life of the discovery of the untimely and withstand the extreme conditions, withstand the sudden increase in the load, the cable breakage may occur.

In 2001, the Nanmen Bridge in Yibin City, Sichuan Province, in 2007, the Mezcala Cable-stayed Bridge in Mexico, in 2019, the Nanfang'ao Bridge in Taiwan, and in 2022, a cable-stayed bridge in western India, all suffered from cable breakage, resulting in serious loss of life and property. Therefore, it is important to accurately assess the effects of cable breakage on cable-stayed system bridge structures to ensure the safety of the structures during the design and operation phases.

A. Bieliatynskyi et al. (Eds.): CSTTE 2023, LNCE 603, pp. 379–387, 2024.
https://doi.org/10.1007/978-981-97-5814-2_34

For the study of broken cables, the American Post-Tensioning Association [7] proposed two simulation methods for broken cables: one is the transient dynamic analysis method; the other is the dynamic amplification factor method, and suggested that the dynamic amplification factor DAF should be taken as 2.0. The European Committee for Standardisation [8] suggested that the DAF should be taken as 1.5. The Institute of Roads and Highways of France [9] suggested that the DAF should be taken in the range of 1.5–2.0. Wolff [10] pointed out that the DAF for bending moment of cable-stayed bridge is generally less than 2.0, and that the DAF for tie force and tower bending moment is higher than 2.0. However, most of them have not considered the vehicle load on the bridge deck, and the force situation of different bridges is different, and the impact of cable breakage is also different. Therefore, the discussion of cable breakage for various bridge types should be subdivided and not generalized. This article studies the cable breakage of a single tower cable-stayed bridge with specific tower beam consolidation, and considers the vehicle loads that are likely to exist in reality.

2   Overview of Cable-Stayed Bridge Project and Finite Element Modelling

The example bridge is a single tower asymmetric cable-stayed bridge with tower beam consolidation, with a main span of 196 m and a side span of $101 + 62 = 163$ m. The bridge deck is 29.5 m wide, with a basic cable spacing of 7 m on the main beam, a side span tail cable area of 4.5 m, and a cable spacing of 2.0 m on the tower. A total of 2×52 diagonal cables are arranged in a fan-shaped layout, with a steel wire diameter of 7 mm and a standard strength of 1770 Mpa. The cable numbers and arrangements are shown in Fig. 1. The overall layout of the entire bridge is shown in Fig. 2. For convenience, each cross is named as shown in Fig. 2. There are a total of 52 key nodes in the entire bridge, corresponding to the positions of the main beams at 52 pairs of diagonal cables, numbered from left to right as 1–52.

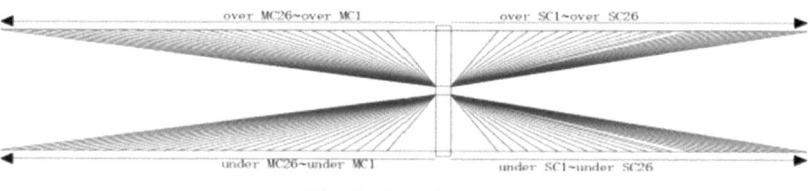

Fig. 1. Rope layout

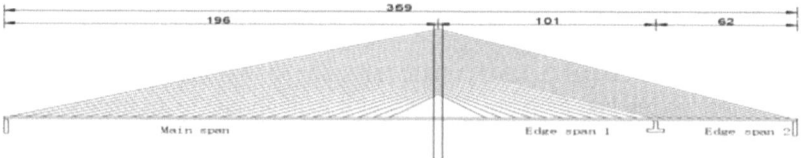

Fig. 2. General layout of the whole bridge

A three-dimensional fishbone beam finite element model of a cable-stayed bridge was established using ANSYS. The main tower was made of C40 concrete, and the main beam was made of C50 concrete. The three-dimensional solid beam element Beam188 was used for both, and the cables were subjected to axial tension or compression using Link10 rod elements. The main beam has a single box three chamber section, and the bridge tower has a rectangular hollow section, both of which use the actual section input form. The section characteristic values are automatically calculated by the software. The bridge tower and foundation are consolidated, the tower beams are consolidated, and both the left and right supports are constrained in vertical and horizontal displacement. The auxiliary piers are constrained in vertical displacement. A finite element model for the calculation of this cable-stayed bridge is established, as shown in Fig. 3.

Fig. 3. Finite element model of cable-stayed bridge

3 Diagonal Cable Fracture Vibration Analysis

To study the dynamic response of the bridge structure of cable-stayed bridges under broken cables, firstly, the deflections of the structure under self-weight, phase II and tension cables are calculated considering the geometric nonlinearity as the initial state of the structure before breaking the cables, and then the dynamic response of the structure after breaking the cables is analysed by removing the tension cables unit in the dynamic time-course calculations.

According to the General Specification for Highway Bridge and Culvert Design (JTG D60-2015) and Design Specification for Highway Cable-stayed Bridges (JTG/T3365-01-2020), the deflection of the main girder of cable-stayed bridge is calculated under the limit state of normal use, and the deformation of the main girder of ANSYS cable-stayed bridge model is shown in Fig. 4, and the maximum value of the main girder downward deflection occurs at the key point 8, with the maximum value of 163 mm, the deflection-to-span ratio is less than L/500 (L is the calculated span diameter) required by the design specification, which meets the specification requirements.

3.1 Selection of Single Rope Breaking Condition

The power response method is used to analyse the vibration impact of different locations of the cable breakage on the whole bridge, the example bridge is an asymmetric structure, the left and right sides of the bridge tower span of the diagonal cable breakage situation is different, it should be considered separately, and choose a total of 10 kinds of single diagonal cable breakage conditions, see Table 1.

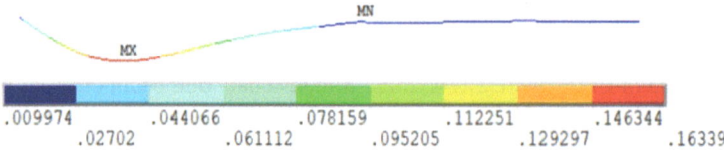

.009974   .044066   .078159   .112251   .146344
        .02702   .061112   .095205   .129297   .16339

Fig. 4. Displacement of main beam under static load

Table 1. Single cable breakage conditions

Working condition number	Explanation of rope breakage conditions	Lasso number
1	Cable break at right end of main span	Under MC1
2	Main span centre breakaway cable	Under MC13
3	Broken cable at the maximum force of the main span	Under MC19
4	Cable break on the left side of the main span	Under MC25
5	Edge span 1 left end broken cable	Under MS1
6	Edge span 1 centre break cable	Under MS6
7	Edge span 1 right end broken cable	Under MS11
8	Edge span 2 left end broken cable	Under MS13
9	Edge span 2 centre break ropes	Under MS20
10	Edge span 2 right end broken cable	Under MS25

3.2 Heavy-Duty Arrangement of Single Broken Ropes

According to the "General Code for Design of Highway Bridges and Culverts" (JTG D60-2015), the standard value of uniformly distributed load on Class I lanes of highways is fully distributed on the same influence line that causes the most adverse effect on the structure. The standard value of concentrated load only acts on the peak value of one influence line in the corresponding influence line. Each working condition applies a uniformly distributed load $q_k = 10.5$ kN/m on the span where the cable is broken, and a concentrated load $p_k = 360$ KN is applied at the critical point of the cable break, which serves as a rough simulation of the vehicle's heavy load effect. The heavy load arrangement for each rope break condition is shown in Table 2.

3.3 Effect of a Single Broken Cable on the Main Girder Line Shape

The maximum displacement generated by the broken cable under heavy load, need to break the cable node vibration time course analysis, the main beam 52 on the diagonal cable, to the cable MC19 and the main beam anchorage at the key point 8, for example, its vibration time course analysis results in working condition 3 in Fig. 5.

From Fig. 5, it can be seen that in the vibration of a single broken cable under heavy load, the maximum displacement after MC19 breaks the cable is 269 mm, and

Table 2. Heavy duty arrangement

Rope break condition number	q_k Layout location	p_k Layout location
1	Main span	Location of main beam at MC1
2	Main span	Main beam location at MC13
3	Main span	Main beam location at MC19
4	Main span	Main beam location at MC25
5	Edge span 1	Main beam location at MS1
6	Edge span 1	Main beam location at MS6
7	Edge span 1	Main beam location at MS11
8	Edge span 2	Main beam location at MS13
9	Edge span 2	Main beam location at MS20
10	Edge span 2	Main beam location at MS25

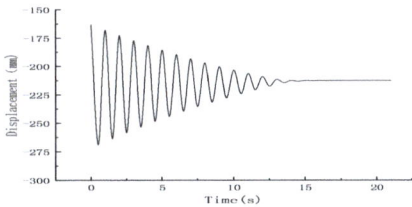

Fig. 5. Vibration time histories at critical point 8 under operating condition 3

the stable displacement after vibration stops is 212 mm. The entire vibration process is in a downward bending state, with the maximum value of 269 mm representing the deflection value of key point 8 after breaking the cable under condition 3. Similarly, for other key points, the peak displacement of the vibration after breaking the cable is taken as the displacement value of the key point after breaking the cable. The displacement values of each key point of the main beam under condition 1–10 of cable breaking are obtained as shown in Fig. 6.

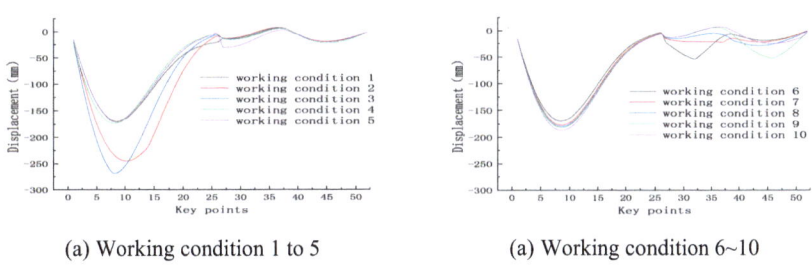

(a) Working condition 1 to 5   (a) Working condition 6~10

Fig. 6. Displacement of key points under a single broken cable

From the upper displacement Fig. 6, it can be concluded that the main girder displacement is the largest in Case 3, and the maximum displacement value occurs at critical point 8, reaching 269 mm. The larger the value of the cable force, the larger the displacement produced by the cable breakage. Therefore, when considering the most unfavourable effect of multi-cord breakage, it is only necessary to increase the broken cords near the maximum displacement of the single-cord breakage result to find out the most unfavourable result of the broken cords.

3.4 Selection of Working Conditions for Multiple Break Ropes

CHEN [11] pointed out that in the case of double-rope fracture, compared with the origin symmetric fracture and transverse symmetric fracture, the longitudinal symmetric fracture has the greatest impact on the line shape, combined with the results of Fig. 5, the location of MC19 (key point 8) was selected for the simulation of multiple-rope fracture, with a total of two types of double-rope fracture conditions and two types of triple-rope fracture conditions, and the specific conditions of rope breakage are shown in Table 3.

Table 3. Multiple tie breakage conditions

Working condition number	Explanation of rope breakage conditions	Lasso number
11	Two broken cables at the maximum force of the main span	Under MC19, Under MC20
12	Two broken cables at the maximum force of the main span	Under MC19, Over MC19
13	Three broken cables at the maximum force of the main span	Under MC18, Under MC19, Under MC20
14	Three broken cables at the maximum force of the main span	Under MC19, Over MC19, Under MC20

The heavy load arrangement for the multi-cord break is the same as for the single cable breakage condition 3.

3.5 Effect of Multiple Broken Cables on the Main Girder Line Shape

The displacement values of each key point of the main girder under the broken rope condition 11–14 are shown in Fig. 7.

As can be seen from the resultant Fig. 7, in the case of multi-cord breakage, the same-side broken cords produce larger displacements than the two-side broken cords, which have a greater impact on the main beam line shape. The maximum displacements

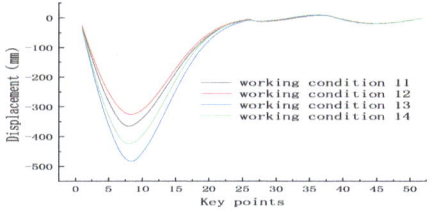

Fig. 7. Displacement map of each key point under multiple broken cables

of the main girder all occurred near the cable MC19 (key point 8), and the maximum value was 366 mm in case 11, the maximum value of the main girder displacement was 323 mm in case 12, the maximum value of the main girder displacement was 481 mm in case 13, and the maximum value of the main girder displacement was 425 mm in case 14. The displacements generated by the 2 conditions of the three cable breakage, 481 mm and 425 mm, have already exceeded the L/500, i.e. 392 mm.

4   Conclusion

In current research, more attention has been paid to the study of the dynamic amplification factor of broken cables, and insufficient attention has been paid to the actual vibration process of broken cables and vehicle loads. Sometimes, the conclusion of a certain bridge is used for all cable-stayed bridges and even cable load-bearing bridges, which is biased towards one sidedness. Different bridge types should be analyzed separately.

This article presents the finite element analysis results of a representative single broken cable, multiple broken cables, the combined action of heavy load and single broken cables, as well as the combined action of heavy load and multiple broken cables in a tower beam consolidated single tower cable-stayed system. The influence of each broken cable on the main beam line shape is obtained, and the conclusions are as follows:

(1) The influence of broken cables on the main girder line shape is different in different locations, the smaller the tie force is and the smaller the bending moment is, the smaller the influence of broken cables on the main girder line shape is, and the larger the tie force is and the larger the bending moment is, the larger the influence of broken cables on the main girder line shape is. In the case of multiple breakage of tie ropes, the displacement produced by the same side breakage is larger than that produced by the two sides breakage, which indicates that the most unfavourable result is produced by the nearest tie ropes breaking together on the same side in the case of multiple breakage of tie ropes.

(2) Considering the cable breakage after heavy load, the maximum displacement caused by a single cable breakage is 269 mm, the maximum displacement caused by two cable breakage is 366 mm, and the maximum displacement caused by three cable breakage is 481 mm, which is 1.65 times, 2.25 times, and 2.95 times that of the original bridge 163 mm, respectively. When three cables break, the displacement exceeds the allowable value in the specifications, indicating that a single tower cable-stayed bridge with tower beam consolidation has a high stiffness and is less affected

by broken cables than a general cable-stayed bridge. For the possible combined effects of heavy loads and broken cables, a single tower cable-stayed bridge with tower beam consolidation can also have a greater safety reserve for broken cables

This article only studied the displacement variation law of the main beam of a single tower cable-stayed bridge after cable breakage, and the conclusion drawn is slightly different from that of a general cable-stayed bridge after cable breakage. It is hoped that in future research, scholars will pay attention to the cable breakage process and vehicle load, and consider the impact effect of vehicle load to participate in time history analysis, so as to continue to discover the variation law of internal forces, cable forces, etc. of other types of cable-stayed bridges, suspension bridges, and suspension arch bridges after cable breakage, Subdivide and summarize the broken rope module.

References

1. Li, Y.B., Li, W., Ni, Y., Yin, L.: Study on fire resistance of large-span suspension bridges under vehicle fire. Fire Sci. Technol. **41**(7), 877–882 (2022)
2. Shen, D.J., Hu, Z.J., Li, Y.: Analysis of blast resistance of diagonal cable during near-field explosion. Vibration Shock **39**(21), 250–257 (2020)
3. Xu, H.S., Wang, Z.H., Yan, D.H.: Experimental study on fatigue life of cable wire with double cracks. Chin. Foreign Highway **43**(6), 127–132 (2023)
4. Wei, H.W., Wu, Z.Z., Huo, X.: Analytical study on damage detection after impact of the cable-stayed cable of Nanjing Yangtze River Second Bridge Nancha Bridge. China Water Transp. **02**, 77–79 (2019)
5. Yang, J.: A review of research on the main disease problems of tension cables of cable-stayed bridges. Western Leather **41**(14), 67 (2019)
6. Chen, Z.B., Lu, N.W.: Study on the reliability of cable-stayed bridge system under corrosion damage of cable-stayed cables. Highway Eng. **45**(01), 6–11 (2020)
7. PTI DC45.1–12. Recommendations for stay cable design, testing and installation, pp. 34–35 (2012)
8. EC1–2. Eurocode 1: Action on structures - Part 2: traffic loads on bridges. British Standard Institution, pp. 134–137S (2006)
9. E.TR.A. Haubans. Recommandations de la commission interministérielle de la précontrainte. Service d'Etudes Techniques des Routes Service d'Etudes Techniques des Routes et Autoroutes, pp. 179–181 (2001)
10. Wolff, M., Starossek, U.: Cable-loss analyses and collapse behaviour of cable-stayed bridges. Bridge Struct. **5**(1), 17–28 (2009)
11. Chen, G.S.: Research on the effect of tie breakage on the line shape of main girder of spatial torsion cable-stayed cable-stayed bridge. Railway Constr. **03**, 17–20 (2016)

Modeling and Simulation in Civil Engineering

The Dynamic Response Characteristics of the Slopes with Different Slope Morphology Under the Seismic Wave Action

Caifeng Hu[1], Feng Xiong[1(✉)], and Xiangkai Zhang[2]

[1] College of Civil Engineering, Hefei University of Technology, Hefei 230009, China
cvexf@hfut.edu.cn
[2] Anhui Vocational and Technical College, Hefei 230011, China

Abstract. The seismic landslides are common natural hazards in the mountainous areas, and they can cause a large number of casualties and property losses directly. The dynamic response characteristics of slope under the action of the seismic wave are the primary problem in evaluating the slope seismic stability. In this paper, the research is carried out on the effects of slope angle, morphology, and seismic wave frequency on the slope dynamic responses. Firstly, the slopes with different shapes (straight, concave, and convex slopes) are modeled. Secondly, the slope acceleration amplification factors and shear strain increment are investigated and analyzed. Finally, the dynamic response mechanisms are revealed using the ray analysis method. The results show that the acceleration response of each part of the slope is different for different frequencies, and the value of the amplification factor at the slope surface center is greater than that at the slope top. Slope morphology has significant effects on the dynamic response of the slope, and the slope surface centroid amplification factors change with different slope morphology. Slope angle has prominent impacts on the slope dynamic responses; the amplification factor is greatest at the slope angle of 30°. This study deepens the understanding of the slope seismic dynamic responses, which is vital for predicting slope instability and disaster reduction.

Keywords: seismic action · dynamic response · topographic effects · slope shape

1 Introduction

The seismic landslides are common seismic hazard phenomena in mountainous areas and have caused many casualties and property losses. Local site conditions have significant effects on ground motion characteristics and earthquake intensity [1–3], including topography (slopes, ridges, and canyons) and geology (sedimentation, basins, and faults).

The earthquakes are one of the most common natural disasters that cause ground movement and damage. The slope shapes may have significant topographic and soil effects during earthquakes, exacerbating the damage caused by earthquakes. Therefore, it is crucial to study the influence of slope morphology on topographic and soil effects. There has been a lot of abroad researchers' attention to this problem, and some results

© The Author(s) 2024
A. Bieliatynskyi et al. (Eds.): CSTTE 2023, LNCE 603, pp. 391–403, 2024.
https://doi.org/10.1007/978-981-97-5814-2_35

have been achieved. For example, a study by R. Tripe et al. [4] shown that the topographic and soil effects were interactive and should not be treated separately. Zhang et al. [5] shown that the amplification factors at the slope angle of 32.3° were maximum. Ashford et al. [6] proposed that the slope dynamic response can be subdivided into topographic amplification, site amplification, and surface amplification effects. They also derived the corresponding calculations that the topographic amplification factors reach their maximum value at $H/\lambda = 0.2$, about 1.5. George et al. [7] shown that the vertical motion components generated within the slopes may be as large as the horizontal. It was due to the reflection of seismic waves on the slope surface.

The domestic scholars have also conducted a lot of dynamic response research on the influence of slope morphology. Through shaking table and numerical analysis methods, Qi Shengwen et al. [8] obtained that the slopes had two types of dynamic response: high slope and low slope typology. They proposed the concept of the critical height of the slopes. The results of the study by Yan Zhixin et al. [9] showed that the rocky slopes' dynamic responses with different shapes had different deformation damage sites, which was closely related to the slope shapes. Zhang Yingbin et al. [10] analyzed three factors: the slope heights, angles, and shapes. The results showed that different slope shapes caused different acceleration response patterns, and the concave slopes' acceleration amplification effects were slightly minor. Zhang Yihao et al. [11] analyzed the different slope types' stabilities and damages of the landslides. The test results shown that the convex slopes required the lowest lifting angles for the same slope gradient and height, and their stabilities were the worst; the concave slopes were the highest, and their stabilities were the best. In addition to this, graded starts were more likely to occur when convex slopes break down.

Although previous studies have yielded some results regarding the dynamic response characteristics of slope under the action of the seismic wave, few studies have been conducted to explore the influence of slope morphology. In this paper, based on the previous work, we used finite difference software to establish three kinds of slope models (straight, concave, and convex slopes) with the same elevation for numerical simulation. And the amplification factors of the slope surface centroid amplification were observed. The laws of the slope dynamic response were summarized when the slope angle, morphology, and seismic wave frequency changed. The mechanism was then analyzed by combining the ray analysis and Snell's law formula derivation.

2   Numerical Simulation

2.1   Soil Parameters

A series of numerical simulations were carried out using the finite difference code FLAC 5.0 [12] to describe the dynamic response characteristics of slope surface centroid amplification under the action of the seismic wave. In the calculation models, the bedrock's upper was a homogeneous, linearly elastic soil layer. A local damping of 0.05 was used to improve the computational efficiency. The soil layer's primary material parameters are shown in Table 1.

Table 1. The physical and mechanical parameters of the soil layer

Density, ρ (Mg/m^3)	Modulus of elasticity, E (MPa)	Shear wave velocity, V_s (m/s)	Poisson's ratio	Local damping
2.0	1333	500	1/3	0.05

2.2 Model Dimensions and Boundary Conditions

Different slope models are needed to correspond to different working conditions to study the slope dynamic responses under the seismic action. The shear waves (SV waves) come into the slope vertically from the bottom in this study, and the slope models were modeled using the fixed slope height $H = 50$ m, width 1000 m ($L_1 = L_2$), and depth of 40m. The 2D models with different slope shapes are shown in Figs. 1, 2 and 3.

Typically, observation points are located at the model free-field boundaries' left or right, which are used as reference points [4, 13] to calculate the topographic amplification factors. Therefore, this study used the ratio of the peak acceleration at the 2D models' slope surface center to the 1D models at the same height to calculate the amplification factors. The 1D models were constructed with the same material, depth, load, and boundary conditions as the 2D models. To ensure the computational efficiency and accuracy of the results, the quiet and free-field boundaries (Fig. 1) were used to absorb the approach borders' waves, preventing any waves from reflecting in the models. The dimensions of the grid (Δl) were chosen according to the recommendations of Kuhlemeyer and Lysmer [14]:

$$\Delta l \leq \frac{\lambda}{10} \tag{1}$$

where λ is the wavelength corresponding to the highest frequency of the input seismic wave.

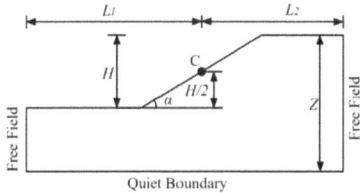

Fig. 1. Schematic diagram of the 2D straight slope models

2.3 The Input Seismic Wave

The input of the seismic wave was obtained using the Gabor wave [15] with a peak ground acceleration of 0.5 m/s^2, given by Eq. (2). To mimic the accumulation and attenuation of the seismic waves during an actual seismic event, the modified Gabor

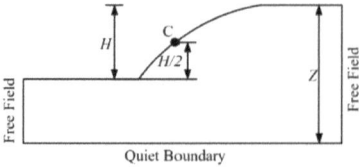

Fig. 2. Schematic diagram of the 2D convex slope models

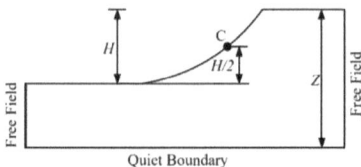

Fig. 3. Schematic diagram of the 2D concave slope models

wave at the slopes' bottom was considered. The constant in Eq. (2) was varied so that the acceleration reached 1.0 m/s² (Eq. 3), and the same number of cycles ($N = 12$) was used for the different frequencies considered. The adjusted seismic acceleration time history profile was shown in Fig. 4.

$$a(t) = \sqrt{\alpha e^{-\beta t^\gamma}} \sin(2\pi f t) \qquad (2)$$

$$a(t) = 2\sqrt{\alpha e^{-\beta t^\gamma}} \sin(2\pi f t) \qquad (3)$$

where f is the frequency of the input seismic wave; t is the time; α, β, and γ are parameters regulating the shape and amplitude of the time history envelope of the accelerated seismic wave. Details of the values taken were shown in Table 2.

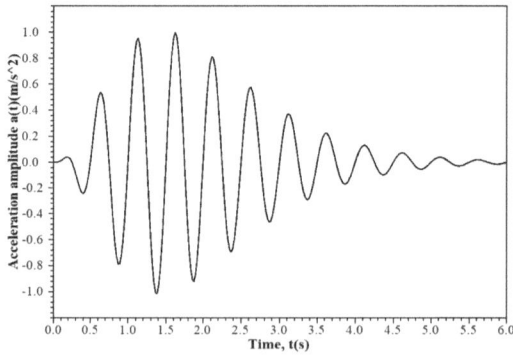

Fig. 4. Time history diagram of ground shaking acceleration ($f = 2$Hz)

Table 2. Parameters controlling the shape of the acceleration time history

cycles	1	2	4	6	12
α	2800000	63000	3100	200	6.5
β	35	22	22	7	3.5
γ	5.7	5	5	5	5

2.4 Observation Points and Working Conditions

In order to monitor the input seismic wave accuracy, the frequencies (including the soil intrinsic frequency) were selected to allow the calculation models to produce topographic and soil layer effects. The natural frequencies were calculated by Eq. (4):

$$f_n = \frac{(2n + 1)V_s}{4Z}, n = 0, 1, 2, ...\infty \tag{4}$$

Observation points were set at the slope models' bottom, surface center, and crest. And evaluated the slope dynamic response characteristics for the operating conditions listed in Table 3.

3 Numerical Simulation Results and Analysis

The a_{max} was the peak acceleration at the 2D slope model surface center, and the $a_{h,ff}$ was at the same height in the 1D model.

Topographic amplification factor:

$$A_{\max} = a_{\max}/a_{h,ff} \tag{5}$$

Horizontal topographic amplification factor:

$$A_{h\,\max} = a_{h\,\max}/a_{h,ff} \tag{6}$$

3.1 Verification of Numerical Simulation Results

As shown in Fig. 5, the monitored horizontal topographic amplification factors A_{hmax} at the slope crest (the straight slope of $Z = 125$ m) were compared with the literature [7]. Although the calculation results were not the same (due to the different model sizes, soil parameters, and input seismic wave types), the overall trend was consistent with the previous results (the amplification factors all reach the maximum value of about 1.5 or so when $H/\lambda = 0.2$), which verified the input seismic wave reasonableness.

Table 3. Numerical simulation cases

Slope morphology	Thickness of soil layer Z (m)	slope angle i (°)	H/λ
Straight slope	125	75°,45°,30°	0.05,0.1,0.2,0.3,0.5,1
	250	75°,45°,30°	0.05,0.1,0.2,0.3,0.5,1
	500	75°,45°,30°	0.05,0.1,0.2,0.3,0.5,1
Concave slope	125	75°,45°,30°	0.5
	250	75°,45°,30°	0.5
	500	75°,45°,30°	0.5
Convex slope	125	75°,45°,30°	0.5
	250	75°,45°,30°	0.5
	500	75°,45°,30°	0.5

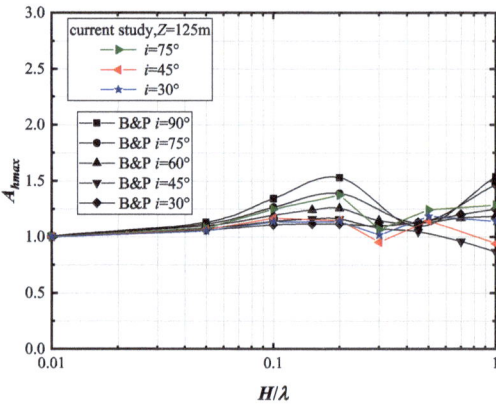

Fig. 5. Topographic amplification factors at the slope crest for $Z = 125$ m and results from Bouckovalas and Papadimitriou.

3.2 Effects of the Seismic Wave Frequency

Figure 6 shown the trend of topographic magnification factors with H/λ at the straight slopes' crest and surface center at different slope angles (45°). When $H/\lambda = 0.2$, the amplification factors at the slope crest were the largest, and the maximum value reached about 1.5; when $H/\lambda = 0.5$, the amplification factors at the slope surface center were the largest, with the maximum value of about 2.5. This indicates that with different frequencies, the acceleration response of each part of the slope is different and characterized by rhythmic changes along the slope. The amplification factors' maximum value at the slope surface center was larger than the crest in this study. So, it is necessary to investigate how the slope surface centroid amplification factors change under the action of the seismic wave.

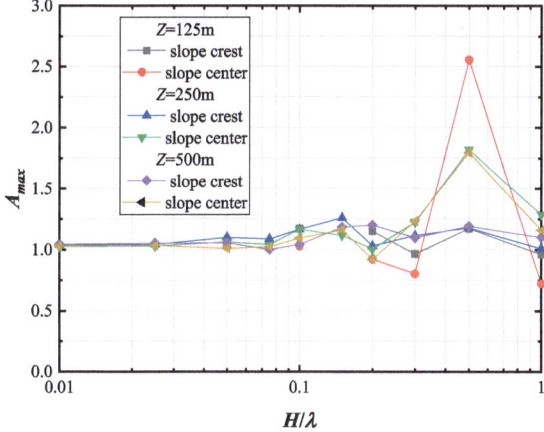

Fig. 6. Amplification factor of ground shaking at the slope crest and the slope center

3.3 Effects of the Slope Angle

As shown in Fig. 7, when the straight slopes' elevation $Z = 125$ m ($H/\lambda = 0.5$), the size relationship between the amplification factors at different slope angles was $30° > 75° > 45°$. This result is consistent with the conclusion of the literature [5] that the straight slope topographic amplification factors are maximum at the slope angle of $30°$.

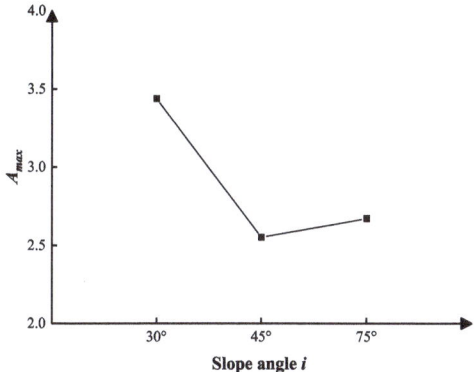

Fig. 7. The straight slope surface centroid amplification factors

3.4 Effects of the Slope Morphology

Figure 8 was the line graphs of the concave, straight, and convex slopes' amplification factors ($H/\lambda = 0.5$). From Fig. 10, it could be seen that for the slope angle of $75°$, the amplification factors were concave > convex > straight slopes; for the slope angle of $45°$: convex > straight > concave slopes; for the slope angle of $30°$: straight >

convex > concave slopes. On the other hand, the size relationship of the concave slope amplification factors at different slope angles was obviously different from the straight slopes (the convex slopes: 30° > 75° > 45°; the concave slopes: 75° > 30° > 45°). These results show that the amplification factors fluctuate as the slope morphology changes.

4 Discussions

4.1 Theoretical Analysis

Based on the wave propagation, reflection, and diffraction physical phenomena. When an incident SV wave impinges on the free surface, reflected SV and P waves will be generated. This is due to the coupling effects between the P and SV waves caused by the displacement and continuity conditions on the free surface [16]. From Fig. 9, it can be seen that the θ_0 (=α) is the angle of the SV waves incidence, and the θ_1 and θ_2 are the angles of the SV and P waves reflection, respectively. In order to satisfy the boundary displacements and stress conditions on the free surface, the straightforward relation can be derived from the Snell's law as follows [16]:

$$\sin \theta_0 = k^{-1} \sin \theta_2 \tag{7}$$

$$\theta_0 = \theta_1 \tag{8}$$

where k is a material constant defined by the ratio between the P and SV wave velocity:

$$k = \frac{V_P}{V_S} = \left[\frac{2(1-\nu)}{1-2\nu} \right]^{\frac{1}{2}} \tag{9}$$

where ν is the Poisson's ratio, and $k = 2$ when $\nu = 1/3$.

In this paper, the numerical analysis was carried out for the working conditions with the slope angle $\alpha \leq 75°$. When the SV waves' reflection angle is $\theta_1 \leq 75°$, the SV waves reflected on the inclined plane cannot reach the upper surface; when the reflected P waves' angle is $\theta_2 > 90° - \theta_0$, the reflected P waves will reach the upper surface. For $\theta_2 = 90° - \theta_0$, the oblique angle can be calculated as 26.6° using Eq. (8); for $\theta_2 \geq 90°$, the reflected P waves propagate along the inclined plane, and the critical oblique angle is 30° when $\theta_2 = 90°$. Thus, when 26.6° < α < 30°, the reflected P waves can reach the upper surface directly. It can superimpose with the incoming SV and Rayleigh waves. The amplitude of the reflected P waves is determined by the slope angle, as shown in Eq. (10), and A_0 was the amplitude value of the incident SV waves.

$$A_1 = -A_0 \frac{k \sin 4\alpha}{\sin 2\theta_2 \sin 2\alpha + k^2 \cos^2 2\alpha} \tag{10}$$

Substituting Eqs. (7) (8) (9) into the above equation yielded that:

$$\frac{A_1}{A_2} = -\frac{k \sin 4\alpha}{4 \sin \alpha \sin 2\alpha \sqrt{1 - 4 \sin^2 \alpha} + k^2 \cos^2 2\alpha} \tag{11}$$

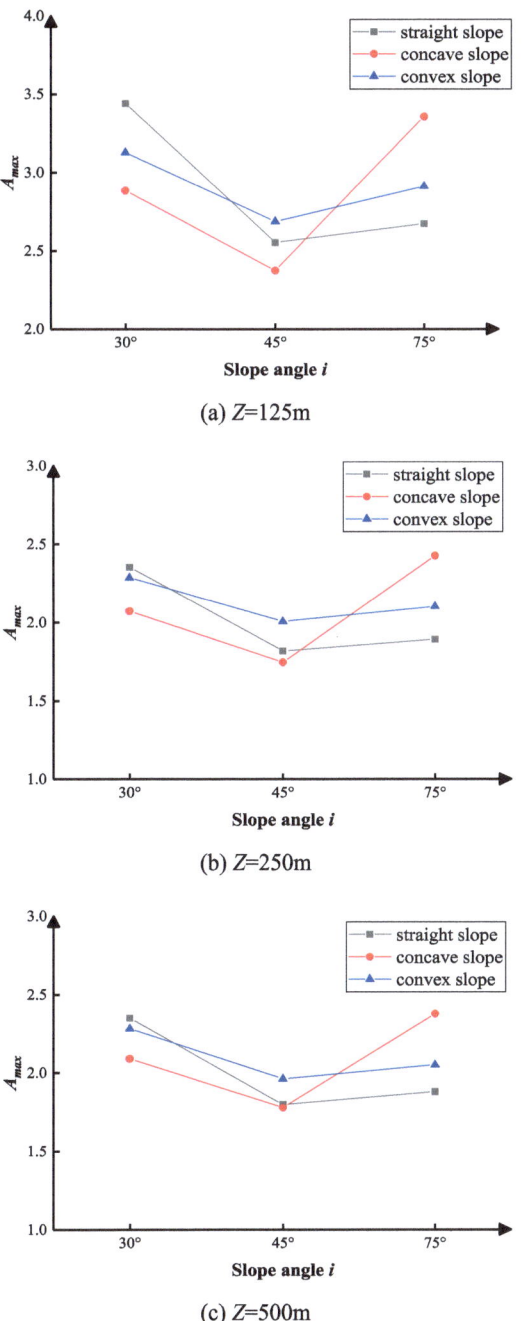

(a) Z=125m

(b) Z=250m

(c) Z=500m

Fig. 8. The slope surface centroid amplification factors ($H/\lambda = 0.5$)

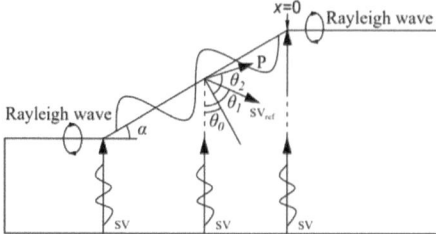

Fig. 9. Propagation of the incident SV, reflected SV and P waves, and the diffracted Rayleigh waves in the slope model [5].

From the variation of $\frac{A_1}{A_0}$ with slope angle in Fig. 10, it can be seen that when the slope angle was less than $30°$, the amplitude of the reflected P waves increased with the slope angle, and the amplification factors increased as well. As long as the slope angle was kept above $30°$, the amplitude of the reflected P waves was close to zero. Therefore, the slope surface centroid amplification factors were reduced somewhat. This is because when the slope angle exceeds $30°$, the reflected P waves are transformed into a surface wave propagating along the free surface. It explains why the straight slope amplification factors are maximum at the slope angle of $30°$.

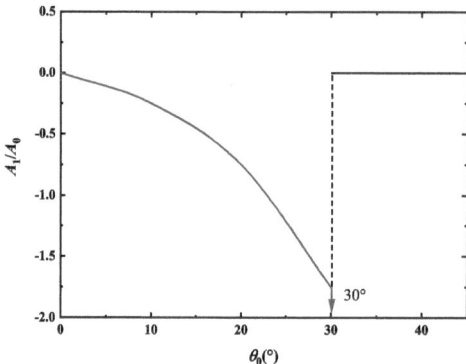

Fig. 10. Variation of the amplitude ratio of the reflected P wave with the angle of incidence. (The arrow points to the critical angle of $30°$)

4.2 Ray Analysis

Under the action of micro-vibration, according to the fluctuation theory of the seismic wave propagation, the rays are used to indicate the seismic waves' propagation direction, and the sparseness of the wave rays can reflect the convergence or dispersion of the seismic waves propagation energy. From the analysis of Snell's law [16] in Sect. 4.1. A, the distribution of the seismic fluctuation energy in an isotropic uniform slope is determined by the incident seismic wave angle, independent of the vibration direction of the waves. The following will analyze and verify the numerical simulation results based on the seismic wave ray path distribution.

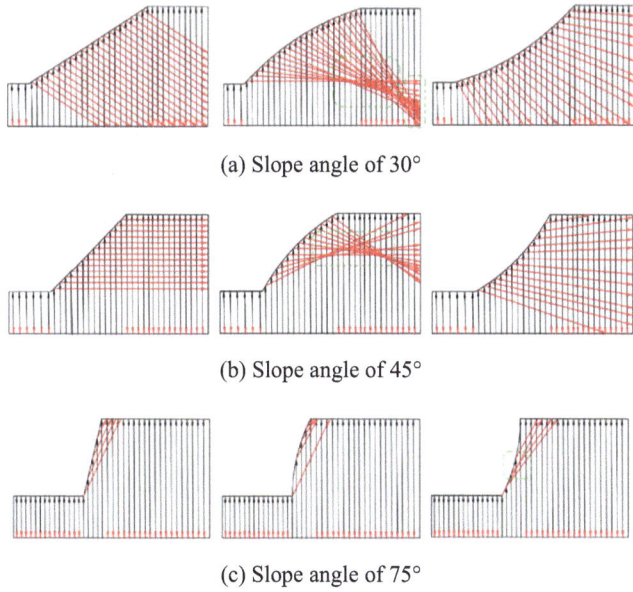

(a) Slope angle of 30°

(b) Slope angle of 45°

(c) Slope angle of 75°

Fig. 11. Schematic diagrams of the straight, concave, and convex slope ray paths

It can be seen from Fig. 11(a) that the convex slopes with the angle of 30° had stronger energy in the slopes' middle and lower regions and weaker energy in the upper part; the straight slopes had more vital energy near the slope surface; and the concave slopes were evanescent to the reflected waves. This was consistent with the numerical simulation results. The amplification factors of the straight slopes were most significant when the slope angle was 30°. It can be seen through Fig. 11(b) that the energy convergence regions caused by convex slopes moved to the slope middle and front. The energy was stronger at the slope surface center. Consistent with the simulation results at the slope angle of 45°, the convex slope amplification factors were the maximum. It can be seen from Fig. 11(c) that when the slope angle was 75°, due to the concave slopes' dispersion effects on the reflected waves and the influence of the high slope angle, the energy convergence phenomenon in the areas near the slope middle and lower part surface was partially caused. In contrast, the convex slopes' convergence effects on the reflected waves were mainly reflected behind the slope crest, and there was no apparent convergence phenomenon near the slope surface center. It was also consistent with the above numerical simulation results for the slope angle of 75°, and the concave slope amplification factors were the maximum.

The above results of ray analysis and numerical simulation corroborated each other, indicating that the slope shapes had significant effects on the slope dynamic response. In summary, the influence of slope shapes on the slope dynamic response is not generalized in practical slope problems. It depends on the slopes' specific conditions.

5   Conclusions

Based on the above numerical simulation and theoretical analysis results, the effects of the slope angle, slope morphology, and seismic wave frequency on the effects on the slope dynamic responses are revealed. The conclusions are drawn as followed.

(1) The acceleration responses are different at the various parts of the slope for different frequencies, and the amplification factor shows a rhythmic phenomenon along the elevation.
(2) The amplification factors are different for the slope with different slope aspect shape. The slope morphology has a significant effect on the dynamic response of the slope.
(3) If the slope elevation is the same and the H/λ equals to 0.5, the amplification factor of straight slope is biggest for the slope angle of 30° and it is smallest for the slope angle of 45°. The slope angle has greatly influences the slope amplification factor.

Funding. This research was funded by the Natural Science Foundation of Anhui province (Grant No. 2108085ME190) and the National Natural Science Foundation of China (Grant No. 42107173).

References

1. Assimaki, D., Gazetas, G., Kausel, E.: Effects of local soil conditions on the topographic aggravation of seismic motion: parametric investigation and recorded field evidence from the 1999 Athens earthquake. Bull. Seismol. Soc. Am. **95**, 1059–1089 (2005)
2. Geli, L., Bard, P.-Y., Jullien, B.: The effect of topography on earthquake ground motion: a review and new results. Bull. Seismol. Soc. Am. **78**, 42–63 (1988)
3. Spudich, P., Hellweg, M., Lee, W.: Directional topographic site response at Tarzana observed in aftershocks of the 1994 Northridge, California, earthquake: implications for mainshock motions. Bull. Seismol. Soc. Am. **86**, S193-208 (1996)
4. Tripe, R., Kontoe, S., Wong, T.: Slope topography effects on ground motion in the presence of deep soil layers. Soil Dyn. Earthq. Eng. **50**, 72–84 (2013)
5. Zhang, Z., Fleurisson, J.A., Pellet, F.: The effects of slope topography on acceleration amplification and interaction between slope topography and seismic input motion. Soil. Dyn. Earthq. Eng. **113**, 420–431 (2018)
6. Ashford, S.A., Sitar, N., Lysmer, J., Deng, N.: Topographic effects on the seismic response of steep slopes. Bull. Seismol. Soc. Am. **87**, 701–709 (1997)
7. Bouckovalas, G.D., Papadimitriou, A.G.: Numerical evaluation of slope topography effects on seismic ground motion. Soil Dyn. Earthq. Eng. **25**, 547–558 (2005)
8. Qi, S., He, J., Zhan, Z.: A single surface slope effects on seismic response based on shaking table test and numerical simulation. Eng. Geol. **306**, 106762 (2022)
9. Yan, Z.-X., Shi, S., Dang, B., Li, B.: Influence of slope morphology on rocky slope stability under earthquake. J. Shandong Univ. Sci. Technol. (Nat. Sci. Ed.) **32**(2): 43–48+78 (2013)
10. Zhang, Y.-B., Liu, J., Tang, Y.-B., Xiang, C.-L.: Seismic dynamic response analysis considering slope topography effect. J. Earthquake Eng. **43**(1), 142–153 (2021)
11. Zhang, Y.-H., Xu, K., Zheng, S.-M., Liu, Z.-Y., Zhu, T.: Stability analysis of landslides with different slope patterns. Energy Environ. Protect. **44**(12), 73–79 (2022)
12. Itasca, Fast Lagrangian Analysis of Continua (FLAC), Version 5.0, Itasca Consulting Group Inc., Minneapolis, Minnesota (2011)

13. Lenti, L., Martino, S.: The interaction of seismic waves with step-like slopes and its influence on landslide movements. Eng. Geol. **126**, 19–36 (2012)
14. Kuhlemeyer, R.L., Lysmer, J.: Finite element method accuracy for wave propagation problems. Technical note. J. Soil Mech. Found. Div. **99**(5), 421–427 (1973)
15. Gabor, D.: Theory of communication. I. The analysis of information. J. Inst. Electr. Eng. **93**, 429–441 (1946)
16. Achenbach, J.: Wave Propagation in Elastic Solids. Elsevier, Amsterdam (2012)

Research on the Accuracy of Lateral Force Coefficient of Pavement Anti-skid

Lei Chen[1], Da Li[2], Jile Jiang[3], and Lu Liu[1(✉)]

[1] Research Institute of Highway Ministry of Transport, Beijing 100088, China
`1076146575@qq.com`
[2] Yitong Engineering Testing Co., LTD., Hohhot 010000, China
[3] Beijing Tesidi Semiconductor Equipment Co., LTD., Beijing 101300, China

Abstract. In order to improve the accuracy of road surface anti slip detection, the six component force testing method is applied to optimize the measurement method of the single wheel lateral force coefficient tester used for road surface anti slip detection, and to eliminate the interference of static factor uncertainty, the comparison of experimental data has proven the rationality and feasibility of the method, providing a basis for further improving the accuracy of pavement anti slip detection.

Keywords: Transverse force coefficient · detection · calibration · six-component

1 Introduction

In addition to meeting the basic conditions of traffic, the road should have good safety. Among them, the anti-skid performance of the road surface can provide good adhesion for high-speed vehicles and reduce the hidden danger of traffic accidents. The safety performance of pavement is expressed by the anti-skid performance of pavement, and the friction coefficient of pavement is the most important index to evaluate its anti-skid performance, so it is necessary to detect and evaluate the friction coefficient of pavement accurately and efficiently.

Due to the limitation of detection range, detection speed, sampling number, etc., the friction coefficient vehicle test method is often used for detection on high-grade roads. The SCRIM, for example, not only represents the braking distance of the vehicle, but also reflects the ability of the road surface to prevent the vehicle from sideslip [1–3]. The equipment is installed on the chassis of a large truck and equipped with a large-capacity water tank, which can be continuously tested for long distances without affecting traffic. It has obvious advantages in technical characteristics, working efficiency and equipment working stability, so it is widely used. SCRIM method is used to measure the transverse force coefficient of pavement under the most unfavorable conditions [4–5]. Since the test wheel is equipped with a fixed counterweight and is set at an Angle of $20°$ with the driving direction, when the test wheel touches the ground, the measuring mechanism is subject to the axial transverse force, which is more in line with the condition of the car when it is driving on the road [5–7].

© The Author(s) 2024
A. Bieliatynskyi et al. (Eds.): CSTTE 2023, LNCE 603, pp. 404–413, 2024.
https://doi.org/10.1007/978-981-97-5814-2_36

The research of this project aims to address the limitations and shortcomings of existing pavement anti slip performance testing, enhance the evaluation ability of pavement friction characteristics, and provide scientific basis for road safety and traffic operation [8].

The current method of quantity traceability does not require key measurement technical parameters such as indication error of friction coefficient. In addition, the calibration technology is based on the dynamic repeated measurement method of coefficient of variation, using the road section method. Due to the limitations of the site, it cannot meet the efficiency requirements of the measurement standard implementation, which increases the difficulty of tracing the value of lateral force friction coefficient, making it difficult to grasp the accuracy of lateral force coefficient testing [9].

Research on the traceability method of lateral force friction coefficient can form technical specifications for ensuring the accuracy of pavement anti slip performance measurement and provide important data support for decision-making [10].

2   Study on the Measurement Source Method of Single-Wheel Friction Coefficient Tester for Lateral Force

2.1   Principle Analysis

The transverse force coefficient (SFC) is assumed that the standard test tire has a certain deflection Angle (20°) with the forward direction of the car traveling at a certain speed, and the contact with the wet road causes it to produce a lateral friction resistance perpendicular to the plane of the test wheel. The ratio of this force to the constant vertical load borne by the test wheel is the SFC value of the transverse force coefficient, as shown in Fig. 1. The lateral friction resistance can be obtained from the force sensor, the vertical load can be obtained according to the known value, the transverse force system value can be obtained, and the road friction coefficient is positively correlated. It can reflect the vertical and lateral friction coefficient of the comprehensive index, is characterized by the ability to measure the emergency braking or sharp turns provided by the road to the vehicle handling performance.

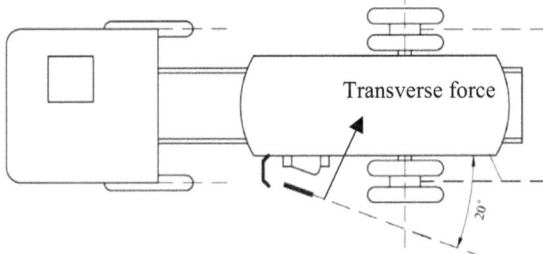

Fig. 1. Schematic diagram of transverse force coefficient testing principle

The vertical load of the tire has little influence on the transverse force when the side deflection Angle is small, but when the side deflection Angle of the tire is large to almost reach the transverse force saturation state, the vertical load will play a dominant role. This is because the effect of vertical load only occurs in the area where there is relative slide between the tire and the ground, when there is more of this relative slide in the ground, that is, when the side Angle is large, the effect of vertical load will be obvious.

In the pavement lateral force coefficient measurement system, the static vertical load is a certain value (China's standard is 2kN), and the measured value is close to the maximum transverse force under the condition that the side deflection Angle is 20°. In some measuring systems, a constant static vertical load is used to calculate the transverse force coefficient, while in some measuring systems, the static vertical load is only a reference standard for equipment calibration, and the actual vertical load measured in real time is used to calculate the transverse force coefficient.

The horizontal load is also involved in the calculation of SFC, so the specified value of its error is relatively strict, especially when the static SFC parameter is calibrated, the calibration is extremely complex.

According to the requirements of technical indicators in JJG (Traffic) 113–2014 "Single-wheel lateral Force Coefficient Tester", the technical indicators are measured and calibrated. Through mastering the measurement methods, some influencing factors in the test are understood and relevant problems are found. At present, the test parameters mainly include the deviation Angle of the test wheel, the vertical load of the test wheel, the horizontal load of the test wheel, the distance sensor, the temperature sensor and the repeatability index of the transverse force coefficient output by the dynamic test. None of these indicators is a static influence factor, so the accuracy of the evaluation of skid resistance under dynamic conditions does not meet the conditions.

In order to eliminate the influence factors on the accuracy of SFC parameter calibration, the static SFC value and dynamic SFC value are compared and tested, and the leading parameter of SFC value is horizontal load, and the axial horizontal load is obtained through the six-component force sensor.

The safe driving performance of road vehicles depends on whether the longitudinal motion (braking, traction control), lateral motion (steering control) and vertical motion (road fluctuation) are normal. If these loads and torque are not normally loaded on the car body, it will cause some of its components to produce additional loads, and in bad circumstances, it may damage the parts and even the whole vehicle. The performance of the above various vehicle motion loads is the most concerned by various supervisory departments and enterprises. The six component test system can provide test and data support for the above concerned projects.

At the same time, the six-component test system can provide accurate road spectrum acquisition function, and provide comprehensive road spectrum signal for vehicle testing in the laboratory. At the same time, since the load collected by the six-component test system is the external load of the vehicle, the load of each component in the vehicle body model can be calculated through the external load, so as to obtain the fatigue life of all components. These calculation data have important guiding significance when the body is still in the modeling stage, and problems can be found in the design stage, rather than slowly improving after the whole vehicle is built. The six-component force test method is different from the previous method of measuring the load inside the vehicle. Because the internal load is not able to calculate the load of other components.

Six component force sensor measurement system, the test index meets the requirements of SAE (Fig. 2) coordinate system. Can quickly measure the three-direction force and three-direction torque of the wheel. The sensor is mounted between the hub adapter and the wheel adapter, the tire is mounted on the modified wheel, and the hub adapter is mounted on the axle. In this way, when the wheel is forced on the transmission hub, it will pass through the sensor first. Six independent Bridges measure force and torque, with low dryness between channels, free from temperature changes and electromagnetic interference. Because the sensor measures force and torque in a rotating coordinate system, the output of the radial channel changes as the wheel rotates. Before applying the acquired data, they must first be converted into the vehicle coordinate system. Slip rings with integrated optical encoders transmit the signal to the stator, and the encoder provides the angular position and wheel rotation speed.

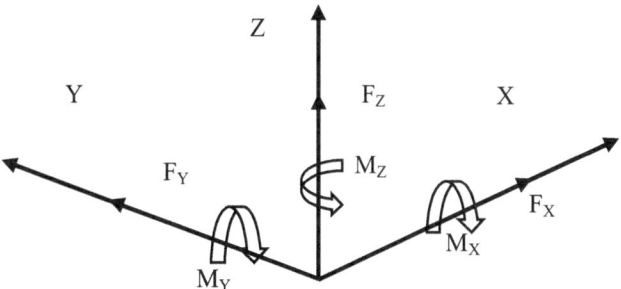

Fig. 2. Wheel six component SAE coordinate system

3 Accuracy of Lateral Force Coefficient

The overall test scheme is to convert the SFC value of the evaluation index into the most original force value, and compare the error value of the standard SFC value obtained through the conversion of the force value with the SFC collected by the single wheel lateral force friction coefficient tester.

3.1 Test Verification

At present, the general calibration method for horizontal load error is as follows:

1) Park the bearing vehicle on the horizontal hard test platform site and turn on the system power supply;
2) Place the afterforce plate of the vehicle horizontally under the test wheel, buckle the reaction device and lower the test wheel to the fully unloaded state, and connect the standard dynamometer with the horizontal loading device in series;
3) After the tester is preheated for 10 min, start the verification state, and start to apply force with the afterburner;

4) When the standard dynamometer is 0, 200N, 400N, 600, 800N, 1000N, 1200N, 1400N, 1600N, 1800N and 2000N, the corresponding response measurement value of the tester is read, the corresponding force value is obtained, and the difference between the two is calculated respectively, and the maximum difference is taken.

For the test of horizontal load error, the six-component force sensor is used to compare the truth value of the error (Fig. 3). At static state, the force standard is applied, and the pressure is 0, 200N, 400N, 600, 800N, 1000N, 1200N, 1400N, 1600N, 1800N, 2000N. The data of single-wheel lateral force friction coefficient tester and force standard sensor were recorded respectively, and the data of six-component force sensor and Mark signal were collected.

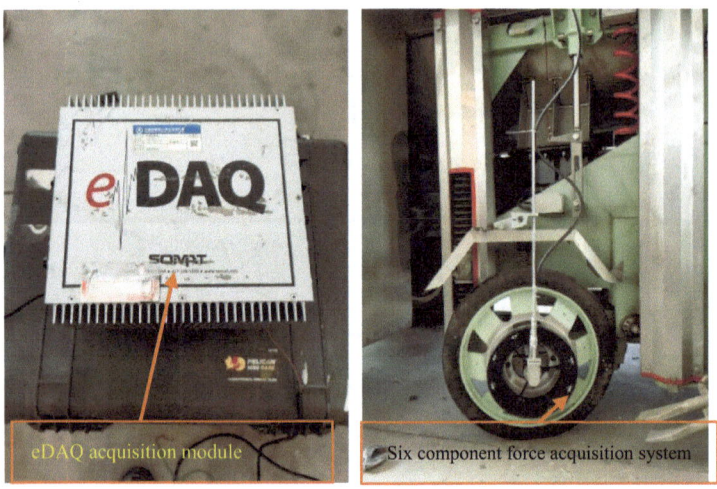

(a) eDAQ Data Acquisition System   (b) Friction collection data system

Fig. 3. eDAQ collects data

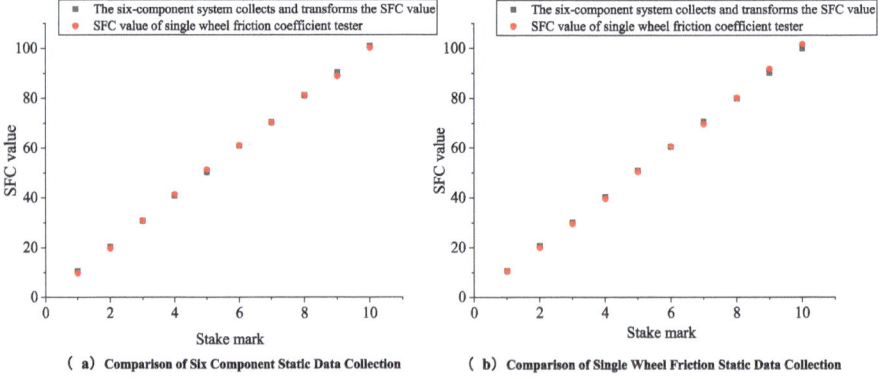

(a) **Comparison of Six Component Static Data Collection**   (b) **Comparison of Single Wheel Friction Static Data Collection**

Fig. 4. Schematic diagram of ten loading data

It can be seen from the Fig. 4 that the SFC value collected by the single-wheel lateral force friction coefficient tester and the force value data collected by the six-component force sensor are basically in the same linear trend after conversion.

3.2 SFC Repeatability Check

The friction coefficient of the single wheel transverse force tester is measured several times at different speeds on the standard 500 m test road, and the data of the single wheel lateral force friction coefficient tester and the six-component force sensor are recorded respectively (Fig. 5).

Fig. 5. SFC repeatability verification

It can be seen from the Fig. 6 that the dynamic SFC values collected by the single wheel lateral force friction coefficient tester at different speeds and the dynamic SFC values converted by the force value data collected by the six-component force sensor are basically the same correlation coefficient.

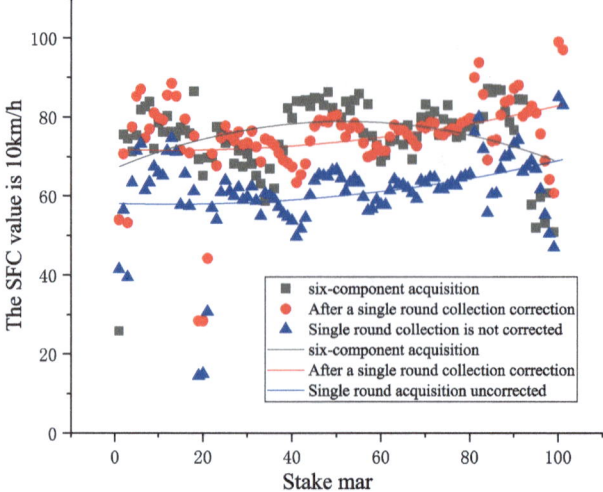

（a）Comparison of 10km/h six component force data collection and single wheel friction data collection

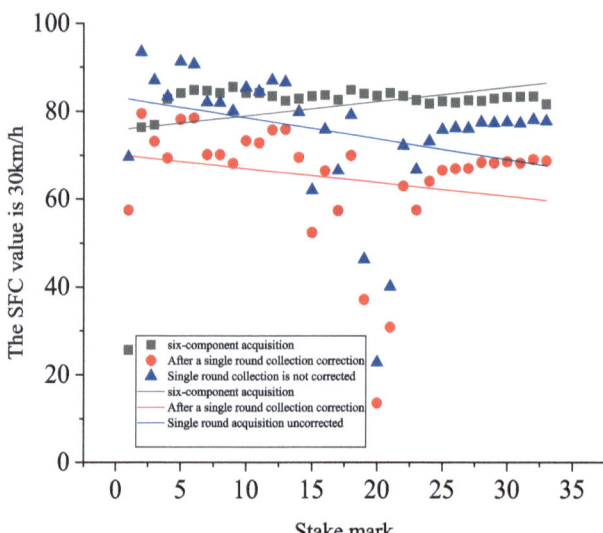

（b）Comparison of 30km/h six component force data collection and single wheel friction data collection

Fig. 6. Dynamic test analysis

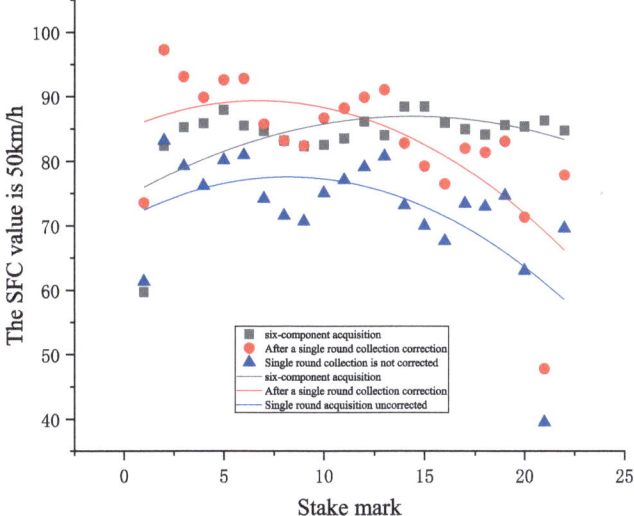

(c) **Comparison of 50km/h six component force data collection and single wheel friction data collection**

Fig. 6. (*continued*)

4 Conclusions

The lateral force coefficient is used to test the anti-skid performance of the pavement. The influencing factors affecting the accuracy of SFC are the main reference factors for the establishment of measurement technology. The external factors in the testing process of the lateral force coefficient are different, so the test results are also related to the pressure of the test wheel, the thickness of the water film and the paving direction of the asphalt surface layer. The tire pressure is less than the standard tire pressure, and the measured result is too large and needs to be corrected. Therefore, the tire pressure should be checked before the test. When the tire tread of the test wheel is in contact with the mixture of the surface layer, under different water film thickness, the anti-skid performance of the tire tread and the road surface will also be affected to a certain extent. The research shows that the thicker the water film thickness, the smaller the measured SFC value. Especially after the rain, the road is in a wet state, the measured data is small.

Internal factors such as temperature, test speed, etc. The influence degree of temperature test value is larger, the temperature is higher and lower, the measured SFC value is different from the actual anti-sliding ability reflected, and this difference will be more obvious on the old road; When the test speed is less than the standard speed (50 km/h), the measured SFC value is larger, reflecting the anti-skid performance preference. When the speed is greater than the standard speed, the SFC value decays significantly.

The six-component test is used to eliminate the influence factors, especially the influence of tire wear state on the test results during the test.

Improve the accuracy of pavement detection results, provide reliable data for detection and maintenance personnel, and improve the quality of engineering construction. Reduce due to inaccurate road anti-skid test data, resulting in detection, maintenance personnel to make wrong remedial judgments to reduce unnecessary traffic accidents caused by road slip, damage to personal and property safety. The independent research of calibration devices has realized the replication of multiple sets of calibration devices in various provincial metrological technical institutions nationwide at the same time, and promoted the establishment of metrological service network.

This experiment starts from the macro perspective of road friction, without studying the microstructure of the road surface. Looking at other industries, most have already entered the micro research of friction coefficient. The micro characterization of road friction coefficient has the advantages of good accuracy and high stability. In the future, relevant research will be carried out on the micro level definition of friction reference materials and road friction coefficient standards.

References

1. Wu, L.-l., Wu, A.-n., Li, J.-s.: Research of the effect of road bump on the sideway force coefficient test system. J. High. Transp. Res. Dev. **22**, 51–53 (2005). https://kns.cnki.net/kns8/defaultresult/index
2. Liang, B.: Research on detection method and evaluation standard of pavement skid-resistance. Shanxi Sci. Technol. Commun. 1, 21–24 (2020). https://kns.cnki.net/kns8/defaultresult/index
3. Zhou, H., Wang, G., Zhen, J., Chen, X.: Numerical analysis method for friction characteristics of tire-pavement under wet slip condition. J. Mech. Eng. **56**, 177–185 (2020). https://doi.org/10.3901/JME.2020.21.177
4. Huang, X-m., Zheng, B-s.: Research status and progress for skid resistance performance of asphalt pavements. China J. Highw. Transp. **32**, 32–49 (2019). https://doi.org/10.19721/j.cnki.1001-7372.2019.04.003, https://doi.org/10.13873/j.1000-97872008.12.025
5. Wang, R., Yong, Z., Yang, Y.: Research progress of tire wear and tire dynamic performance change. China Rubber Ind. **68**, 140–145 (2020). https://kns.cnki.net/kcms2/article/abstract?v=3uoqIhG8C44YLTlOAiTRKibYlV5Vjs7iy_Rpms2pqwbFRRUtoUImHUNKuq_ADummANi-EeBLH_BQ54XgN1XLW5q8leb2xNX0&uniplatform=NZKPT
6. Peng, H., Guo, R., Xu, L., Si, Y.: Research on pavement texture and skid-resistance performance based on CEEMD and SVD. J. Wuhan Univ. Technol. (Transp. Sci. Eng.) **45**, 1140–1150 (2021). https://kns.cnki.net/kcms2/article/abstract?v=3uoqIhG8C44YLTlOAiTRKibYlV5Vjs7iJTKGjg9uTdeTsOI_ra5_XYOfC2X_K5OYwFF5B33atclofg4Pa2U_iORA6jd8La5o&uniplatform=NZKPT
7. Sun, R., Zhang, Y., Wu, A., Xi, L., Li-He, Y., Liu, C.: Data acquisition and analysis system design based on wheel six-component. J. Chongqing Jiaotong Univ. (Nat. Sci.) **39**, 15–21 (2020). https://kns.cnki.net/kcms2/article/abstract?v=3uoqIhG8C44YLTlOAiTRKibYlV5Vjs7i8oRR1PAr7RxjuAJk4dHXooZ5F9j_u2mWhE2yToTJ8o9qqolF7NkRNmDKE2_G6QFd&uniplatform=NZKPT
8. Guo, K., Yang, Y., Xu Nan Chen Ping Zhang Lihao. (2014). Forces and Moments Computation, Calibration and Optimization of New Tire Test Rig. Nongye Jixie Xuebao **45**, 8–15. https://doi.org/10.6041/j.issn.1000-1298.2014.05.002
9. Wang, X.G.:Analysis and countermeasures on reason of influence on accuracy of cornering ratio testing system MU METER MK6. Bei Fang Jiao Tong 1, 4–6 (2013). https://doi.org/10.15996/j.cnki.bfjt.2013.01.003
10. Wang, J., Zhan, F., Mao, P.-j, Lei, L.: Design of measurement system for six-component of plow load based on virtual instrument technology. Transd. Microsyst. Technol. **27**, 67–69 (2008). https://doi.org/10.13873/j.1000-97872008.12.025

Mitigating Underestimation in Infiltration Cooling Load in Metro Stations: A Simulation Method for Determination

Bowen Guan[(⊠)], Haobo Yang, and Xinke Wang

School of Human Settlements and Civil Engineering, Xi'an Jiaotong University, Xi'an 710054, Shaanxi, China
guanbw@xjtu.edu.cn

Abstract. Accurate and rational estimation of cooling loads is crucial for the energy-efficient design of ventilation and air-conditioning systems in metro stations, contributing significantly to the low-carbon development of metro systems. This study conducted a comparative analysis between the cooling load estimates from design drawings and the actual measurements obtained on-site. The results revealed a significant underestimation, ranging from 81.0% to 87.5%, of the cooling load attributed to air infiltration through entrances during the design phase. To address this misestimation, a simulation method is proposed, utilizing a combined application of STESS software and Fluent software. The findings demonstrate that employing this simulation method can effectively mitigate the misestimation, reducing it to a range of 8.3% to 50.2%.

Keywords: Metro station · CFD simulation · Infiltration cooling load · Piston wind

1 Introduction

As the backbone of urban public transportation, the metro system boasts the advantages of speed, efficiency, and high passenger capacity. However, inevitably, this also brings about significant energy consumption and carbon emission [1]. Ventilation and air-conditioning (VAC) systems dominate the energy consumption of metro stations. Accurate and rational estimation of the air conditioning system's cooling load is crucial for energy-efficient and low-carbon design of the VAC system, as well as ensuring the thermal and humidity environment in the station [2].

Due to the presence of multiple interconnected entrances between metro stations and the external environment, it is challenging to effectively separate the public areas from the outside during actual operation. This inevitably leads to the infiltration of outdoor air into the public areas of the metro station through entrances [3, 4]. To the best knowledge of the authors, there are few studies on the infiltration airflow at the entrances and exits of subway stations. Liu et al. study the natural ventilation of entrances [5]. Zhang et al. use IDA software to establish a one-dimensional model to study the ventilation of entrances [6, 7]. Based on the current design and operational conditions of metro stations,

A. Bieliatynskyi et al. (Eds.): CSTTE 2023, LNCE 603, pp. 414–420, 2024.
https://doi.org/10.1007/978-981-97-5814-2_37

there is a lack of comprehensive understanding and research on the infiltration airflow through entrances. Consequently, the impact of infiltration airflow through entrances is not adequately considered in practical system design. This results in erroneous estimates of the infiltration airflow, ultimately affecting the load estimation and system design of VAC systems in metro stations. To address the research gap, a simulation method utilizing the coupled application of STESS software and Fluent software is proposed for mitigating the misestimation. This study is expected to serve as a valuable reference for the rational estimation of cooling loads and energy-efficient design of VAC systems in metro stations.

2 Methodology

2.1 Station Information

Four typical underground double-layered metro stations in a metro line in North China Plain are surveyed in this study. Figure 1 illustrates the station structure. The station hall is located on the underground first floor, while the platform is situated on the underground second floor. The platform follows a typical island configuration, allowing trains to pass on either side. Platform screen doors (PSDs) are installed between the track area and the platform. For public spaces, the thermal and moisture sources primarily include building envelopes, occupants, equipment, lighting, etc. Due to the thermal inertia of the soil, the heat transfer load from the building envelope is generally considered negligible in the design of underground station VAC systems. The occupant load can be calculated based on passenger flow conditions and the heat generated by individuals. Equipment and lighting are typically calculated using unit area indicators or specific heat emission criteria for individual equipment. The station hall is connected to the external environment through entrances. Permeating airflows induced by these entrances may significantly impact the design of the station's VAC system. The accurate estimation of infiltration cooling loads through entry and exit points is crucial for effective control of the thermal and moisture environment in metro stations and the design of the VAC system.

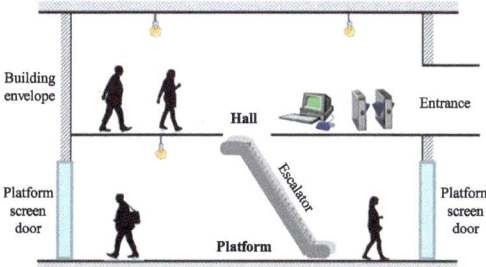

Fig. 1. Schematic diagram of indoor thermal and humid load composition in an underground metro station with PSD system

2.2 Simulation Method

In this study, a simulation method utilizing the coupled application of STESS software and Fluent software is proposed. Besides the well-known CFD software Fluent, STESS is a metro thermal environment simulation and analysis software developed by Tsinghua University. Based on extensive on-site testing at metro stations and theoretical research, it has the capability to predict airflow, air volume, pressure, as well as long-term (monthly) and short-term (hourly) temperature variations of metro systems under various conditions, including different regions, climates, system configurations, passenger densities, train operation schedules, and locations (tunnels and platforms) [8].

In this simulation method, the logical relationship between STESS and Fluent software is illustrated in Fig. 2. A STESS simulation model is developed based on the actual conditions of the surveyed metro line, with the model outputting air pressure variations over time at PSD locations. The air pressure variations serve as inputs for Fluent, providing boundary conditions for simulating the airflow within the metro station. The Fluent simulation model, depicted in Fig. 2, ultimately yields the output of air infiltration volume through entrances. The utilization of the STESS simulation model's output as input for Fluent simulation enhances the fidelity of the simulation boundary, thereby significantly enhancing calculation accuracy.

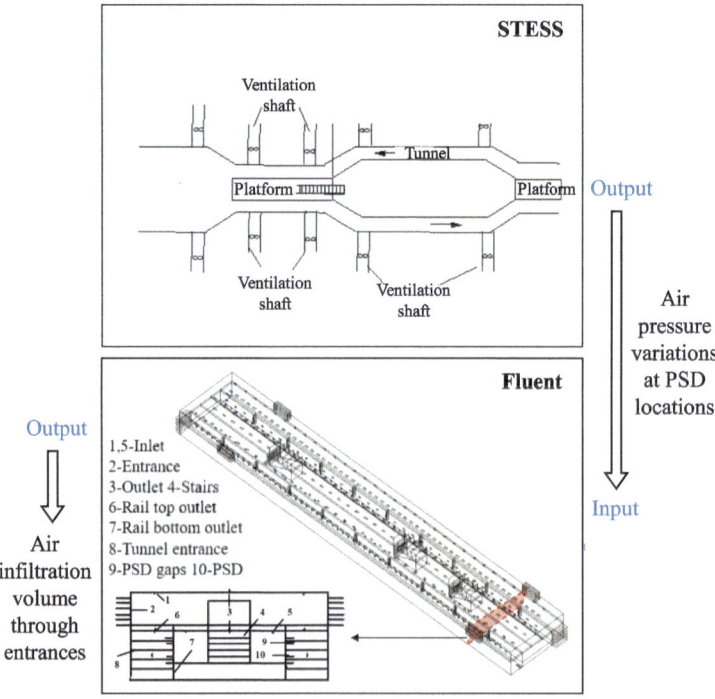

Fig. 2. Simulation method

To verify the grid independence and time step independence of the Fluent simulation, grid and time step refinement were conducted. In the base model, the grid count was set at 0.94 million with a time step of 0.1 s. The grid was refined to 1.23 million, and the time step was reduced to 0.05 s. The comparative simulation results are presented in Fig. 3. The impact of grid and time step refinement on the simulation results is minimal, indicating that the current base model satisfies the independence requirements for both grid and time step.

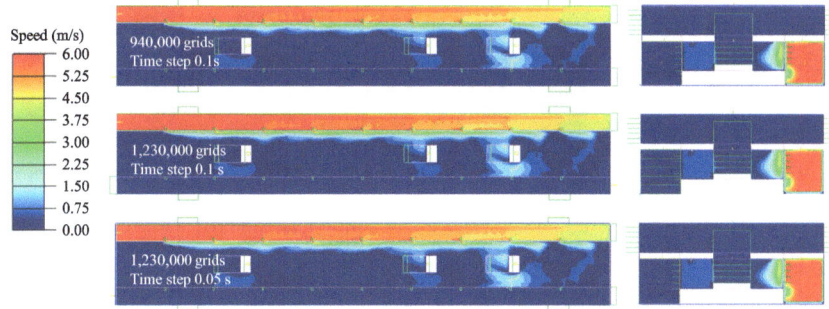

Fig. 3. Grid Independence and Time Step Independence Testing

3   Results and Discussion

3.1   Underestimation in Air Infiltration Through Entrances

To ascertain whether the metro station design phase accurately estimated the cooling load induced by infiltration through entrances, this section involves a comparison between the design drawings and on-site test results. It contrasts the designed infiltration load with the actual load. The on-site test was conducted in August, during which the outdoor parameters closely resembled the design conditions. The comparative results are illustrated in Fig. 4. Examining Fig. 4(a), it is evident that air infiltration through entrances constitutes 1.7%–2.7% in the four metro stations, A, B, C, and D. However, the proportion rises to 15.0%–19.8% in Fig. 4(b). This suggests that the conventional design substantially underestimates the cooling load attributed to infiltration through entrances, posing challenges for effective thermal and humidity environment control in metro stations. In summary, while the cooling load resulting from air infiltration through entrances constitutes a significant portion of the total load, the proportion of this load is substantially underestimated during the schematic design phase.

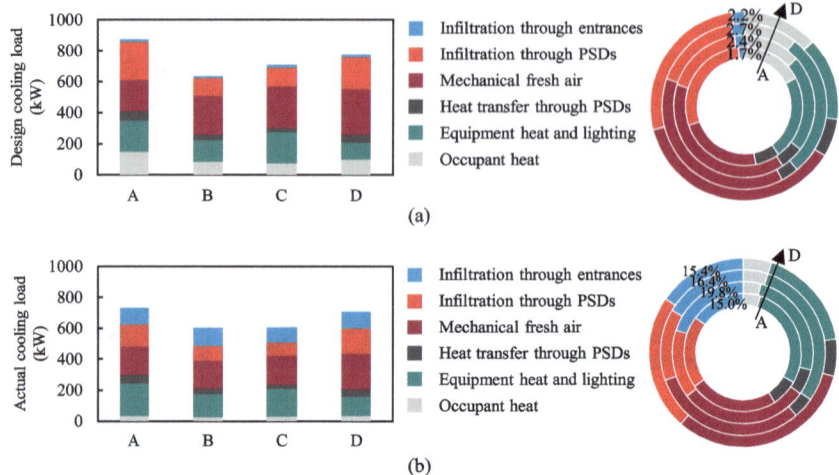

Fig. 4. A breakdown of cooling load: (a) design cooling load, and (b) actual cooling load.

3.2 Underestimation Mitigation with the Proposed Method

The simulation results of air infiltration through entrances will be introduced in this section. For the known four stations, the infiltration volume is influenced by various factors, including train frequency and whether trains in both directions enter or exit the station simultaneously. The simulation results indicate that the air infiltration volume resulting from one pair of trains entering and exiting the station ranges from 940 to 1,930 m^3. The train frequency varies throughout the hours, with the air infiltration volume ranging from 1.0 to 1.8 × 10^4 m^3/h. Based on the simulated air infiltration volume, the corresponding cooling load can be estimated.

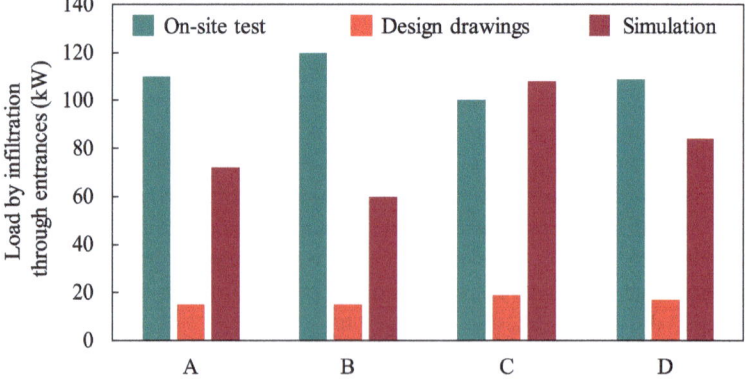

Fig. 5. Underestimation mitigation by simulation

Figure 5 illustrates the mitigation of underestimation using the simulation method under design conditions. The results demonstrate a significant reduction in underestimation when employing the proposed simulation method. The estimation error rate can be decreased from 81.0%–87.5% to 8.3%–50.2% with the application of the proposed simulation method. By using this method, the estimation of station load can be more accurate, and the design of VAC system can be more reasonable and efficient.

4   Conclusions

This study conducted a comparative analysis between the cooling load estimates from design drawings and the actual measurements obtained on-site. The results reveal that, while the cooling load resulting from air infiltration through entrances constitutes a significant portion of the total load, the proportion of this load is substantially underestimated during the schematic design phase.

Moreover, a simulation method is proposed, utilizing a combined application of STESS software and Fluent software. The findings demonstrate that employing this simulation method can effectively mitigate the misestimation, reducing it from 81.0%–87.5% to 8.3%–50.2%.

The simulation method proposed in this paper combines STESS software and Fluent software, which requires high computing power. And the current research object is only underground subway stations. The follow-up research will focus on simplifying the simulation and promoting the application scenarios.

Acknowledgments. This work was supported by the National Key Research and Development Programs of China (2022YFC3802505 and 2023YFC3807001).

References

1. Ahn, J., Cho, S., Chung, D.H.: Development of a statistical analysis model to benchmark the energy use intensity of subway stations. Appl. Energy **179**, 488–496 (2016). https://doi.org/10.1016/j.apenergy.2016.06.065
2. Yu, Y.Z., You, S.J., Zhang, H., et al.: A review on available energy saving strategies for heating, ventilation and air conditioning in underground metro stations. Renew. Sust. Energ. Rev. **141**, 110788 (2021). https://doi.org/10.1016/j.rser.2021.110788
3. Xue, P., You, S.J., Chao, J.Y., Ye, T.Z.: Numerical investigation of unsteady airflow in subway influenced by piston effect based on dynamic mesh. Tunn. Undergr. Space Technol. **40**, 174–181 (2014). https://doi.org/10.1016/j.tust.2013.10.004
4. Faugier, L., Marinus, B.G., Bosschaerts, W., et al.: CFD model for airflow in a subway station compared to on-site measurements: the challenges of as-built environment. Tunn. Undergr. Space Technol. **140**, 105248 (2023). https://doi.org/10.1016/j.tust.2023.105248
5. Liu, Y., Hu, Y., Xiao, Y., et al.: Effects of different types of entrances on natural ventilation in a subway station. Tunn. Undergr. Space Technol. **105**, 103578 (2020). https://doi.org/10.1016/j.tust.2020.103578
6. Zhang, X., Ma, J., Li, A., et al.: Train-induced unsteady airflow effect analysis on a subway station using field experiments and numerical modelling. Energy Build. **174**, 228–238 (2018). https://doi.org/10.1016/j.enbuild.2018.06.014

7. Ma, J., Zhang, X., Li, A., et al.: Analyses of the improvement of subway station thermal environment in northern severe cold regions. Build. Environ. **143**, 579–590 (2018). https://doi.org/10.1016/j.buildenv.2018.07.039
8. Wang, Y., Li, X.F.: STESS: subway thermal environment simulation software. Sust. Cities Soc. **38**, 98–108 (2018). https://doi.org/10.1016/j.scs.2017.12.007

A Prediction Model for Loess Collapse Coefficient Based on Genetic Algorithm Back Propagation Neural Network

Yuetong Zhang[1]([✉]), Wuwei Zhu[3], Shenglong Jin[1], Jianhui Niu[2], Qize Wang[2], Zhixun Xie[2], Xiao Yang[3], and Yanchao Yue[1]

[1] School of Human Settlements and Architectural Engineering, Xi'an Jiaotong University, Xi'an, China
2856401378@qq.com, yuey@xjtu.edu.cn

[2] Shaanxi Construction Engineering Holdings Group Future Urban Innovation Technology Co., Ltd./Future Urban Construction and Management Innovation Joint Research Center Xi'an, Xi'an, China

[3] Shaanxi Academy of Building Sciences Co., Ltd., Xi'an 710049, Shaanxi, China

Abstract. As an important property of loess, the collapsibility coefficient is commonly used in engineering to evaluate the collapsibility of loess. The main methods for predicting the collapsibility coefficient of loess currently include fuzzy algorithm, principal component analysis, data mining, etc. Due to the numerous indicators that affect the collapsibility of loess and their mutual influence, current research has problems such as incomplete consideration of indicators or insufficient data used to fit prediction models, resulting in insufficient prediction accuracy. Using factor analysis method, factor analysis is conducted on the selected indicator data based on the theory of collapsible deformation structure. Water content, porosity, and plasticity indicators are selected as independent main factors, and a Back Propagation (BP) neural network prediction model is constructed to predict the collapsible coefficient. Genetic algorithm is used to optimize the initial network parameters of the model. Compared to the BP neural network prediction model, the model optimized by genetic algorithm has varying degrees of improvement in accuracy, high prediction accuracy, and better practicality in engineering.

Keyword: Collapse coefficient · Factor analysis · BP neural network · genetic algorithm

1 Introduction

Loess is a loess-like accumulation formed by a combination of factors under arid or semi-arid conditions during the Quaternary period [1]. Loess is less compressible and has a certain degree of strength under natural moisture conditions. However, when loess is under the joint action of water and vertical stress, its structure will be destroyed rapidly, and wet subsidence occurs.

© The Author(s) 2024
A. Bieliatynskyi et al. (Eds.): CSTTE 2023, LNCE 603, pp. 421–431, 2024.
https://doi.org/10.1007/978-981-97-5814-2_38

In recent years, with the rapid development of microscopic technology, scholars at home and abroad use microscopic technology to study the microstructure of soil. Lei Xiangyi, Dibben S C et al. [2, 5] used microscopic techniques to evaluate the relationship between pore type and wet subsidence of loess. Lin Guanghua, Tovey et al. [3–6] used microscopic techniques to study the microstructure of loess and rated the effect of labile salts, pore characteristics etc. on the wetting properties of loess.

For the wet subsidence model, the main construction method in China is to fit the empirical formula with the data obtained from the wet subsidence test. For example, the relationship between the physical property index of loess and the coefficient of wet subsidence was constructed by using the least squares method [7], linear fitting [11] and other ways. In order to further improve the prediction effect of empirical formulas, Lingxia Gao and Ping Li [12] constructed the multifactor regression equations between loess wet subsidence coefficient and water content, plasticity index, and pore ratio by using principal component analysis. Meanwhile, with the development of computer performance, the high computing power provided by computers is used for data mining to further improve the prediction accuracy of the wet subsidence model. For example, BP neural network [8] and fuzzy algorithm [9, 10] are used to predict the coefficient of loess wet subsidence, which further reduces the prediction error of the model and makes it more valuable for engineering applications.

In order to make the selected physical property indexes have strong independence and improve the prediction accuracy of the model, this paper adopts factor analysis to screen the main factors and uses genetic algorithm to optimize the initial network parameters of the BP neural network, to achieve the purpose of improving the prediction accuracy.

2   Factor Analysis Based on Physical Properties Indicators of Loess

Due to the complexity of loess wet subsidence mechanism, the physical indicators affecting the loess wet subsidence coefficient are not completely independent. If the correlation of the selected indexes is too high, the BP neural network model constructed will be overfitting; and if the selected indexes are too few, the accuracy of the model will be reduced. Therefore, this part will be based on the theory of wet subsidence deformation of loess, select the characteristic indexes, use principal component analysis to screen out the relatively independent physical property indexes, then build the BP neural network, and use genetic algorithm to optimize the weights and thresholds of the BP neural network in order to achieve the purpose of improving the accuracy.

2.1   Selection of Variable

According to the "Wet Loess Area Construction Standard" (GB50025-2018), when the depth of the overlying soil layer is less than 10 m, the vertical stress is set at 200 kPa to test the wetting coefficient; when it is greater than 10 m, the wetting coefficient is tested according to the saturated self-weight stress of the overlying soil body. In this paper, the wetting coefficient under the vertical stress of 200 kPa is selected, i.e., the vertical stress is considered as an invariant value, and the data with wetting coefficient less than

0.015 are excluded. According to the relevant literature, the analysis may be related to the following factors: Pore ratio and dry density and Initial moisture content and degree of saturation and Plastic limit, liquid limit and plasticity index and so on.

2.2 Factor Analysis

To standardize the data in the database to achieve the consistency of the physical indicators in terms of scale and order of magnitude, and then calculate the correlation coefficient matrix of the physical indicators to determine whether there is a strong correlation between the physical indicators with the help of the correlation coefficient matrix of the physical indicators. The number of factors is selected according to the variance contribution rate and cumulative variance contribution rate of each factor, and if the correlation between the obtained factors and the physical indicators is not very strong, it is necessary to rotate the factors to get the main factors with high correlation with the physical indicators. This paper is based on SPSS software to factor analyze the data of physical indicators, the specific steps are shown below.

Determination of the Correlation Matrix

By organizing the test data under the vertical stress of 200 kPa in this project with the data collected in the literature with a total of 120 sets of test data [9–12], and 60 sets of data were selected for correlation analysis:

Based on the standardized physical property index data set, the correlation coefficients between the variables were calculated, as shown in Table 1. According to the correlation coefficient matrix, there is a good correlation between moisture content and saturation; a strong correlation between dry density and pore ratio; and a strong correlation between liquid limit, plastic limit, and plasticity indexes.

Table 1. Matrix of correlation coefficients among physical property indicators

	dry density γ_d	porosity ratio e	moisture content ω	Saturation Sr	liquid limit wL	plastic limit wp	Plasticity index Ip
dry density γ_d	1	−0.995	0.178	0.715	−0.222	−0.222	−0.221
porosity ratio e	−0.995	1	−0.199	−0.726	0.234	0.234	0.233
moisture content ω	0.178	−0.199	1	0.805	0.086	0.09	0.082
Saturation S_r	0.715	−0.726	0.805	1	−0.077	−0.074	−0.08
liquid limit wL	−0.222	0.234	0.086	−0.077	1	1	0.999
plastic limit wp	−0.222	0.234	0.09	−0.074	1	1	0.998
Plasticity index Ip	−0.221	0.233	0.082	−0.08	0.999	0.998	1

Factor Determination

Based on the correlation analysis of the collected data, factors are established to downsize the original data structure. The importance of each factor was measured based on the variance contribution rate and cumulative variance contribution rate, and then the factors were selected, as shown in Table 2. The number of factors was determined to be 3, and the cumulative variance contribution rate of the factors was 99%.

Table 2. Cumulative variance contribution of factors

factor	Initial eigenvalue	
	Variance contribution/%	Cumulative variance contribution/%
1	48.82	48.82
2	36.166	84.987
3	14.787	99.773
4	0.149	99.923
5	0.048	99.971
6	0.029	100
7	0	100

Factor Rotation

In this paper, the maximum variance rotation method is used to rotate the factor loading matrix. Through the factor analysis after the rotation of Table 3, the factors that have a strong influence on the first factor are liquid limit, plastic limit and plasticity index; the second factor is mainly related to the pore ratio and dry density; and the third factor has a strong connection with water content and pore ratio. Through factor analysis, this paper selects plasticity index, pore ratio and water content to replace all the physical property indexes.

Table 3. Factor loads before and after rotation

Physical indicators	Pre-rotation factor loading			Post-rotation factor loading		
	1	2	3	1	2	3
liquid limit w_L	0.825	0.550	−0.124	0.994	−0.103	0.024
plastic limit w_p	0.825	0.552	−0.120	0.994	−0.102	0.019
Plasticity index I_p	0.824	0.554	−0.116	0.994	−0.104	0.028
porosity ratio e	0.715	−0.558	0.417	−0.122	0.984	0.123
dry density γ_d	−0.703	0.557	−0.439	0.135	−0.979	−0.143
moisture content ω	−0.571	0.772	0.268	0.069	0.062	0.994
Saturation S_r	−0.211	0.632	0.743	−0.031	0.626	0.776

3   Wet Trapping Modeling

Based on the principal component analysis of the dataset, the water content, pore ratio and plasticity indexes are taken as the principal components, to achieve the relative independence among the indexes as well as the purpose of reducing the dimensionality of the data structure.

3.1   Construction and Optimization of Wet Depression Model

Introduction to BP Neural Networks

Compared with other traditional linear regression methods, BP neural networks can use data to train the network parameters to establish the mapping relationship between variables and predicted values, and there is no need to describe this relationship beforehand, so BP neural networks are very suitable for nonlinear regression neural networks.

Nonlinear mapping ability: BP neural network essentially achieves a mapping function from input to output, and mathematical theory proves that a three-layer neural network can approximate any nonlinear continuous function with arbitrary accuracy. This makes it particularly suitable for solving problems with complex internal mechanisms, where BP neural networks have strong nonlinear mapping capabilities.

Self learning and adaptive ability: During training, BP neural networks can automatically extract "reasonable rules" between output and output data through learning, and adaptively memorize the learning content in the network's weights. BP neural network has high self-learning and adaptive capabilities.

Generalization ability: The so-called generalization ability refers to the ability of the BP neural network to apply learning results to new knowledge when designing a pattern classifier, which not only considers whether the network can correctly classify the required classification objects, but also whether the network can correctly classify unseen patterns or patterns contaminated with noise after training.

Fault tolerance: The BP neural network does not have a significant impact on the global training results when its local or partial neurons are damaged, which means that even if the system is damaged locally, it can still work normally. The BP neural network has a certain degree of fault tolerance.

Establishment Steps

In this paper, a BP neural prediction model based on database [9–12] is developed with reference to experimental data and collected data, and the specific steps are shown below:

Input layer, output layer establishment: this paper will take the water content, pore ratio, plasticity index as the characteristic value of the input layer, so the number of nodes in the input layer is 3. There is a strong independence between the indexes, and it can comprehensively evaluate the whole data structure. The output layer is one layer, and the number of nodes is 1.

Hidden layer establishment: in general, increasing the number of layers of the hidden layer can improve the accuracy of the evaluation results, but the increase in the number of hidden layers will lead to the network training time is too long, so in this paper, we will establish a layer of hidden layer, the number of nodes in the hidden layer is determined by the following empirical formula 1:

$$n_1 = \sqrt{n+m} + \alpha \tag{1}$$

where: n_1 - number of nodes in the hidden layer; n - number of nodes in the input layer; m - nodes in the output layer; α - node number modifier, chosen between [1, 10].

Indicator Data Normalization: In order to keep the input data within the specified range to ensure that the neural network can converge and reduce the training time, the selected indicator data needs to be normalized before modeling. In this paper, the sample data will be normalized so that the range is between [−1, 1].

Network training: through the weights and thresholds between different layers, connect the input layer, hidden layer, and output layer, train the network through forward transfer, utilize the backward propagation of error, continuously update the weights and thresholds between layers, make the prediction results of the network closer to the expected value, and finally use part of the data for the validation of neural network prediction model.

3.2 GA Optimization Modeling

Genetic Algorithm (GA) is an intelligent optimization algorithm proposed by American professor J. Holland in 1975, which is a global optimization algorithm simulated based on biological genetic evolution.

Genetic algorithms refer to the phenomenon of selection inheritance, gene crossover, and gene mutation in Mendelian genetic theory, which has the characteristics of global optimization, parallel processing, and high generality. Firstly, a set of optimal solutions are selected not to be eliminated during each iteration, and then the solutions of the fitness function are used to select the excellent individuals, and the selected individuals are reorganized using the genetic operator to form a new population. The new population formed has evolved compared to the previous generation, and the iteration can be terminated after reaching the preset goal through one iteration.

Due to the stochastic nature of the initial parameters of the BP neural network, the optimization of the network parameter configurations using genetic algorithms can minimize the prediction set error.

Initial Population Creation

Firstly, all the parameters are flattened into the form of chromosomes, and all the values and thresholds in the BP neural network are selected to form an N-dimensional vector (N = input vector dimension × number of implied layers + number of implied layers + number of output vectors × number of implied layers + number of output vectors) as an individual chromosome. The initial population size is set to M = 10, which in turn randomly generates an M × N matrix as the initial population.

In the loess wet subsidence model of this paper, the input vector dimension is 3, which are water content, pore ratio, and plasticity index, and the output vector dimension is 1, which is the wet subsidence coefficient.

Heredity, Crossover, Variation
In this paper, the roulette algorithm is selected as the selection operator. The roulette algorithm is able to pass on individuals with higher fitness function values to the next generation, so in each iteration, firstly, the fitness function value of each individual in the current population needs to be calculated; then the best and the worst fitness individuals are saved into the preset space, and the selection probability of the other individuals in the population is calculated; finally, the individual selection is performed according to the selection probability of each individual, and the chosen individuals to replace the individuals in the original population. After completing the crossover and mutation, the fitness of everyone in the current population is calculated, and the individual with the best fitness in the previous generation is used to replace the individual with the worst fitness in the current population.

In order to maintain a diverse population, two individual structures in the previous generation need to be partially replaced and reorganized to form a new individual, a process called crossover. In this paper, the two-point crossover method is chosen as the crossover operator, and the crossover probability is set at 0.8.

In order to make the genetic algorithm have the ability of local search and maintain the diversity of the population, the genetic algorithm introduces the mutation algorithm to generate new individuals. In this paper, Gaussian variation is chosen as the variation algorithm. The mutation probability is 0.2.

Termination Conditions
The algorithm terminates when the optimal individual fitness reaches a minimum error of 0.00001, with a preset number of generations to terminate at 100.

3.3 Model Accuracy Validation and Result Analysis

In this paper, the model program will be established based on MATLAB 2020b software, and the BP neural network will be written together with the genetic algorithm program.

Relative Error Analysis
The range of the number of hidden layer nodes of the BP neural network is [3, 11], Training in the selected range of hidden layers, when the number of hidden layer nodes is 12, the genetic algorithm optimized model has the highest accuracy, and its relative error is 5.81%, currently, the relative error of the BP neural network prediction model is 32.71%. The accuracy of the prediction model of loess wet subsidence coefficient optimized by genetic algorithm is improved by 30%, and the error is less than 20%, which achieves the expected effect. When the hidden layer node is equal to 12, the specific prediction data are shown in Table 4.

From Table 4, the BP neural network for the medium wetted loess prediction accuracy is poor, but after the optimization of the genetic algorithm, the error is within 12.19%, the prediction accuracy of the reliability of the higher.

Table 4. Prediction data table for hidden layer node number of 12

Serial No	Water content ω/%	porosity ratio e	Plasticity index Ip	Measured value	BP Predictive Value	relative error/%	GA optimization model	relative error/%
1	18.2	0.706	10.2	0.011	0	100.00%	0.0113	3.02%
2	23.5	1.138	15	0.056	0.040	28.09%	0.0494	11.76%
3	24	1.092	14.5	0.045	0.04	10.30%	0.0491	9.14%
4	23.4	1.109	14.1	0.049	0.042	14.70%	0.0506	3.26%
5	21.3	0.906	12.6	0.023	0.036	54.44%	0.0202	12.19%
6	20.5	0.905	12.4	0.026	0.035	36.04%	0.0265	2.08%
7	21.7	1.14	12.9	0.058	0.046	21.22%	0.058	0.01%
8	18.4	1.09	13.2	0.06	0.057	5.65%	0.0587	2.14%
9	18.8	1.169	13.6	0.073	0.056	23.96%	0.0666	8.78%

When the number of hidden layer nodes is 7, the prediction accuracy of the BP neural network prediction model is the highest, and the average relative error is 19.49%, currently, the average relative error of the prediction results of the genetic algorithm optimization model is 9.39%. The specific prediction data is shown in Table 5. From Table 5, the relative error of the BP neural network prediction values is all within 34.62%, and the prediction model has a large error, but after the genetic algorithm optimization of the BP neural network prediction model, excluding the first group, the other data prediction error is within 17.83%, and the prediction accuracy of the model optimized by the genetic algorithm is improved by 10%.

When the hidden layer node is too low, the prediction accuracy is poor for non-wetted loess, which is excluded in this paper to study the wetting coefficient when wetting occurs in wetted loess. After the exclusion of the GA optimization model prediction set of the prediction value error within 17%, the error is less than 20%, the accuracy reaches the expected accuracy.

Table 5. Prediction data table for hidden layer node number of 7

Serial No	Water content $\omega/\%$	porosity ratio e	Plasticity index Ip	Measured value	BP Predictive Value	relative error/%	GA optimization model	relative error/%
1	18.2	0.706	10.2	0.011	0.0109	1.13%	0.0138	25.35%
2	23.5	1.138	15	0.056	0.0442	21.07%	0.0587	4.82%
3	24	1.092	14.5	0.045	0.0365	18.98%	0.0530	17.83%
4	23.4	1.109	14.1	0.049	0.0371	24.32%	0.0534	9.01%
5	21.3	0.906	12.6	0.023	0.0310	34.62%	0.0247	7.39%
6	20.5	0.905	12.4	0.026	0.0321	23.63%	0.0260	0.11%
7	21.7	1.14	12.9	0.058	0.0422	27.32%	0.0536	7.56%
8	18.4	1.09	13.2	0.06	0.0486	19.00%	0.0600	0.08%
9	18.8	1.169	13.6	0.073	0.0769	5.40%	0.0640	12.37%

3.4 Analysis of Results

For the BP neural network optimized by genetic algorithm, the optimal initial network parameters can be selected for training in the iterative process. Therefore, with the increase of the number of hidden layer nodes, the accuracy of the optimized neural network will be further improved.

The prediction analysis shows that the genetic algorithm can make up for the decrease in prediction accuracy due to the randomness of the initial network parameters. When the number of hidden layer nodes is 12, the relative error of the prediction model is within 12.19%, and all kinds of errors are reduced to different degrees, the prediction accuracy is reliable, and it has a good engineering practical value.

4 Conclusions

(1) Using the factor analysis method, the water content, pore ratio, and plasticity indexes were selected as factors based on the variance contribution rate and cumulative variance contribution rate, and the prediction model of the wet subsidence coefficient of loess and physical property indexes was established by using BP neural network.

(2) As the genetic algorithm can optimize the initial network parameters of the BP neural network, the GA-BP neural network prediction model has a higher prediction accuracy compared with the BP neural network during the training process, which is closer to the real value, and has a strong engineering practicability.

(3) The GA-BP neural network optimization model created in this paper has the highest accuracy when the number of nodes in the input layer is 3, the number of nodes in the output layer is 1, and the number of nodes in the hidden layer is 12, and the average relative error of the predicted data is 5.81%.

(4) Due to the ability of genetic algorithm to optimize the initial network parameters of BP neural network, during the data training process, the GA-BP neural network prediction model has higher prediction accuracy compared to BP neural network,

and can approach the true value more closely, which has strong engineering practicality. The modeling steps for the surrounding rock and soil layers of the reference tunnel, as well as the modeling results of the surrounding terrain and topography, are consistent with the internal volume method, and will not be repeated here.

Acknowledgment. During the writing of this paper, we are grateful for the financial support from Shaanxi Academy of Building Research Ltd: Research Project on Microstructural Analysis and Wet Depression Mechanism of Loess, and from Xindian Comprehensive Survey and Design Research Institute Ltd. (Project No. 2019-DKY-N01) Future Urban Construction and Management Innovation Joint Research Center (20211177-ZKT14) and Shaanxi Province Four Body One Joint Soil Center (2022ZY2-CXJD-09) and Shaanxi Natural Science Foundation Research Program (2024SF-YBXM-537) and Key R&D Plan Project for Shaanxi Province in 2024: Research on Dynamic Characteristics and Design Theory of New Prefabricated Steel Concrete Wind Power Tower (2024SF-YBXM-625).

References

1. Xie, W., Wang, Y., Ma, Z.: Current status and development trends of research on loess collapse mechanism. Geoscience **2**, 397–407 (2015)
2. Fan, W., Wei, Y., Yu, B.: Current status and development trends of microscopic mechanism research on loess collapse. Hydrogeol. Eng. Geol. **49**(05), 144–156 (2022)
3. Lin, G., Liu, B., Zhang, A.: The influence of soluble salts on the collapsibility of Yili loess. J. Xi'an Technol. Univ. **39**(02), 167–171 (2019)
4. Romero, E., Simms, P.H.: Microstructure investigation in unsaturated soils: a review with special attention to contribution of mercury intrusion porosimetry and environmental scanning electron microscopy. Geotech. Geol. Eng. **26**(6), 705–727 (2008)
5. Wang, H., Cheng, Y., Lu, Z.: A study on constitutive model of the cohesive soil considering soil-structure interactions. In: Proceedings of the 7th International Conference on Environmental Science and Civil Engineering (ESCE 2021), pp. 1172–1180. IOP (2021)
6. Johnson, P.S., Eberl, M., McBride, M., Aguila, R.E.: Using dynamic image analysis as a method for discerning microdebitage from natural soils in archaeological soil samples. Lithic Technol. **46**(2) (2021)
7. Huang, J., Li, X., Teng, H.: Evaluation model for loess collapsibility based on partial least squares method. Disaster Sci. **36**(2), 60–64 (2021)
8. Jing, Y., Wu, Y., Cui, Z.: Intelligent evaluation of loess collapsibility. Bull. Soil Water Conserv. **26**(1), 53–56 (2006)
9. Li, R., Gu, T., Wang, J.: Evaluation of loess collapsibility based on fuzzy information optimization technology. J. Xi'an Univ. Archit. Technol. (Nat. Sci. Edn.) **41**(2), 213–218 (2009)
10. Zhang, Q., Wang, J., Li, B.: A prediction method for loess collapsibility based on fuzzy similarity priority ratio. Yangtze River **1**(90–93), 104 (2015)
11. Liu, H., Ni, W., Yan, B.: Preliminary exploration of the relationship between the strength of loess structure and its collapsibility. Soil Mech. **29**(3), 722–726 (2008)
12. Gao, L., Luan, M., Yang, Q.: Evaluation of loess collapsibility based on principal component analysis of microstructural parameters. Soil Mech. **33**(7), 1921–1926 (2012)

Numerical Simulation Study on the Effect of Height to Width Ratio on the Force and Deformation of Netting

Liulin Yang, Zhaocai Wang, and Zhenhan Chu[✉]

School of Civil Engineering and Architecture, Shandong University of Science and Technology,
Qing Dao, Shan Dong, China
17863813615@163.com

Abstract. In this paper, based on the basic principle of finite element, the structure of metal mesh is simulated by beam unit in ABAQUS software, the contact parts of mesh are simulated by adding connection unit, and the calculation is carried out after adding steady-state oceanic flow characteristics by ABAQUS/Aqua module, and the mathematical model is verified by citing the experimental results of previous people. On this basis, the total area of the mesh and the flow velocity are guaranteed to be constant, and the height and width of the mesh are changed to analyze the overall mesh, the change law of the long and short sides of the stress and the deformation of the mesh. The results of the numerical simulation show that the total force of the mesh remains unchanged under each working condition; the force of the mesh is mainly transferred along the short side in the uniform incoming flow; the maximum deformation of the mesh is positively related to the mesh height and width ratio under the same flow velocity.

Keywords: Metal mesh coat · Finite element method · Water flow · Aspect ratio · Hydrodynamic characteristics

1 Introduction

The shortcomings such as poor durability and stability of the fiber nets used in inshore netting aquaculture industry under the action of long wave currents have seriously affected the marine environment and the living space of fish, resulting in the reduction of fish quantity and quality. The deep-water metal mesh box has become a popular mode of aquaculture development with the advantages of wave current resistance, high stability, smooth and durable, and low pollution. Metal netting is an important part of deep-water netting and one of the most complex and major components under stress, which is directly related to the safety and energy saving of fishery production and quality improvement, and is an important branch of aquatic science.

More research has been conducted by scholars in various countries on the hydrodynamic characteristics of net coats, and Wang Yintao et al. [1] studied the hydrodynamic characteristics of nets through experiments, and calculated the wave force of nets and

© The Author(s) 2024
A. Bieliatynskyi et al. (Eds.): CSTTE 2023, LNCE 603, pp. 432–440, 2024.
https://doi.org/10.1007/978-981-97-5814-2_39

nets frames. Cao Xuerui et al. [2] used the "nine-point coordinate method" to calculate and analyze the effects of flow velocity and counterweight on the preparation of netting. Chen, Cheng et al. [3] investigated the interaction of the mesh coat with regular and focused waves using porous medium theory. Wang, Wen et al. [4] investigated the water resistance characteristics of metal rhombic mesh coat under the action of water flow by model experiments. Huke et al. [5] based on the finite element principle, a computational model that can be used to calculate the hydrodynamic force of the mesh coat under the action of water flow was prepared by combining the empirical equations of the mesh plane. Shi Xinghua [6] Based on a new mesh grouping method, numerical simulation is used to analyze the deformation and force of the mesh coat. Huang Xiaohua et al. [7, 8] numerical simulation of mesh based on concentrated mass method to analyze the effect of counterweight on the force, deformation and motion characteristics of mesh under the action of water flow. Lader et al. [9] assumed the mesh coat as a flexible body composed of micro-element mesh connected by nonlinear springs, and studied the force and deformation of the circular mesh coat under the action of water flow by numerical methods. Although many scholars have studied various aspects of the hydrodynamic properties of mesh coats, most of them are analyzed for fixed-size mesh coats, and the analysis of the hydrodynamic laws of mesh coats with different height and width ratios is rare.

Metal diamond chain link mesh is a small-scale porous elastic mesh structure, and its connection method is more special compared with traditional fiber-based flexible mesh clothing. In this paper, we adopt the finite unit method [10] ABAQUS software is used to numerically calculate the hydrodynamic properties of a zinc-aluminum alloy rhombic metal mesh [3, 4]. The hydrodynamic characteristics of a zinc-aluminum alloy rhombic metal mesh under the action of water flow are numerically calculated, and the relationship between the maximum deformation and axial stress of the mesh and the height/width ratio of the mesh is analyzed on the basis of the results of good fit with the previous experimental results, changing the preparation direction and height/width ratio of the mesh, and the relationship between the overall mesh, the water resistance of the long and short sides and the force transfer on the fixed side of the mesh, as well as the maximum deformation of the mesh. The results of the study will provide design basis for the optimization design of the mesh jacket and the maximization of energy saving benefits.

2   Numerical Model

2.1   Geometric Model of the Mesh Coat

A three-dimensional wire model is used to model the metal rhombic mesh coat, and the vertical single mesh wire is regarded as a component, for which operations such as vertical array are performed to simulate the transversely woven metal mesh coat. The contact is simulated by adding a connection unit at the intersection of the two wires, and the relative degrees of freedom in the three displacement directions at the contact are constrained during the calculation, and the relative degrees of freedom of rotation are not constrained [5]. The relative rotational degrees of freedom are not constrained.

Figure 1 shows the model diagram of a metal diamond-shaped chain link mesh with mesh size m = 45 mm, wire diameter 3.2 mm, horizontal side x and vertical side y.

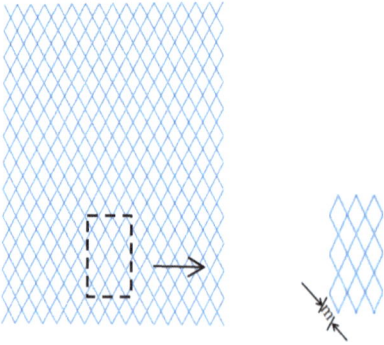

Fig. 1. Model diagram of metal diamond-shaped chain link network

2.2 Numerical Methods

Force Analysis

Gravity and Buoyancy

Because the whole net coat is submerged below the water surface, the gravity and buoyancy are considered at the same time and the circular section rod [11] of gravity and buoyancy are calculated as follows:

$$P_z = \rho_f gV - \rho_w gV = (\rho_f - \rho_w)g\frac{\pi D^2}{4}l \tag{1}$$

where: P_z is the combined force of gravity and buoyancy on the circular rod; ρ_f and ρ_w are the densities of the metal mesh and water, respectively; D is the cross-sectional diameter of the circular rod; l is the length of the rod; V is the volume of the rod.

Water Flow Force

The water flow can be approximated as a steady plane flow, so the interaction of the water flow with a circular bar [12] can be expressed by the equation of planar flow and lead cylindrical load.

$$F_c = \frac{1}{2}\rho_w C_d Dv^2_{c\,max} \tag{2}$$

where: F_c is the water flow load per unit length of the circular rod; C_d is the drag coefficient; v^2_{cmax} is the maximum possible velocity of the fluid.

Calculation Method

The mesh coat is simulated using a two-node beam unit. Each node has three displacement degrees of freedom in three dimensions, and its discretized equation of motion [4]:

$$M\ddot{u} + C\dot{u} + Ku = Q \tag{3}$$

where: M, C, K and Q are the mass matrix, damping matrix, stiffness matrix and nodal load vector of the system nodes, respectively; $\ddot{u}$ and $\dot{u}$ and u are the acceleration vector, velocity vector and displacement of the system nodes, respectively.

$$
\begin{aligned}
M &= \sum_e M^e \quad M^e = \int_{V_e} \rho N^T N dV \\
C &= \sum_e C^e \quad C^e = \int_{V_e} \eta N^T N dV \\
K &= \sum_e K^e \quad K^e = \int_{V_e} B^T D B dV \\
Q &= \sum_e Q^e \quad Q^e = \int_{V_e} N^T f dV + \int_{S_e} N^T T ds
\end{aligned}
\tag{4}
$$

where: M^e, C^e, K^e, and Q^e are the mass matrix, damping matrix, stiffness matrix, and load vector of the unit nodes, respectively; N is the interpolation function matrix; B is the strain matrix; D is the elasticity matrix; f is the nodal load matrix composed of the unit volume forces; T is the nodal load matrix composed of the unit surface forces; V_e is the unit volume; S_e is the unit area; ρ is the mass density; and η is the damping coefficient. The discretized equations of motion are solved using the implicit algorithm in the ABAQUS/Standard direct integration method.

3   Results and Discussion

3.1   Total Resistance to Water Flow of the Mesh Coat

In this paper, the material parameters of zinc-aluminum alloy, mesh size 45 mm and wire diameter 3.2 mm are used in the validated literature. Keeping the total area of the mesh coat the same, the aspect ratio is varied to $L_y/L_x = 1/2.3$, $1/2$, $1/1.7$, $1/1.4$, 1, 1.4, 1.7, 2, 2.3, and calculated at 0.2 m/s, 0.4 m/s, 0.6 m/s, 0.8 m/s and 1.0 m/s different flow velocity conditions of the net clothing water resistance and the relationship between the height and width ratio.

Figure 2 shows the comparison of the water resistance of the mesh coat under the above conditions.

From Fig. 2, it can be seen that the water resistance of the mesh coat is independent of the height to width ratio. This is due to the fact that under the premise of identical environmental conditions, the magnitude of the force on the mesh is related to the projected area of the mesh in the direction of the water flow. In this paper, the total area of the mesh, mesh size, mesh diameter and the direction of incoming flow are determined to be unchanged, and the weave direction and height/width ratio of the mesh are changed to calculate the constant force on the mesh.

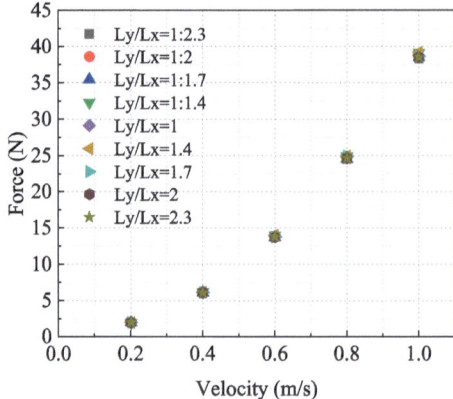

Fig. 2. Variation of water resistance of netting with height to width ratio at different flow rates

3.2 The Force Transmission Law of the Fixed Edge of the Net Clothing

Figure 3 shows the relationship between the ratio of the force on the x(y) side of the mesh and the total force on the mesh for different aspect ratios of 0.2 m/s, 0.4 m/s, 0.6 m/s, 0.8 m/s, and 1 m/s flow velocity $\lambda_{A,x}$ or $\lambda_{A,y}$ and the aspect ratio L_y/L_x, respectively.

From Fig. 3, the ratio $\lambda_{A,x}$ of the horizontal side of the mesh to the total force decreases as the aspect ratio L_y/L_x increases; the ratio $\lambda_{A,y}$ of the vertical side to the total force increases as the aspect ratio L_y/L_x increases. This reflects that under the effect of uniform incoming flow, the forces on the mesh are mainly transmitted along the short side.

The maximum value of the difference between the distribution coefficients of horizontal and vertical side forces all appear in the $L_y/L_x = 1/2.3$ working condition; when the aspect ratio is less than 1, the flow velocity has little effect on the force transfer of different aspect ratios of net clothing; when the aspect ratio is greater than 1, the two curves gradually approach with the increase of flow velocity, and the changes are more obvious; the maximum values of $\lambda_{A,y}$ are 19%, 21%, 26%, 32% and 36% under different flow velocity conditions from low to high.

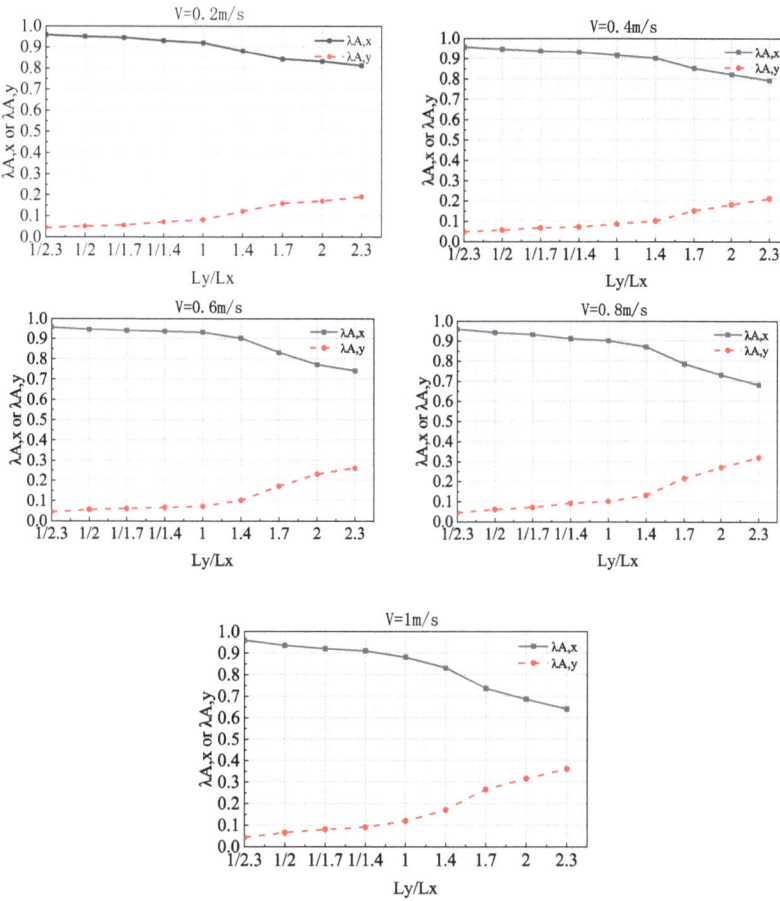

Fig. 3. Relationship between the reaction force distribution coefficient on the x(y) side of the mesh and the aspect ratio Ly /Lx at different flow rates

3.3 Relationship Between Mesh Deformation and Mesh Height/Width Ratio

Figure 4 shows the comparison of the maximum deformation produced by the mesh coat under the above working conditions.

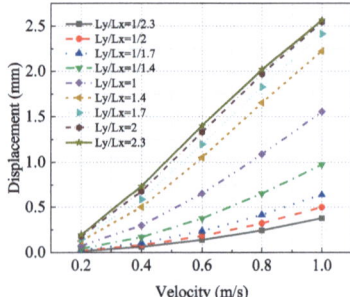

Fig. 4. Variation of the maximum deformation of the mesh with the aspect ratio at different flow rates

Figure 5 shows the deformation diagram of the mesh coat with different aspect ratios under the flow velocity of 1 m/s at 100 times magnification.

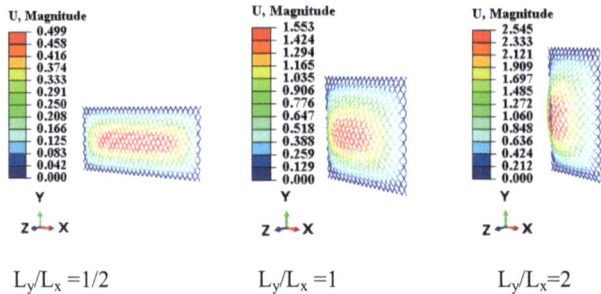

$$L_y/L_x = 1/2 \qquad L_y/L_x = 1 \qquad L_y/L_x = 2$$

Fig. 5. Deformation of mesh coat with different aspect ratios at flow velocity 1 m/s with 100 times magnification

As can be seen from Fig. 4, the maximum deformation produced by the mesh coat is positively correlated with the mesh coat height to width ratio. This is because the intertwined mesh knots at the connection of the diamond-shaped metal mesh make the transverse and longitudinal bending stiffness of the mesh differ, and the larger the aspect ratio is, the more hinges the mesh can rotate longitudinally, and the more bending moments are transmitted transversely at the same flow rate, and the larger the displacement of the mesh can produce. As can be seen from Fig. 5, the middle of the mesh is the largest deformation, and the deformation gradually decreases when it spreads around, and the mesh is deformed in a mesh pocket inside the mesh.

4 Conclusion

Based on the basic principle of finite element, this paper investigates the overall mesh, force transfer law and mesh deformation under different aspect ratios under the action of water flow through numerical simulation. The force on one side is related to the aspect ratio of the mesh, and the force on the vertical side increases gradually with the increase of aspect ratio, while the force on the horizontal side decreases gradually with the increase of aspect ratio; the force on the mesh is mainly transmitted along the short side in the uniform incoming flow. The maximum deformation produced by the mesh coat is positively related to the mesh coat height to width ratio. Therefore, in the production of the physical mesh box, attention should be paid to the reinforcement on the longer fixed edges of the mesh, and the optimal aspect ratio should be determined according to the incoming flow velocity to make the mesh have better safety and stability.

References

1. Wang, Y.T., Wang, Q., Guo, X.Y.: Analysis of hydrodynamic characteristics of aquaculture nets under the action of regular waves. J. Fish. Modern. **50**(02), 40–49 (2023). https://doi.org/10.3969/j.issn.1007-9580.2023.02.006
2. Cao, X.R., Ma, H., Xu, M.L.: Numerical simulation study on the factors influencing the deformation of flexible mesh coat structure. J. Ocean Univ. China (Nat. Sci. Ed.) **53**(2), 69–76 (2023). https://doi.org/10.16441/j.cnki.hdxb.20210334
3. Chen, C., Song, W., Xie, Z.L.: Research on hydrodynamic characteristics of farming equipment net clothing based on porous media model. J. Fish. Modern. **49**(05), 115–126 (2022)
4. Liu, H.F., Chen, C.P., Zheng, Y.N.: Numerical simulation study of water resistance characteristics of a kind of cultured metal mesh coat under the action of water flow. J. Fish. Modern. **44**(06), 73–79 (2017)
5. Hu, K., Geng, B.L., Zhao, X.: Numerical method for hydrodynamic study of fishing nets under the action of water currents. J. Waterway Harbor **40**(2), 199–205 (2019). 10. 3969/j.issn.1005-8443.2019.02.012
6. Shi, X.H., Zhou, Y., Qian, J.Q.: Numerical simulation method of mesh clustering of netting based on hydrodynamic performance. J. Fish. Modern. **48**(3), 74–79+96 (2021). 10. 3969/j.issn.1007-9580.2021.03.010
7. Huang, X.H., Guo, G.X., Hu, Y.: Numerical simulation of force and motion deformation of deep-water nets under the action of wave flow. J. China Fish. Sci. **18**(2), 443–450 (2011)
8. Huang, X.H., Guo, G.X., Hu, Y.: Numerical simulation of water action net coating process. J. Southern Fisheries Sci. **7**(03), 56–61 (2011). https://doi.org/10.3969/j.issn.2095-0780.2011.03.010
9. Lader, P.F., Fredheim, A.: Dynamic properties of a flexible net sheet in waves and current-A numerical approach. J. Aquacult. Eng. **35**(3), 228–238 (2006). https://doi.org/10.1016/j.aquaeng.2006.02.002
10. Zhao, Y.P., Bi, C.W., Dong, G.H.: Three-dimensional numerical simulation of the flow field around a planar mesh coat. J. Hydrodyn. Res. Prog. A Ser. **26**(5), 606–613 (2011). 10. 3969/j.issn1000-4874.2011.05.012
11. Yu, J.Z., Zhang, X.T., Li, X.: Hydrodynamic characteristics of planar net-clothing structures under the action of focusing waves. J. Mar. Eng. **40**(05), 98–110 (2022). https://doi.org/10.16483/j.issn.1005-9865.2022.05.011
12. Choo, Y.I., Casarella, M.J.: Hydrodynamic resistance of towed cables. J. Hydronaut. **5**(4), 126–131 (1971)

A Method for Identifying Public Transportation Super Spreaders Considering Community Structure

Jun Chen[✉], Zaiqi Li, Zixuan Zhang, and Xiaowei Li

School of Civil Engineering, Xi'an University of Architecture and Technology, No. 13, Yanta Road, Beilin, Xi'an, China
chenjuntom@163.com

Abstract. Due to variations in passengers' travel behaviours, not all passengers exhibit the same epidemiological transmission ability when they are infected. Public transportation super spreaders are passengers who can cause more extensive infections when they are infected. This study utilizes multi-source public transit data to construct a weighted passenger contact network and proposes the Gravity Hub Bridge method (GHB) for node identification based on the gravity model and the community structure. Compared to other identification methods, GHB exhibits the largest transmission range difference at low, medium, and high epidemiological levels. In other words, the public transportation super spreaders identified by GHB possess a higher epidemiological transmission ability.

Keywords: Complex networks · Public health · Passenger contact networks · Identification method

1 Introduction

Efficient urban public transportation plays a crucial role in facilitating human mobility and sustaining economic development. However, the shared and confined high-density travel spaces create favourable conditions for the transmission of diseases among passengers [1]. Infected passengers may potentially transmit the disease to other non-infected passengers during travel. Due to variations in passenger behaviours, some individuals have the capacity to cause more extensive infections when they are infected, identifying them as public transportation super spreaders [2]. Therefore, it is essential to identify these super spreaders. This identification process can assist relevant authorities in implementing targeted measures, such as travel restrictions, vaccination programs, health monitoring, and other disease control strategies.

Currently, the identification methods [3–5] for super spreaders are not fully developed and do not take into account the robust community structure within the passenger contact network [6, 7]. In complex networks, community structures facilitate the spread within communities while limiting the spread between communities, thus impacting the speed and extent of transmission. Therefore, in situations characterized by a strong

© The Author(s) 2024
A. Bieliatynskyi et al. (Eds.): CSTTE 2023, LNCE 603, pp. 441–450, 2024.
https://doi.org/10.1007/978-981-97-5814-2_40

community structure, node identification methods that consider community structure prove more effective than traditional identification methods. Existing approaches for identifying public transportation super spreaders often overlook the community structure of the passenger contact network. This oversight results in a lack of insights into relationships within communities, leading to an inaccurate representation of epidemiological interactions among passengers. Consequently, these methods fail to precisely capture the transmission abilities of individual passengers, which, in turn, affects the final identification of public transportation super spreaders. This discrepancy may introduce biases into decision-making regarding epidemiological control strategies. Considering the community structure within the passenger contact network when identifying public transportation super spreaders may offer a more accurate identification of passenger groups with higher transmission capabilities.

2   Research Method

2.1   GHB Method

Passenger Contact Network Construction
The passenger contact network is defined as a weighted undirected graph $G = \{V, E, W, M\}$, where $V = \{v_i | i = 1, 2, \cdots, n\}$ is the set of nodes, with each node v_i representing a passenger using public transportation. $E = \{e_{ij} | i, j = 1, 2, \cdots, n, i \neq j\}$ is the set of edges, where e_{ij} represents the connection between two passengers, v_i and v_j, who are simultaneously present in the same vehicle during their travel. $W = \{w_{ij} | i, j = 1, 2, \cdots, n, i \neq j\}$ is the set of edge weights, where the weight of each edge e_{ij} is denoted as w_{ij}, representing the duration of the passengers v_i and v_j in the same vehicle. $C = \{C_k | k = 1, 2, \cdots, m\}$ is the set of communities obtained through community detection algorithms, with each community C_k containing N_k nodes.

Community Division
To identify public transport super spreaders while considering the community structure of the passenger contact network, it is necessary to employ a community detection algorithm. This algorithm partitions the network into several communities based on the network's topology, aiming for strong connections within communities and weak connections between them. The Infomap algorithm is efficient, stable, and applicable to community detection in large networks. Therefore, this study utilizes the Infomap algorithm to perform community detection on the weighted passenger contact network.

The Infomap algorithm, based on the minimization of code length, employs a random walk approach to identify communities with the shortest path encoding length [8]. The description length $L(M)$ for the random walk paths generated by partitioning the network's n nodes into m communities using partition method M is represented by Eq. (1). Initially, the Infomap algorithm treats each node as an individual community and progressively merges adjacent communities to maximize the reduction of the objective function $L(M)$ until the reduction becomes negligible.

$$L(M) = q_\curvearrowright H(Q) + \sum_{i=1}^{m} p^i_\circlearrowleft H(P^i) \tag{1}$$

$$H(X) = -\sum_{l}^{n} p_i \log p_i \tag{2}$$

where $q_{\curvearrowright}$ is the probability that a certain step in the random walk will be converted to other communities at any node; $H(Q)$ is the information entropy of random walks among different communities; $H(P^i)$ is the information entropy of random walks within the community; $p^i_{\circlearrowright}$ is the sum of the probability of visiting each node in the community i and the probability of exiting the community i.

Weighted Interconnection Density Calculation

As the community structure of the network plays a role in promoting transmission within communities while inhibiting transmission between communities, the level of connectivity between communities has significant implications. When a community has fewer connections to other communities, the nodes within that community primarily transmit the disease to their neighboring nodes within the same community. Their impact on other communities is relatively minimal, making the hub nodes within the community particularly crucial. On the other hand, when a community has numerous connections to other communities, the nodes within that community possess the ability to spread the disease to neighboring communities. In this scenario, identifying bridge nodes between communities becomes essential. Therefore, there is a need to quantify the interaction between communities and distinguish the contribution of nodes to transmission within their own community and transmission to other communities. For community C_k, in combination with all edges involving nodes within and outside C_k, along with their strengths, the weighted interconnection density of community C_k is defined as follows:

$$\rho_{C_k} = \frac{\sum\limits_{v_i \in C_k} S_{in}(i)/(S_{in}(i) + S_{out}(i))}{N_k} \tag{3}$$

where $S_{in}(i)$ represents the internal community weight, which is the sum of edge weights between node v_i and neighboring nodes within the community. $S_{out}(i)$ represents the external community weight, which is the sum of edge weights between node v_i and neighboring nodes outside the community.

Node GHB Value Calculation

The relationships between internal and external connections within communities, as well as the impact of community size, are reflected through weighted interconnection density and the number of community nodes. If node v_i belongs to community C_k and its neighboring community is C_l, considering the reciprocal of edge weights as the distance between nodes, along with the internal and external weights of node v_i and its neighboring node v_j, as well as the weighted interconnection density of their respective communities and community size, we calculate the centrality of node v_i using the calculation method of the gravity model. The GHB (Gravity Hub Bridge) value of node v_i is calculated as follows:

$$GHB(i) = \rho_{C_k} H(i) + (1 - \rho_{C_l}) B(i) \tag{4}$$

$$H(i) = N_k \sum_{v_j \in I(i), v_j \in C_k} \frac{S_{in}(i)S_{in}(j)}{1/w_{ij}^2} \tag{5}$$

$$B(i) = \sum_{v_j \in I(i), v_j \notin C_k} N_l \frac{S_{out}(i)S_{in}(j)}{1/w_{ij}^2} \tag{6}$$

where $I(i)$ is the set of neighboring nodes of node v_i.

2.2 Baseline Method

Existing research has selected degree, strength, and k-shell decomposition as methods for identifying super spreaders in public transportation. However, as the passenger contact network is a weighted network, the s-shell decomposition method [9] is an extension of k-shell decomposition for weighted networks. In this study, we have chosen strength (NS) and the s-shell decomposition method (s-shell) as baseline methods for comparison. Additionally, we have selected weighted betweenness centrality (WBC) [10], weighted eigenvector centrality (WEC) [11], and weighted gravity model (Gravity) [12] as baseline methods for comparative analysis alongside GHB.

1) NS: The NS of node v_i is the sum of the edge weights connected to it.

$$NS(i) = \sum_{j \in I(i)} w_{ij} \tag{7}$$

2) WBC: The WBC of node v_i is the number of shortest paths passing through this node in the weighted network, reflecting the hub of node propagation in the network.

$$WBC(i) = \sum_{i \neq s, i \neq j, s \neq j} \frac{d_{sj}(i)}{d_{sj}} \tag{8}$$

where d_{sj} is the number of all shortest paths from node v_s to node v_j in the weighted network, and $d_{sj}(i)$ is the number of shortest paths passing through node v_i in d_{sj}.

3) WEC: The WEC evaluates the importance of the neighboring node by using the information of the neighboring nodes and calculates the weighted adjacency matrix corresponding to the complex network.

$$WEC(i) = \lambda^{-1} \sum_{j=1}^{N} w_{ij} e_j \tag{9}$$

where λ is the largest eigenvalue of the weighted adjacency matrix W, and the eigenvector corresponding to W is denoted as $e = (e_1, e_2, \ldots, e_n)^T$.

4) s-shell: Remove the node with the lowest NS in the network, all the removed nodes have an s-shell value of 1, denoted as $wk_s = 1$, and remove the remaining sub-networks, and assign an s-shell value of 2, denoted as $wk_s = 2$, and repeat this step until there are no nodes in the network.

5) Gravity: In [12], the degree of each node is regarded as its mass, and the shortest path distance between two nodes is regarded as the distance between them, and an

index of gravitational centrality is proposed to identify influential spreaders in complex networks. For the weighted network, this paper replaces the indicators in the formula with the indicators in the weighted network.

$$Gravity(i) = \sum_{d_{ij} \in \Psi_i} \frac{NS(i) * NS(j)}{d_{ij}^2} \qquad (10)$$

where ψ_i is the set of neighborhoods whose distance to node v_i is less than or equal to the given value, set to 3 in [12].

2.3 Evaluation Method

Weighted SIR Model
We use the SIR Model to capture the transmission capability of the super-spreaders. The SIR model categorizes nodes into three health states: susceptible (S), infected (I), and recovered (R). Infected nodes transmit the disease to their neighbors with an infection probability λ, and infected nodes recover and gain immunity at a recovery rate β (in this paper, $\beta = 1$). The calculation formula for the infection probability λ is as follows:

$$\lambda_{ij} = m\lambda_t w_{ij} \qquad (11)$$

where m controls the spread of the epidemic. In this paper, m is set to represent low (0.2), medium (0.5), and high (0.8) levels of epidemiological transmission ability. According to reference [5], $\lambda_t = 8.17 \times 10^{-4}$ h^{-1}, where w_{ij} represents the edge weight between node v_i and node v_j.

For each identification method, selecting the first p percent nodes as the initial infected nodes, applying the SIR model to simulate the propagation of the network 100 times, and calculating R_m which represents the average final number of recovered nodes for the tested identification methods. This value is considered as the transmission capability of the super spreaders identified by this method.

Transmission Range Difference
To facilitate the comparison of the transmission capability among different methods, we selected NS, a widely used and easily comprehensible metric, as a reference method. We calculated the transmission range difference, denoted as r, between the other five methods and NS.

$$r = \frac{R_m - R_s}{R_s} \qquad (12)$$

where R_s stands for the average final number of recovered nodes for the NS method.

3 Algorithm Experiment

We used MySQL to capture passenger contact relationships and python to build passenger contact networks and community division.

3.1 Passenger Contact Network Construction

We collected raw data for one week of bus routes in a city, including smart card data, GPS data, transit operation records, and bus stop coordinates. The passenger boarding stations were determined by linking these data sources. Using the assumption of the next trip, the last trip, and the return trip, the alighting stations were determined using the trip chain method. Transfer behaviours were identified using an independent threshold-based public transit transfer model. The methods for determining boarding and alighting points, as well as transfer judgments, are detailed in references [13, 14]. Individual passenger trips with the same travel purpose were combined into public transit Origin-Destination (OD) pairs, and data cleaning was performed to obtain the weekly public transit OD data. Based on the passenger's travel chain data, the algorithm for determining contact among passengers within the same train carriage is designed, as illustrated in Fig. 1.

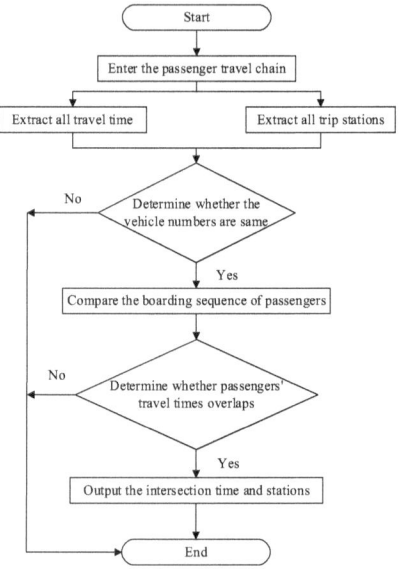

Fig. 1. Flow chart of passenger contact judgment

3.2 Community Division Result

The passenger contact network is divided into communities and the modular is shown in Table 1. It can be observed that the passenger contact network exhibits a strong community structure. However, a minority of passengers travel on less popular routes during off-peak hours, resulting in the presence of some independent communities.

Table 1. Community division results

Date	Modular
Monday	0.719
Tuesday	0.724
Wednesday	0.684
Thursday	0.703
Friday	0.698
Saturday	0.705
Sunday	0.721

3.3 Comparison of Methods

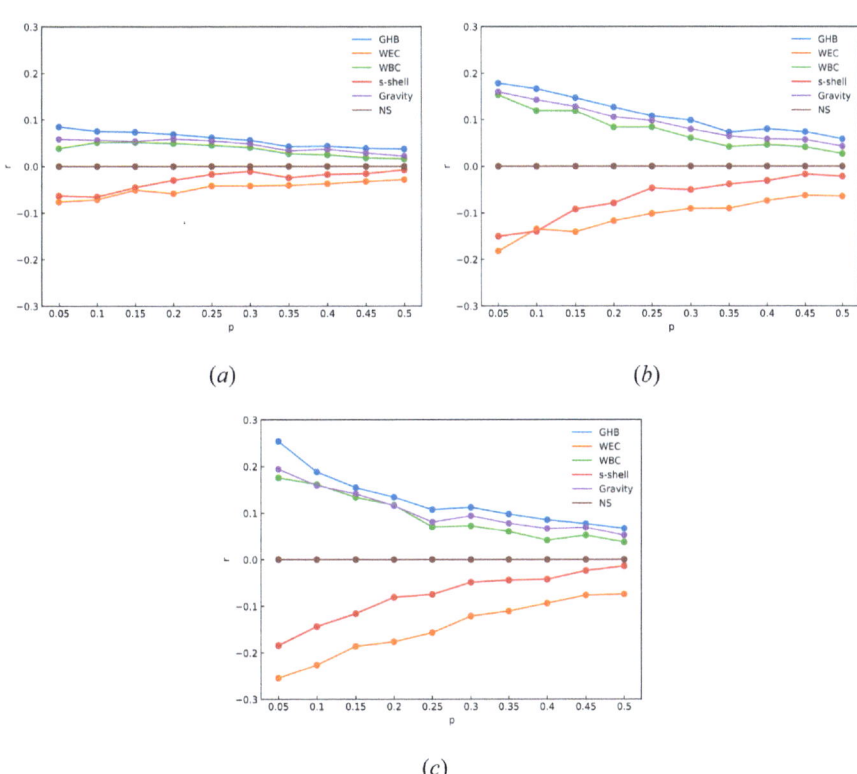

Fig. 2. Transmission range difference plot at different epidemiological levels (**a**) low level, (**b**) medium level, (**c**) high level

Table 2. Transmission range difference table

P method	0.05	0.1	0.15	0.2	0.25	0.3	0.35	0.4	0.45	0.5
Low level										
GHB	0.085	0.075	0.073	0.069	0.062	0.056	0.042	0.043	0.039	0.037
Gravity	0.058	0.056	0.053	0.059	0.054	0.048	0.033	0.036	0.028	0.022
WBC	0.038	0.051	0.051	0.049	0.045	0.040	0.027	0.024	0.018	0.019
WEC	−0.077	−0.072	−0.051	−0.059	−0.042	−0.042	−0.041	−0.038	−0.033	−0.029
s-shell	−0.064	−0.065	−0.045	−0.030	−0.017	−0.011	−0.025	−0.018	−0.016	−0.008
NS	0	0	0	0	0	0	0	0	0	0
Medium level										
GHB	0.178	0.166	0.147	0.127	0.108	0.098	0.072	0.080	0.073	0.058
Gravity	0.160	0.143	0.128	0.105	0.098	0.079	0.064	0.058	0.056	0.043
WBC	0.153	0.119	0.119	0.083	0.084	0.060	0.042	0.046	0.041	0.026
WEC	−0.182	−0.135	−0.141	−0.117	−0.102	−0.091	−0.091	−0.074	−0.063	−0.065
s-shell	−0.150	−0.139	−0.092	−0.079	−0.047	−0.051	−0.039	−0.031	−0.017	−0.022
NS	0	0	0	0	0	0	0	0	0	0
High level										
GHB	0.254	0.188	0.154	0.134	0.107	0.112	0.097	0.085	0.077	0.067
Gravity	0.194	0.159	0.141	0.116	0.081	0.094	0.078	0.066	0.069	0.052
WBC	0.175	0.161	0.134	0.117	0.070	0.072	0.060	0.041	0.052	0.037
WEC	−0.255	−0.226	−0.186	−0.176	−0.157	−0.122	−0.111	−0.094	−0.077	−0.075
s-shell	−0.185	−0.144	−0.116	−0.081	−0.075	−0.049	−0.044	−0.043	−0.024	−0.014
NS	0	0	0	0	0	0	0	0	0	0

Using different super spreader identification methods to analyze the passenger contact network will yield different node ranking results. Therefore, it's essential to compare the transmission ranges caused by the identification results of each method. To facilitate comparisons, the transmission range difference for each method is calculated using Eq. (12). Taking Monday as an example, the transmission range difference of super spreaders selected by each method at different epidemiological levels in different proportions is shown in Table 2. As the intensity is the control method, the transmission range difference is 0. The transmission range difference is shown in Fig. 2, where the horizontal axis represents the super spreader ratio p, and the vertical axis indicates the method's relative transmission range difference r. A positive value suggests that the method's identification of public transportation super spreaders has a higher transmission capability compared to the results obtained with NS.

Combining Table 2 and Fig. 2, it can be observed that the GHB method demonstrates the most substantial disparity in transmission range among the identified super spreaders in public transportation. This signifies that the results produced by the GHB method can lead to a broader reach, indicating that GHB is more proficient in identifying public transportation super spreaders when compared to other methods. For epidemics at low, medium, and high levels, the range of transmission range difference for GHB falls within 0.037–0.085, 0.058–0.178, and 0.067–0.254, respectively. The smallest transmission range difference for GHB occurs during low-level epidemics with a recognition rate p of 0.5, which is 0.037. Conversely, in high-level epidemics with a recognition rate p of 0.05, the highest transmission range difference for GHB reaches 0.254. As the recognition rate p of super spreaders decreases, and the epidemiological level increases, the extent of the transmission range difference for GHB also increases. This implies that the identified super spreaders possess a more robust transmission capability. Moreover, as the recognition rate p of super spreaders increases, the results obtained by various methods significantly overlap. In scenarios characterized by lower epidemiological levels, the transmission ability of super spreaders is more constrained. In these cases, the GHB method proves more effective, although the difference is not particularly pronounced.

Each method has its own characteristics, so it has different performance. Among the identification methods, it is observed that the WEC method yields relatively poor transmission capability in its identification results. This is primarily due to the presence of nodes with exceptionally high degree in the network, resulting in a phenomenon where the centrality scores tend to concentrate around these high-degree nodes. As a consequence, the discriminative power among scores for other nodes becomes significantly reduced. Similarly, the s-shell method also demonstrates limited transmission capability in its identification results. This can be attributed to a shared drawback with the k-shell method, namely, the inability to precisely partition nodes within the same shell. Consequently, the s-shell method falls short in providing a nuanced quantification of the transmission capabilities of different nodes located within the same shell. In contrast, the WBC method evaluates the significance of nodes as pivotal transmission points in a weighted network. A higher WBC value implies a greater likelihood of disease transmission occurring through that particular node. The Gravity method takes into account both the NS of nodes and their neighboring nodes, as well as the distance between them. However, it neglects the network's community structure and does not achieve the desired identification results. Conversely, the GHB method takes into account the community structure inherent in the passenger contact network. This approach offers a more accurate representation of the epidemiological interactions among passengers. As a result, its identification results exhibit significantly enhanced transmission ability at different epidemiological levels.

4 Conclusion

The GHB method identifies public transport super spreaders with higher epidemiological transmission ability. As the identification proportion of super spreaders decreases and the epidemiological level increases, the GHB method becomes more effective.

However, this paper does not consider further infection caused by the virus spreading inside the bus after the infected passenger has exited the bus, or exposure caused by the infected passenger while waiting on the platform. More comprehensive pathways of epidemic spread will be considered in our future studies.

References

1. Morawska, L., Cao, J.: Airborne transmission of SARS-CoV-2: The world should face the reality. Environ Int. **139**, 105730 (2020). 2020-06-01
2. Liu, Y., et al.: Characterizing super-spreading in microblog: An epidemic-based information propagation model. Physica A **463**, 202–218 (2016). 2016-12-01
3. Kang, L., Ling, Y., Zhanwu, M., Fan, Z., Juanjuan, Z.: Investigating physical encounters of individuals in urban metro systems with large-scale smart card data. Physica A: Statist. Mecha. Appl. **545** (2020). 2020-05-01
4. Mo, B., et al.: Modeling epidemic spreading through public transit using time-varying encounter network. Trans. Res. Part C: Emerg. Technol. **122**, 102893 (2021)
5. Qian, X., Sun, L., Ukkusuri, S.V.: Scaling of contact networks for epidemic spreading in urban transit systems. Scientific Reports **11** (2021). 2021-02-23
6. Hajdu, L., Bóta, A., Krész, M., Khani, A., Gardner, L.M.: Discovering the hidden community structure of public transportation networks. Netw. Spat. Econ. **20**, 209–231 (2020)
7. Kumar, P., Khani, A., Lind, E., Levin, J.: Estimation and mitigation of epidemic risk on a public transit route using automatic passenger count data. Trans. Res. Record **2675** (2021). 2021-05-01
8. Rosvall, M., Bergstrom, C.T.: Maps of random walks on complex networks reveal community structure. Proceedings of the National Academy of Sciences - PNAS, vol. 105, pp. 1118–1123 (2008). 2008-01-01
9. Eidsaa, M., Almaas, E.: s-core network decomposition: a generalization of k-core analysis to weighted networks. Physical review. E, Statistical, nonlinear, and soft matter physics **88**, 062819 (2013). 2013-01-01
10. Opsahl, T., Agneessens, F., Skvoretz, J.: Node centrality in weighted networks: generalizing degree and shortest paths. Social Networks **32**, 245–251 (2010). 2010-01-01
11. Newman, M.E.: Analysis of weighted networks. Phys. Rev. E Stat. Nonlin. Soft. Matter. Phys. **70**, 056131 (2004). 2004-11-01
12. Ma, L., Ma, C., Zhang, H., Wang, B.: Identifying influential spreaders in complex networks based on gravity formula. Physica A **451**, 205–212 (2016)
13. Zhao, J., Rahbee, A., Wilson, N.H.M.: Estimating a rail passenger trip origin-destination matrix using automatic data collection systems. Comp.-Aided Civil and Infrastr. Eng. **22**, 376–387 (2007). 2007-01-01
14. Chen, J., Yang, D.: Estimating smart card commuters origin-destination distribution based on APTS data. J. Transport. Sys. Eng. Info. Technol. **13**, 47–53 (2013). 2013-08-15

Analysis of Parking Lots Site Selection for the 2024 Chengdu World Horticultural Exposition

Guangjun Zhan[1] and Qin Liu[2(✉)]

[1] Chengdu Municipal Engineering Design and Research Institute Co., Ltd., 269 Sanse Road, Jinjiang District, Chengdu 610023, Sichuan, China
[2] Sichuan University Jinjiang College, 1 Jingjiang Avenue, Pengshan District, Meishan 620860, Sichuan, China
`liuqin@scujj.edu.cn`

Abstract. 2024 Chengdu World Horticultural Exposition will hold from April 26th to October 28th, lasting for 184 days, estimated tourist volume reaching 5.5 million. It predicts that average daily tourists of weekdays will be approximately 29,000 and on the peak days will be expected to 97,000. The aim of this study is to plan and lay out the parking lots of the 2024 Chengdu World Horticultural Exposition in advance, in order to promote the construction of parking lot infrastructure. By analyzing the proportion of travel modes, parking lot demand, and land use around the Expo Park, a total of 7 parking lots, including a total of 8457 private parking spaces and 570 tourist/dedicated buses, are planned and arranged to meet the rigid parking demand on weekdays and weekends. The surrounding buildings and the surrounding roads would be reasonably used to set about 7500 flexible parking spaces to meet the parking demand during extreme peak days. The results of this study will provide useful information for the traffic organization in the later stage of the 2024 Chengdu World Horticultural Exposition and will greatly facilitate the smooth development of traffic organization in the later stage.

Keywords: Parking Lot · Site Selection · Travel Mode · World Horticultural Exposition

1 Introduction

The International Horticultural Expo, latest issue held in Chengdu in 2024, is the international highest level professional exhibition of landscape and horticulture, which can promote the exchange and development of economy, culture, science and technology around the world. The 2024 Chengdu International Horticultural Expo, with the theme of "Park City, Beautiful Habitat", will be held from April 26th to October 28th, lasted for 184 days, is a continuous large-scale exhibition activity.

According to the previous International Horticultural Expo in China, the exhibition period will attract a large amount of visitors, which generated high intensity of transportation demand, will pose serious challenges to the organization and management of

A. Bieliatynskyi et al. (Eds.): CSTTE 2023, LNCE 603, pp. 451–460, 2024.
https://doi.org/10.1007/978-981-97-5814-2_41

exhibition transportation [1, 2]. Previous studies on the transportation of continuous large-scale exhibition mainly focus on the analysis of visitor characteristics or traffic organization experience or daily visitor volume forecast [2–4]. Wang and Wan (2015) summarize transportation organization experience of Qingdao International Horticultural Exposition, which indicate that the overly dispersed layout of parking lots will bring significant inconvenience in terms of use and management [5]. Based on the basic parking data of the 2019 Beijing International Horticultural Exposition, Liu et al. (2022) analysis the parking characteristics, including the daily parking distribution characteristics, the average parking time on weekends and holidays, and the distribution of car entry and exit times, propose the parking supply and organizational control countermeasures, including advance booking ticket, graded response, remote information induction and near end traffic control, provide certain reference for the car parking demand prediction, parking facilities supply planning [6]. Karri Sowmya and Meera M. Dhabu show that the parking charge can efficiently manage parking occupancy during peak and off-peak hours [7]. Selcuk D. et al. using a GIS-based fuzzy analytical hierarchy process (AHP) approach analyze the problem of parking lot site selection in mega cities based on the three major standards of land, finance, and transportation, and indicate that transportation had the greatest weight while finance had the least weight [8].

Parking lots are an important infrastructure for exhibition activities. By properly solving the parking problem, the exhibition services, resources and facilities would be Effective fully utilized, which will maximize the benefits of the exhibition. However, there is a serious shortage of research on exhibition parking lot planning before the exhibition. Therefore, this article will focus on the research of parking lot planning and site selection schemes, which will be the foundation for improving the driving experience and enhancing the comprehensive evaluation of the Expo and provide guidance and basis for the construction of parking lots.

2 Parking Demand Analysis

2.1 Location and General Layout of the Expo

The main venue of the World Horticultural Exposition is located in the core area of Airport New City in the Eastern New District of Chengdu. Holding the Expo will drive the improvement of the urban infrastructure, aggregate a great number of the population, and improve the urban space morphology. The Expo covers an area of 242 hectares, surrounded by 4 main urban roads, and has five entrances and exits (as shown in Fig. 1). The NO. 1 and NO. 2 Entrance/Exit are the main entrances and exits of the Expo, mainly used to serve tourists, while other entrances and exits are used as functional Entrance/Exit to ensure conference affairs, or as backup entrances and exits which are only activated in emergency situations.

Fig. 1. The Entrance/Exit Layout of 2024 Chengdu World Horticultural Exposition

2.2 Tourist Number Forecast

Based on the effective tourist area of the Expo, the maximum daily tourist capacity can be calculated, which is defined as:

$$P_{daily} = \frac{A_{valid}}{A_{per-capita}} \times \frac{T_{valid}}{T_{per-capita}} \tag{1}$$

The P_{daily} denote the maximum daily tourist capacity, the A_{valid} denote the effective tourist area, the $A_{per-capita}$ denote the index of per capita tourist area, the T_{valid} denote the daily effective opening hours of the Expo, and $T_{per-capita}$ denote the Average travel time of per capita tourist. The effective tourist area of the Expo includes exhibition gardens, roads, and squares.

Table 1. The calculation parameters of maximum daily tourist capacity.

CALCULATION PARAMETERS			
A_{valid}	hectare	$A_{per-capita}$	m^2/person
Roads and Squares	36	Roads and Squares	5
Exhibition Garden	30	Exhibition Garden	100

Using the corresponding parameters listed in Table 1, the maximum daily tourist capacity is calculated as 112.5 thousand people per-day. Considering the comfort of tourists, taking 86% of the calculated maximum daily tourist volume as the final maximum daily tourist volume, the final determined value is 96 thousand people per-day, which will be the tourists of extreme peak days.

By analogy with the experience of other city expo events in China (as shown in Table 2), it is predicted that the total number of tourists during the 2024 Chengdu Expo period will be approximately 5.5 million, with approximately 29 thousand tourists on weekdays, 49 thousand tourists on weekends, and approximately 97 thousand tourists on extreme peak days. The extreme peak days mainly refer to the long holidays during the 2024 Chengdu World Horticultural Exposition period, including May Day and National Day holidays.

Table 2. Analysis of total tourists on different days.

Item	2014 Qingdao	2016 Tangshan	2019 Beijing	2024 Chengdu (Prediction)
Area of the Expo	241 hectares	540 hectares	960 hectares	242 hectares
Total Tourists of the Expo	6.0 million	5.0 million	9.3 million	5.5 million
Weekday Tourists	33 thousand	29 thousand	57 thousand	29 thousand
Weekend Tourists	56 thousand	45 thousand	87 thousand	49 thousand
Extreme Peak Day Tourists	80 thousand	73 thousand	123 thousand	97 thousand

2.3 Travel Model Split

As shown in Fig. 2, the 2024 Chengdu World Horticultural Exposition Park is relatively far from the surrounding major core cities, and is located in the Eastern New Area, which is also an urban new area under construction and has not yet formed scale. According to location relationship and urban development situation, tourists mainly come from nearby major cities, especially Chengdu with an estimated proportion of over 70%. Therefore, private car travel, as a convenient, comfortable, free and the suitable travel mode for long-distance, will be one of the main travel modes during the Expo.

As shown in Fig. 3, the Metro Line 18 has been built near the Expo, and the nearest metro station, Sancha station, is approximately 1.7 km away from the Expo in a straight line. The Metro Line 19 is planned to be put into operation by the end of 2023. During the operation of the Expo, the Expo can be connected to Tianfu International Airport, Shuangliu International Airport, and the city center of Chengdu through taking metro and then transferring to bus. Therefore, metro travel will be another major travel mode during the Expo.

Based on the above analysis, during the exhibition period, the proportion of travel modes is mainly composed of private cars (40%) and metro transit (35%), supplemented by tourism buses (12%) and dedicated bus (7%) (as shown in Fig. 4).

Fig. 2. Major Cities Around the 2024 Chengdu World Horticultural Exposition

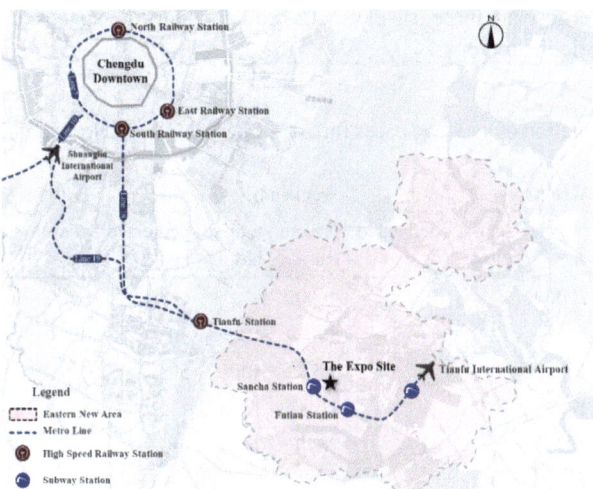

Fig. 3. The Subway Resources for 2024 Chengdu World Horticultural Exposition

2.4 Parking Demand

Based on the tourist flow and travel mode split during the exhibition period, the total number of parking spaces demand under different dates, such as on weekdays, weekends, or extreme peak days, can be calculated using the following formula.

$$Ps_{num} = P_{daily} \times Td_{pro} \times Veh_{cap} \div Tu_{rate} \tag{2}$$

The Ps_{num} denote the total number of parking spaces, the P_{daily} denote the maximum daily tourist capacity, the Td_{pro} denote the proportion of travel modes, the Veh_{cap} denote the average number of passengers per vehicle, the Tu_{rate} denote the daily turnover rate of

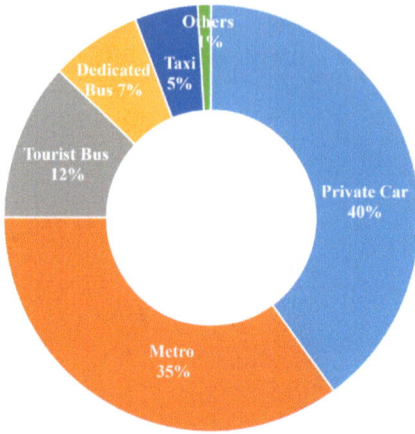

Fig. 4. The Travel Mode Split of 2024 Chengdu World Horticultural Exposition

parking spaces. There are three travel modes requiring parking spaces, including private car, tourist bus, dedicated bus.

Table 3. Calculation of parking spaces for 2024 Chengdu World Horticultural Exposition.

Item	Weekdays			Weekends			Extreme Peak Days		
	private car	tourist bus	dedicated bus	private car	tourist bus	dedicated bus	private car	tourist bus	dedicated bus
Ps_{daily} (Person)	29000			49000			97000		
Td_{pro} (%)	40	12	7	40	12	7	40	12	7
Veh_{cap} (Person)	2.5	40	50	2.5	40	50	2.5	40	50
Tu_{rate}	1	1	1	1	1	1	1	1	1
Ps_{num} (Vehicle)	4640	87	41	7840	147	69	15520	291	136

From Table 3, we may conclude that there is actually a big difference of parking spaces between different days. Taking private car parking spaces as an example, the demand for parking spaces on weekdays is 4640, and on weekends it increases to 7840, while on extreme peak days it increases to 15520. Therefore, at the planning level, we need to differently and precisely provide parking spaces according to different operating conditions.

3   Parking Lots Site Selection

3.1   Parking Site Selection Principles

During the 2024 Chengdu World Horticultural Exposition of 186 days, there are totally 112 weekdays, 60 weekend days, and 14 extreme peak days. Therefore, the normalized parking spaces demand is mainly for weekdays and weekends. The parking lot site selection process is detailed in Fig. 5. Based on different operating exposition days, we propose three overall principles for selecting parking lot sizes, respectively including rigid supply workdays and weekends, flexible supply extreme peaks, time-sharing response and operations. Rigid supply parking spaces refer to fixed parking spaces located near the Entrance/Exit of the Expo Park, which can supply the parking demand of tourists on weekdays and weekends in a normalized and convenient manner. Flexible supply parking spaces refer to fully utilize the parking spaces of surrounding buildings of the Expo Park and fully excavate parking spaces of surrounding roads, in order to meet the parking demand of the not many extreme peak days. Due to the uneven daily tourist flow and fluctuating parking demand during the exhibition period, in order to reduce the operating costs and burden of parking lots, time-sharing response and operations refer that priority is given to using nearby parking lots on weekdays, weekends, and small and long holidays, while remote parking lots are used during extreme peak hours.

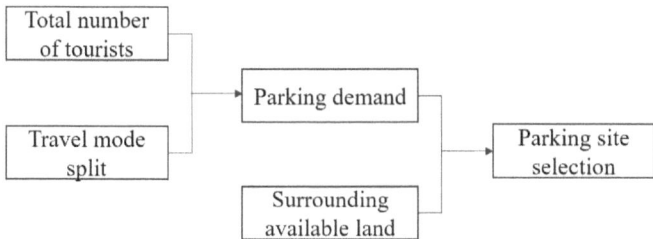

Fig. 5. Parking site selection general processes

3.2   Parking Site Selection Land

There will be a large demand for the use of parking lots during the exhibition period, however, after the exhibition period the parking demand generated by the Expo Park will sharply decrease. Therefore, the parking lots for the Expo should be divided into two types: temporary parking lots during the exhibition period and permanent parking lots after the exhibition. When planning the layout of the parking lot, the actual situation of land use and the type of parking lot should be comprehensively considered to achieve a combination of permanent and temporary using function.

The permanent parking space should be combined with the comprehensive layout planning of the park, and should be as close as possible to the entrance and exit, and use the internal land of the park to facilitate the parking needs of tourists after the exhibition period. Temporary parking lots should avoid permanent basic farmland, which shouldn't change its use under any circumstances. Therefore, temporary parking lots outside the park should be selected for urban construction land that has not yet been planned, and after the exhibition, the nature of the land should be restored according to the plan and priority should be given to construction.

3.3 Parking Lot Site Selection

According to the principles of selecting parking lots for the Expo and the surrounding land use, a total of 7 parking lots have been arranged, which will set a total of 9027 parking spaces (as shown in Table 4). The specific distribution of the parking lots are shown in Fig. 6. Among them, parking lots P1 to P5 are private car parking lots, while parking lots P6 and P7 are bus parking lots.

Fig. 6. Parking lot site selection layout of 2024 Chengdu World Horticultural Exposition

The parking lot P6 is an exclusive parking lot for tourist buses and dedicated buses, with a quantity that can meet the parking demand of any period. To avoid interweaving with the traffic organization of private car tourists, tourist buses and dedicated buses first arrive at the No. 2 Entrance/Exit of the Expo Park to put down tourists, and then park at the P6 parking lot. According to the needs of tourists, tourist buses and dedicated buses can pick up tourists at the No. 1 Entrance/Exit or No. 2 Entrance/Exit to leave the Expo Park. P7 is a VIP exclusive bus parking lot that can only be used when needed.

The parking lots P1 and P2 are nearby the main Entrance/Exit, which are the main parking lots during the exhibition period. On normal working days, opening up the use of parking lots P1 and P2 can meet the parking demand of private cars. When on weekends, parking lot P3 will be activated to meet the increasing demand for private car parking.

Table 4. Basic information of the parking lots for 2024 Chengdu World Horticultural Exposition.

Parking lots	Area *(Hectares)*	Number of parking spaces	Land use nature
P1	12.5	4297	Planned construction land
P2	10.7	2260	Planned construction land
P3	7.5	1446	Land use within the Expo Park
P4	1.3	237	Land use within the Expo Park
P5	0.7	217	Land use within the Expo Park
P6	6.4	530	Planned construction land
P7	0.7	40	Land use within the Expo Park

Parking lots P4 and P5 are located inside the park and serve as exclusive parking spaces for staff during the exhibition period, and will be retained as permanent parking spaces after the exhibition.

The surrounding buildings and the surrounding roads can be reasonably used to set about 7500 parking spaces, and reasonable traffic organization measures should be taken out to meet the parking demand during extreme peak days. The urban roads around the park are all urban trunk roads, with no less than 6 lanes in both directions, providing a basic guarantee for the traffic organization during the Expo period. In the future, special design will be carried out for traffic organization based on the functional layout of the park and the layout of parking lot location.

Due to the temporary parking lots outside the Expo Park only used for the exhibition period of 186 days, the construction of the temporary parking lots should minimize investment as much as possible. Therefore, the temporary parking lots outside the park should be arranged in accordance with the terrain and topography as much as possible to reduce engineering earthwork. And some high-quality natural native trees within the site can be retained according to the actual situation, forming a natural landscape.

4   Conclusion

This article analyzes the number of tourists at different days during the 2024 Chengdu International Horticultural Exposition, and combines the public transportation resources around the Expo Park to predict the proportion of travel modes. Based on this, the parking space demand of private cars, tourist buses and dedicated buses on weekdays, weekends and extreme peak days are detailed analyzed and then calculated out. According to the distribution of entrances and exits of the Expo Park and the surrounding land use, a total of 7 parking lots, including a total of 8457 private parking spaces and 570 tourist/dedicated buses, are planned and arranged to meet the rigid parking demand on weekdays and weekends. The surrounding buildings and the surrounding roads would be reasonably used to set about 7500 flexible parking spaces to meet the parking demand during extreme peak days. By planning and laying out the parking lots for the Expo in advance, it will help solidify the land use for parking facilities and carry out land requisition and consolidation in advance. Setting the parking lot location in advance will be an important prerequisite for the special design of traffic organization in the later stage of the 2024 Chengdu World Horticultural Exposition and will greatly facilitate the smooth development of traffic organization in the later stage. The research has not integrated into the construction stage and is relatively not in-depth enough, further research can be conducted based on the actual tourist flow and parking lot operation of the Expo for in-depth analysis.

References

1. Su, A.T., Cheng, C.K., Lin, Y.J.: Modeling daily visits to the 2010 Taipei international flora exposition. Urban Forest. Urban Green. **13**(2014), 725–733 (2014). https://doi.org/10.1016/j.ufug.2014.07.001
2. Jiao, Y.T., Wu, L., Yao, G.Z.: Traffic protection program research of world expo Beijing. Logisticis Sci-Tech. **8**(2019), 96–99 (2019). https://doi.org/10.13714/j.cnki.1002-3100.2019.08.025
3. Zhang, Y.: Daily visitor volume forecasts for Expo 2010 Shanghai China. In: 2011 14th International IEEE Conference on Intelligent Transportation Systems (ITSC) (2011). https://doi.org/10.1109/itsc.2011.6082994
4. Xu, M., Yang, L., Cao, M.: Strategies for managing complex visitor flow issues at Shanghai World Expo 2010. In: 2009 IEEE/INFORMS International Conference on Service Operations, Logistics and Informatics (2009). https://doi.org/10.1109/soli.2009.5204003
5. Wang, T.T., Wan, H.: Analysis of traffic organization experience of the 2014 Qingdao international horticultural exposition. China Transp. Rev. **37**(01), 79–84 (2015)
6. Liu, Y., Yao, G.Z., Cai, C.C., Zhao, L.C.: Parking characteristics and parking management strategy of Beijing expo. Traffic Transp. **38**(01), 1–5 (2022)
7. Karri, S., Meera, M.D.: Model free reinforcement learning to determine pricing policy for car parking lots. Exp. Syst. Appl. **230**, 120532 (2023). https://doi.org/10.1016/j.eswa.2023.120532. ISSN 0957-4174
8. Selcuk, D., Melih, B., Alev, T.G.: Selection of suitable parking lot sites in megacities: a case study for four districts of Istanbul. Land Use Policy **111**, 105731 (2021). https://doi.org/10.1016/j.landusepol.2021.105731. ISSN 0264-8377

Thermal Analysis Simulation of High-Speed Train Brake Disc Based on Fluid-Solid-Thermal Coupling

Shize Zheng[1] and Jianyong Zuo[1,2(✉)]

[1] Institute of Rail Transit, Tongji University, Shanghai 201804, China
zuojy@tongji.edu.cn
[2] Shanghai Key Laboratory of Rail Infrastructure Durability and System Safety, Shanghai 201804, China

Abstract. With the improvement of the high-speed trains, brake discs are facing a series of dangers caused by temperature. This article uses fluid-solid-thermal coupling simulation to simulate the changes in the thermal temperature of the brake disc itself under high-speed braking conditions, and studies the changes in the temperature of the surrounding environment. The surface temperature of the brake disc is mainly affected by heat conduction and heat convection. Under emergency braking conditions with an initial speed of 400 km/h, the temperature of the brake disc reaches a maximum of 829.93 °C.

Keywords: brake disc · finite element simulation · fluid-solid-thermal coupling · heat convection

1 Introduction

The braking system is the safety guarantee for the operation of the train. Among all the braking methods, disc braking is widely used in high-speed train braking due to its advantages of safety, stability and good heat dissipation performance. During braking, the braking system converts kinetic energy of the train into thermal energy, thus slowing or stopping the train [1]. Experimental research has shown that for every doubling of the train speed, the required braking power increases eight times [2]. During emergency braking, more kinetic energy of the train is converted into thermal energy of the brake disc through friction, and the temperature of the brake disc rises rapidly. As one of the important devices to ensure the safety of trains, the braking system will bring great potential safety hazards to the operation of the train if the heat generated by the brake disc cannot be dissipated in a timely and effective manner [3, 4].

Therefore, this article starts from the temperature changes of the brake disc during the entire braking process in a flow field environment, and studies the simulation method of heat dissipation. This article uses the ANSYS-FLUENT software to simulate it. Assuming that the brake disc is in an open flow field environment, and studies the temperature distribution law of the brake disc surface and the heat dissipation of surface under emergency braking conditions with an initial speed of 400 km/h, and obtains the factors that affect the temperature rise of the brake disc.

A. Bieliatynskyi et al. (Eds.): CSTTE 2023, LNCE 603, pp. 461–469, 2024.
https://doi.org/10.1007/978-981-97-5814-2_42

2   Simulation Model

2.1   3D Model and Grid Division

This article analyzes the shaft-mounted brake disc used in high-speed trains. Using SOLIDWORKS software for modeling, and in order to save computational costs and facilitate the generation of meshes, the brake disc is reasonably simplified, ignoring subtle features that have a small impact on the simulation results. The main dimensions of the brake disc are shown in the following Table 1.

Table 1. Main dimensions of brake disc

Outer diameter of brake disc	Inner diameter of brake disc	Thickness of brake disc	thickness of friction ring	Diameter of cooling rib
640 mm	350 mm	80 mm	22 mm	20 mm

The actual brake disc model is completely symmetrical and consists mainly of friction rings, heat dissipation ribs, etc. During train braking, the friction ring surface comes into direct contact with the brake pads, generating strong frictional resistance that slows or stops the train. Although the train and other structures are also neglected here, the brake disc is not placed separately in the flow field for calculation [5], which will result in the brake disc being close to the axis region of the inner ring, that is, the axle position is also filled with air, making the original leeward area of the disc surface become windward. Based on the above considerations, when establishing the solid calculation domain model of the brake disc, the disc claw and a section of the axle connected to the brake disc are retained. Figure 1 is a three-dimensional solid model of the brake disc for calculation, where the right side is a section model of the brake disc cut along the symmetrical section. As can be seen from the figure, the interior of the brake disc mainly contains several cylindrical heat dissipation ribs with equal diameter and uniform arrangement. The uniform arrangement of the heat dissipation ribs can make the temperature rise of the brake disc friction surface consistent under the action of friction heat generation, and the thermal stress on the friction ring is equal [6].

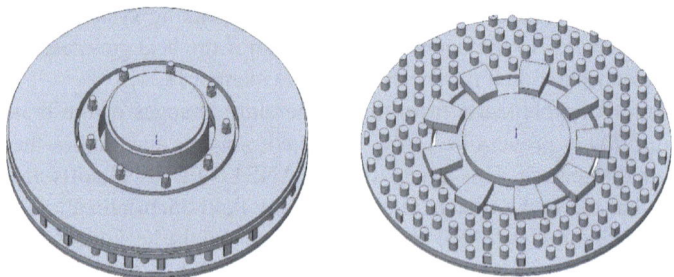

Fig. 1. Three-dimensional model of solid computational domain

Based on the three-dimensional model of the brake disc, a fluid-solid coupling model containing a fluid calculation domain is created. In fluid calculation, in order to prevent the boundary from affecting the research object, it is required to have an infinite flow field, but it is difficult to achieve in actual grid generation [7]. The fluid region established in this article is mainly to simulate the physical environment in which the brake disc works. However, because the single brake disc heat dissipation simulation calculation is mainly used for the principle verification analysis of brake disc heat transfer, to study the heat transfer characteristics of the brake disc, and to compare and verify with the model simulation of the complex structure of the high-speed train head, in order to save computational costs, it is assumed that the brake disc is in an open and unobstructed fluid environment, that is, the fluid domain is simplified to a rectangular region with a size of 7 m * 3.5 m * 3.5 m. The calculation model is shown in the Fig. 2(a). The processed model is meshed, and the meshed grid is shown in the Fig. 2(b).

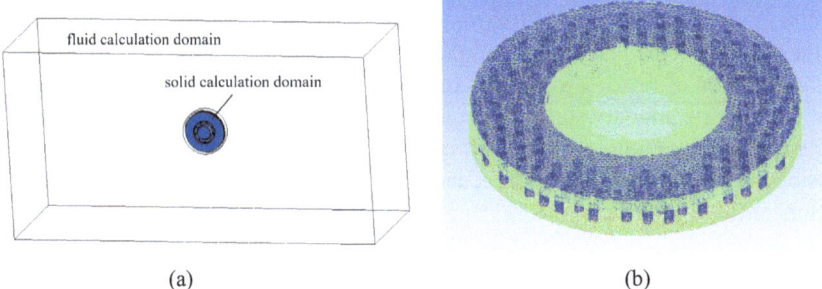

(a)   (b)

Fig. 2. (a). Fluid-solid coupling calculation model (b). Grid division of brake disc

2.2 Working Conditions and Material Parameters

The braking condition used in the simulation is that the initial braking speed is 400 km/h. The deceleration during braking is as Table 2. The parameters of the brake disc materials used in the simulation are shown in Table 3. The physical parameters of air are shown in Table 4.

Table 2. Braking condition

Velocity (km/h)	Deceleration (m/s^2)
400-200	−0.98
200-0	−1.25

Table 3. Parameters of brake disc material

specific heat capacity[J/(kg·k)]	density[kg/m³]	thermal conductivity[W/(m· k)]
489.9	7980	30.9

Table 4. Parameters of air

specific heat capacity[J/(kg·k)]	density[kg/m³]	viscosity[kg/m·s]	thermal conductivity[W/(m·k)]
1006.43	Change with temperature	1.7894e-5	0.0242

The loading of thermal flow on the surface of the brake disc is using an equivalent friction heat calculation method. The emergency braking of a train with a initial speed of 400 km/h takes a total of 101 s from the application of braking to the complete stop of the train. The calculated set duration is 120 s, which includes the entire process from the application of braking to the end of braking at the time of 101 s, as well as the 19 s of the train's stationary state after completing braking. The simulation results mainly focus on the temperature variation law of the brake disc and the dynamic convection heat dissipation of the brake disc surface, and study the temperature rise law and influencing factors of the brake disc.

3 Fluid-Solid-Thermal Coupling Simulation Results

3.1 Law of Brake Disc Solid Temperature Rise

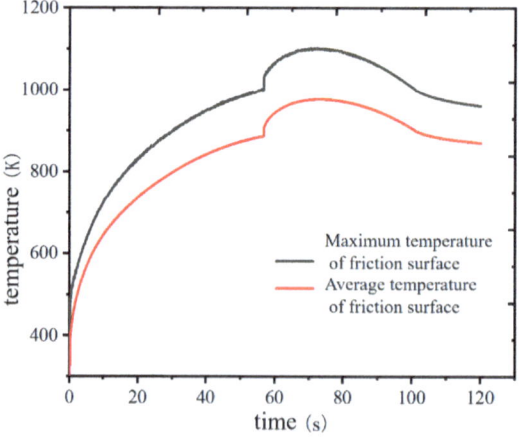

Fig. 3. Temperature curve of brake disc friction surface

The surface of the brake disc comes into direct contact with the outside air, so there is both thermal conduction between solids and convection between fluids on the surface. As shown in Fig. 3, the maximum and average temperatures of the brake disc friction surface vary with time. It can be clearly seen from the figure that the disc surface temperature rises first and then falls with time. This is because the heat flow input at the initial node is greater than the output, resulting in an increase in the node temperature. At around 60 s, the temperature curve rises abruptly, which is due to changes in the deceleration of the train and the work done by the friction force, resulting in a sudden change in the input heat flow. When the heat flow input and output reach equilibrium, the node temperature reaches its maximum, and then the heat flow input at the node is less than the output, and the temperature begins to decrease. This calculation condition reaches its maximum disc surface temperature of 1103.08 K at around 72 s.

Figure 4 shows the temperature cloud maps of the brake disc surface at different times. It can be seen from the figure that the temperature distribution of the brake disc friction surface along the circumferential direction has little difference, and it generally shows an upward trend along the radial direction. The surface temperature of the heat dissipation ribs gradually increases with time, mainly due to the heat conduction between solids, where heat is transferred from the higher temperature friction surface to the lower temperature area, especially through the comparison of the temperature cloud maps at different times. It can be seen that with the increase of time, the temperature difference between the friction surface and the heat dissipation ribs decreases significantly.

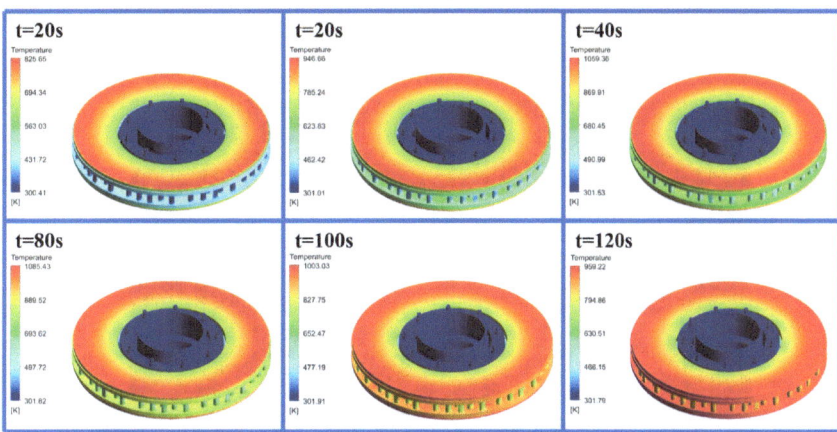

Fig. 4. Cloud map of brake disc surface temperature at different times

3.2 Convection Heat Dissipation on the Surface of the Brake Disc

Distribution of Air Velocity Flow Field Around the Brake Disc

During emergency braking, the running speed of the train decreases continuously as the braking time increases. At the same time, the rotational motion of the brake disc will interfere with the surrounding air flow, causing the flow velocity of the air to change continuously, which in turn leads to a corresponding change in the convective heat transfer coefficient of the brake disc surface. Therefore, air convection, as a major mode of heat dissipation of the brake disc, analyzing the distribution of the air flow field around the brake disc plays an important role in studying the temperature rise law of the brake disc [8].

In this simulation, the distribution of air flow around the brake disc is relatively simple. As time changes, the flow field speed gradually decreases, but the distribution law of the flow field around the brake disc remains basically unchanged. Therefore, only the flow field at the time of emergency braking for 20 s is analyzed here. Figure 5 shows the instantaneous velocity distribution cloud map of the air flow field at the z-x symmetric section position of the brake disc, which is similar to the temperature distribution of the air flow field.

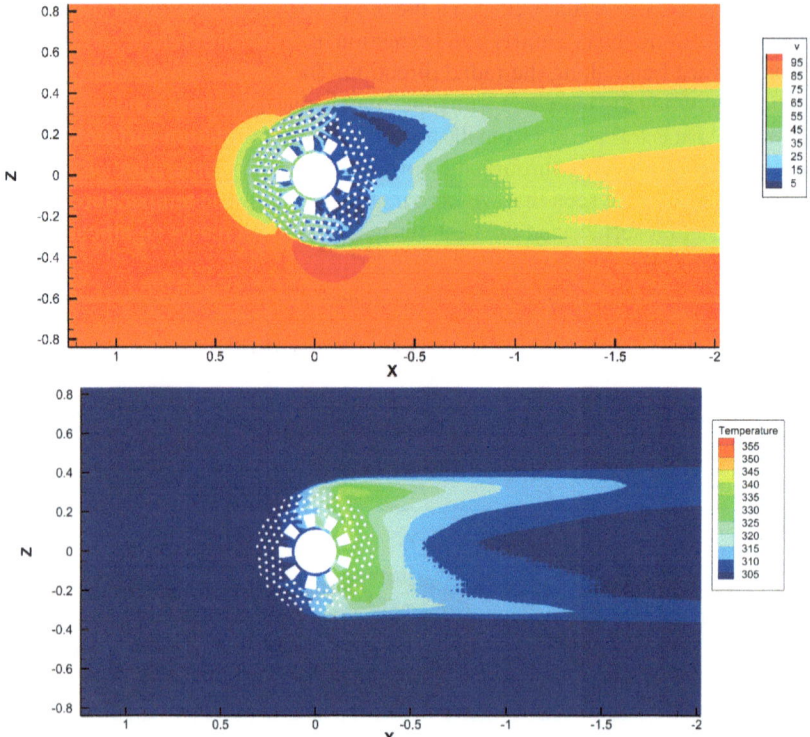

Fig. 5. (a). Cloud map of air velocity; (b). Temperature cloud map of the air domain

From the air velocity cloud map around the brake disc, it can be seen that during the braking process of the train, the air blown in from the front interacts with the airflow generated by the rotation of the brake disc, reducing the air flow velocity in front of the brake disc. The air flow velocity above and below the brake disc increases due to rotation. However, behind the brake disc, the air flow velocity decreases due to the obstruction of the brake disc.

In order to better analyze the flow pattern of air around the brake disc and in the channels near the cooling ribs under two different conditions: train operation and brake end, the distribution of air velocity around the brake disc and streamline diagram are output. Figure 6(a) shows the results during the braking process (time 20 s), when the train is still running. Correspondingly, Fig. 6(b) shows the results when the train is at a standstill (time 120 s).

From the velocity streamline diagram at time 20 s, it can be seen that some of the air flowing in from the inlet enters a complex channel consisting of heat dissipation ribs, and the flow velocity of the air inside the brake disc decreases significantly. Due to the multi-scale complex motion formed by the rotation of the brake disc and the horizontal travel of the train, as well as the complex structure of the brake disc, multiple flow phenomena with vortex structures appear around the brake disc, including near the heat dissipation ribs.

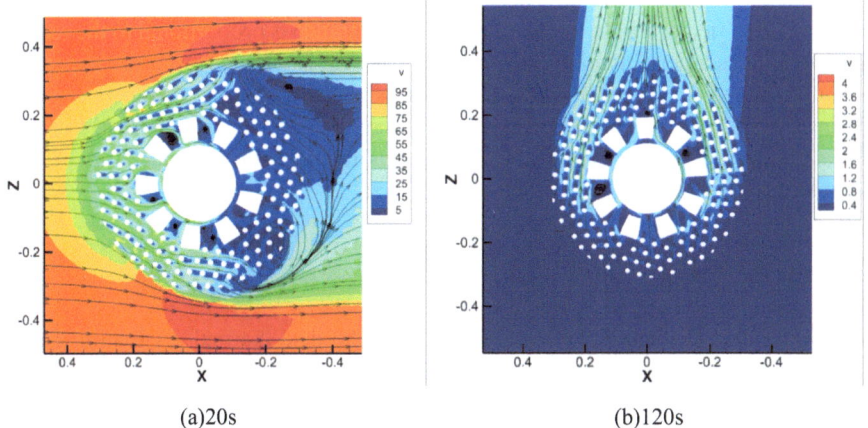

(a)20s   (b)120s

Fig. 6. Velocity streamline diagram around the brake disc

From the velocity streamline diagram at time 120 s, it can be seen that after the braking is completed, the train and the brake disc are in a stationary state, and the air is not completely static or in irregular motion, but generally shows an upward trend, which is due to natural convection. Natural convection is a spontaneous heat transfer process that occurs due to the spontaneous motion of fluids participating in heat transfer due to their different densities [9]. From the previous analysis, it can be seen that after the emergency braking of the train is completed, the temperature of the solid brake disc is high, and the air around it will also be affected. Although the train is in a stationary state, due to uneven temperature distribution, the air around the brake disc has a low density of high temperature air, and the density difference causes fluid motion under the influence of gravity.

Coefficient of Convective Heat Dissipation on the Surface of Brake Disc

Thermal convection can be divided into forced convection and natural convection. The heat dissipation of trains under braking and running conditions belongs to forced convection, while the heat dissipation after the completion of braking belongs to natural convection [10]. In this simulation, the shape of the brake disc remains unchanged, and the convective heat transfer coefficient is mainly related to the flow state of the fluid. During the braking process, the spatial position of each point on the brake disc changes continuously and periodically, which causes the magnitude of the convective heat transfer coefficient at each point on the surface of the brake disc to also vary periodically. When a point on the disc surface rotates to a position with high air flow velocity in the flow field, the convective heat transfer coefficient at that point increases accordingly. Conversely, when a point on the surface rotates to a position with low air flow velocity in the flow field, the convective heat transfer coefficient at that point decreases accordingly.

By extracting the heat dissipation power of each part of the brake disc, it can be found through comparison that the heat dissipation power of the heat dissipation ribs is significantly lower than that of the brake disc friction surface, and the decrease in heat dissipation power of the cooling ribs compared to the brake disc friction surface has a significant lag. This is because, except for the contact surface, the input heat flow in the axial direction is mainly from the heat conduction of the upper nodes, and according to the principle of heat transfer, the heat conduction distance is proportional to the required time, so it takes a certain amount of time for the heat conducted from the upper layer to be transmitted to the lower nodes. Therefore, during braking, the temperature of the cooling ribs is always lower than that of the friction surface, and the numerical difference in heat dissipation power between the cooling ribs and the friction surface is mainly due to the temperature difference.

4   Conclusions

This article analyzes the simulation results of brake disc thermal dissipation under a simplified flow field environment, and obtains the general rules of brake disc temperature rise and heat dissipation process:

(1) The surface of the brake disc is greatly affected by the heat flux density and air convection. The surface temperature of the brake disc is mainly affected by heat conduction and heat convection. The temperature of the friction surface rises first and then falls with time. Under emergency braking conditions with an initial speed of 400 km/h, the temperature of the brake disc reaches a maximum of 829.93 °C at time 72 s;

(2) The convective heat transfer coefficient of the disc surface changes dynamically with time and space during braking, the air flow field distribution around the brake disc can be roughly divided into two different situations: the train running stage and the train stationary state. During the running process, the air flow velocity around the brake disc can be regarded as the superposition of the brake disc rotation and the air inlet velocity, and the flow field at the cooling rib is complex. After braking, the air flow is mainly caused by air convection generated by temperature difference in the stationary state of the train.

Acknowledgements. The research work is supported by National Key R&D Program of China (2021YFB3703805), Science and Technology Research and Development Programme Topics of China State Railway Group Co., Ltd. (Grant No. K2023J005, Grant No. P2023J040-4) and Shanghai Collaborative Innovation Research Center for Multi-network & Multi-modal Rail Transit.

References

1. Li, H., Lin, H.: Design and research on foundation brake system of high speed train. China Railway Sci. **02**, 11–16 (2003)
2. Lei, F.M.: Analysis of temperature field and thermal stress field of CRH2 EMU brake disc based on ANSYS. Southwest Jiaotong University (2017)
3. Zhou, S.-X., et al.: Study on heat dissipation of high-speed train brake disc based on phase change heat storage principle. J. Mech. Eng. (6), 202–210 (2022)
4. Jin, X.: Study on temperature field of disc brake. Hefei University of Technology (2007)
5. Ting, Y., Chao, X., Qing, J., et al.: Application of Reynolds stress turbulence model in numerical simulation of automobile external flow field, Chongqing, China (2021)
6. Xin, W., Guoquan, W., Yong, C.: Research on heat dissipation of heat sink structure of high-speed train disc brake. Locomotive Electr. Drive (03), 94–99 (2021)
7. Zhu, W.: Research and application of high-speed train flow field simulation technology coupled with track model. Beijing Jiaotong University (2016)
8. Du, X., Yang, Z., Li, Q., et al.: Study on the cooling characteristics of a passenger car brake disc. J. Tongji Univ. (Nat. Sci. Edn.) **44**(05), 787–793 (2016)
9. Zhao, Z., Wei, A.: Research on two strengthening measures of natural convection heat dissipation of vertical heat sink. Energy-Saving Technol. (3), 200–205 (2019)
10. Zhang, J., Yu, D., Lin, P.: Temperature field simulation of train disc brake based on flow field analysis. Railway Veh. **56**(03), 8–13 (2018)

XFEM Composite Failure Criterion and Slope Failure Simulation Based on ABAQUS

Zhiluo Li, Shijie Cheng, and Peng Yu$^{(\boxtimes)}$

School of Civil and Hydraulic Engineering, Ningxia University, Yinchuan 750021, China
yupeng111@163.com

Abstract. Conventional numerical methods face major challenges in simulating the complex failure process of soil slopes effectively and accurately. This paper introduces a tension-shear composite failure criterion that elucidates the compound failure mechanisms of soil slopes, realized in the ABAQUS software through the secondary development of a user subroutine and simulated via the extended finite element method (XFEM) module. This method is utilized to simulate the process of soil slope failure under conditions that include heaping load at the crest and excavation at the toe, accompanied by analyses of the failure patterns. The methodology's validity and accuracy are substantiated through comparison with experimental data. The proposed approach adeptly captures the initiation, propagation, and ultimate penetration of cracks during the slope failure process, offering an effective method for simulating the entire slope failure.

Keywords: tension-shear composite failure criterion · ABAQUS · XFEM · Slope failure mode · crack propagation

1 Introduction

Landslides represent a prevalent geological hazard, capable of inflicting substantial destruction. They have consistently been a focal point of scholarly inquiry within geotechnical engineering. Conventional analyses of slope stability are predominantly founded on the principles of limit equilibrium, augmented by assorted techniques for identifying potential sliding planes. Yet, these methodologies are contingent upon a rigid-plastic assumption, implying that the shear strength of the sliding surface attains its threshold simultaneously, and they neglect the nonlinear stress-strain relationship within the soil, as well as the integral procession of slope failure, encompassing the failure initiation, evolution, and ultimate rupture [1]. Contrastingly, the finite element method (FEM) is favored for its comprehensive capture of the deformation and the failure mechanics of the slopes, accomplished by integrating the constitutive relationship of the soil mass. FEM refines the assessment of slope stability by incorporating a strength reduction factor to modulate the strength parameters of the slope. As the slope approaches instability, the nonlinear finite element computation ceases to converge, with the reduction coefficient concurrently serving as the slope's safety factor [2]. Slope failures seldom manifest as a sudden incident; rather, the onset of discontinuity originates

© The Author(s) 2024
A. Bieliatynskyi et al. (Eds.): CSTTE 2023, LNCE 603, pp. 470–485, 2024.
https://doi.org/10.1007/978-981-97-5814-2_43

at the most vulnerable juncture or nature flaw, proceeding through a succession of stages within which the stress at the crack tip transfer to adjacent area and the discontinuity grows, culminating in the slope's eventual total penetrative slip failure [3]. However, neither the limit equilibrium approach nor the strength reduction method can authentically and precisely encapsulate the entire progression from the inception of the slip surface to its ultimate penetration [4, 5].

In the application of the traditional finite element method for simulating discontinuities, such as cracks and shear bands, the mesh boundary must coincide with the discontinuity. As the discontinuity progresses, the mesh requires ongoing refinement to adapt the discontinuity geometry, which costs heavy. The extended finite element method (XFEM) enhances the finite element approach by integrating local enrichment functions, primarily the Heaviside function and the near tip asymptotic functions, into the FE approximation, based on the theory of partition of unity (PU). The interpolation of the enriched degrees of freedom reproduces the local enrichment functions within the element, thus permitting the simulation of a crack's arbitrary growth path without remeshing [6–11]. The corpus of research dedicated to the utilization of XFEM in the depiction of soil's discontinuous deformation and failure mechanisms is expanding. Yu et al. [12] devised a nonlinear analysis system that model the discontinuous internal boundaries within nonlinear materials employing the XFEM. A contact algorithm for the XFEM enriched soil discontinuities was established, incorporating a cohesive crack model and Willner's theory to simulate the adhesive, sliding, and separation states at the contact interface [13]. Wang et al. [14] formulated a methodology for assessing soil cracking through elemental stress analysis and stress backtracking, and utilized a sector control domain for the weighted mean stress evaluation and cracking direction determination, enhancing the precision and adaptability in ascertaining the nature, timing, and orientation of soil failure. Yu et al. [15] advanced a hybrid integration strategy within the extended finite element method, amalgamating Gaussian integration's precision with the expediency of single-point integration. Building upon this, they simulated the expansion of an arbitrary three-dimensional failure surface characterized by tensile-shear composite failure [16].

Since the release of version 6.9, ABAQUS, a widely utilized premier commercial CAE software, has integrated the essential functionalities of the extended finite element method. Nonetheless, the incorporation of more sophisticated and recently proposed methodologies necessitates the implementation via user-defined subroutines [17–20]. Giner et al. [17] initially employed a user subroutine to facilitate the implementation of the extended finite element method within ABAQUS in a two-dimensional framework, subsequently analyzing crack propagation in the context of a fretting fatigue problem. Cheng et al. [21] studied the desiccation shrinkage and cracking of soils by a 3D hydromechanical model established with ABAQUS via the subroutine of UMAT and the XFEM tools in ABAQUS. Cruz et al. [19] employed a user subroutine to surmount the constraints of ABAQUS on multiple cracks or crack intersection within a single element, thereby facilitating the simulation of crack intersection in porous rock formations. To replicate the progression of hydraulic fracture propagation, Tawfik et al. [22] notched plain and reinforced concrete beams were investigated numerically to study their flexural response using the contour integral technique (CIT), the extended finite element method

(XFEM), and the virtual crack closure technique (VCCT) in ABAQUS. Teimouri [23] amalgamated the virtual crack closure technique (VCCT) with the extended finite element method to model the delamination growth of Type I fatigue fractures in composites, employing the direct cycling method within Abaqus.

From a mechanical perspective, the discontinuities in soils can be bifurcated into two distinct types: tensile cracks induced by tensile forces, and shear slips induced by shear forces. Presently, XFEM-based simulations predominantly address the formation of tension cracks, with simulations about the sliding phenomena under shear in soils being notably sparse. Specifically, the implementation of shear and tension-shear compound failure cracks on the ABAQUS platform remains unobserved. This study integrates an XFEM crack propagation criterion tailored for soil tension-shear composite failure via the ABAQUS user subroutine interface. In conjunction with ABAQUS's robust nonlinear computing capabilities, it presents an innovative approach for simulating the tension-shear composite failure process in soil slopes.

The latter part of this paper arranged as follows: Sect. 2 encapsulates the foundational principles of the extended finite element method (XFEM) and delineates the procedure for simulating crack propagation in ABAQUS utilizing its XFEM module. In Sect. 3, the tension-shear compound failure criterion is introduced and implemented via the UDMGINI subroutine. Concurrently, the weighted average stress is deployed to ascertain the crack propagation direction. The fourth section illustrates the application of this method in simulating centrifugal model experiments of soil slope failure under two disparate operational conditions (heaping load at the crest of the slope and excavation at the toe), corroborating the method's rationality and accuracy.

2 Numerical Method

2.1 Nodal Enrichment Functions

XFEM introduces the asymptotic crack-tip functions [6] to capture the singularity of the crack tip and the Heaviside step function to characterize the displacement jump at the crack [24]. The displacement vector, inclusive of both displacement enrichment functions, is articulated as follows:

$$\mathbf{u} = \sum_{I=1}^{n} N_I(x) \left[\mathbf{u}_I + H(x)\mathbf{a}_I + \sum_{\alpha=1}^{4} F_\alpha(x)\mathbf{b}_I^\alpha \right] \tag{1}$$

where, x is the coordinates, $N_I(x)$ is the shape function, $\mathbf{u}_I$ is the node displacement vector describing continuous deformation, $\mathbf{a}_I$ is the enriched degree of freedom scaling the displacement jump, $H(x)$ is the Heaviside step function, $\mathbf{b}_I^\alpha$ is the crack tip enriched degree of freedom scaling the displacement singularity, $F_\alpha(x)$ is the asymptotic crack-tip function.

The displacement description adheres to the partition of unity principle and is capable of reproducing the discontinuity and singularity of the crack displacement field within the element. The Heaviside step function is defined as follows:

$$H(x) = \begin{cases} 1, & \text{if } (\mathbf{x} - \mathbf{x}^*) \cdot \mathbf{n} \geq 0 \\ -1, & \text{otherwise} \end{cases} \tag{2}$$

where, $\mathbf{x}^*$ is the point on the crack closest to $\mathbf{x}$, and $\mathbf{n}$ is the unit normal vector on the crack.

The asymptotic crack-tip functions consist of four functions, which are as follows:

$$F_\alpha(x) = \left[\sqrt{r} \sin\frac{\theta}{2}, \sqrt{r} \cos\frac{\theta}{2}, \sqrt{r} \sin\theta \sin\frac{\theta}{2}, \sqrt{r} \sin\theta \cos\frac{\theta}{2} \right] \tag{3}$$

where, (r, θ) is the polar coordinate system with the crack tip as the origin, and $\theta = 0$ is the crack tangent direction.

2.2 Phantom Node Method

To facilitate the enrichment of degrees of freedom and delineate the discontinuities within cracked elements, phantom nodes (overlaid atop the original authentic nodes) are embedded within the elements of the enriched region in the extended finite element module of ABAQUS. For the sake of coherence, each phantom node is entirely synchronized with its respective genuine node when the element remains intact, as illustrated in Fig. 1. As the crack penetrates the element, the element is cleaved into two parts, each associated with actual nodes at one side and phantom nodes at the other side. At this point, the phantom nodes are decoupled from their real counterparts, allowing for independent movement. By generating phantom nodes and managing their independent movement, the displacement description within ABAQUS's XFEM module behaves good flexibility and accommodates the variations in discontinuities during simulating.

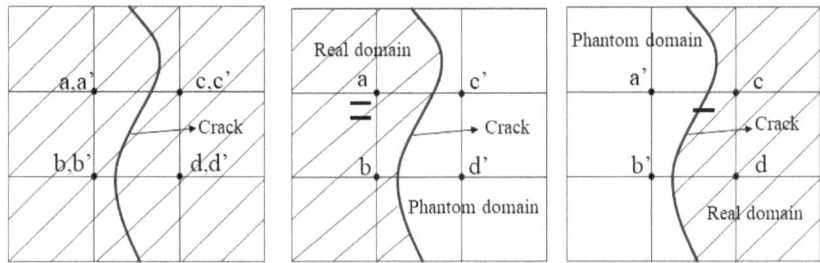

Fig. 1. Phantom nodes of the cracking element during crack propagation.

2.3 Damage Initiation Criteria

Within the XFEM module of ABAQUS, six damage initiation criteria are built in: the maximum principal stress criterion, the maximum principal strain criterion, the maximum nominal stress criterion, the maximum nominal strain criterion, the quadratic separation interaction criterion and the quadratic traction interaction criterion. These criteria are pertinent to tensile failures and do not precipitate failure under compressive conditions. Yet, there exists no equivalent criterion for the prevalent shear failure in soils. To more accurately replicate the failure of a soil slope, secondary development through a user subroutine is indispensable. This entails the integration of a shear failure criterion within the purview of the XFEM module.

Within the ABAQUS' XFEM module, the damage initiation criteria are amenable to customization, permitting users to define these thresholds via the subroutine [25]. Considering that soil slopes may concurrently undergo shear and tensile failures under varying stress conditions, any newly formulated failure criterion must encompass both eventualities. This research adopts the maximum tensile stress criterion for tensile failures and the Mohr-Coulomb criterion for shear failures. By the combination of these two criteria, the particular failures modes are determined and composite failure mechanism is revealed, depending on the stress conditions.

2.4 Damage Evolution Mechanism

Upon fulfillment of the pertinent damage initiation criterion, the damage evolution mechanism delineates the degradation rate of the bond stiffness. D is included to indicate the average global damage at the junction of the crack surface and the edge of the crack element. During the damage evolution process, D incrementally increases from 0 to 1. The normal and tangential stresses of the element, as influenced by the damage, are articulated as follows:

$$t_n = \begin{cases} (1-D)T_n, & T_n \geq 0 \\ T_n, & T_n < 0 \end{cases} \tag{4}$$

$$t_s = (1-D)T_s \tag{5}$$

$$t_t = (1-D)T_t \tag{6}$$

where, D is the damage variable, t_n, t_s, t_t are the normal stress component and two tangential stress components respectively, T_n are the normal stress component without damage, T_s and T_t are the first and second tangential stress components without damage respectively. For user defined damage initiation criterion, corresponding damage evolution criterion must be defined simultaneously [26]. The cumulative effect of normal and tangential traction displacements is typically quantified through effective traction displacement, which characterizes the ensuing damage evolution. The effective traction displacement is defined as follows:

$$\delta_m = \sqrt{\langle \delta_n \rangle^2 + \delta_s^2 + \delta_t^2} \tag{7}$$

where, δ_n is normal traction displacement, δ_s and δ_t is the first and second tangential traction displacement.

3   Initiation and Propagation of Cracks

3.1   Crack Initiation

ABAQUS, by default, employs the stress or strain at the centroid of the element preceding the crack tip as the foundational metric for determining whether the damage initiation criteria are met. This method is both precise and efficient given a sufficiently refined mesh. However, when the precision of the mesh near the crack tip is coarse relative to the stress or strain field, the accuracy of stress or strain measurements based on the centroid's position diminishes. To remedy this discrepancy, ABAQUS offers an option to shift the damage initiation assessment from the centroid of the element to the position of the crack tip itself, thus leveraging the stress or strain at the crack tip for a more accurate evaluation, as illustrated in Fig. 2.

When tensile or shear failure occurs, the element is regarded to have met the damage initiation requirement, and damage evolution occurs, with the damage variable increasing from 0 to 1, resulting in the creation and spread of crack. To identify the failure criteria, the tensile stress level T_l and shear stress level S_l are set in this study for the tensile failure and shear failure, respectively. If $T_l > 1$, the element is considered to have tensile failure. Shear failure occurs in the judgment element when $S_l > 1$. The tensile stress level T_l and shear stress level S_l is defined as follows:

$$T_l = \left| \frac{\sigma_3}{f_t} \right| \quad (\sigma_3 < 0) \tag{8}$$

$$S_l = \frac{\sigma_1 - \sigma_3}{\sin \varphi (\sigma_1 + \sigma_3 + 2c / \tan \varphi)} \tag{9}$$

where σ_1 is the maximum principal stress, σ_3 is the minimum principal stress, f_t is the tensile strength of the soil, φ is the internal friction angle of the soil, and c is the cohesion of the soil. When the stress level at the crack tip reaches 1, the appropriate type of failure is regarded to have occurred.

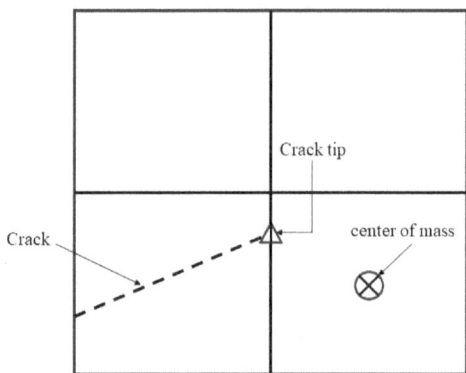

Fig. 2. Location of the crack initiation.

3.2 Propagation of Crack

The mode of failure dictates the trajectory of crack expansion. Tensile failure transpires when the tensile stress level, T_l, exceeds unity, aligning the tensile failure plane with the normal direction of the minimum principal stress. Similarly, shear failure ensues when the shear stress level, S_l, surpasses unity, positioning the shear failure plane at an angle of $45° + \varphi/2$ relative to the direction of the minimum principal stress.

It is essential to recognize that the stress estimation at the centroid or crack tip is a localized computation for an individual element. In cases of inferior mesh quality, a non-local averaging method may be employed to more accurately evaluate the stress or strain field preceding the crack tip, thus refining the precision of the estimated crack propagation direction.

To precisely govern the range of non-local averaging and smoothing for the crack growth direction, a semi-circular local zone is established with the crack tip as the epicenter and a predefined influence radius r_c positioned anterior to the crack tip, designated as the control domain (illustrated in Fig. 3). Within this control domain, non-local averaging and smoothing are exclusively conducted. The average stress within the control domain is determined as the weighted mean stress of all integration points within that domain, and the Gaussian weighting function is employed to assign greater significance to the integral points in proximity to the crack tip. The Gaussian weighting function is delineated as follows:

$$\omega(r) = \frac{1}{(2\pi)^{3/2}r_c^3} \exp\left(\frac{-r^2}{2r_c^2}\right) \tag{10}$$

where, r is the distance between an integral point in the control domain and the crack tip, r_c is the control domain's influence radius, and the default is three times the enrichment element feature length.

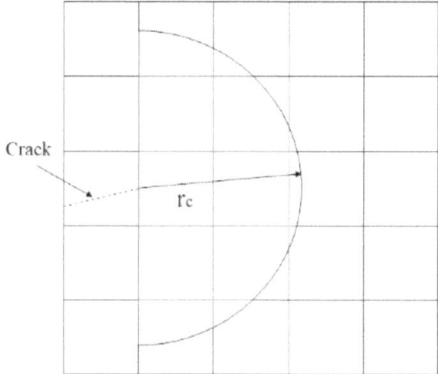

Fig. 3. Scope of control domain.

3.3 The Realization of Compound Failure Criterion

For the management of crack growth, the UDMGINI subroutine in ABAQUS can be configured to report the conditions of crack initiation and the orientation of crack propagation. Crafted in Fortran, the UDMGINI subroutine is capable of real-time extraction of stress, strain, coordinate positions, and temporal data from the element integration points. Utilizing these data, the subroutine computes the magnitude of the index value pertinent to the prevailing damage initiation criterion via a bespoke algorithm, thereby discerning whether, at a given moment, the element is experiencing cracking and the specific direction of such cracking.

The implementation of the tension-shear composite failure criterion within the UDMGINI subroutine involves the following steps:

1) The UDMGINI subroutine retrieves the stress tensor from the integration point of the element.
2) The principal stress magnitude and its directional orientation at the element's integration point are ascertained utilizing ABAQUS's SPRIND function.
3) The principal stress is then input into the formulas for tensile stress level (Eq. 8) and shear stress level (Eq. 9) for evaluation. An element is deemed to have undergone tensile failure if its tensile stress level exceeds the prescribed threshold, while the shear stress level remains below it. Conversely, an element is considered to have experienced shear failure if its shear stress level surpasses the threshold, irrespective of the tensile stress level. Should both tensile and shear stress levels meet or exceed their respective thresholds, a comparison of the two levels is conducted, and the failure mode corresponding to the higher stress level value is selected.
4) After ascertaining the mode of failure, the trajectory of crack propagation is established by converting the local coordinates to global coordinates, using the orientation of the minimum principal stress as a benchmark. In the event of tensile failure, the crack propagates in alignment with the normal to the minimum principal stress; conversely, during shear failure, the crack advances at an angle of $45° + \varphi/2$ relative to the normal to the minimum principal stress.

4 Verification by Numerical Examples

4.1 Simulation of Soil Slope Failure Process Under the Condition of Heaping Load on Top of Slope

Regueiro et al. [27] probed the behavior of pronounced discontinuous fields in pressure-responsive plastic materials by employing the embedded discontinuity approach in conjunction with the augmented Drucker-Prager elastoplastic model. Within their paper, they developed a computational model depicting soil slope instability induced by a heaping load on the top of the slope, as represented in Fig. 4, wherein the slope ultimately succumbed to a comprehensive slide under the influence of the load. It is significant to note that the paper revealed the figuration of the slip surface (Fig. 5) but refrains from detailing the processes of crack initiation and propagation.

The parameters employed in this analysis are derived from the data presented in the paper [27]. Encompassing an elastic modulus of 10 MPa, Poisson's ratio of 0.4, soil cohesion of 40 kPa, a friction angle of 10°, and a unit weight of 20 kN/m3. To enhance the comparability of the calculated results, these parameters align with those specified in the computational examples [27]. Additionally, these values fall within the typical spectrum of soil properties and do not undermine the objectives or conclusions of this study, thereby affirming the significance of the research findings. The model configuration is as follows: a rigid foundation is emplaced atop the slope, intimately integrated with the

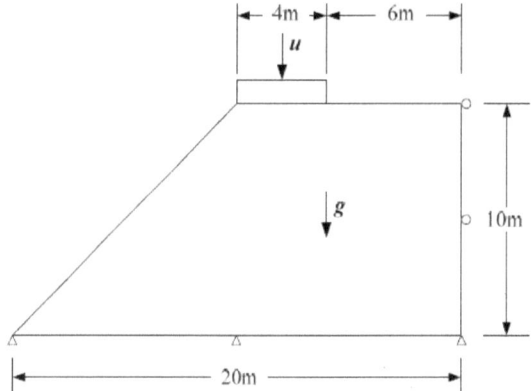

Fig. 4. Instability problem of pushing slope.

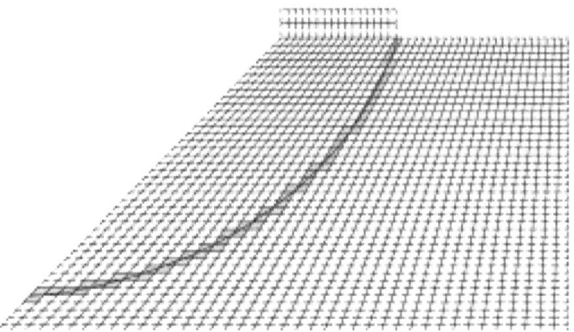

Fig. 5. Crack track diagram after slope failure.

subjacent soil; the slope's gravity stress is factored into the calculation and a downward displacement u is exerted upon the center of the foundation.

As the loading displacement reaches u=4 cm, the soil in proximity to the bottom right corner of the robust foundation at the summit of the slope starts to fracture, as shown in Fig. 8; the tensile crack only cuts two layer of element before it turned to shear failure. Upon escalation to u=5 cm, the shearing slip plane proliferates swiftly, with its depth extending to 8.18m, as shown in Fig. 9. Then, the inclination of the slip plane moderates and the pressure induced by the loading amplifies the friction along the fracture surface, decelerating the slip plane's propagation. Complete penetration of the slope by the cracks transpires at a displacement load of u = 10 cm, as shown in Fig. 10. With the increase of the displacement load, the slope undergoes a process of uniform deformation, nonuniform deformation, formation of the slip surface and finally overall failure. At this time, The vertical and horizontal displacement cloud maps of the slope are shown in Fig. 6 and Fig. 7. The analytical outcomes reveal that the configuration and location of the identified shear zone are congruent with the findings documented in the literature [27]. Plastic deformation is localized within the elements of the shear zone, while the elements beyond this zone exhibit predominantly elastic deformation.

The results corroborate the ability of the methodology applied in this investigation to meticulously monitor the initiation, progression, and ultimate penetration of cracks and shear zones in soil slopes subjected to top slope loading. This research elucidates the morphology and progression patterns of the shear zone with greater clarity and precision than the embedded discontinuity approach.

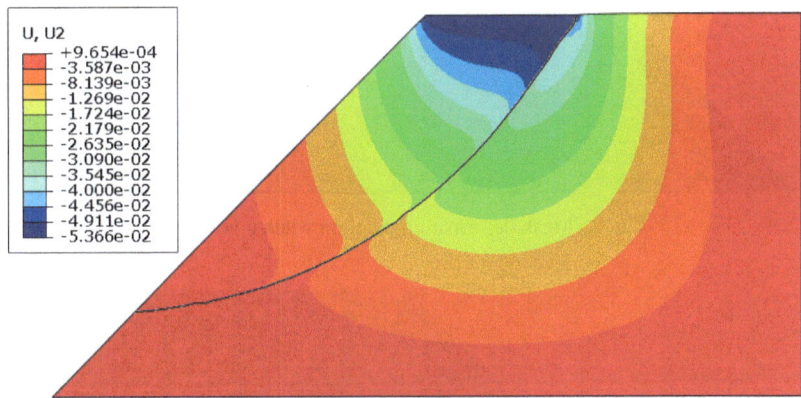

Fig. 6. Vertical displacement diagram.

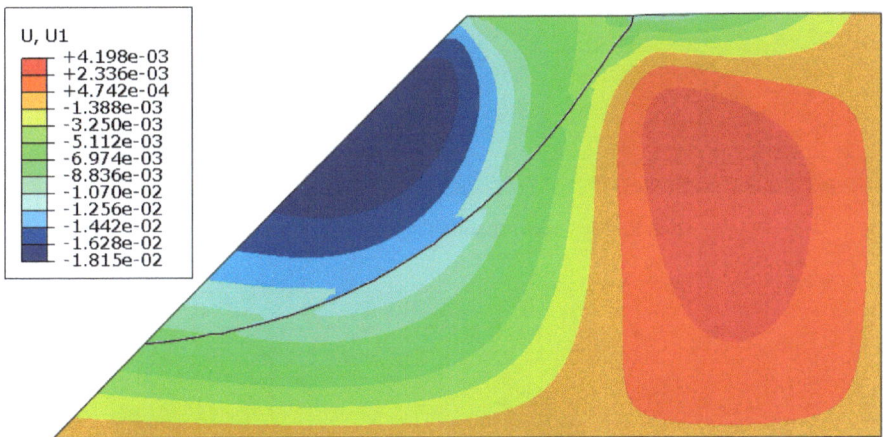

Fig. 7. Horizontal displacement diagram.

4.2 Simulation of Soil Slope Failure Process Under Excavation at the Toe of the Slope

Excavation-induced landslides, as a common mode of soil slope failure, have garnered considerable focus. Li [28] conducted a thorough investigation into the processes and mechanisms underpinning the instability and collapse procession of such landslides

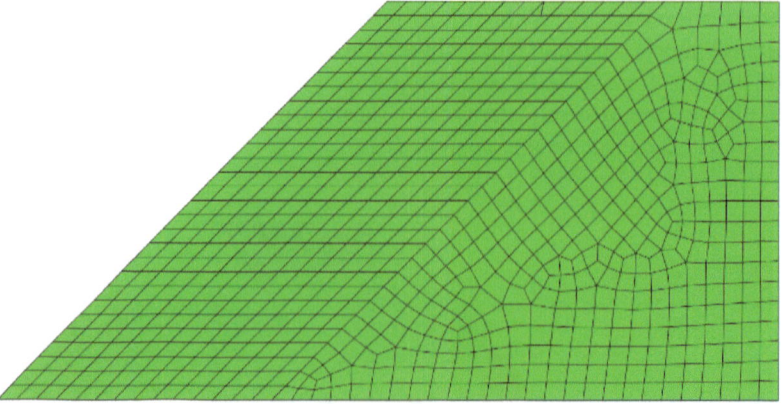

Fig. 8. Mesh deformation diagram when u is 4 cm.

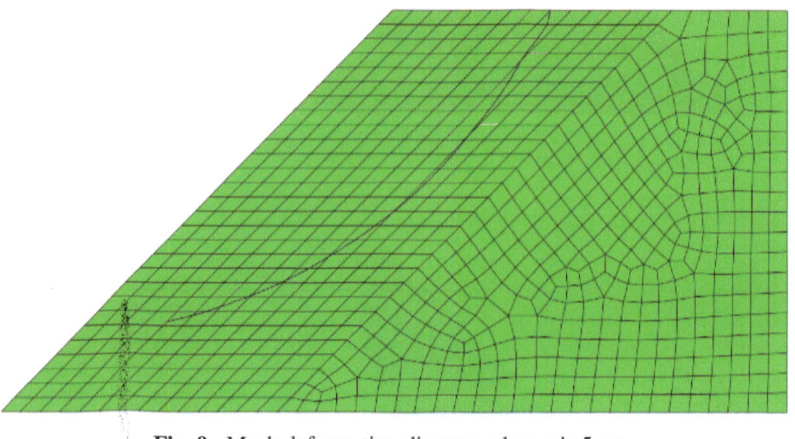

Fig. 9. Mesh deformation diagram when u is 5 cm.

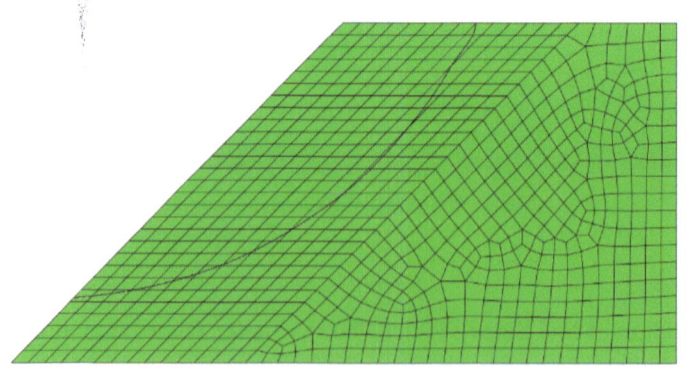

Fig. 10. Mesh deformation diagram when u is 10 cm.

through centrifugal model tests. This study employs the C1 experiment as a paradigm to simulate and analyze the inception and propagation of the slip failure of the tested slope.

The experimental model features a slope with a height of 25 cm, and an excavation depth at the toe of the slope measuring 8 cm. Following the attainment of 50g during centrifugation and the sample achieving relative stability, the excavation at the slope's toe is executed in one continuous process.

Figure 11 illustrates the failure surface post-landslide occurrence within the test specimen, while Fig. 12 presents the trajectory of the associated crack patterns. The specimen exhibits a principal slip plane near the slope's surface and two tension cracks at the crest of the slope. For simplicity purpose, this study primarily examines the emergence and progression of a principal crack.

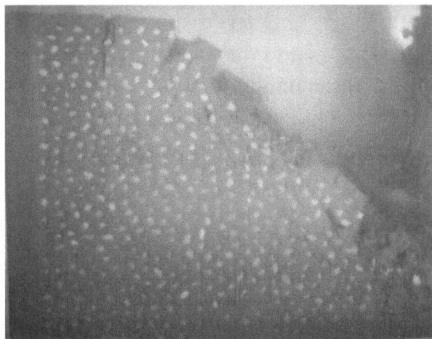

Fig. 11. A photo of the test sample after it was destroyed.

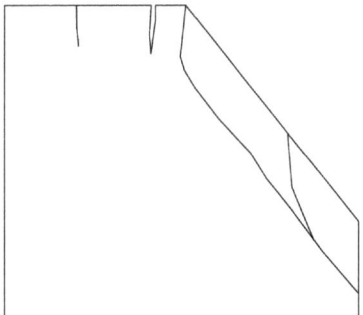

Fig. 12. Crack path diagram.

In the course of the centrifuge experiment, Li [28] observed that the primary slip plane on the slope originated at the excavation front and progressively ascended. Plastic deformation is localized within the shear zone elements, whereas elements outside this zone display characteristics of near-elastic deformation. The slip surface, to an extent, serve as demarcations between the zones of elastic and plastic deformation. Table 1 details the specific physical and mechanical properties of the test soil specimen.

Table 1. Physical and mechanical parameters of C1 test soil1.

Sample number	Grade of side slope	Moisture capacity /%	Dry density /(g/cm^3)	Cohesion /KPa	The angle of internal friction /(°)	Elastic modulus /KPa	Poisson ratio	Tensile strength /KPa
C1	1.2:1	18	1.5	26	24.5	3200	0.3	17

The specific simulation process is as follows:

1) Three steps of the analysis are established. Step 1 applies a self-weight load to the entire model and conducts geo-stress equilibrium, yielding a model that sustains the self-weight load without undergoing any displacement.
2) Centrifugal load of 50g is applied to the model gradually with 5g each increment.
3) The birth-and-death element technique is utilized to simulate the slope excavation process, with Step 2 initiating the birth-and-death elements prearranged in the excavation zone, effectively rendering the regional elements inactive. The self-weight load is preserved, and calculations for crack propagation are carried out.

Figure 15 illustrates the simulation results from this study. At Step Time 0.3178 in STEP-2, shear failure manifests in the element at the base of the excavation face (Fig. 15a). This shear failure continues, reaching a height of 3.2 cm at Step Time 0.9431 (Fig. 15b). Subsequently, the crack height swiftly escalates to 24.3 cm, transitioning into a tensile failure, which persists until it completely cleaves through the slope's summit, culminating in a breach (Fig. 15c). At this time, The vertical and horizontal displacement cloud maps of the slope are shown in Fig. 12 and Fig. 13. The simulated trajectory and ultimate form of the principal crack correlate with the experimental findings, affirming the effectiveness of the proposed methodology. Specifically, it confirms the precision with which the position of slope fissures can be pinpointed and the accuracy of the predicted direction of their expansion.

The numerical simulation indicates that the shear crack within the slope predominantly propagates within the strain localization zone observed in the centrifuge test, ascending gradually over time. However, these fissures are situated considerably lower than the corresponding areas shown in Fig. 12. Moreover, the crosspoint of shear and tension cracks near the slope's apex is markedly higher than illustrated in Fig. 12. This variance could stem from two potential factors: Initially, the physical model test may be subject to three-dimensional effects, in contrast to the numerical simulation, which is constrained to plane strain conditions, potentially leading to significant discrepancies between the two. Secondly, the excavation face's surface area might have neared a critical state, and the centrifugal test model could have fractured due to a minor, spontaneously occurring local crack on the excavation face (Fig. 14).

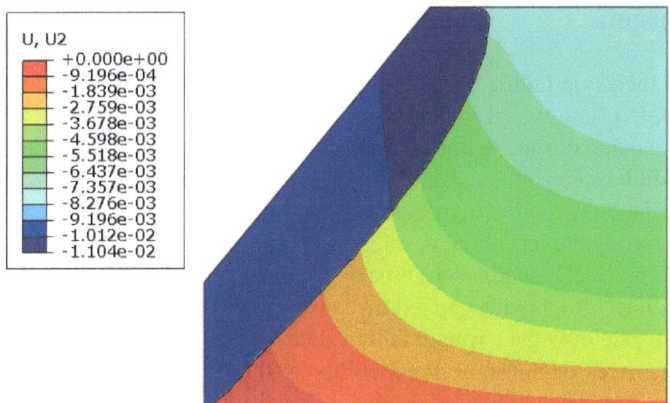

Fig. 13. Vertical displacement diagram.

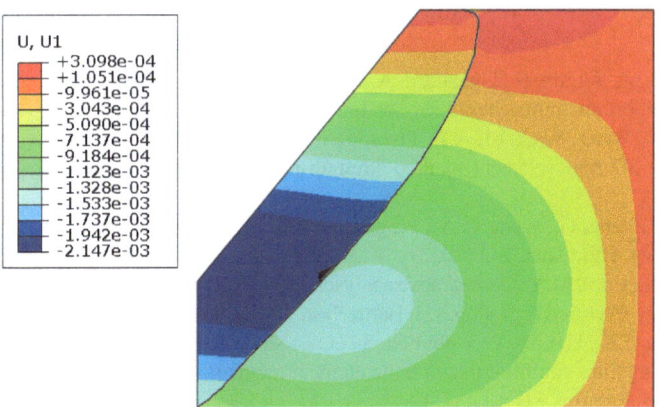

Fig. 14. Horizontal displacement diagram.

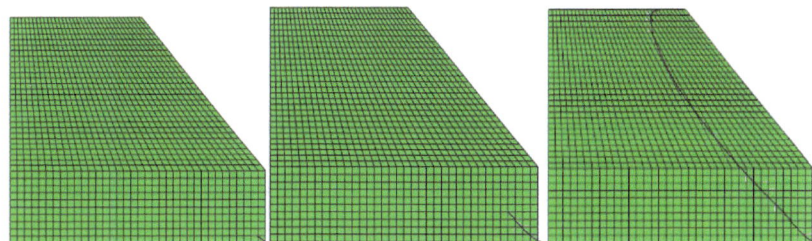

(a) Step-2, Step Time=0.3178 (b) Step-2, Step Time=0.9431 (c) Step-2, Step Time=0.9432

Fig. 15. Crack track diagram.

5 Conclusion

To replicate the slope failure, the tension-shear composite failure criterion is encoded into a UDMGINI subroutine and incorporated into the ABAQUS extended finite element module. Employing this methodology, the failure patterns of the slope under two distinct loading conditions are simulated and meticulously analyzed.

The simulation outcomes corroborate that the methodology delineated in this study can vividly portray the inception, development, and terminal failure state of the slope's shear zone. It precisely identifies the type of onset, the timing, and the trajectory of crack propagation, thus bridging the gap left by conventional slope failure simulations, which typically illustrate only the terminal failure state without the evolutionary process. The two cited instances demonstrate the viability of this approach in emulating complex soil slope failures. The simulation results are readily interpretable and instrumental in elucidating the causation and progression of soil slope collapse.

References

1. Azarafza, M., Akgün, H., Ghazifard, A., et al.: Discontinuous rock slope stability analysis by limit equilibrium approaches–a review. Int. J. Digital Earth **14**(12), 1918–1941 (2021)
2. Lin, H.D., Wang, W.C., Li, A.J.: Investigation of dilatancy angle effects on slope stability using the 3D finite element method strength reduction technique. Comput. Geotech. **118** (2020)
3. Liu, S.J., Zhang, G., Wang, A.X.: Progressive failure mechanism of structuralized cemented slopes. Eng. Fail. Anal. **143** (2023)
4. Koca, T.K., Koca, M.Y.: Comparative analyses of finite element and limit-equilibrium methods for heavily fractured rock slopes. J. Earth Syst. Sci. **129**(1) (2020)
5. Mebrahtu, T.K., Heinze, T., Wohnlich, S., et.al.: Slope stability analysis of deep-seated landslides using limit equilibrium and finite element methods in Debre Sina area, Ethiopia. Bull. Eng. Geol. Environ. **81**(10) (2022)
6. Jiang, Y., Dong, J., Nie, D. F., et.al.: XFEM with partial Heaviside function enrichment for fracture analysis. Eng. Fract. Mech. **241** (2021)
7. Deng, H.C., Yan, B., Koyanagi, J.: Improved XFEM for 3D interfacial crack modeling. Mech. Mater. **186** (2023)
8. Crusat, L., Carol, I., Garolera, D.: XFEM formulation with sub-interpolation, and equivalence to zero-thickness interface elements. Int. J. Numer. Anal. Meth. Geomech. **43**(1), 45–76 (2019)
9. Li, H., Li, J.S., Yuan, H.: A review of the extended finite element method on macrocrack and microcrack growth simulations. Theoret. Appl. Fract. Mech. **97**, 236–249 (2018)
10. Yu, P., Wang, X.N., Yu, J.L., et.al.: XFEM Simulation of Soil Crack Evolution Process Considering the Stress Concentration and Redistribution at the Crack Tip. Int. J. Geomech. **22**(9) (2022)
11. Yu, P., Hao, Q.S., Yu, J.L., et al.: XFEM-based investigation on sliding regularities of soil slopes. Chin. J. Geotech. Eng. **44**(08), 1416–1424 (2022)
12. Yu, J.L.: Mechanism investigation and numerical simulation of Evolution of Shear Band in Soil. Qinghua University (2009)
13. Yu, J.L., Yu, Y.Z., Zhang, B.Y., et al.: A contact algorithm based on extended finite element method. Eng. Mech. **28**(04), 13–17 (2011)
14. Wang, X.N., Li, Q.M., Yu, Y.Z., et al.: Simulation of the failure process of landslides based on extended finite element method. Rock and Soil Mech. **40**(06), 2435–2442 (2019)

15. Yu, P., Hao, Q.S., Wang, X.N., et.al.: Mixed Integration Scheme for Embedded Discontinuous Interfaces by Extended Finite Element Method. Frontiers in Earth Science, **9**, (2022)
16. Yu, P.: Numerical simulation of evolution of discontinuous surfaces in soil by 3D XFEM. Qinghua University (2022)
17. Giner, E., Sukumar, N., Denia, F.D., et al.: Extended finite element method for fretting fatigue crack propagation. Int. J. Solids Struct. **45**(22–23), 5675–5687 (2008)
18. Giner, E., Sukumar, N., Tarancón, J.E., et al.: An Abaqus implementation of the extended finite element method. Eng. Fract. Mech. **76**(3), 347–368 (2009)
19. Cruz, F., Roehl, D., Vargas, E.D.: An XFEM implementation in Abaqus to model intersections between fractures in porous rocks. Comput. Geotech. **112**, 135–146 (2019)
20. Abaqus: V 6.14 Users Manual. RI:Hibbitt, Karlsson & Sorensen Inc (1999)
21. Cheng, W.Q., Bian, H.B., Hattab, M., et al.: Numerical modelling of desiccation shrinkage and cracking of soils. Eur. J. Environ. Civ. Eng. **27**(12), 3525–3545 (2023)
22. Tawfik, A.B., Mahfouz, S.Y., Taher, S.E.F.: Nonlinear ABAQUS Simulations for Notched Concrete Beams. Mater. **14**(23) (2021)
23. Teimouri, F., Heidari-Rarani, M., Aboutalebi, F.H.: An XFEM-VCCT coupled approach for modeling mode I fatigue delamination in composite laminates under high cycle loading. Eng. Fract. Mech. 249 (2021)
24. Moës, N., Dolbow, J., Belytschko, T.: A finite element method for crack growth without remeshing. Int. J. Numer. Meth. Eng. **46**(1), 131–150 (1999)
25. Liu, D.C., Cao, D.F., Hu, H.X., et al.: Numerical study on failure behavior of open-hole composite laminates based on LaRC criterion and extended finite element method. J. Mech. Sci. Technol. **35**(3), 1037–1047 (2021)
26. Jin, G.L.: Numerical analysis of fatigue crack growth in asphalt pavement based on extended finite element model. Southeast University (2016)
27. Regueiro, R.A., Borja, R.I.: Plane strain finite element analysis of pressure sensitive plasticity with strong discontinuity. Int. J. Solids Struct. **38**(21), 3647–3672 (2001)
28. Li, M.: Research on Failure Mechanism and Deformation Rules of Cohesive Slopes under Excavation Conditions. Qinghua University (2009)

Calculation of Active Earth Pressure on a Circular Retaining Wall Based on Energy Method

Senlin Jia[1], Guigui Zhou[2], Tao Qin[1], Yifan Mei[1], Jiahui Li[1], and Kunlin Lu[1(✉)]

[1] School of Civil and Hydraulic Engineering,
Hefei University of Technology, Hefei 230009, China
lukunlin@hfut.edu.cn
[2] Scivic Engineering Corporation, Luoyang 471000, China

Abstract. For an axisymmetric circular retaining wall, a calculation method of active earth pressure on the circular retaining wall based on energy method is being proposed. The analysis of the soil behind the wall employs the Coulomb failure mechanism. Assuming the soil to be a completely rigid plastic body, the calculation formula of active earth pressure on the circular retaining wall is being established according to the associated flow law and virtual work principle. The results are indicating that the active earth pressure on the circular retaining wall is lower than that predicted by the Coulomb solution for plane retaining walls. This active earth pressure is increasing as the radius-to-height ratio of the wall increases, eventually aligning with the Coulomb solution for plane walls. Additionally, the inclination angle of the slip surface is being observed to increase with the circumferential stress coefficient, while the active earth pressure is decreasing correspondingly; For non-cohesive soil, when the radius-to-height ratio approaches infinity, the calculation method in this paper is being consistent with Coulomb theory. This method can calculate the active earth pressure on circular retaining wall with cohesive soil behind the wall. As the cohesion increases, the inclination angle of the slip surface rises, and the active earth pressure decreases. The result is being confirmed by model tests. It can provide a reference for the engineering design of circular retaining structures.

Keywords: circular retaining wall · active earth pressure · energy method · circumferential stress · virtual work principle

1 Introduction

The circular retaining structure, known for its good stress and deformation characteristics, is widely used in various applications. These include retaining walls of circular reservoirs, turns in mountain roads, river bends, and other areas, as well as in practical projects like coal mine shafts. A critical issue in the design of circular retaining walls is the development of a robust and precise methodology for calculating earth pressure. Accurately computing the earth pressure exerted on circular retaining walls necessitates

© The Author(s) 2024
A. Bieliatynskyi et al. (Eds.): CSTTE 2023, LNCE 603, pp. 486–499, 2024.
https://doi.org/10.1007/978-981-97-5814-2_44

the consideration of three-dimensional spatial effects, a domain where current theoretical research remains notably lacking. The classical Rankine earth pressure theory and Coulomb earth pressure theory have been widely used because of their simple calculation. After Rankine and Coulomb, Paik and Salgado (2003), Lu (2010), Zhu (2018), Deng (2023), and numerous scholars have progressively advanced methods for calculating earth pressure. However, for circular retaining structures, a comprehensive theoretical foundation for earth pressure calculation remains elusive. Consequently, the design of such structures often relies on the conventional calculation methods applicable to plane retaining walls. For example, the plane static earth pressure is used to calculate the earth pressure of the diaphragm wall in the circular foundation pit project of the wagon tipper house of the fourth phase of Qinhuangdao Coal Port Terminal (Liu 1995), and the classical Rankine earth pressure theory is used to calculate the earth pressure of the circular diaphragm wall support of the water pump in Huangjuezhuang Power Plant (Li 1994). However, the existing model tests earth pressure on the circular retaining wall is less than the theoretical value of the plane earth pressure (Tran et al. 2014, Cho et al. 2015, Tan et al. 2018 and Tangjarusritaratorn et al. 2022). Continuing to apply the plane earth pressure theory for calculating the earth pressure on circular retaining walls can lead to conservative designs, resulting in unnecessary costs and resource wastage. Therefore, it's crucial to explore and develop a method for calculating the active earth pressure on circular retaining walls. Such research not only deepens theoretical understanding but also provides essential technical support for optimizing the design of these structures, thus holding significant theoretical and practical engineering value.

The issue of calculating earth pressure on circular retaining walls is a spatially axially symmetric problem, which has been theoretically explored by various scholars. Derezantzev (1958) using the slip-line method, proposed a theory for the limit earth pressure of granular materials under axial symmetry conditions. According to this theory, the resulting earth pressure distribution is nonlinear and tends to be lower than the values derived using the plane earth pressure calculation method for retaining walls. Xiong et al. (2019, 2020), Xiong and Wang (2020) further improved the earth pressure calculation method of circular retaining wall based on the slip-line method, and achieved some research results. Prater (1977) introduced the hoop stress coefficient to account for the influence of hoop stress. Using the limit equilibrium method, he derived a formula for calculating earth pressure that considers the ratio of the wall's radius to its height. This approach, grounded in clear mechanical principles, effectively captures the nonlinear distribution of earth pressure along the height of the wall. Kim et al. (2013) assumed that the sliding surface of the soil is Rankine sliding surface and used the lateral earth pressure coefficient, as derived by Paik and Salgado (2003), to calculate the lateral earth pressure on the circular shaft. It was pointed out that the lateral earth pressure on circular retaining walls, estimated by Rankine theory, significant overestimates the earth pressure, a conclusion supported by test results. Numerical modeling is also an approach used to clarify the mechanical behaviors of the surrounding soil and the cylindrical shaft after excavation. Prior to the current study, most numerical simulations of cylindrical shaft problems were conducted using 2D axisymmetric simulations (Meftah et al., 2018). However, recognizing the significance of three-dimensional effects in cylindrical shaft simulations, Chehadeh et al. (2019) and Meftah et al. (2022) employed a 3D finite element analysis to reveal the distribution of earth pressure on a cylindrical shaft.

Current research on the earth pressure affecting circular retaining walls remains limited and not fully developed. Addressing this gap, this paper employs the energy method as its foundational approach. It posits that soil exhibits completely rigid-plastic behavior and follows the associated flow rule at the point of failure. By applying the principle of virtual work, a new formula for calculating the active earth pressure on circular retaining walls is derived and subsequently compared with results from existing model tests.

2 Basic Principle of Energy Method

2.1 Flow Rule

The soil is considered an ideal plastic body, with yield stress at each point in plastic flow state denoted as σ_n and τ. The yield stress is defined as the stress that causes the soil to enter into a plastic flow state. The flow represents the relationship between the yield stress and the plastic strain rate in a plastic body. Mises proposed that the flow rule can be expressed by a plastic potential function f, which satisfies the following equation (GU 2005):

$$\frac{\varepsilon_\tau}{\varepsilon_n} = \frac{\partial f / \partial \tau}{\partial f / \partial \sigma_n} \tag{1}$$

where ε_n is the normal strain rate and ε_τ is the shear strain rate.

For a soil subject to the Mohr-Coulomb strength condition, the yield condition is the strength condition, and the potential function f is:

$$f = \tau - c - \sigma_n \tan \varphi \tag{2}$$

where c is the cohesion of the soil; φ is the internal friction angle of the soil; τ is the shear stress; σ_n is the normal stress on the shear plane.

When the soil reaches the yield state or the plastic flow state $\tau = \tau_f$, the plastic potential function $f = 0$, and τ_f is the shear strength. This flow rule determined by the yield condition is called the associated flow rule. Substituting the yield Eq. (2) into Eq. (1), we can get:

$$\frac{\varepsilon_\tau}{\varepsilon_n} = -\frac{1}{\tan \varphi} \tag{3}$$

Equation (3) shows that when the soil is in the state of shear sliding or plastic flow, the angle between the strain velocity vector v on the sliding surface and the sliding surface is φ.

2.2 Principle of Virtual Work

The principle of virtual work states that, for a continuous deformable body, the external virtual work done by a statically admissible stress field on a kinematically admissible displacement field equals the internal virtual work. The principle of virtual power involves replacing the displacement and strain from the principle of virtual work with velocity and strain rate. It is necessary to take into account the energy dissipation on the velocity discontinuity when constructing the dynamic allowable stress field with the velocity discontinuity. If the velocity discontinuity is regarded as a thin layer in which the velocity changes sharply and continuously, the shear strain rate will tend to infinity when the thickness of the thin layer tends to zero, which indicates that the velocity discontinuity must be a slip surface. The internal energy dissipation power on the discontinuity is:

$$\dot{W} = \int_S (\tau v_t - \sigma_n v_n) dS \tag{4}$$

where: S is the velocity discontinuity; v_t is the tangential component of the velocity discontinuity; v_n is the normal component of the velocity discontinuity.

Substituting Eq. (2) into Eq. (4), we can get:

$$\dot{W} = \int_S c v_t dS \tag{5}$$

Equation (5) is the internal energy dissipation rate on the discontinuity surface.

3 Derivation of the Active Earth Pressure Formula for Circular Retaining Wall

The plan and sectional views of the circular retaining wall are shown in Fig. 1.

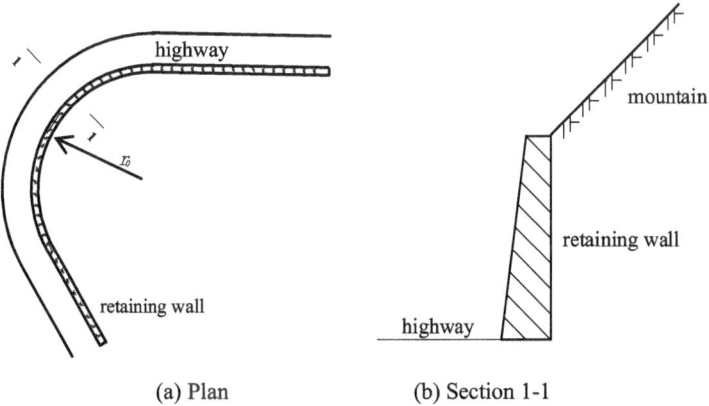

(a) Plan   (b) Section 1-1

Fig. 1. Plan and section of circular retaining wall

For an axisymmetric circular retaining wall, the Coulomb failure mechanism is used to analyze the soil behind the wall. When the wall moves away from the soil to reach the state of active limit equilibrium, the sliding surface of the soil is a curved surface passing through the heel of the wall, and its generatrix is a straight line, as shown in Fig. 2. A circular retaining wall has a radius of r_0 and a height of H.

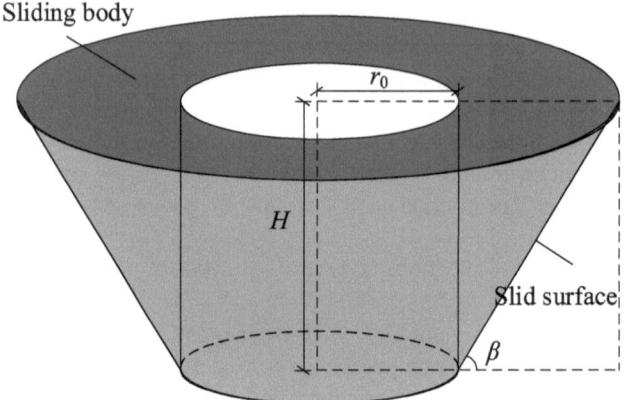

Fig. 2. Analysis model of active earth pressure on circular retaining wall

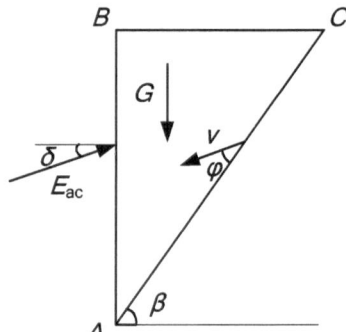

Fig. 3. Calculation diagram of energy method

Due to the axial symmetry of the model, any vertical section can be selected for analysis, as illustrated in Fig. 3. It is assumed that the wall is vertical, the wall back is rough, and the displacement mode of the wall is translation. It is assumed that the wall moves centripetally at a velocity of $v\cos(\beta - \varphi)$, and the relative velocity on the sliding surface is v. δ is the wall-soil-friction angle, φ is the internal friction angle of the soil behind the wall, β is the angle between the generatrix of the sliding surface of the soil behind the wall and the horizontal plane, that is, the inclination angle of the sliding surface of the soil, and G is the gravity of the soil behind the wall.

The gravity and volume of the sliding soil mass are:

$$G = \gamma V \tag{6}$$

$$V = \int_0^H \left[\pi \left(r_0 + \frac{z}{\tan \beta} \right)^2 - \pi r_0^2 \right] dz = \frac{\pi H^3}{3 \tan^2 \beta} + \frac{\pi r_0 H^2}{\tan \beta} \tag{7}$$

The gravity work power is:

$$\dot{W}_G = \gamma V v \sin(\beta - \phi) = \gamma v \sin(\beta - \phi) \left(\frac{\pi H^3}{3 \tan^2 \beta} + \frac{\pi r_0 H^2}{\tan \beta} \right) \tag{8}$$

The reaction of the retaining wall to the sliding soil is E_{ac}, which is equal to the active earth pressure acting on the back of the wall in magnitude and opposite in direction. The work power of the retaining wall on the sliding soil is:

$$\begin{aligned} \dot{W}_{E_{ac}} &= -2\pi r_0 E_{ac} (\cos \delta) v \cos(\beta - \phi) - 2\pi r_0 E_{ac} (\sin \delta) v \sin(\beta - \phi) \\ &= -2\pi r_0 E_{ac} v \cos(\beta - \phi - \delta) \end{aligned} \tag{9}$$

The cohesive force of cohesive soil is c, and according to the associated flow rule, the internal energy dissipation rate on the velocity discontinuity surface (slip surface AC) is:

$$\begin{aligned} \dot{W}_{AC} &= cv(\cos \phi) S_{AC} = cv(\cos \phi) \left(\frac{2\pi r_0 H}{\sin \beta} + \frac{\pi H^2}{\tan \beta \sin \beta} \right) \\ &= cv \left(\frac{2\pi r_0 H}{\tan \beta} + \frac{\pi H^2}{\tan^2 \beta} \right) \end{aligned} \tag{10}$$

As the circular retaining wall moves away from the soil, the soil behind the wall will be compressed in the circumferential direction, and the internal energy dissipation power caused by the work done by the circumferential stress is:

$$\dot{W}_\theta = \int_V \sigma_\theta \dot{\varepsilon} dV \tag{11}$$

Introducing the hoop stress coefficient λ:

$$\sigma_\theta = \lambda \sigma_1 = \lambda \gamma z \tag{12}$$

There are others:

$$dV = 2\pi r dr dz \tag{13}$$

$$\dot{\varepsilon} = \frac{v \cos(\beta - \varphi)}{r} \tag{14}$$

Substitute Eqs. (12)–(14) into Eq. (11) to obtain:

$$\dot{W}_\theta = \int_V \sigma_\theta \dot{\varepsilon} dV = \int_0^H \int_{r_0}^{r_0 + \frac{H-z}{\tan \beta}} \lambda \gamma z \frac{v \cos(\beta - \phi)}{r} 2\pi r dr dz$$

$$=2\pi\lambda\gamma v\cos(\beta-\phi)\frac{H^3}{6\tan\beta} \tag{15}$$

According to the principle of virtual work, for sliding soil, the power external by external forces is equal to the power of internal energy dissipation. This can be expressed as:

$$\dot{W}_G + \dot{W}_{E_{ac}} = \dot{W}_\theta + \dot{W}_{AC} \tag{16}$$

Substituting Eqs. (8)–(10) and (15) into Eq. (16) gives:

$$\gamma v\sin(\beta-\phi)\left(\frac{\pi H^3}{3\tan^2\beta} + \frac{\pi r_0 H^2}{\tan\beta}\right) - 2\pi r_0 E_{ac}v\cos(\beta-\phi-\delta)$$
$$= 2\pi\lambda\gamma v\cos(\beta-\phi)\frac{H^3}{6\tan\beta} + cv\left(\frac{2\pi r_0 H}{\tan\beta} + \frac{\pi H^2}{\tan^2\beta}\right) \tag{17}$$

Eliminate the velocity v and sort it out to get:

$$E_{ac} = \frac{\sin(\beta-\phi)}{\cos(\beta-\phi-\delta)}\left(\frac{\gamma H^3}{6r_0\tan^2\beta} + \frac{\gamma H^2}{2\tan\beta}\right)$$
$$-\frac{\cos(\beta-\phi)}{\cos(\beta-\phi-\delta)}\frac{\lambda\gamma H^3}{6r_0\tan\beta} - \frac{cH(2r_0\tan\beta + H)}{2r_0\tan^2\beta\cos(\beta-\phi-\delta)} \tag{18}$$

In Eq. (18), the inclination angle β of the slip surface is arbitrarily assumed, and only the slip surface with the maximum value of E_{ac} is the most dangerous slip surface. E_{ac} is a function of β. According to the condition of $dE_{ac}/d\beta = 0$, the angle β at the maximum value of E_{ac} is obtained, and then the active earth pressure of the circular retaining wall can be obtained by substituting Eq. (18).

Equation (18) is the expression of the resultant force of the active earth pressure of the circular retaining wall. For the determined circular retaining wall, the ratio of the radius to the wall height r_0/H is constant, and the earth pressure intensity at the depth H can be expressed as:

$$\sigma_h = \frac{dE_{ac}}{dh} = \frac{\sin(\beta-\phi)}{\cos(\beta-\phi-\delta)}\left(\frac{H}{6r_0\tan^2\beta} + \frac{1}{2\tan\beta}\right)\gamma h$$
$$-\frac{\cos(\beta-\phi)}{\cos(\beta-\phi-\delta)}\frac{\lambda H}{6r_0\tan\beta}\gamma h - \frac{c}{2\tan^2\beta\cos(\beta-\phi-\delta)}\left(2\tan\beta + \frac{H}{r_0}\right) \tag{19}$$

It can be seen from Eq. (19) that the earth pressure intensity is distributed in a triangle. Compared with the Coulomb earth pressure for a plane retaining wall, the earth pressure for a circular retaining wall is also related to the radius-height ratio r_0/H and the hoop stress coefficient λ.

4   Model Test Verification

4.1   Model Test 1

Tran and Meguid (2014) conducted an indoor model test out of a shaft. The model parameters are: $r_0 = 0.075$ m, $H = 1$ m, $\gamma = 14.7$ kN/m^3, $c = 0$, $\varphi = 41°$, $\delta = 0$. The results of this method are compared with those from Tran and Meguid (2014).

Figure 4 displays the results, which are compared with those derived from Coulomb's earth pressure theory. T5, T6, and T7 in the figure represent the test results from Tran and Meguid (2014).

The earth pressure values obtained by the method in this study closely match the upper portion of the retaining wall's measured data from Tran and Meguid (2014). However, a deviation is observed in the lower third of the wall height, where the calculated values differ from the measurements. It due to due to the reduced earth pressure at the base of the wall caused by friction, a factor not reflected in the linear distribution model of this study. Notably, the earth pressure strengths calculated using the classical Coulomb earth pressure theory are substantially greater than those in the findings of Tran and Meguid (2014).

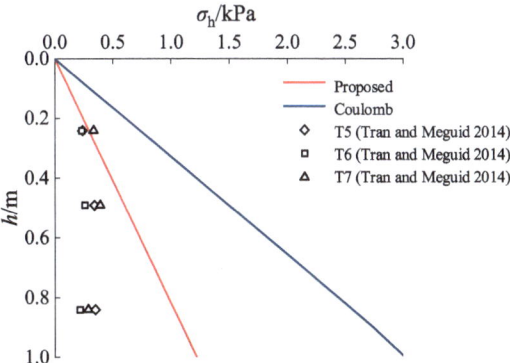

Fig. 4. Comparison of the proposed method and the experimental data of Tran and Meguid (2014)

4.2 Model Test 2

Cho et al. (2015) conducted a centrifuge model test to investigate the lateral earth pressure on a circular shaft, applying a centrifugal acceleration coefficient of $N = 75$ g. The vertical shaft of the model, made of aluminum alloy, featured a hollow circular section. It had an embedment height of 0.2 m and an external diameter of 0.08 m. The model is equivalent to a reinforced concrete shaft with a height of 15 m and an outer diameter of 6 m. Model test parameters are: $\gamma = 12.66$ kN/m^3, $c = 0$ kPa, $\varphi = 36.95°$, Cho et al. (2015) does not give the value of wall-soil friction angle, which is taken in this paper $\delta = \varphi/2$. The earth pressure values obtained from the model test of Cho et al. (2015) are shown in Fig. 5.

Figure 5 compares the calculation results in this paper with the model test results in Cho et al. (2015) and Coulomb's earth pressure theory. The results indicate that the calculated values more closely align with the model test of Cho et al. (2015). The earth pressure values derived from Coulomb's earth pressure theory are higher than those observed in Cho et al. (2015). The enhanced precision of the calculation results of this method, closely aligning with the test earth pressures, can be attributed to the methodological consideration of both the diameter-to-height ratio of the circular retaining wall and the effects of hoop stress.

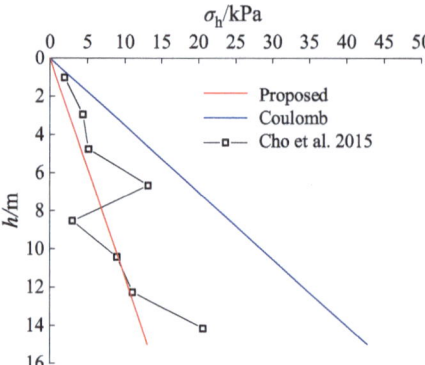

Fig. 5. Comparison of the proposed method and the experimental data of Cho et al. (2015)

5   Analysis of Cohesionless Soil Law

5.1   Dip Angle of Slip Surface

For a circular retaining wall with cohesionless backfill, the variation of the slip plane inclination angle β with the ratio of diameter to height r_0/H is shown in Fig. 6. The dip angle of the slip plane calculated by the classical Coulomb earth pressure theory is also shown in the figure.

Figure 6 indicated that the inclination angle of the sliding surface of the soil in the active limit state of the circular retaining wall is greater than that of the Coulomb solution of the plane retaining wall, and the inclination angle of the sliding surface of the soil in the active limit state of the circular retaining wall decreases with the increase of the diameter-height ratio r_0/H, and gradually tends to the Coulomb solution of the plane retaining wall. The greater the internal friction angle, the larger the inclination angle of the sliding plane becomes. The higher the hoop stress coefficient, the larger the inclination angle of the slip surface becomes.

5.2   Active Earth Pressure

For cohesionless soil, when the diameter-height ratio r_0/H of a circular retaining wall tends to infinity, Eq. (18) can be transformed in:

$$E_a = \frac{1}{2}\gamma H^2 \frac{\sin(\beta - \varphi)}{\tan \beta \cos(\beta - \varphi - \delta)} \tag{20}$$

Equation (20) is the active earth pressure expression of plane retaining wall calculated according to the classical Coulomb earth pressure theory. It can be seen that when the ratio of diameter to height r_0/H tends to infinity, the calculation method is consistent with Coulomb's theory.

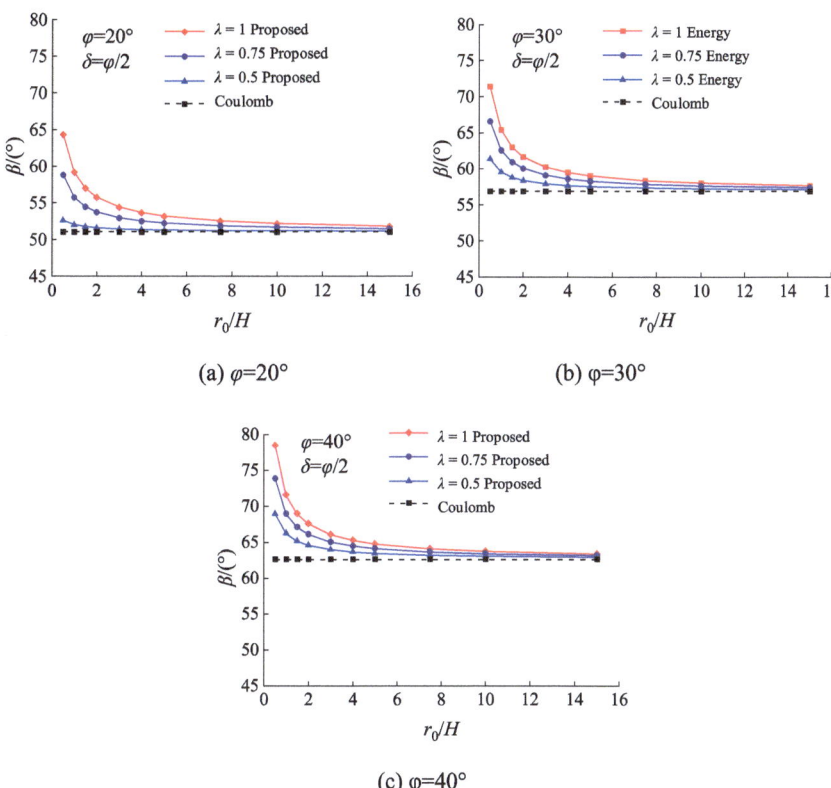

Fig. 6. Variation curve of slip surface inclination with the ratio of diameter to height

For a circular retaining wall, the variation of the active earth pressure with the ratio of diameter to height r_0/H obtained by the energy method is shown in Fig. 7. The Coulomb earth pressure of plane retaining wall is also given in Fig. 7. The ordinate $2E_{ac}/\gamma H^2$ in Fig. 7 represents the magnitude of the normalized resultant active earth pressure.

Figure 7 shows that the active earth pressure of the circular retaining wall is less than the Coulomb solution of the plane retaining wall, and the active earth pressure of the circular retaining wall increases with the increase of the diameter-height ratio r_0/H, and gradually tends to the Coulomb solution of the plane retaining wall. The greater the value of the circumferential stress coefficient, the lower the active earth pressure of the circular retaining wall.

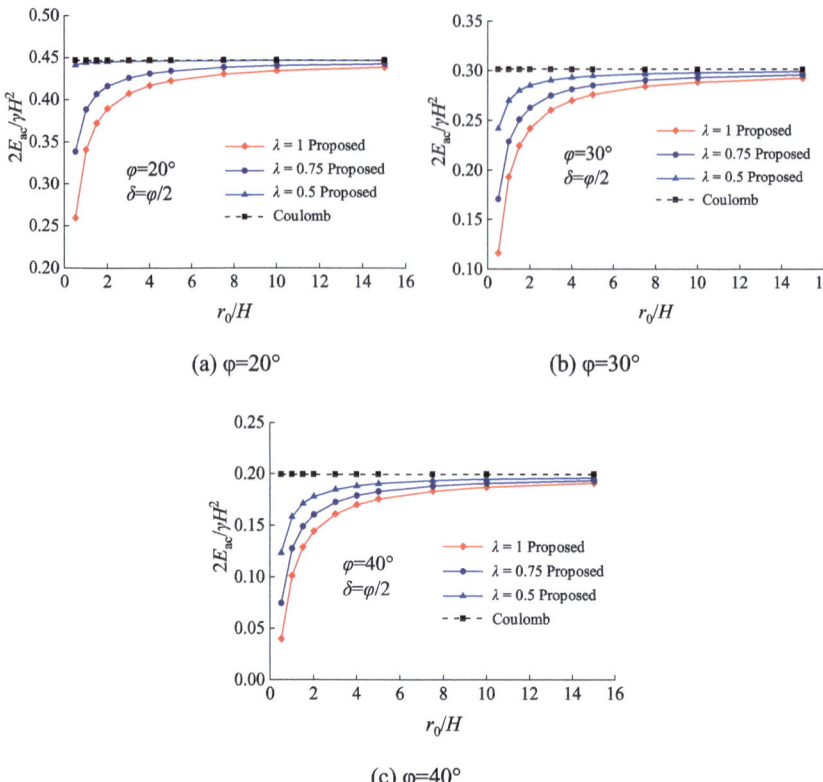

Fig. 7. Variation of earth pressure with the ratio of diameter to height

6   Analysis of Clayey Soil Law

The calculation and discussion are based on the cohesionless soil. Equation (18) is derived from the energy method, it can also be used to calculate the active earth pressure of the circular retaining wall with cohesive soil behind the wall.

To investigate the impact of cohesion on the sliding surface and active earth pressure of circular retaining walls, the following parameters have been chosen as reference values for the calculations. Theses parameters include: soil unit weight $\gamma = 18 \text{ N/m}^3$, circular retaining wall radius $r_0 = 10$ m, wall height $H = 10$ m, soil internal friction angle $\varphi = 30°$, wall-soil friction angle $\delta = 1/2\varphi$, and cohesion $c = 10$ kPa.

6.1   Dip Angle of Slip Surface

The relationship between the inclination angle of the slip surface and the cohesion is shown in Fig. 8. It can be seen from Fig. 8 that the inclination angle of the sliding surface of the circular retaining wall increases with the increase of the cohesion. A higher internal friction angle corresponds to a greater inclination angle of the sliding plane.

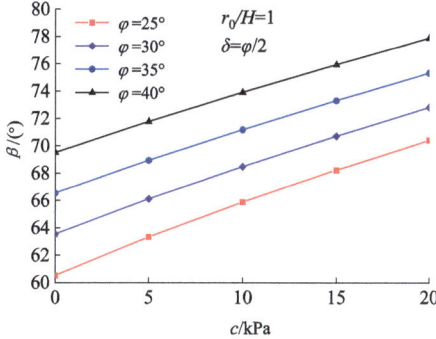

Fig. 8. Variation of dip angle of slip plane with cohesion

6.2 Active Earth Pressure

The relationship between the active earth pressure and the cohesion of the circular retaining wall is shown in Fig. 9. It can be seen from Fig. 9 that the active earth pressure of circular retaining wall decreases with the increase of cohesion. An increase in the internal friction angle results in a reduction of the active earth pressure.

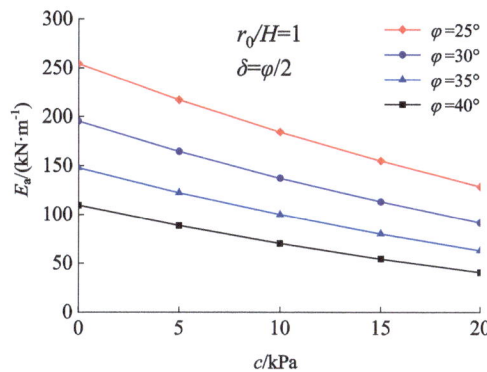

Fig. 9. Variation of active earth pressure with cohesion

7 Conclusion

(1) Utilizing the energy method, this study establishes a calculation method for the active earth pressure on circular retaining walls. It takes into account the internal energy dissipation resulting from hoop stress. The efficacy and applicability of this proposed method have been corroborated by results obtained from model tests.

(2) The active earth pressure of the circular retaining wall is found to be less than that of plane retaining walls. As the ratio of diameter to height (r_0/H) increases, the active earth pressure on circular walls gradually approaches the Coulomb solution

for plane walls. Additionally, a higher hoop stress coefficient results in lower active earth pressure for circular walls.

(3) The proposed method in this paper is applicable for calculating the active earth pressure on circular retaining walls with both cohesionless and cohesive soils. For cohesionless soils, when the ratio of diameter to height r_0/H tends to infinity, the calculation method in this paper is consistent with Coulomb's theory.

(4) The earth pressure calculation method obtained in this paper is only a theoretical solution, and there is a certain deviation from the actual value. In future research, we need to consider more influencing factors, such as the uniformity of the soil, the load on the soil surface, the underground water conditions, and the time effect.

References

Paik, K.H., Salgado, R.: Estimation of active earth pressure against rigid retaining walls considering arching effects. Geotechnique **53**(7), 643–653 (2003)

Lu, K.L., Zhu, D.Y., Yang, Y.: Calculation method of active earth pressure under non-limit state considering soil arching effects. China J. Highway Transp. **23**(1), 19–25 (2010)

Zhu, Y.L., Yu, J., Zhou, J.F., et al.: Calculation of earth pressure on rigid retaining walls with considerations to the seismic load and soil stress-deflection. J. Vibroengineering **20**(3), 1488–1500 (2018)

Dang, F., Wang, X., Cao, X., et al.: Calculation method of earth pressure considering wall displacement and axial stress variations. Appl. Sci. **13**(16), 9352 (2023)

Liu, S.X.: Design for round diaphragm wall of car dumper shed of the fourth stage project of coal terminal, Qinhuangdao Port. Port Eng. Technol. **4**, 28–35 (1995)

Li, Y.L.: Circular underground continuous wall support for water pumping station of Huangjuezhuang power plant. Water Power **3**, 15–17 (1994)

Tran, V.D.H., Meguid, M.A., Chouinard, L.E.: Discrete element and experimental investigations of the earth pressure distribution on cylindrical shafts. Int. J. Geomech. **14**(1), 80–91 (2014)

Cho, J., Lim, H., Jeong, S., et al.: Analysis of lateral earth pressure on a vertical circular shaft considering the 3D arching effect. Tunn. Undergr. Space Technol. **48**(4), 11–19 (2015)

Tan, Y., Lu, Y., Xu, C., et al.: Investigation on performance of a large circular pit-in-pit excavation in clay-gravel-cobble mixed strata. Tunn. Undergr. Space Technol. **79**(9), 356–374 (2018)

Tangjarusritaratorn, T., Miyazaki, Y., Sawamura, Y., et al.: Numerical investigation on arching effect surrounding deep cylindrical shaft during excavation process. Underground Space **7**(5), 944–965 (2022)

Derezantzev, V.G.: Earth pressure on the cylindrical retaining walls. In: Brussels Conference on Earth Pressure Problems, II, pp. 21–27 (1958)

Xiong, G.J., Wang, J.H., Chen, J.J.: Theory and practical calculation method for axisymmetric active earth pressure based on the characteristics method considering the compatibility condition. Appl. Math. Model. **68**(4), 563–582 (2019)

Xiong, G.J., Chen, J.J., Li, M.G., et al.: General axisymmetric active earth pressure obtained by the characteristics method based on circumferential geometric condition. Sci. China Technol. Sci. **63**(2), 341–356 (2020)

Guojun, X., Jianhua, W.: A rigorous characteristic line theory for axisymmetric problems and its application in circular excavations. Acta Geotech. **15**(2), 439–453 (2018). https://doi.org/10.1007/s11440-018-0697-7

Prater, E.G.: An examination of some theories of earth pressure on shaft linings. Can. Geotech. J. **14**(1), 91–106 (1977)

Kim, K.Y., Lee, D.S., Cho, J., et al.: The effect of arching pressure on a vertical circular shaft. Tunn. Undergr. Space Technol. **37**(8), 10–21 (2013)

Meftah, A., Benmebarek, N., Benmebarek, S.: Numerical study of the active earth pressure distribution on cylindrical shafts using 2D finite difference code. J. Appl. Eng. Sci. Technol. **4**(2), 123–128 (2018)

Chehadeh, A., Turan, A., Abed, F., et al.: Lateral earth pressures acting on circular shafts considering soil-structure interaction. Int. J. Geotech. Eng. **13**(2), 139–151 (2019)

Meftah, A., Benmebarek, N., Benmebarek, S.: Active earth pressure acting on circular shafts using numerical approach. Civil Eng. J. **8**(4), 734–750 (2022)

Gu, W.C.: Manual of Earth Pressure Calculation for Retaining Walls, pp. 223–225. China Building Materials Press, Beijing (2005)

Numerical Analysis of Large-Diameter Shield Tunneling Disturbance Considering Stratum Strength Anisotropy

Hao Jiang[1], Wu Liu[2(✉)], Jin Cheng[2], Huayan Yao[2], Renjie Li[1], and Jinhang Shang[2]

[1] Shandong Electric Power Engineering Consulting Institute Co., Jinan 250013, China
{jianghao,lirenjie}@sdepci.com
[2] College of Civil Engineering, Hefei University of Technology, Hefei 230009, China
{liuwu168,yaohuayan}@hfut.edu.cn

Abstract. The anisotropic characteristics of natural strata could have a significant effect on the tunneling disturbance of shield tunnels. A shield tunneling disturbance simulation method considering the effect of stratum strength anisotropy is proposed in this study. The proposed method adopts the microstructure tensor theory to characterize the anisotropy effects of the stratum material cohesion c and the internal friction angle φ, and finely simulates the shield tunneling processes, including the shield shell advancement, the lining installation, and the shield tail grouting. The shield tunneling refined simulation method is validated by simulating the tunneling process in a certain section of a super-large-diameter shield tunnel in Wuhu City, Anhui Province, China, with good agreement between the simulated surface settlements on the axial and transverse profiles of the excavated tunnel after tunneling to different lining-ring numbers and the monitoring data. On this basis, according to the anisotropic compression strength property of the layered stratum in other sections, the effect of the stratum shear strength anisotropy on the shield tunneling disturbance is analyzed. The shield tunneling-induced surface settlement under the case considering stratum shear strength parameters anisotropy is obviously greater than that without accounting for the anisotropy, demonstrating the importance of considering strata anisotropy caused by geological structures during shield tunneling disturbance modeling.

Keyword: shield tunnel · surface settlement · refined simulation · stratum anisotropy

1 Introduction

With the rapid growth of the economy in recent years, the transportation channels have expand rapidly, increasing the number and size of transportation tunnel projects. The shield method, as an important method of tunnel construction, is widely used in engineering [1]. Geological and engineering factors play a crucial role in determining the extent of surrounding rock disturbance caused by shield tunneling. Geological factors include stratum stress field, groundwater seepage field, stratum structure, etc., while

A. Bieliatynskyi et al. (Eds.): CSTTE 2023, LNCE 603, pp. 500–513, 2024.
https://doi.org/10.1007/978-981-97-5814-2_45

engineering factors mainly include shield machine type, tunnel shape and size, tunnel burial depth, etc. In addition, the influence of tunneling parameters, such as palm face pressure and grouting pressure, also has a great impact on tunnel settlement. The evaluation of tunneling-induced disturbance in shield tunnel projects is of paramount importance, extensive research has been conducted. For instance, Qin [2] utilized Plaxis software to simulate the tunneling process of a shield machine under three distinct stratigraphic conditions, and studied the influence law of palm face pressure, shrinkage, and grouting pressure on the ground surface settlement and deformation. Ma et al. [3] found that increasing the excavation chamber pressure and synchronous grouting pressure for diversion tunnel engineering can reduce the maximum settlement. However, the present research often focuses on studying the influences of engineering factors, tunneling parameters and seepage process on the tunneling disturbance, research that considers the influence of strata anisotropy has not been paid enough attention.

Affected by the complex development of geological structures, natural strata often exhibit significant anisotropic characteristics. Many methods have been proposed to consider the influence of material anisotropy on its strength. Shi et al. [4] proposed a novel method for predicting the anisotropic strength of layered rocks based on the principle that weak surfaces within the rocks cause a reduction in the compressive strength. Based on the theory of a single weak surface, Bao et al. [5] derived a functional relationship between the compressive strength of slate and the inclination angle of the discontinuity. Pietruszczak et al. [6] proposed to incorporate the microstructure tensor and the relevant mixed invariants to characterize the anisotropic strength of material. Characterizing the influence of material anisotropy on strength based on microstructure tensor approach has a clear physical significance. Inspired by the work done by Pietruszczak et al. [6], the microstructure tensor theory has widely been used in material anisotropy characterization [7, 8].

In this study, a refined shield tunneling disturbance simulation method based on the microstructure tensor theory is proposed to reveal the influence of stratum strength anisotropy induced by the complex development of natural geological structures on the shield tunneling disturbance. The proposed method finely considered the process of shield shell advancement, lining installation, and shield tail grouting. The microstructure tensor theory is used to characterize the material anisotropy effects on its shear strength parameters. Based on a large shield tunnel project located in Wuhu City, Anhui Province, China, the proposed method is used to simulate the shield tunneling disturbance and verified with monitoring data, and the influence of strata strength anisotropy on the tunneling disturbance is analyzed.

2 Theoretical and Simulation Methods

In order to better simulate the shield tunneling disturbance with the consideration of stratum strength anisotropy induced by geological structures. The methods for refined shield tunneling simulation and stratum strength anisotropy characterizing are elaborated in this section.

2.1 Shield Tunneling Refined Simulation Method

The shield tunneling disturbance is simulated by a refined simulation method based on the structural dimensions and operation mode of the shield machine. The method accurately considers the influences of shield shell advancement, lining installation, and shield tail grouting, through continuously altering the attributes of the shield shell and lining elements and adjusting the positions of applied palm face pressure and grouting pressure according to the shield tunneling process in FLAC3D software.

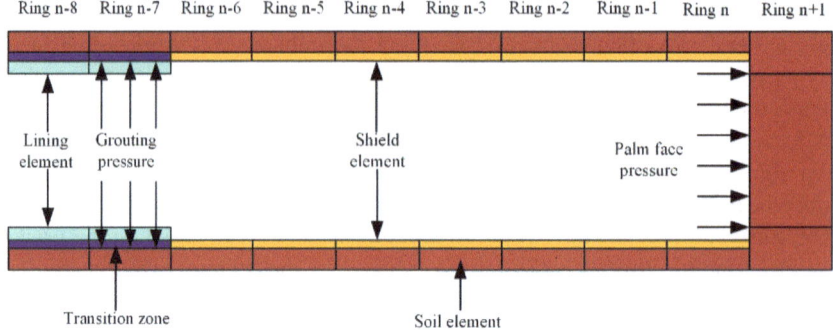

Fig. 1. Schematic diagram of the refined simulation of shield tunneling

Based on the shield tunneling process of the shield tunnel project selected in this study, the shield shell is modeled with seven rings of lining segment, as shown in Fig. 1, in which the shield is assumed to tunnel to the nth ring of lining segment. The transition zone is used to simulate the shield tail grouting, considering the size difference between the diameter of the shield and the outer ring diameter of the lining. The length of the grouting pressure acting in the transition zone at the tail of the shield is one ring of lining segment. After finely building the shield tunneling calculation model based on the tunneling size of each ring, the specific simulation of shield tunneling is divided into two stages as follows:

(1) In the first stage, the shield shell gradually enters the stratum, there is no shield tail lining installation process. Whenever the shield tunneling forwards a ring distance, excavate stratum elements inside the tunnel, activate the shield shell elements and apply palm face pressure on the excavation face until the shield shell fully enters the stratum (the shield tunneling forwards seven-ring distance).

(2) In the second stage, the whole shield shell is in the stratum, and the processes of the lining installation and shield tail grouting should be considered. After the completion of the previous stage of tunneling simulation, when the shield tunneling forwards to the 8th ring, excavate stratum elements inside the tunnel, activate the shield shell elements in the 8th ring, and apply palm face pressure on the excavation face. Meanwhile considering the lining installation and shield tail grouting process, change the shield shell elements in the 1st ring as the transition zone elements, activate the lining elements in the 1st ring, and apply grouting pressure in these elements. Afterwards, the subsequent simulation of shield tunneling is conducted

according to the above simulation process, by continuously-regularly activating the shield elements, applying palm face pressure on the excavation face, adjusting the shield shell element in the tail to the transition zone elements and applying grouting pressure, until the end of the simulation.

2.2 Theory of Stratum Strength Anisotropy Analysis

Affected by the dominant-orientation development of geological structures, the stratum strength may exhibit significant anisotropic characteristics [7]. In order to consider the influence of stratum strength anisotropy during shield tunneling modeling, the microstructure tensor theory proposed by Pietruszczak et al. [6] is adopted to characterize the anisotropy of the shear strength parameters cohesion c and internal friction angle φ, with the following expressions:

$$c = c_0\left[1 + A_{ij}l_il_j + b_1\left(A_{ij}l_il_j\right)^2\right] \tag{1}$$

$$\varphi = \varphi_0\left[1 + B_{ij}l_il_j + b_2\left(B_{ij}l_il_j\right)^2\right] \tag{2}$$

where A_{ij} and B_{ij} are traceless symmetric tensors describing the bias in the spatial distribution of cohesion and internal friction angle [4]; c_0 φ_0 b_1 and b_2 are material parameters; l_i are the components of the generalized loading vector associated with the principal triad of the microstructure tensor [8].

For the layered stratum, cut by only one set of parallel structural planes, one obtains, $A_{ij}l_il_j = A_1(1 - 3l_3^2)$ $B_{ij}l_il_j = B_1(1 - 3l_3^2)$ in which $l_3^2 = \frac{tr(\mathbf{m}\otimes\mathbf{m}\mathbf{\Sigma}^2)}{tr(\mathbf{\Sigma}^2)}$, $\mathbf{m}$ denotes the unit normal vector of the parallel structural planes, represents the stress tensor. Then Eqs. (1)– (2) 1can be simplified as:

$$c = c_0\left[1 + A_1\left(1 - 3l^2\right) + b_1\left(A_1\left(1 - 3l_3^2\right)\right)^2\right] \tag{3}$$

$$\varphi = \varphi_0\left[1 + B_1\left(1 - 3l_3^2\right) + b_2\left(B_1\left(1 - 3l_3^2\right)\right)^2\right] \tag{4}$$

3 Modeling Analysis of Large-Diameter Shield Tunneling Disturbance

A large-diameter shield tunnel under construction, located at the "Big Bend" of the Wanjiang section of the Yangtze River in Wuhu City, Anhui Province, China is selected as the research object in this paper. The length of the shield tunnel is about 3.9 km, starting from the Jiangbei work shaft with two mud-water balanced shields, tunneling southward crossing the Yangtze River, and then reaching the Jiangnan work shaft. The right-line shield tunneling first, with the diameter of the shield shell is 15.07 m. After tunnel excavation, a reinforced concrete lining with a thickness of 0.6 m is utilized, with a length of 2 m for each ring of the lining segment. The outer diameter of the lined tunnel is 14.5 m, while the inner diameter is 13.3 m. Taking the tunneling section of the right-line shield tunnel from 60 m to 122 m in front of the Jiangbei working shaft for study, a 3D finite element calculation model is established, as shown in Fig. 2. The actual geological strata and tunnel structure are finely considered by the calculation model, with a total of 208530 elements and 218816 nodes. The Mohr-Coulomb model in the FLAC3D software is utilized to characterize the mechanical behavior of strata.

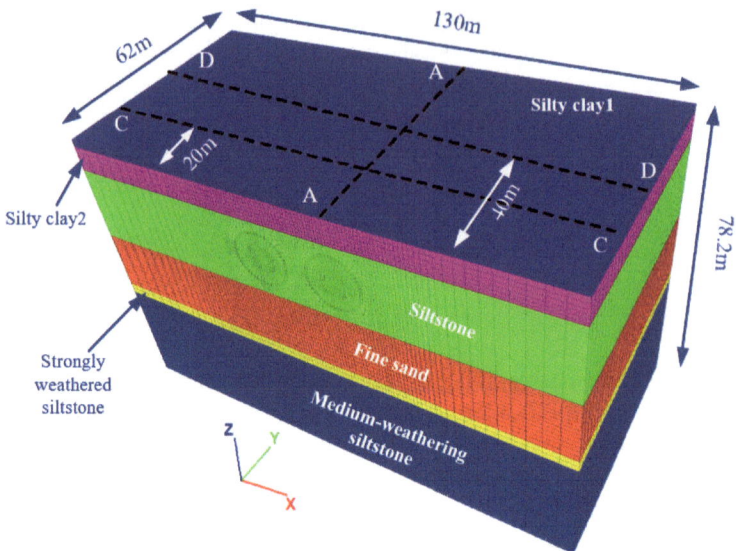

Fig. 2. Schematic diagram of the 3D finite element calculation model of the tunnel

According to the engineering geological investigation report, the material parameters for shield tunneling simulation are shown in Table 1. In order to truly reflect the weight of all components inside the shield, the equivalent weight of the shield shell elements should be converted based on its thickness [9]. The calculated parameters for the shield shell are Elastic modulus of 200 GPa, Poisson's ratio of 0.3, and specific weight of 247 kN·m^{-3}. The palm surface pressure is taken as 0.25 MPa, and the grouting pressure is 0.2 MPa. The assumptions for calculation are: (1) the strata are homogeneous and continuous; (2) the possible temporary loads on the ground are neglected; (3) the stiffness reduction effect of the connection between lining segments is not considered; (4) the grouting pressure is evenly distributed along the surface of the pipe segment. The initial and boundary conditions used are: Zero-displacement normal constraints are applied on the bottom and lateral boundaries of the calculation model; The initial stress field is determined by conducting strata self-weight stress calculation.

Table 1. Material parameters for shield tunneling simulation

Material	Elastic modulus (MPa)	Poisson's ratio	Specific weight (kN·m^{-3})	Cohesion (kPa)	Internal friction angle (°)
Silty clay	9.6	0.20	17.3	40	25
Clay	10.74	0.34	18.4	20	14.63
Silty sand	6.33	0.32	17.8	15	11.9
Fine sand	51	0.26	19.4	12	20
Strongly weathered siltstone	100	0.26	20.5	200	34.73
Moderately weathered siltstone	2000	0.24	23	300	28
Reinforced concrete lining	30000	0.20	27.0	/	/
Transition zone	1200	0.25	23.0	/	/

4 Results Analysis for the Shield Tunneling Disturbance

To verify the effectiveness of the refined shield tunneling simulation method, the simulation analysis of the shield tunneling disturbance is first conducted and compared with monitoring data. Figure 3 presents the vertical displacement distribution on the A-A profile (location shown in Fig. 2) induced by shield tunneling forward to different lining rings. When tunneling to the 7th ring, the whole shield shell just enters into the stratum, the induced longitudinal settlement range is about 38 m, and the maximum surface settlement is 7.5 mm located right above the excavation surface. As the shield tunnels forward, the settlement at the top of the tunnel increases. When tunneling to the 25th ring, the maximum surface settlement increases to 14.4 mm, which is about 92% higher than the result obtained when tunneling to the 7th ring. After shield tunneling, there are locally large settlement areas at the top and excavated surface of the tunnel, while a significant uplift occurred at the bottom, which is mainly caused by the unloading effect of tunnel excavation.

Figure 4 shows the vertical displacement distribution on the C-C profile (location shown in Fig. 2) induced by shield tunneling forward to different lining ring numbers. After shield tunneling, the ground surface settlement on the transverse profile is symmetrically distributed along the tunnel's central axis, and the range of the settlement area is about 3 times the diameter of the tunnel. The induced settlement above the tunnel on the C-C profile increases with the forward tunneling of the shield. When tunneling to the 7th ring, the maximum settlement on the surface is 6.7 mm, and obviously increases to 12.2 mm after tunneling to the 25th ring, with an increment of 82%.

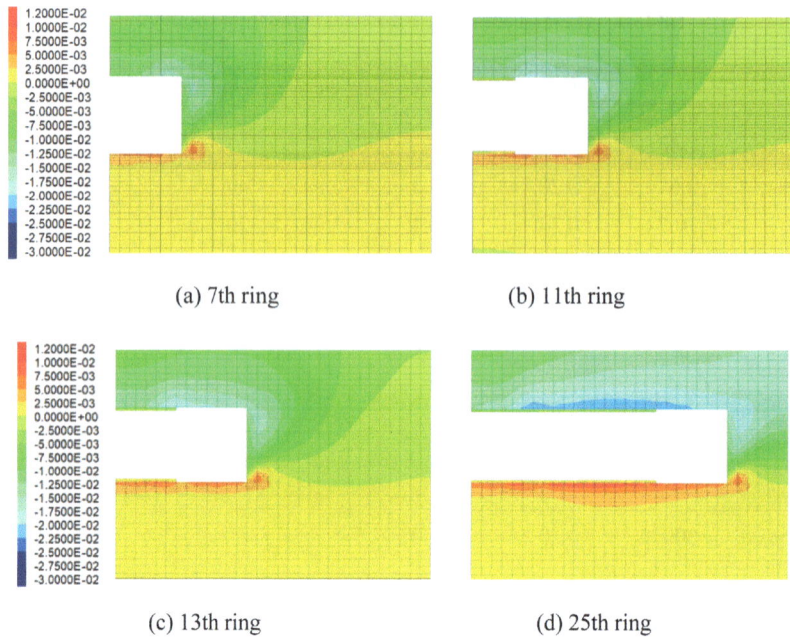

(a) 7th ring   (b) 11th ring

(c) 13th ring   (d) 25th ring

Fig. 3. Vertical displacement distribution on A-A profile after tunneling to different ring numbers (m)

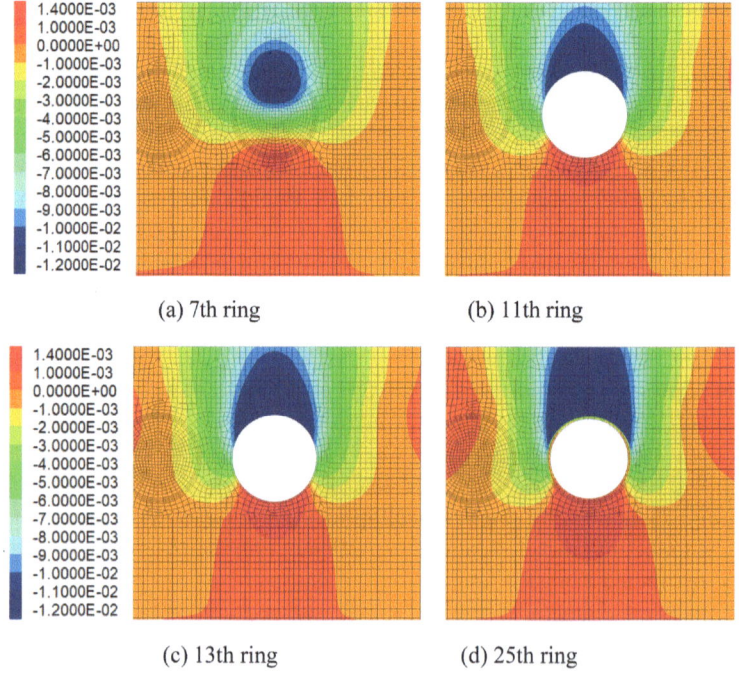

(a) 7th ring   (b) 11th ring

(c) 13th ring   (d) 25th ring

Fig. 4. Vertical displacement distribution on C-C profile after tunneling to different ring numbers (m)

The comparisons between the simulated and measured settlement values on the A-A and C-C profiles after tunneling to the 7th ring, 11th ring, 13th ring and 25th ring are respectively shown in Figs. 5 and 6. With the forward tunneling of the shield, the ground surface settlement gradually increases. The surface settlement in front of the excavation surface decreases with the increasing distance from the excavation surface. The simulated surface settlements after tunneling to different lining ring numbers agree well with the monitoring data on the A-A and C-C profiles, demonstrating the effectiveness and reliability of the refined shield tunneling simulation method considering the process of shield shell advancement, lining installation, and shield tail grouting.

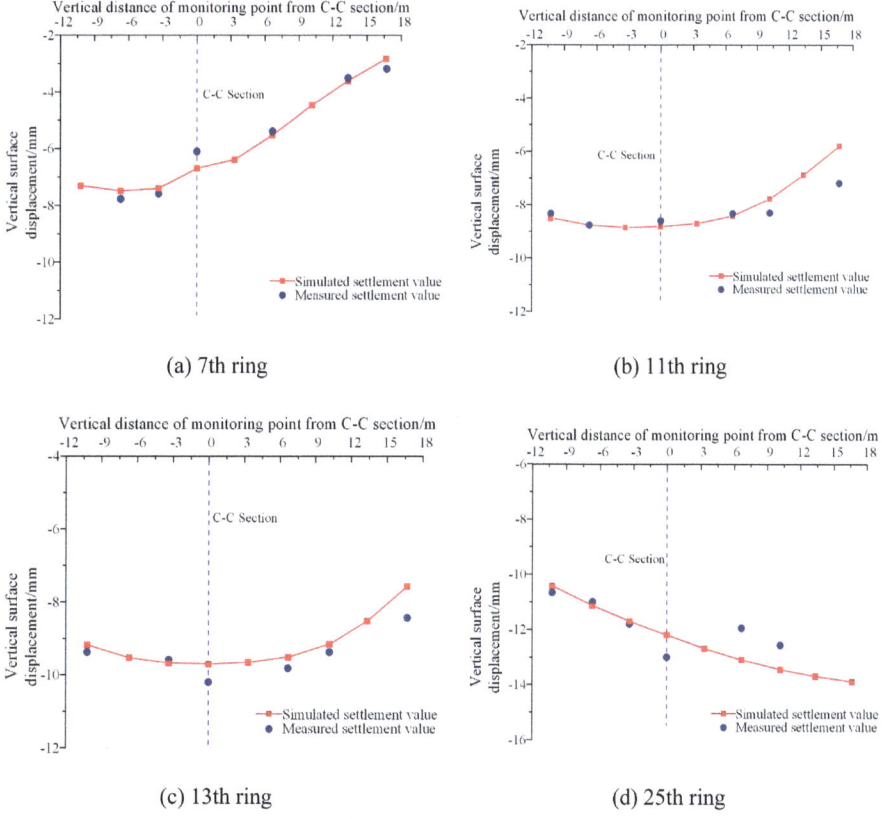

Fig. 5. Comparison of simulated and monitored surface settlements on A-A profile after tunneling to different ring numbers

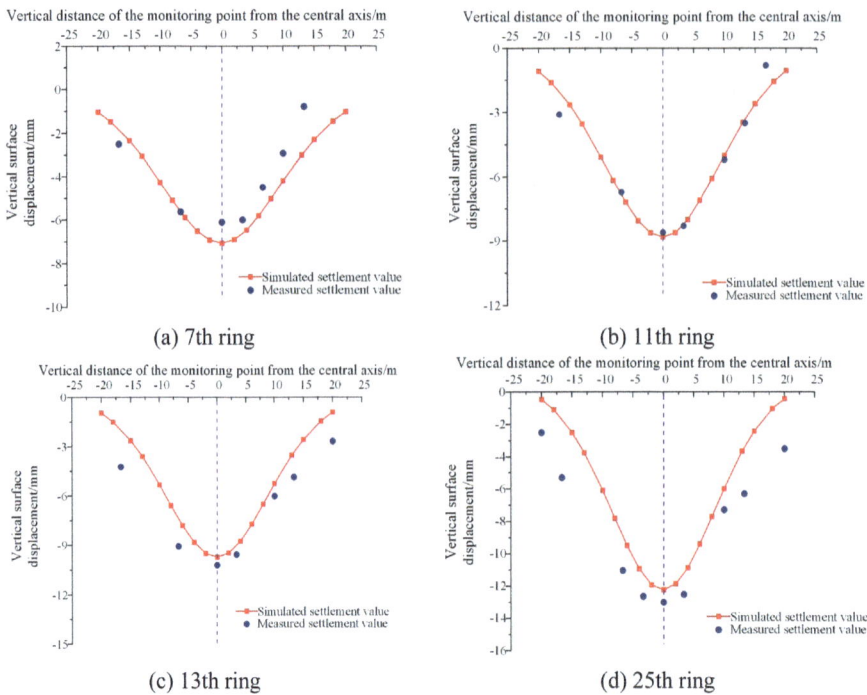

Fig. 6. Comparison of simulated and monitored surface settlements on C-C profile after tunneling to different ring numbers

5   Influence of Stratum Strength Anisotropy on the Tunneling Disturbance

To analyze the influence of stratum strength anisotropy on the shield tunneling disturbance, the experimental data of anisotropy compression strength on layered stratum in other sections of the large-diameter shield tunnel projects is used to characterize the anisotropy of the stratum where the shield located. By best fitting the triaxial compression strengths under the confining pressure of 0.5 MPa, 1.5 MPa and 3.0 MPa with different dip angles of the layered structures, the anisotropic expressions for cohesion and internal friction angle can be obtained as.

$$c = 5.33\left[1 + 0.6\left(1 - 3l_2^2\right) + 1.81\left(0.6\left(1 - 3l_2^2\right)\right)^2\right] \tag{5}$$

$$\varphi = 15.65\left[1 + 0.172\left(1 - 3l_3^2\right) + 3.7\left(0.172\left(1 - 3l_3^2\right)\right)^2\right] \tag{6}$$

Based on Eqs. (5)–(6), the comparison of the predicted and experimented strengths for layered stratum with different dip angles of layered structures are shown in Fig. 7. One can find from Fig. 7 that the strength of layered stratum is closely related to the dip angle of the layered structures, and the predicted anisotropic strengths for layered stratum based on the microstructure tensor theory agree well with the experimental results. By applying the expressions for the anisotropy of cohesion and internal friction angle in the refined shield tunneling simulation, the influence of stratum strength anisotropic can be studied.

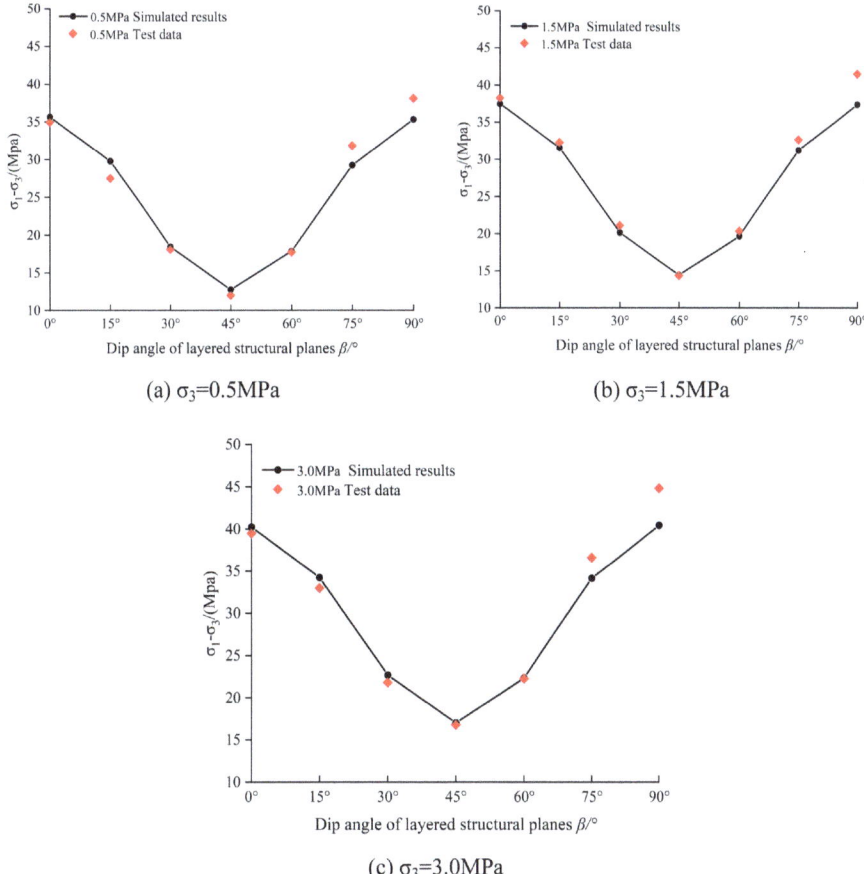

(a) σ_3=0.5MPa   (b) σ_3=1.5MPa

(c) σ_3=3.0MPa

Fig. 7. Predicted and experimented strengths for layered stratum with different dip angles of layered structures

Previous studies have shown that the anisotropic strength characteristic of layered strata plays a significant impact on the excavation disturbance of underground engineering [10]. In this section, the layered stratum with anisotropic strength characterized by Eqs. (5)–(6) is used to simulate the stratum where the tunnel located in the calculation model (shown in Fig. 2), and the dip angle of layered structural planes is set to be 45°. The simulation cases for studying the influences of shear strength parameters anisotropy on the shield tunneling disturbance are described in Table 2, in which case 1 and case 2 respectively only consider the cohesion and internal friction angle anisotropy, case 3 considers the anisotropy of cohesion and internal friction angle, case 4 does not consider the anisotropy influence.

Table 2. Calculation parameters for different considerations of stratum anisotropy

Cases	Elastic modulus (MPa)	Poisson's ratio	Specific weight (kN·m⁻³)	Cohesion (kPa)	Internal friction angle (°)
Case 1	51	0.26	19.4	Equation (5)	20
Case 2				12	Equation (5)
Case 3				Equation (6)	Equation (6)
Case 4				12	20

After shield forward tunneling to the 25th ring, the distributions of vertical displacement on profiles of A-A and C-C under different considerations of shear strength parameters anisotropy are shown in Figs. 8–9. The settlement above the tunnel increases obviously with the consideration of cohesion and internal friction angle anisotropy, and the surface settlement caused by the anisotropy of internal friction angle is obviously bigger than that of cohesion anisotropy.

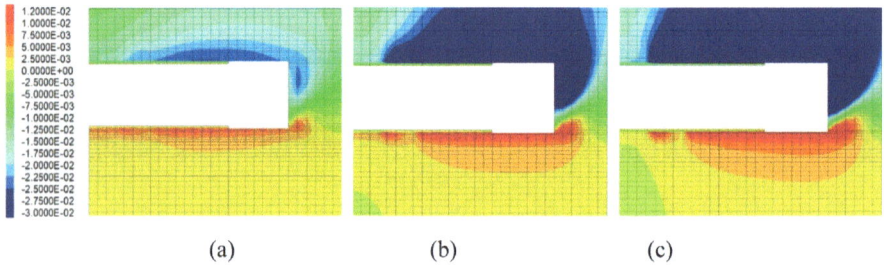

(a)   (b)   (c)

Fig. 8. Vertical displacement distribution on A-A profile under different considerations of strength anisotropy after tunneling to the 25th ring (m): (a) cohesion anisotropy (b) internal friction angle anisotropy (c) anisotropy of cohesion and internal friction angle

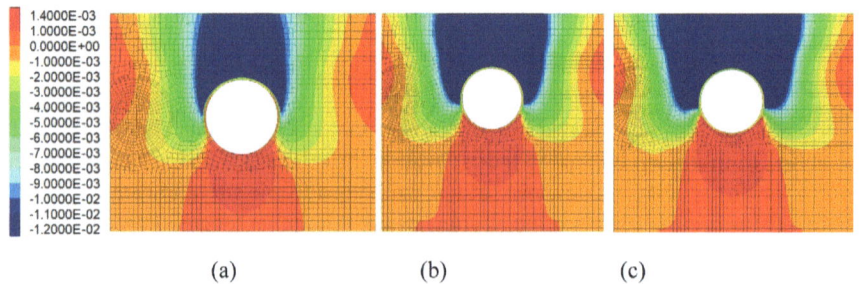

(a)   (b)   (c)

Fig. 9. Vertical displacement distribution on C-C profile under different considerations of strength anisotropy after tunneling to the 25th ring (m): (a) cohesion anisotropy (b) internal friction angle anisotropy (c) anisotropy of cohesion and internal friction angle

Figure 10 shows the surface settlement curves on C-C profile under different considerations of strength anisotropy after tunneling to the 25th ring. The surface settlement curves caused by shield tunneling all show a "U-shaped" curve, and the maximum settlement is basically on the central axis of the excavation tunnel. The difference between the surface settlement curves calculated by only considering cohesion anisotropy and the case without considering strength anisotropy is small. The surface settlement calculated by considering the anisotropy of the internal friction angle is obviously larger than that without considering strength anisotropy, with the maximum surface settlement increasing from 10.2 mm to 24.1 mm. When considering the anisotropy of both cohesion and internal friction angle, the shield tunneling-induced surface settlement increases further, the maximum settlement value is 38.0 mm. Therefore, the anisotropic strength characteristics of the stratum can play a significant effect on the shield tunneling disturbance. It's dangerous to ignore the effect of stratum strength anisotropy, and its impact should be accurately considered in the design and construction of shield tunnels based on the actual anisotropy characteristics of the stratum.

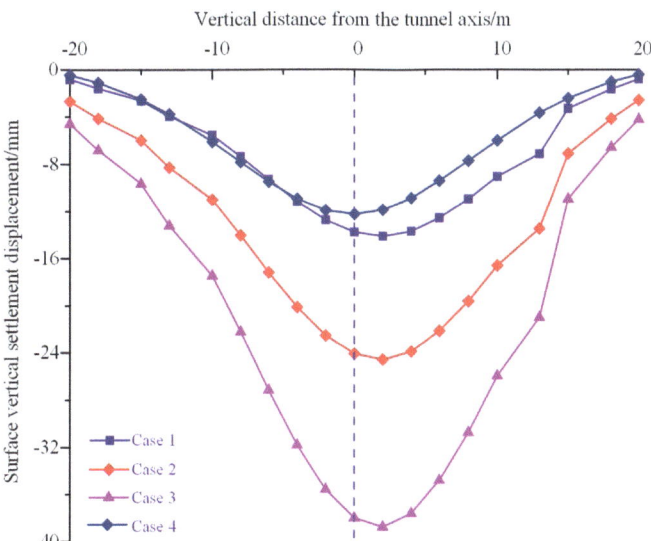

Fig. 10. Surface settlement curves on C-C profile under different considerations of strength anisotropy after tunneling to the 25th ring

6   Conclusions

In this study, a shield tunneling refined simulation method considering stratum strength anisotropy is established and utilized to simulate the tunneling disturbance in a certain section of a large shield tunnel project, and the influence of shear strength parameters' anisotropy is analyzed. Following conclusions are drawn:

(1) The processes of shield shell advancement, lining installation, and shield tail grouting are effectively considered by the refined shield tunneling simulation method, which has successfully modeled the tunneling disturbance in a certain section of a super-large-diameter shield tunnel in Wuhu City, Anhui Province, China. The "U-shaped" distribution of surface settlement induced by shield tunneling is in a range of about three times the diameter of the tunnel, with the maximum value located on the central axis of the excavated tunnel and increases as the shield tunneling forward. The simulated surface settlements on the axial and transverse profiles of the excavated tunnel after tunneling to different lining-ring numbers agree well with the monitoring data.

(2) The shield tunneling disturbance is greatly influenced by the anisotropy of stratum shear strength caused by the dominant-orientation development of geological structures. The shield tunneling-induced surface settlement under the case considering the anisotropy of layered stratum material's cohesion and internal friction angle is much larger than the result under the case of isotropic shear strength. Compared to the influence of stratum cohesion anisotropy, the effect of internal friction angle anisotropy has a greater impact on the shield tunneling disturbance. Therefore, attention should be paid to considering the influence of shear strength anisotropy on the shield tunneling disturbance in actual anisotropic strata, and it's dangerous to ignore its effect.

It is worth noting that in the simulation studies of this paper, only the influence of anisotropy caused by the development of structures in a single direction on shield tunneling is considered, and the anisotropic characteristics of actual strata may be more complex, it will be considered in our future research.

Acknowledgment. The financial supports from the Natural Science Foundation of Anhui Province (2208085ME153), the University Synergy Innovation Program of Anhui Province (GXXT- 2022-020) and the National Natural Science Foundation of China (51709072) are gratefully acknowledged.

References

1. Dai, H., Ji, Y.: Statistical analysis of Chinese large-diameter shield tunnel and state-of-art and prospective of comprehensive technologies. Tunnel Constr. **42**(5), 757–783 (2022)
2. Qin, J.: Study on parameters control of shield tunneling in composite strata in Hefei area. Tunnel Constr. **40**(3), 435–443 (2020)
3. Ma, Z., Zhang, J., Ren, X., et al.: Influence of soil parameter sand shield operation parameters on surface settlement of shield tunnel in water-rich sand layer. Water Resources Power **41**(10), 149–153 (2023)
4. Shi, X., Yang, X., Meng, Y., et al.: An anisotropic strength model for layered rocks considering planes of weakness. Rock Mech. Rock Eng. **49**(9), 3783–3792 (2016)
5. Bao, H., Yang, C.: Study on effects of discontinuities on mechanical characters of slate. Chin. J. Rock Mech. Eng. **24**(20), 53–58 (2005)
6. Pietruszczak, S., Mroz, Z.: On failure criteria for anisotropic cohesive-frictional materials. Int. J. Numer. Anal. Meth. Geomech. **25**(5), 509–524 (2001)

7. Liu, W., Zheng, L., Zhang, Z., et al.: A micromechanical hydro-mechanical-damage coupled model for layered rocks considering multi-scale structures. Int. J. Rock Mech. Min. Sci. **142**, 104715 (2021)
8. Chen, Y.F., Wei, K., Liu, W., et al.: Experimental characterization and micromechanical modelling of anisotropic slates. Rock Mech. Rock Eng. **49**, 3541–3557 (2016)
9. Wang, L., Zhong, J., Zhu, B.: Numerical analysis on ground settlement induced by shield tunnelling with adjacent end shaft works. Mod. Tunn. Technol. **59**(1), 124–132 (2022)
10. Zhu, Z., Sheng, Q., Mei, S., et al.: Improved ubiquitous-joint model and its application to underground engineering in layered rock masses. Rock Soil Mech. **30**(10), 3115–3121 (2009)

Simulation Research on Emergency Evacuation of Subway Station Based on Latent Class Model

Zhen Cao, Chenggong Tang$^{(\boxtimes)}$, Yuping Wang, Juan Geng, and Zhun Tian

School of Civil Engineering, Xi'an University of Architecture and Technology, Xi'an 710055, China

`1187446950@qq.com, {wangypjt,xjdtmgj}@xauat.edu.cn`

Abstract. In order to deeply study the influence of pedestrian psychological behavior on route choice in the emergency evacuation process of subway stations, and quantify the influence of pedestrian psychological heterogeneity on evacuation efficiency, this study establishes an emergency evacuation simulation model based on latent class model to describe the decision-making behavior of pedestrians. Firstly, the stated preference and revealed preference (SP&RP) survey method is used to investigate the psychological behavior of pedestrian groups and the behavior selection mode under different scenarios, and the latent class analysis (LCA) model is established to obtain the latent categories of pedestrians. Then, with the path length in the SP survey as the dependent variable, the distribution probability is obtained by using ordered multiple logistic regression. Finally, the evacuation environment is established in Anylogic for simulation and compared with the questionnaire results. The results show that the LCA model is in good agreement with the real observed values, and can better reflect the path choice behavior of evacuees in reality. There are differences in the influence degree of people with different psychological characteristics on evacuation efficiency. The calm type has the greatest influence on evacuation efficiency, followed by the autonomous type, and the influence of the mass type on evacuation efficiency can be ignored.

Keywords: Subway stations · Emergency evacuation · LCA model · Multiple logistic regression · Anylogic simulation · Efficiency of evacuation

1 Introduction

As an important part of the urban public transportation system, the subway shares a large number of ground passenger flow and effectively reduces the situation of urban ground traffic congestion due to its advantages of large passenger volume, fast running speed, punctuality, short departure interval and low energy consumption. The subway station is one of the main places for crowded activities in the city for pedestrians to wait, transfer and rest for a short time. With the rapid growth of the operating mileage, the working day, weekend and holiday in many cities have shown a normal large passenger flow. The vast majority of subway stations are located underground, with closed space, complex

© The Author(s) 2024
A. Bieliatynskyi et al. (Eds.): CSTTE 2023, LNCE 603, pp. 514–531, 2024.
https://doi.org/10.1007/978-981-97-5814-2_46

structure, few entrances and exits, and long distance walking. When emergencies come, people will inevitably be intertwined, bringing more risks and challenges to emergency evacuation.

At present, domestic and foreign scholars have carried out a lot of research on emergency evacuation in subway stations and achieved good results. Some of them study the path choice behavior of passengers from the pedestrian decision-making level. The other part of the study starts from the simulation method level, based on cellular automata, social force and agent micro-simulation models, to explore the impact of pedestrian behavior on emergency evacuation. Based on census data and questionnaire survey results, Li Feng et al. analyzed the significant influencing factors of each type of passengers through multiple logit models, and estimated the proportion of different types of passengers in subway stations [1]. Wang Heng et al. used the conditional logit model and the random parameter logit model [2] to demarcate the utility coefficients of four factors, distance, exit density, passenger flow direction and exit visibility, on emergency evacuation of subway stations, and analyzed the heterogeneity of passenger decision preference based on the results of the questionnaire survey [3]. Xu Huizhi et al. used a questionnaire combined with stated preference (SP) survey [4] and revealed preference (RP) survey [5] to investigate passengers' choice of evacuation behavior in different situations. The results showed that: The difference in the proportion of blindly following, panicked, autonomous and impulsive passengers in subway stations has a significant impact on evacuation efficiency [6]. Lin Xiaofei et al. designed 8 evacuation methods by using orthogonal method, and analyzed the influence of 4 factors by using range method based on Pathfinder and taking evacuation time as index [7]. Guo Haixiang et al. improved the social force model [8] by considering the impact of pedestrian awareness on pedestrian's expected speed when an emergency occurred, and the results showed that pedestrian's expected speed increased with the increase of pedestrian awareness, thus reducing the evacuation time [9]. Based on the social force model, Meng et al. simulated the emergency evacuation of subway transfer stations in case of emergencies, and proposed evacuation methods to shorten the evacuation time by analyzing and comparing the evacuation time of passengers in peak periods [10]. Hu Mingwei et al. built a simulation model of subway station water invasion evacuation based on AnyLogic platform to study the impact of emergency evacuation strategies on evacuation efficiency under water invasion conditions. After optimizing evacuation strategies, the evacuation efficiency of the model increased by 11.4% overall, and the safe evacuation ratio reached 92.2%. Wang Lixiao et al. constructed a latent class model (LCM) considering pedestrian psychological heterogeneity [11] to characterize individual decision-making processes. The simulation results show that: The latent class model is a better reflection of the path choice behavior of pedestrians in reality and is more effective and reasonable than multinomial logit (MNL) model [12–14]. According to previous studies, most evacuation simulation models only simulate the phenomenon of crowd evacuation, and the comprehensive analysis of evacuation environment and psychological behavior is relatively scarce. The built simulation models only describe the emergency response of pedestrians from the appearance, and the models are generally studied from the perspective of group decision-making rather than individual decision-making. In addition, the pedestrian exit decision rules embedded in most simulation models are only based on the shortest path decision rules under the assumption of complete pedestrian rationality and complete information grasp, or the discrete choice model rules based on stochastic utility maximization theory, which is inconsistent with reality.

In view of this, firstly, this study adopts SP&RP survey method to investigate the influencing factors (exit distance, exit density, exit congestion, exit visibility) of pedestrian group's psychological state and behavior. Secondly, the influence of pedestrian psychological heterogeneity on path selection behavior is explored by establishing LCA model. Finally, the model of pedestrian groups with different psychological categories is used to simulate, which effectively improves the reliability of the simulation process and results. This study simulates the problem of pedestrian emergency evacuation in subway stations, considers the influence of pedestrian group's psychological behavior on route selection, and takes Nanhuomen Station in Xi 'an City (China) as an example to conduct modeling simulation. The results have certain reference value and can provide theoretical support for emergency evacuation management in subway stations.

2 Data Source

2.1 Passenger Flow Data

Based on the AFC data of Nanhuomen Metro transfer station in Xi 'an, the distribution characteristics of passenger flow are analyzed. The morning and evening peak traffic of working days, weekends and holidays is examined, and the flow level of large passenger flows is estimated according to the statistics. On the day of the survey, the passenger volume of Metro Line 2 was 917,000 people, the passenger flow of Nanhuamen station was 56,000 people, the maximum inbound passenger flow in 15min was 2078 people, and the outbound passenger flow was 1787 people. Nanzhaomen Station is a single transfer channel, passengers can transfer between lines 2 and 5 through the transfer channel, and the shortest time for walking to complete the conversion of the station hall is 96 s. In this study, a total of 3865 people were evacuated, including 1787 on the platform floor and 2078 on the station hall floor.

2.2 Questionnaire Survey

In the process of emergency evacuation, the psychological state of pedestrians will continue to change, and they are easy to make a series of decisions influenced by their own emotions and cognition as well as others. Under normal circumstances, the pedestrian evacuation path is composed of key nodes such as stairs or escalators between the platform and the station hall, ticket offices, security channels, gates, transfer channels, entrances and exits. However, when encountering an emergency, the psychological state of pedestrians is more fragile, and they are more susceptible to the influence of the surrounding environment and others, thus losing their rational judgment. When encountering obstacles, large crowd density and other situations, they will choose to decircuit or force crossing. Usually, waiting for service in front of a gate in the direction of an exit causes a large number of pedestrians to jam. The inconsistency of pedestrian flow velocity between stations or platforms leads to stampede events in the interweaving process. Pedestrians who lack the ability of independent judgment give up changing the evacuation path because of crowd behavior. Panic pedestrians repeatedly change the evacuation path and reverse flow of people repeatedly detour.

In order to confirm the correlation between pedestrian psychological behavior and emergency evacuation decision-making level, the influence of the above behavioral factors on pedestrian path selection was analyzed through experimental investigation, and the psychological heterogeneity of the influencing factors was quantified during pedestrian decision-making. RP survey [15], also known as behavioral survey, is a survey of actual action or completed selection behavior, and RP survey can reflect the real choice behavior of the survey object in the existing scene. SP survey [16], also known as intention survey, is a survey of choice intention to select how the subject chooses and how to consider it under hypothetical conditions. SP survey can reflect the choice behavior of the survey object for the things that have not happened in the set scene. In this study, RP&SP survey method was used to collect individual pedestrian data, so as to realize scenario selection survey based on similar scenario selection. The questionnaire is divided into three parts: the first part is about individual attributes and travel attributes of pedestrians, the second part is about psychological latent variables and preference characteristics, and the third part is about emergency evacuation choice behavior. 240 valid questionnaires were collected, and the survey results are shown in Table 1 and Table 2.

Table 1. Questionnaire on emergency evacuation of subway pedestrians

Sort	Question	Options (%)
a. Personal attributes and travel attributes	a1. Your gender is?	1. Male {53.6%} 2. Female {46.4%}
	a2. Your age is?	1. ≤ 18{18.6%} 2.18–59{54.0%} 3. ≥ 60{27.4%}
	a3. Your education level is?	1. Junior high school and below {9.8%} 2. High school and technical secondary school {22.2%} 3. Bachelor degree or above {68.0%}
	a4. Have you participated in emergency evacuation safety education training?	1. Have recently attended {5.2%} 2. Have been involved before {52.2%} 3. Have never been involved {42.6%}
	a5. The frequency of your visit to the station?	1. Come almost every day {29.3%} 2. Three or five times a week {36.3%} 3. Come once in a while {34.4%}
	a6. The purpose of your trip is?	1. Commuting {44.3%} 2. Transfer {19.4%} 3. Leisure and entertainment {23.9%} 4. Other {12.4%}
	a7. Travel time?	1.7:00–9:00{38.3%} 2.17:00–19:00{27.8%} 3. Other time period {33.9%}
	a8. Nature of travel date?	1. Holidays and festivals {18.9%} 2. Working day {54.2%} 3. Day off {26.9%}
	a9. Do you have luggage with you?	1.Yes{30.7%} 2.No{69.3%}
	a10. Whether there was a companion?	1.0{72.3%} 2.1–2{19.8%} 3. ≥ 3{7.9%}

(continued)

Table 1. (*continued*)

Sort	Question	Options (%)
b. Emergency evacuation behavior and preference characteristics	b1. I have a good sense of direction and can locate myself on a map?	1.Agree {62.3%} 2.General{22.6%} 3.Disagree{15.1%}
	b2. I didn't know what to do when the emergency happened?	1.Agree{25.0%} 2.General{45.3%} 3.Disagree{29.7%}
	b3. I walk with most people?	1.Agree{31.2%} 2.General{57.2%} 3.Disagree{11.6%}
	b4. I will follow the direction of the staff?	1.Agree{68.4%} 2.General{20.1%} 3.Disagree{11.5%}
	b5. I stand in line patiently when there is a crowd?	1.Agree{54.7%} 2.General{32.3%} 3.Disagree{13.0%}
	b6. I am good at using plans, signs and other signs?	1.Agree{38.8%} 2.General{46.6%} 3.Disagree{14.6%}
	b7. I know the transfer route?	1.Agree{47.7%} 2.General{32.9%} 3.Disagree{19.4%}
	b8. I know the locations of entrances and escalators?	1.Agree{48.2%} 2.General{30.8%} 3.Disagree{21.0%}
	b9. I think the station space layout is reasonable?	1.Agree{34.9%} 2.General{50.4%} 3.Disagree{14.7%}
	b10. I will take the less crowded path?	1.Agree{60.9%} 2.General{20.8%} 3.Disagree{18.3%}
	b11. I am used to evacuating through familiar tunnels?	1.Agree{58.3%} 2.General{11.5%} 3.Disagree{30.2%}
	b12. I will take the nearest exit?	1.Agree{50.6%} 2.General{24.5%} 3.Disagree{24.9%}
	b13. Compared with the exit visible, I think the exit distance?	1.Important{23.1%} 2.Equal{47.2%} 3.Unimportance{29.7%}
	b14. Compared to the exit visible, I think the queue time?	1.Important{67.5%} 2.Equal{23.0%} 3.Unimportance{9.5%}
	b15. Compared to the exit visible, I think the evacuation guidelines?	1.Important{17.7%} 2.Equal{31.8%} 3.Unimportance{50.5%}
	b16. Compared to the distance to the exit, I think the queuing time?	1.Important{38.9%} 2.Equal{50.3%} 3.Unimportance{10.8%}
	b17. Compared to the distance to the exit, I recognize the evacuation guidelines?	1.Important{7.1%} 2.Equal{25.8%} 3.Unimportance{67.1%}
	b18. Compared to the queuing time, I think the evacuation guidelines?	1.Important{27.1%} 2.Equal{25.0%} 3.Unimportance{47.9%}

Table 2. Questionnaire on emergency evacuation of subway pedestrians

Scene (c.)	Exit	Exit visibility	Evacuation guidance	Queuing time	Exit distance	Your choice (%)
c1	A	Low	Yes	>3 min	200m	35.1%
	B	Normal	Yes	1–3 min	400m	24.6%
	C	Low	Yes	1–3 min	300m	11.8%
	D	High	No	<1 min	400m	28.5%
c2	A	Low	No	<1 min	300m	22.3%
	B	Low	Yes	1–3 min	400m	32.9%
	C	Low	Yes	>3 min	200m	22.6%
	D	Normal	No	<1 min	400m	22.2%
c3	A	Low	No	<1 min	200m	42.6%
	B	High	No	>3 min	400m	16.8%
	C	Low	Yes	1–3 min	300m	32.8%
	D	High	No	<1 min	400m	7.8%
c4	A	Normal	No	1–3 min	300m	31.9%
	B	High	Yes	>3 min	400m	14.1%
	C	Low	Yes	1–3 min	300m	16.8%
	D	High	No	<1 min	400m	37.2%
c5	A	High	No	>3 min	200m	15.8%
	B	Normal	Yes	<1 min	400m	31.2%
	C	Low	Yes	1–3 min	300m	40.7%
	D	High	No	<1 min	400m	12.3%
c6	A	High	Yes	1–3 min	200m	32.5%
	B	Normal	No	>3 min	400m	14.6%
	C	Low	Yes	1–3 min	300m	25.8%
	D	High	No	<1 min	400m	27.1%

3   Behavioral Decision Model

Anylogic software is an effective tool for studying and simulating pedestrian behavior. Since it allows the customization of individual evacuees' attributes and behaviors and simulates multiple evacuees at the same time, Anylogic simulation platform can explore the mechanism of pedestrian psychological behavior and emergency evacuation decision from the perspective of individual decision making. The model defines the basic attributes of individual evacuees according to the questionnaire survey results, and sets the path decision mechanism of individual evacuees based on utility theory and latent category model, so as to simulate the path choice behavior of evacuees.

3.1 Establishment of Utility Function

Utility theory is a decision-making theory used by individuals to choose decision schemes. Decision-makers are often affected by subjective consciousness. In decision-making problems, the consideration of individual benefits and losses is called utility. The Utility theory is based on the Random Utility Maximization model (RUM) which is completely rational for decision makers. In this study, it is assumed that individuals choose the maximum utility path for evacuation according to their personal attributes, travel attributes (variables a_i, $i = 1, 2, ..., 10$), emergency evacuation behaviors and preference characteristics (variables b_i, $i = 1, 2, ..., 18$). The utility function formula consists of directly observable (V_{ij}) and unobservable (ε_{ij}) items. The utility (U_{ij}) of the decision maker's choice scheme is shown in formulas (1) and (2).

$$U_{ij} = V_{ij} + \varepsilon_{ij} \tag{1}$$

$$V_{ij} = \alpha_i S_{ij} + \beta_i M_{ij} + r_i \tag{2}$$

where: S_{ij} is the personal attribute and travel attribute vector that influence decision maker's choice of path, and M_{ij} is the emergency evacuation behavior and preference feature vector (i.e., psychological latent variable vector) that influence decision maker's choice of path; α_i and β_i is the parameter to be estimated of the corresponding attribute vector, reflecting the sensitivity of the attribute vector to the path. The regression analysis is performed based on the SP&RP questionnaire survey data and subsequent latent category analysis results. If its value is positive, it has a positive impact on the path selection; if its value is negative, it has a negative impact on the path selection; r_i is a constant term. The basic form of the decision model is derived based on the assumption that it has independent and identical distribution characteristics and follows Gumbel distribution. The expression of the probability density function of the random error term is shown in Eq. (3), and the probability of the decision maker choosing the path is shown in Eq. (4).

$$f\left(\varepsilon_{ij}\right) = e^{-\varepsilon_{ij} - e^{-\varepsilon_{ij}}} \tag{3}$$

$$P_{ij} = \frac{e^{(V_{ij})}}{\sum_{k=1}^{I} e^{(V_{kj})}} = \frac{e^{\alpha_i S_{ij} + \beta_i M_{ij} + r_i}}{\sum_{k=1}^{I} e^{\alpha_k S_{kj} + \beta_k M_{kj} + r_k}} \tag{4}$$

where: I is the set of other paths, V_{kj} is the utility observable item of the decision maker's choice of path, S_{kj} is the personal attribute and travel attribute vector that influence the decision maker's choice of path, M_{kj} is the emergency evacuation behavior and preference feature vector that influence the decision maker's choice of path; α_k and β_k is the parameter to be estimated of the corresponding attribute vector; r_k is a constant term.

3.2 Latent Category Model

LCA [17, 18] is a statistical method for estimating parameters based on individuals' response patterns on explicit indicators, i.e. different joint probabilities, clustering into different latent classes based on posterior probabilities. In this study, emergency evacuation behavior and preference characteristics (variables b_i, $i = 1, 2, ..., 18$) are used as observable explicit variables for potential category analysis. The potential category model includes two model parameters, potential category probability and conditional probability. When there are multiple explicit variables (survey data), the number of options for each variable is i_i. At the same time, there is also latent variable X with T latent categories after the explicit variable, and each explicit variable in each category of X has local independence, then the basic equation of the LCA model can be obtained, as shown in Eq. (5).

$$\pi_{i_1 i_2,...,i_{18}}^{b_1 b_2,...,b_{18}} = \sum_{t=1}^{T} \pi_t^X \pi_{i_1 t}^{\overline{b_1}X} \pi_{i_2 t}^{\overline{b_2}X} \cdots \pi_{i_{18} t}^{\overline{b_{18}}X} \tag{5}$$

where: $\pi_{i_1 i_2,...,i_{18}}^{b_1 b_2,...,b_{18}}$ represents the joint distribution probability of the explicit variables estimated by the latent class model, and π_t^X represents the conditional probability that the observed data belongs to a certain latent class of i_i.

3.3 Parameter Calibration

Mplus 7.4 software was used to carry out LCA on pedestrian's emergency evacuation behavior and preference characteristics (variable b_i, $i = 1, 2, ..., 18$) attributes. The indicators [19] used in this study include: Log-L, Akaike Information Criterion (AIC), Bayesian information criterion (BIC), Adjusted Bayesian information Criterion (aBIC), Entropy, likelihood ratio test (LMR), Bootstrap-based likelihood ratio test (BLRT). The results show that the smaller the values of Log-L, AIC, BIC and aBIC, the better the fitting effect is. The higher the Entropy value, the higher the classification accuracy, LMR and BLRT values were significant (p-value&lt, 0.05), indicating that the model of C categories is better than that of C-1 categories. By comparing the relevant fitting indicators of each category, the best model is selected, and the fitting results are shown in Table 3.

Table 3. Goodness of fit of LCA model and model selection

Class	Log-L	AIC	BIC	aBIC	Entropy	LMR p-value	BLRT p-value	Minimum category proportion (%)
1-C	−4042.808	8453.615	9094.053	8510.818	-	-	-	100
2-C	−4070.237	8434.473	8946.127	8480.173	0.750	***	***	12.5
3-C	−4106.034	8432.067	8814.937	8466.264	0.788	0.017	***	16.7
4-C	−4136.894	8419.787	8673.874	8442.482	0.847	0.025	0.037	18.3
5-C	−4174.089	8420.177	8545.480	8431.369	0.796	0.261	0.104	7.5

Note: - indicates no number of values, *** indicates < 0.05

According to the data of various models in Table 3, when the number of categories is 4, all indicators are optimal, so the category 4 model is selected to divide the population. Further, the posterior conditional probability and class probability of the four-class model are analyzed to determine the mental activity characteristics of various other groups. In the part of psychological latent variables and preference characteristics of this RP&SP questionnaire, option 1 means sensitive to the problem, option 2 means normal, option 3 means insensitive to the problem, and the conditional probability of option 1 is shown in Fig. 1.

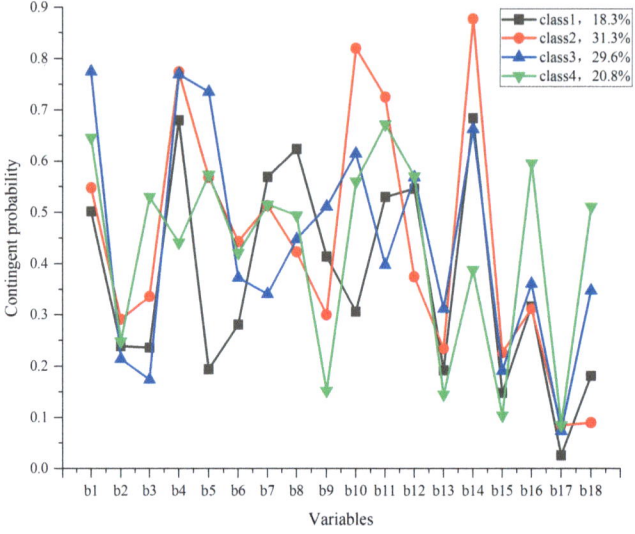

Fig. 1. Conditional probability that the 4-category model option is 1

For people in category 1, the category probability is 18.3%, showing that they are more sensitive to "following the command of staff", "knowing the location of entrances and escalators" and "knowing the transfer route", indicating that they have clear spatial cognition ability and follow the suggestions of others, so they can be named "calm" people. Among the people in category 2 (31.3%), "they can find their own location on the map", "they are used to withdrawing from familiar channels" and "choosing less crowded channels" are sensitive, which can be called "autonomous" people. Among the people in category 3 (29.6%), the characteristics of "strong sense of direction", "following the command of staff", "I will patiently queue up when encountering crowds" and "I will choose the passageway with fewer people" are more sensitive, belonging to the "mass" crowd. Category 4 group (20.8%) is the only group that "walks with the majority of people" and shows sensitivity and other behaviors are more general, so it can be regarded as a "conformity" group.

The group category is taken as variable b, option 1 is the calm group, option 2 is the autonomous group, option 3 is the mass group, option 4 is the conformity group, and the personal attributes and travel attributes variables a_i, $i = 1, ..., 10$ are combined into independent variables, and the distance of pedestrians' choice of entrance and exit in SP survey is taken as the dependent variable c, option 1 is the shortest path, option 2 is the longer path. Option 3 is the longest path, and Option 3 is used as a utility reference to perform ordered multivariate logictic regression analysis [20, 21] on the data and solve the regression parameters of the respective variables. The results are shown in Table 4.

Table 4. Results of multivariate Logictic regression model parameter estimation

c1	B	Wald test	Significance	Exp(B)	c2	B	Wald test	Significance	Exp(B)
Intercept	1.746	2.555	0.110	-	Intercept	−0.152	0.015	*	-
[a1 = 1]	0.117	0.105	0.254	1.124	[a1 = 1]	−0.309	0.638	*	0.734
[a2 = 1]	0.106	0.037	0.152	1.112	[a2 = 1]	−0.346	0.333	**	0.707
[a2 = 2]	−0.515	1.412	0.235	0.598	[a2 = 2]	−0.466	1.062	0.303	0.628
[a3 = 1]	0.006	0.000	***	1.006	[a3 = 1]	0.373	0.338	**	1.452
[a3 = 2]	−0.053	0.014	**	0.948	[a3 = 2]	0.641	1.927	0.165	1.899
[a4 = 1]	−0.170	0.050	0.178	0.843	[a4 = 1]	−0.715	0.664	*	0.489
[a4 = 2]	−0.624	2.745	**	0.536	[a4 = 2]	−0.604	2.274	0.132	0.547
[a5 = 1]	0.262	0.322	*	1.300	[a5 = 1]	0.295	0.346	**	1.343
[a5 = 2]	−0.012	0.001	***	0.988	[a5 = 2]	0.175	0.156	***	1.192
[a6 = 1]	0.276	0.251	*	1.318	[a6 = 1]	0.740	1.427	0.232	2.097
[a6 = 2]	0.092	0.022	0.118	1.096	[a6 = 2]	0.771	1.235	0.266	2.163
[a6 = 3]	−0.108	0.032	0.142	0.898	[a6 = 3]	0.863	1.736	0.188	2.371
[a7 = 1]	0.018	0.002	**	1.018	[a7 = 1]	−0.279	0.393	*	0.756
[a7 = 2]	0.307	0.446	*	1.360	[a7 = 2]	0.298	0.382	**	1.347
[a8 = 1]	0.066	0.014	**	1.068	[a8 = 1]	0.227	0.144	**	1.255
[a8 = 2]	−0.536	1.689	0.194	0.585	[a8 = 2]	−0.181	0.162	*	0.835
[a9 = 1]	0.227	0.355	*	1.255	[a9 = 1]	0.097	0.056	*	1.102
[a10 = 1]	−0.776	1.146	0.284	0.460	[a10 = 1]	0.389	0.215	**	1.476
[a10 = 2]	−1.058	1.891	0.169	0.347	[a10 = 2]	−1.155	1.560	0.212	0.315
[b = 1]	0.943	2.401	0.121	2.569	[b = 1]	0.746	1.299	0.254	2.108
[b = 2]	0.293	0.332	*	1.340	[b = 2]	0.789	2.158	0.142	2.201
[b = 3]	−0.419	0.766	**	0.658	[b = 3]	−0.289	0.313	***	0.749

Note: - indicates no number of values, * indicates < 0.1, ** indicates < 0.05, *** means < 0.01

4 Simulation and Result Analysis

4.1 Evacuation Environment Construction

Nanhuomen Station is the transfer station of Xi 'an Metro Line 2 and Line 5, Line 2 station is north-south, Line 5 station is east-west, Nanhuomen station has six entrances: Line 2 has four entrances A, B, C, D, and Line 5 has two entrances F and G. The layout of pedestrian passages and facilities inside the subway station was investigated on the spot, and the plane and three-dimensional diagram of the station were combined to clarify the plane and spatial relationship of each part of the subway station, and the emergency evacuation simulation environment was established, as shown in Fig. 2.

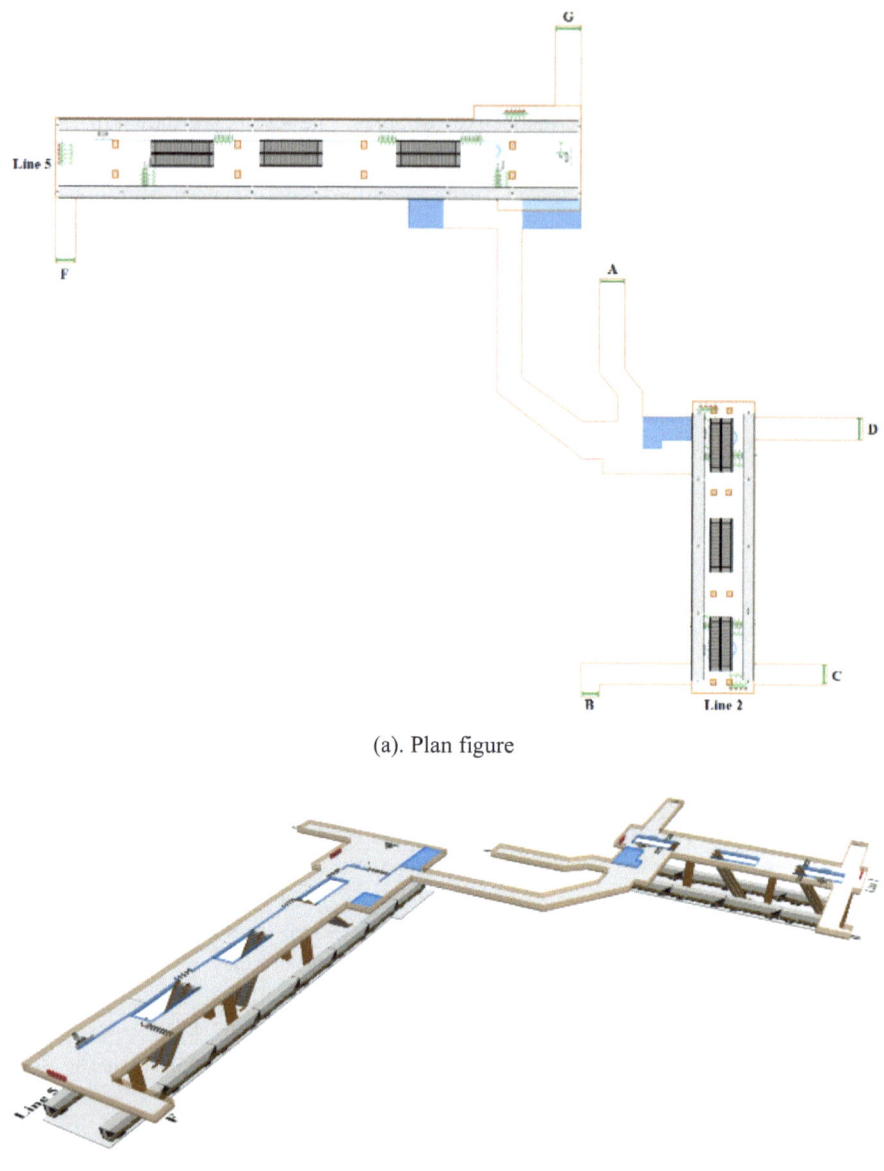

(a). Plan figure

(b). Solid figure

Fig. 2. Emergency evacuation simulation environment construction

4.2 Analysis of LCA Model Results

In this study, passengers who later enter the train and the entrance are not considered in the simulation process, and emergency evacuees are loaded to the station floor and platform floor at the beginning of the simulation. By using the evacuation simulation model established above, the LCA model considering the influence of psychological

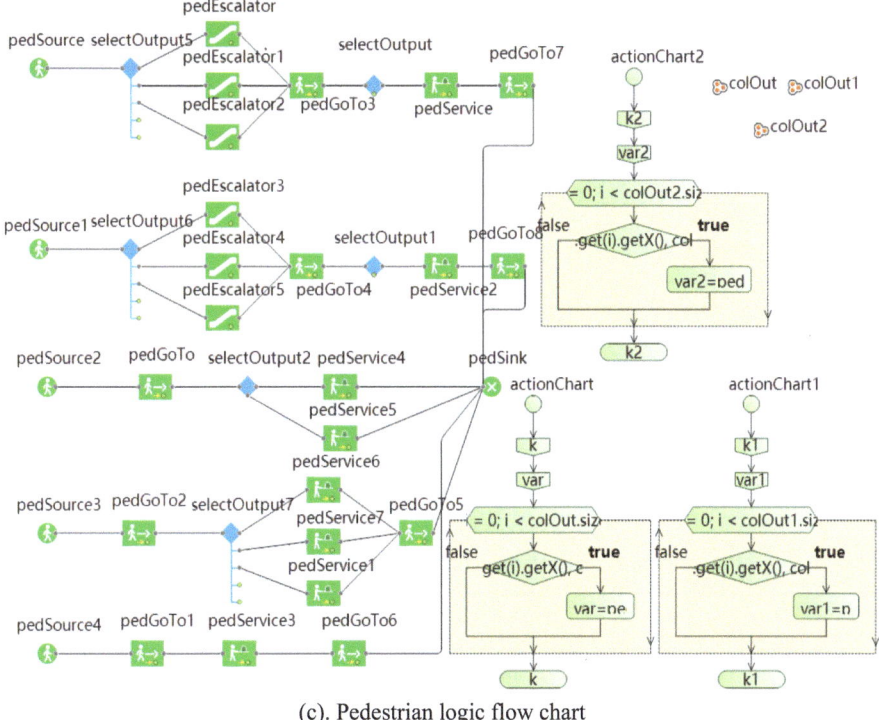

(c). Pedestrian logic flow chart

Fig. 2. (*continued*)

behavior is compared and analyzed with the results of the actual SP survey. To verify the impact of psychological heterogeneity on route selection and evacuation efficiency, the simulation results are shown in Fig. 3 and Fig. 4.

By observing Fig. 3, it can be found that the evacuation time curve based on the LCA model is smooth. With the increase of the number of evacuees, the overall evacuation time increases, which is approximately linear. The fitting line is shown in Eq. (6).

$$y = 0.0857x - 0.634, \ R^2 = 0.995 \tag{6}$$

where: y is the evacuation time and x is the number of evacuees. This is because in the evacuation process, pedestrians based on the distance and queuing time of the two optimal conditions in the gate and the result of the selection of the entrance and exit, therefore, the entire evacuation process is relatively smooth, there is no large crowded period, in the evacuation time curve is no obvious "bump" and "depression". In addition, through the comparison of the number of evacuees at each entrance and exit in Fig. 4, it can be found that the simulation results are roughly consistent with the results of SP survey, with a maximum error of 1.1%. The number of evacuees at entrance C and entrance D is relatively small, accounting for no more than 14.6%, and the number of evacuees at entrance F is the largest, accounting for the largest 22.0%.

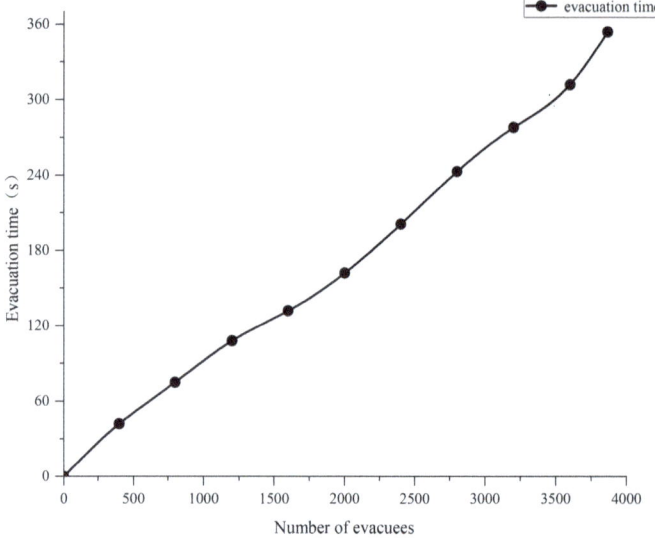

Fig. 3. The evacuation time varies with the number of evacuees

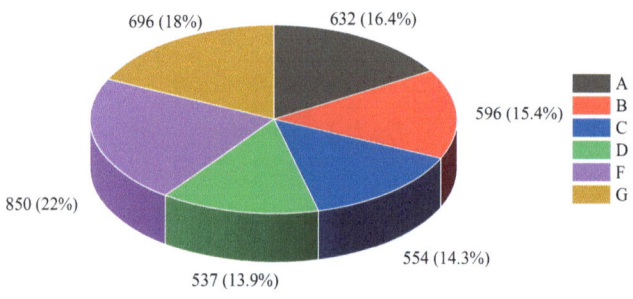

(a). LCA model

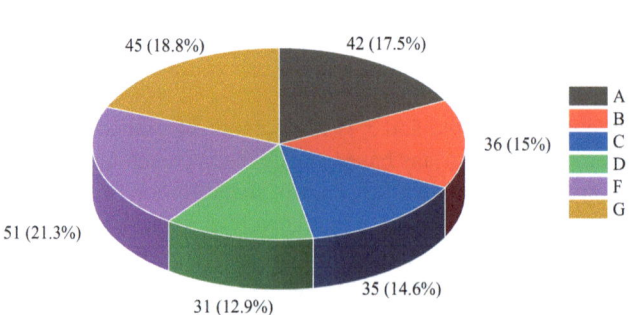

(b). Questionnaire result

Fig. 4. Comparison of the number of evacuees at each entrance and exit

4.3 Comparison of Evacuation Efficiency of Different Proportions of People

The crowd is divided into calm type, autonomous type, mass type and conformity type. In order to explore the differences in the path selection behavior of each latent group, the influence of different groups of people on evacuation efficiency under different proportions is studied. The proportion of conformity type is taken as a variable, and the proportion of autonomous and mass type people is kept unchanged. In order to observe the influence of coolness on evacuation efficiency, the proportion of coolness on evacuation efficiency should be increased while the number of conformity groups should be reduced. The research methods used to observe other groups of people are similar. Evacuation efficiency is expressed by the average number of evacuees, and the results are shown in Fig. 5.

Figure 5(a) shows that in the case of calm crowd with different proportions, the evacuation efficiency increases with the proportion of calm crowd. In the first 5.2% of the increase of crowd proportion, the evacuation efficiency curve changes greatly, and then the change trend gradually slows down with the increase of the proportion. When the proportion reaches the maximum, the maximum evacuation efficiency is 12.0 person/s, and the evacuation efficiency is increased by 10.1%, indicating that the proportion of calm crowd can effectively improve the evacuation efficiency, while the other proportion is not conducive to crowd evacuation.

The initial proportion of autonomous people is relatively large, which is consistent with the actual situation. As can be seen from Fig. 5(b), evacuation efficiency increases steadily as the proportion of autonomous people increases. However, when the proportion of autonomous people increases to the third 5.2%, evacuation efficiency no longer changes, and the maximum evacuation efficiency is 11.8 person/s, which increases by 7.3%. It shows that the proportion of autonomous population can improve the evacuation efficiency to a certain extent, but there is a threshold for the improvement effect.

The mass crowd is similar to the autonomous crowd. Figure 5(c) shows that in the case of the mass crowd with different proportions, the evacuation efficiency changes with the proportion of the crowd. The evacuation efficiency curve changes greatly only in the first 5.2% when the proportion of the crowd increases. Compared with the initial 10.9 person/s, the evacuation efficiency only increased by 0.3, and the evacuation efficiency only increased by 2.8%, indicating that the proportion of autonomous people had a general impact on the evacuation efficiency.

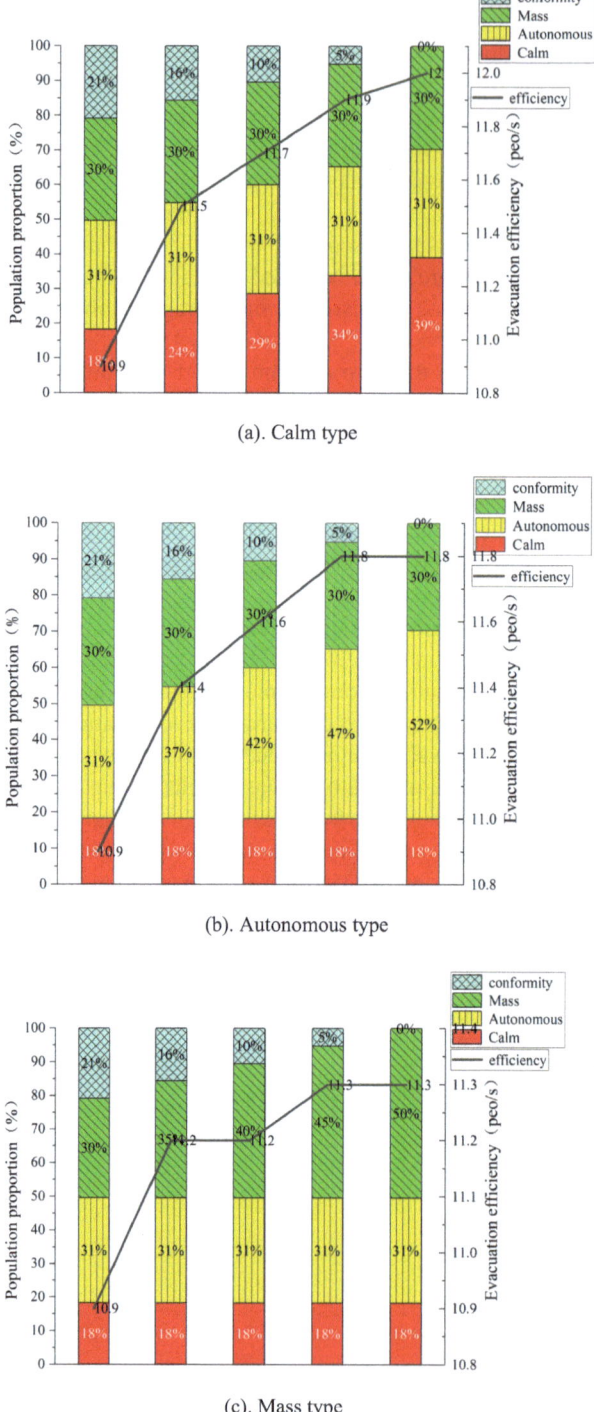

(a). Calm type

(b). Autonomous type

(c). Mass type

Fig. 5. The relationship between different proportion of people and evacuation efficiency

5 Conclusion

This study models the crowd based on LCA, and verifies the necessity of considering the impact of pedestrian psychological behavior in the emergency evacuation process by comparing the simulation results with the SP survey results. Based on this, the impact of different psychological characteristics on the evacuation efficiency is further analyzed. The main research results are as follows:

(1) LCA model can accurately classify people in emergency evacuation. In this study, four categories of model are used, and the classification accuracy rate is as high as 84.7%, which better reflects the path decision-making behavior of pedestrians in the actual scene. The simulation results show that the LCA model is in good agreement with the real observed values in the SP questionnaire scenario setting, and the maximum error is less than 1.1%.
(2) The influence of people with different psychological characteristics on evacuation efficiency exists objectively, and the degree of influence is different. The simulation results show that the calm crowd has the greatest impact on the evacuation efficiency, with the evacuation efficiency increased by 10.1%, the independent crowd has a moderate impact on the evacuation efficiency, with the evacuation efficiency increased by 7.3%, the mass crowd has a little impact on the evacuation efficiency, with the evacuation efficiency only changing by 2.8%, which is basically negligible.
(3) The research results show that the psychological behavior of pedestrians in the process of emergency evacuation has an important impact on route selection, and the more calm crowd and autonomous crowd occupy, the more conducive to emergency evacuation. Therefore, it is necessary for pedestrians to improve their spatial cognition ability, ability to identify signs and maps, and emergency evacuation safety education. The evacuation guidance of the staff also plays an important role in the emergency evacuation process.

Finally, all the data in this study come from the field investigation, but the modeling and simulation method of subway station emergency evacuation for people with different psychological characteristics has a certain universality, which can provide reference for further research on the combined decision model of emergency evacuation.

Author Contributions.   The authors confirm contribution to the paper as follows: study conception and design: Zhen. Cao. Author, Chenggong. Tang. Author; data collection: Chenggong. Tang. Author; analysis and interpretation of results: Chenggong. Tang. Author, Yuping. Wang. Author, Juan. Geng. Author; draft manuscript preparation: Chenggong. Tang. Author, Zhun. Tian. Author. All authors reviewed the results and approved the final version of the manuscript.

References

1. Li, F., Wu, T., Ge, L.: Research on Improvement of Subway station Emergency evacuation Simulation Method under Terrorist Attack. Urban Rail Transit Res. **23**(3), 106–112 (2020)
2. Anwarahm, M., Tieu, K., Gibson, P., et al.: Analyzing the heterogeneity of traveler mode choice preference using a random parameter logit model from the perspective of principal agent theory. Int. J. Logistics Syst. Manage. **17**(4), 447 (2014)

3. Wang, H., Li, F., Jiang, Z., Xu, T.: A quantitative method of Passenger decision behavior Preference in Subway emergency evacuation. J. Tongji Univ. (Natural Science Edition) **50**(04), 571–579 (2022)
4. Wang, F., Chen, J., Zhang, D.: Application of SP survey in Transportation Mode Selection Model. Transp. Syst. Eng. Inf. (5), 90 (2007)
5. Wilson, T.: Monte Carlo analysis of SP-off-RP data. J. Choice Model. **2**, 101 (2009)
6. Xu, H., Wu, X., Lu, P., Wang, L.: Research on Pedestrian emergency Evacuation Simulation based on subway station passenger flow characteristics. Computer Simulation, pp. 1–7. (2023)
7. Lin, X., Yu, X., Hou, Z., Yu, P.: Research on Influencing Factors of Subway emergency evacuation. Sci. Technol. Work Saf. Chin. **16**(S1), 41–45 (2020)
8. Blue, V., Adler, J.: Emergent fundamental pedestrian flows from cellular automata microsimulation. Transp. Res. Rec. **1644**, 29 (1998)
9. Guo, H., Zeng, Y., Chen, W.: Simulation of Multi-exit indoor emergency evacuation based on Social Force model. J. Syst. Simul. 33(3): 721–731 (2021)
10. Meng, Y., Jia, C.: Research and application of metro station evacuation simulation, Proceedings of the 4th (ICISCE).Changsha, China, IEEE, 1123–1125 (2017)
11. Grwwne, W.H., Hensher, D.A.: A latent class model for discrete choice analysis: contrasts with mixed logit. Transp. Res. Part B Methodol. **37**(8), 681–698 (2003)
12. Shi, Z., Chen, Z., Zhou, L., Ling, J.: Cellular automata model for pedestrian evacuation with multiple exit conditions. Syst. Eng. **28**(09), 51–56 (2010)
13. Zhang, Q., Li, R., Ru, H., et al.: Simulation and Analysis of Transportation mode Selection for Crowd Evacuation under Emergency. Chin. Work Saf. Sci. Technol. **19**(07),162–168 (2023)
14. Wang, L., Gai, X., Hao, M.: Simulation of Subway station emergency evacuation considering pedestrian psychological heterogeneity. J. Saf. Environ..**22**(06), 3333–3341 (2022)
15. Wang, W., Xue, F.: Research on capacity matching degree of integrated passenger terminal rail transit system. Transp. Eng. Inform. **19**(03), 111–122 (2019)
16. Zhao, L., Kang, L., Cao, J.: Research on travel mode Selection based on TODIM method. Sci. Technol. Innovation Appl. **13**(22), 77–81 (2023)
17. Pan, Y., Wei, S., Chen, S., et al.: Analysis of potential categories of factors influencing tidal lane usage intention. J. Chongqing Univ. Technol. (Natural Science), **37**(03),183–193 (2023)
18. Liu, S., Zhou, X., He, M., et al.: Heterogeneity of community group-buying behavior in the context of COVID-19: an approach based on latent category analysis. Market Forum. (03),13–20 (2022)
19. Liu, L., Kan, Y., Li, X., et al.: The relationship between latent type of exercise self-efficacy and physical activity stage in patients with type 2 diabetes mellitus. J. Nurs. **38**(04):23–27 (2019)
20. Qiao, Y., Fan, C., Cao, H., et al.: Investigation and analysis of consumers' willingness to consume branded fruits based on multiple ordered Logistic regression model. J. Agric. Sci. Hebei. **26**(05),19–25 (2019)
21. Zhao, F., Wang, J., Chen, J., et al.: Study on the correlation between air quality and residents' subjective well-being based on multiple ordered logistic regression model: A case study of Beijing. J. Capital Normal Univ. (Natural Science Edition) **44**(04), 59–67 (2019)

Flow Test and Simulation of Underground Drainage Pipeline in Substation

Xuezhi Fang[1], Jianhua Xiao[1], Rongwei Liu[1], Xiao Chen[1], Hanbo Yang[1], Jiangshun Yu[2(✉)], and Yun Cai[2]

[1] Guizhou Power Grid Co., Ltd. Kaili Power Supply Bureau, Guizhou Kaili, China
[2] Power China Guizhou Electric Power Engineering Co., Ltd., Guizhou Guiyang, China
875023006@qq.com

Abstract. The flow process of underground water drainage pipeline is of great significance to the flood prevention and drainage of substation, and to ensure the demand of power supply. In this paper, the control test and numerical simulation of the flow process of underground drainage pipeline in substation are carried out. Based on the basic equations of water volume (continuity) and momentum of the full flow in pipelines, combined with the Preissmann slit assumption, the governing equations and their discrete forms of one-dimensional full flow are derived, and a one-dimensional full flow model is established. On this basis, the control experiment is carried out by using the flow process of drainage pipe, and the model is verified by the measured results. The results show that the one-dimensional full-flow model based on Preissmann slit hypothesis can effectively simulate the flow process of underground drainage pipes in substations, and provide technical support for flood prevention and waterlogging early warning of substations in humid areas.

Keyword: Mingmanliu; Pipeline; Water flow process; Substation

1 Introduction

With the development of Chinese economy and society, the electricity load is increasing gradually. It is particularly important to ensure the safe and stable operation of regional substations and improve the ability of disaster prevention and reduction of substations to ensure the power supply demand of our country [1]. At present, the calculation and evaluation of pipe network flood control and drainage mainly include hydrology [2] and hydrodynamics [3]. Among them, hydrology method has a simple calculation process and easy access to required data. However, its description of the physical process of water flow is insufficient, and it is difficult to accurately describe the spatial-temporal variation characteristics of water flow in pipe network system [4]. By solving the Saint-Venant equations (dynamic wave) and its simplified equations (diffusion wave and moving wave), the hydrodynamic method can calculate the flow process in the pipe network [5]. This kind of method can accurately simulate the physical process of water flow with high calculation accuracy, and it has been widely used in the calculation of pipe network water flow [6–8]. Open full flow (i.e. the coexistence or alternate flow of unpressurized flow and pressurized flow) is a water flow phenomenon that often occurs in the process of pipeline drainage, and it is one of the hot spots and difficulties in the analysis and simulation of pipe network flow [9]. However, at present, the analysis and calculation of the open full flow process of the pipe network are mainly focused on the diversion tunnel, the pressure discharge of the reservoir and the urban flood control, and there are few studies on the substation waterlogging prevention and control. Therefore, based on the basic equation of open full flow (continuity) and momentum, combined with Preissmann slit hypothesis, the one-dimensional open full flow governing equation and discrete form are derived in this paper. On this basis, the flow control test and simulation of groundwater drainage pipe in substation are carried out. The research results will provide scientific basis for substation flood disaster loss assessment, environmental risk control, flood control and other planning analysis and formulation.

2 One-Dimensional Bright Full Flow Model and Discretization

2.1 Mathematical Model

The continuity and momentum equation of unsteady flow in one-dimensional open channel is [10, 11]:

$$\frac{\partial Z}{\partial t} + \frac{1}{B}\frac{\partial Q}{\partial x} = 0 \tag{1}$$

$$\frac{\partial Q}{\partial t} + \frac{\partial}{\partial x}(Qv) + gA\left(\frac{\partial Z}{\partial x} + S_f\right) = 0 \tag{2}$$

The continuity and momentum equation of one-dimensional pressurized unsteady flow is:

$$\frac{\partial H}{\partial t} + \frac{c^2}{gA}\frac{\partial Q}{\partial x} = 0 \tag{3}$$

$$\frac{\partial (Q)}{\partial t} + \frac{\partial}{\partial x}(Qv) + gA\left(\frac{\partial H}{\partial x} + S_f\right) = 0 \tag{4}$$

where, Z is the water level, H is the pipe flow head, B is the section width, Q is the flow, S_f is the friction slope, which can be calculated by Manning formula, c is the wave velocity, v is the velocity.

In 1961, Preissmann proposed that the narrow slit method is an effective method for the calculation of pressurized and nonpressurized interfaces. It is assumed that there is an extremely narrow gap at the top of the pipe, which does not increase the area and hydraulic radius of the pipe. When the water depth is higher than the top of the tube, the area of wet circumference and water through section remains unchanged. The narrow slit method is used to treat the pressurized pipe flow equivalent, and the slit width is set as B (Fig. 1), then the slit width can be calculated by the formula $B = \frac{gA}{c^2}$, Where, A is the total area of the section. The water level in the control equation of open channel unsteady flow is represented by the water head H of pressure tube in the control equation of pressure unsteady flow, and $\frac{c^2}{gA} = \frac{1}{B}$, then the governing equation of one-dimensional open channel unsteady flow and one-dimensional pressure unsteady flow can be written in the following form:

$$\frac{\partial Z}{\partial t} + \frac{1}{B}\frac{\partial Q}{\partial x} = 0 \tag{5}$$

$$\frac{\partial Q}{\partial t} + \frac{\partial}{\partial x}(Qv) + gA\left(\frac{\partial Z}{\partial x} + S_f\right) = 0 \tag{6}$$

Considering local head loss and lateral inflow, the governing equation of one-dimensional bright full flow model can be obtained according to Eqs. (5) and (6) as follows:

$$\frac{\partial A}{\partial t} + \frac{\partial Q}{\partial x} = q_0 \tag{7}$$

$$\frac{\partial Q}{\partial t} + \frac{\partial}{\partial x}(Qv) + gA\left(\frac{\partial Z}{\partial x} + S_f + S_L\right) = 0 \tag{8}$$

Fig. 1. Preissmann slit schematic diagram

2.2 One-Dimensional Bright Full Flow Equation Discretization

By means of staggered grid and semi-implicit discretization, the motion equation is discretized on the river segment (pipe segment), and the continuity equation is discretized on the node. This discrete scheme can not only ensure the conservation and stability of the format, but also facilitate the calculation of data input. Figure 2 shows the spatial staggered grid discrete diagram, in which I represents nodes and i represents units. It is assumed that each section of river network or pipe network has (N + 1) nodes and N units.

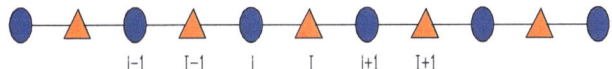

Fig. 2. Equation discretization diagram

Dispersion of Momentum Equation.

Equation (8) is integrated on the unit i, and the obtained formula is semi-implicit discrete, which can be obtained as follows:

$$(Q_i^{t+\Delta t} - Q_i^t)\frac{\Delta x_i}{\Delta t} + v_{I+1}Q_{I+1}^{t+\theta \Delta t} - v_I Q_I^{t+\theta \Delta t} + gA_i(S_{fi} + S_{Li})\Delta x_i \\ = gA_i(Z_I^{t+\theta \Delta t} - Z_{I+1}^{t+\theta \Delta t})$$ (9)

where, Δx_i is the length of the unit i, θ is the implicit coefficient, the format has unconditional stability when $\theta > 5$, S_{Li} is the local head loss, which can be deduced from $S_{Li} = \frac{|Q_i|Q_i}{gC_{cs}A_{cs}\Delta x_i}$, where C_{cs} is the flow coefficient of the building, A_{cs} is the water flow area of the building, S_{fi} is the head loss along the road.

In order to increase the stability of the discrete scheme, a first-order upwind interpolation scheme is used for node flow.

$$v_{I+1}Q_{I+1}^{t+\theta \Delta t} = \begin{cases} v_{I+1}Q_i^{t+\theta \Delta t} & v_{I+1} > 0 \\ v_{I+1}Q_{i+1}^{t+\theta \Delta t} & \text{Other} \end{cases}$$ (10)

$$v_I Q_I^{t+\theta \Delta t} = \begin{cases} v_I Q_{i-1}^{t+\theta \Delta t} & v_I > 0 \\ v_I Q_i^{t+\theta \Delta t} & \text{Other} \end{cases}$$ (11)

Substituting the above two formulas into Eq. (9), we can get:

$$a_i Q_{i-1}^{t+\Delta t} + b_i Q_i^{t+\Delta t} + c_i^k Q_{i+1}^{t+\Delta t} \\ = P_i + \theta g A_i (Z_I^{t+\Delta t} - Z_{I+1}^{t+\Delta t})$$ (12)

Among them:

$a_i = -\theta \max(v_I, 0)$,

$$b_i = \frac{\Delta x_i}{\Delta t} + \theta\big[\max(v_{I+1}, 0) + \max(-v_I, 0)\big] + \frac{gn_i^2|Q_I|\Delta x_i}{A_I R_I^{4/3}}$$

$c_i = -\theta \max(-v_{I+1}, 0)$,

$$P_i = -(1-\theta)\big[\max(v_{I+1}, 0) + \max(-v_I, 0)\big]Q_i + (1-\theta)\max(-v_{I+1}, 0)Q_{i+1} \\ +(1-\theta)\max(-v_I, 0)Q_{i-1} + (1-\theta)gA_i(Z_{Ik} - Z_{I+1}) + Q_i\frac{\Delta x_i}{\Delta t}$$

Discretization of Continuity Equation.

Integrate Eq. (7) on node I, as follows:

$$\Delta x_i B_i \frac{Z_I^{t+\Delta t} - Z_I^t}{\Delta t} + Q_{I-1}^{t+\theta \Delta t} = (q_{mh})_i$$ (13)

where, $\Delta x_I B_I = \frac{\Delta x_{i-1} B_{i-1}}{2} + \frac{\Delta x_i B_i}{2} + A_{mZ,i}$, q_{mh} indicates the node traffic, $A_{mZ,i}$ indicates the node area. Substituted into Eq. (13), we can get:

$$Q_i^{t+\Delta t} - Q_{i-1}^{t+\Delta t} + E_I Z_I^{t+\Delta t} = D_I \tag{14}$$

Among them:

$$E_{Ik} = \frac{1}{\theta \Delta t} \left(\frac{\Delta x_{i-1} B_{i-1}}{2} + \frac{\Delta x_i B_i}{2} + A_{mh,i} \right)$$

$$D_{Ik} = \frac{1}{\theta} \left[-\frac{Z_{Ik}}{\Delta t} \left(\frac{\Delta x_{i-1} B_{i-1}}{2} + \frac{\Delta x_i B_i}{2} + A_{mh,i} \right) - (1-\theta) Q_i^t + (1-\theta) Q_{i-1}^t \right]$$

3 Test and Simulation of Water Flow Control of Underground Drainage Pipe in Substation

In this paper, a 35 kV substation located in Qiandongnan Prefecture, Guizhou Province is selected to carry out the water flow control experiment of underground drainage pipeline. As shown in Fig. 3, the underground water pipe of the substation is L-shaped, with the length of the transverse and longitudinal pipes being 12.2 m and 13.5 m respectively. The pipe is buried in the central position of the transformer station's electric field with a depth of 1 m. Pipe shape and hydraulic parameters are shown in Table 1. During the control test, water was injected from the transverse pipe and outflow was observed in the standpipe (Fig. 3). The water injection was controlled by an automatic pumping device, and the water injection time was 500s. The water injection per unit time presented a linear change, as shown in Fig. 4. An automatic flow monitor is used at the outlet of the standpipe to observe the outflow. The outflow process is shown in Fig. 5.

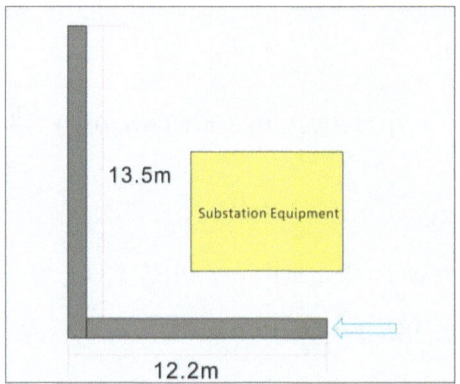

Fig. 3. Substation underground drainage pipeline

The flow process of pipeline was simulated by using the deduced one-dimensional model and discrete model, and the simulation results were evaluated by Nash efficiency

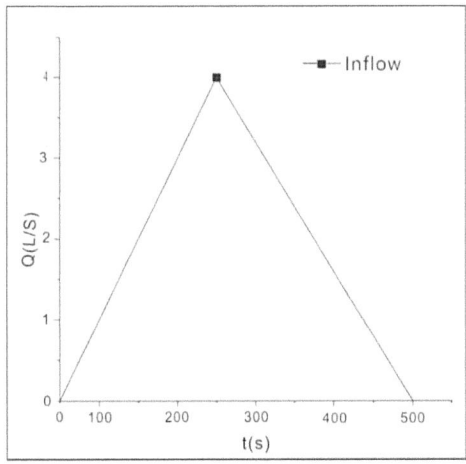

Fig. 4. Control the inflow process of the drainage pipeline in the test

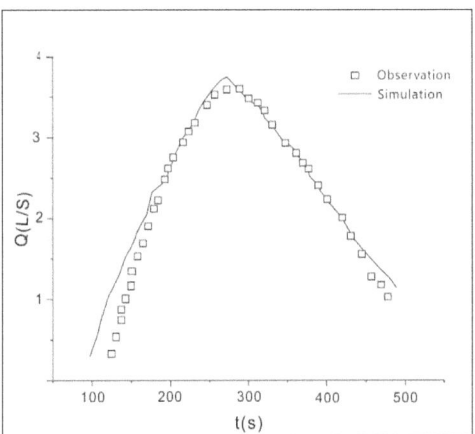

Fig. 5. Process simulation of control test drainage pipeline outflow

Table 1. Shape and hydraulic parameters of underground drainage pipelines

Pipe section	Length (m)	Slope (%)	Roughness	Pipe diameter (m)
Transverse pipe	12.2	0.32	0.01	0.1
Longitudinal pipe	13.5	0.28	0.01	0.1

coefficient (NSE). The results show that the model can well calculate the pipeline flow migration process (Fig. 5), and the NSE is 0.9. The observed peak discharge is 3.6 L/s, and the simulated peak discharge is 3.7 L/s, indicating that the model can accurately capture the flood discharge and provide technical support for the analysis and early warning of flood control and drainage in the substation.

4   Conclusion

Flood control and drainage of substation are of great significance to ensure the demand of power supply. In this paper, drainage of underground drainage pipeline of substation is taken as the research object, and the actual substation pipeline is used to carry out the water flow control test. On this basis, the hydrodynamics method is used to carry out the simulation calculation, and the accuracy of the calculation of pipeline drainage is analyzed and evaluated. Based on the basic equation of open full flow (continuity) and momentum, combined with Preissmann slit hypothesis, the one-dimensional open full flow governing equation and its discrete form are derived, and the one-dimensional open full flow calculation model is established. The model is verified by the test results of water flow control in underground drainage pipe of substation. The results show that the model can simulate the pipeline flow process and flood peak flow well, and provide technical support for flood control and drainage early warning of substation in wet area.

References

1. Zhi, G., Shiqi, Y.: Analysis on the impact of flood disaster on power grid. China's Strategic Emerging Industries **08**, 75 (2018)
2. Haozheng, W., Yiting, Q., Guanyu, H., Qiuyi, W., Feng, Y., Lei, Z.: Construction and application of conceptual model based on muskingum method in urban drainage system. China Water & Wastewater **37**(13), 113–120 (2021)
3. Liu, Q., Sun, C.: Research on flood disaster assessment of substation based on MIKE21. China Ener. Environm. Protect. **44**(12), 86–91 (2022)
4. Xiaona, Z., Jie, F., Fanggui, L.: Development and application of storm flood computation model for urban rain pipe network. Water Resources and Power **26**(5), 40–42 (2008)
5. Li, Q., Liang, Q., Xia, X.: A novel 1D–2D coupled model for hydrodynamic simulation of flows in drainage networks. Adv. Water Resour. **137**, 103519 (2020)
6. Li, D., Hou, C., Shen, R.-z., Gao, X.-j., Huang, M.-s., Ma, Y.: Partition adaptive model of urban rainstorm and flood process based on the simulation concept of plots generalization and road networks fine. Advances in Water Science (2023)
7. Xiaolong, C., Xiaoyu, W., Yonggui, W.: Urban drainage systrm assessment based on land surface - pipe network coupling model. Yangtze Rive **53**(11), 79–85 (2022)
8. Li, Z., Yuhong, C., Chengguang, L.: Numerical simulation of urban waterlogging based on TELEMAC-2D and SWMM model. Water Resources Protection **38**(1), 117–124 (2022)
9. Chen, Y., Yu, G.: One-dimensional numerical simulation of transcritical flows and transition between free surface and pressurized flows. Adv. Sci. Technol. Water Resour. **30**(1), 80–84 (2010)
10. Li, C., Hou, G.: Simulation ON urban flood based on coupling of 1D and 2D hydrodynamic models. Waterconser and Hydropower Technology, 83–85 (2010)
11. Zhang, X., Wang, G., Jin, S.: Study on the numerical solution methods for shallow water equations. Advances In Water Science, 317–323 (2004)

Three-Dimensional Geological BIM Modeling Method Based on Shield Construction Simulation

Ling-pei Chen$^{(\boxtimes)}$ and Tian-cheng Wang$^{(\boxtimes)}$

Department of Civil Engineering, Guangzhou Institute of Science and Technology, Guangzhou 510540, Guangdong, China
75450513@qq.com, mrwangtc@163.com

Abstract. This paper presents a three-dimensional geological Building Information Modeling (BIM) method aimed at addressing safety simulation and rational construction parameter challenges in shield tunneling under varying geological conditions. Incorporating the characteristics of metro shield tunnel modeling, this study utilized both Revit and Dynamo software for modeling. The integration of information on completeness and depth, essential for meeting the requirements of shield tunnel construction, was achieved in the model. This integrated information was then effectively transferred to the finite element simulation model, addressing issues related to communication among participating units and the transmission of information between the BIM model and the finite element model. The research findings demonstrate that the three-dimensional geological BIM model, grounded in shield construction simulation, achieves the automation of shield construction object representation and information updating, thereby enhancing the level of informatization in shield construction.

Keyword: Shield Tunnel Construction · 3D Geological BIM Model · Finite Element Analysis · Information Technology

1 Introduction

The shield tunnel of a metro system is characterized by its considerable length and significant variations in geological conditions along the alignment. Influenced by the scale of the project and intricate geological conditions, the conventional depth of engineering geological exploration proves inadequate. Moreover, the information gathered from sporadically distributed exploration points does not align seamlessly with shield construction conducted on the basis of segment rings. Consequently, this mismatch introduces certain risks into the control of shield construction [1].

BIM model plays an important role in engineering construction because of its 3D visualization, coordination and simulation [2]. Finite element simulation serves as a vital tool for analyzing the interaction between shield tunnel segments and the surrounding rock and soil. It is instrumental in addressing issues related to shield construction parameters and construction control [3, 4].

© The Author(s) 2024
A. Bieliatynskyi et al. (Eds.): CSTTE 2023, LNCE 603, pp. 540–552, 2024.
https://doi.org/10.1007/978-981-97-5814-2_48

The accuracy of finite element simulation in shield tunnel construction is significantly influenced by the completeness and depth of engineering geological information. Establishing a three-dimensional geological model based on the simulation requirements of shield tunnel construction and addressing the information transfer challenge between the geological model and the finite element simulation model constitute key aspects in resolving simulation issues under diverse engineering geological conditions.

The relevant problems was studied by many scholars. In 3D geological BIM modeling and information integration, using the original function of Civil 3D software, combined with the engineering example and the modeling idea, Li Wanhong [5] has created the 3D geological stratigraphic curved surface with the measured and inferred data and the characteristic line, and has established the complete 3D geological model; Using Revit software, combining with Civil3D and Dynamo, the parameterized tunnel engineering model and 3D geological model were established by Yangzhu [6]; Huang di [7] has carried on the 3D geology BIM model extension based on the IFC standard, has established the 3D geology modeling application system and the application frame based on the BIM. While these studies have approached the creation of 3D geological BIM models from various perspectives, their focus has primarily been on modeling methodologies and the integration of geological information. Notably, there has been a lack of further exploration into how BIM models can effectively guide construction processes, leading to a deficiency in the completeness of model information.

In the combination of BIM model and construction simulation, Liu Bei [8] studied the conversion method between BIM model and structural analysis model, and developed the interface program between Revit and ANSYS Using C# language to realize the automatic conversion from BIM model to finite element model; Xie Jisheng [9] established a connection channel between Revit and the finite element software ABAQUS by the file of "*. Sat", and loaded the 3D model in the finite element software successfully; Liu Yujia [10] used BIM technology to build a parameterized 3D geological model, cut the model by Rhino and initially built the grid, which was imported into Kubrix for grid division, the information exchange between BIM model and numerical simulation software was realized, and the accuracy of calculation results was improved. While these studies successfully achieved the transfer of geometric information between BIM and finite element models, they fell short in transferring non-geometric information [11, 12], such as construction details. Consequently, the model depth does not fully meet the requirements of construction simulation.

Against the backdrop of the shield tunnel project from Wanqingsha Station to Hengli Station on Guangzhou Metro Line 18, this paper presents a modeling method and information transmission approach for a 3D geological information model based on BIM. The aim is to address challenges related to safety simulation and the rational selection of construction parameters under varying geological conditions.

2 Project Profile

2.1 Project Profile

The shield tunnel for the Guangzhou Metro Line 18, stretching from Wanjingsha Station to Hengli Station, is situated in Nansha District, Guangzhou City. The design parameters for the main line indicate a starting mileage of ZDK0 + 740.313 and a concluding mileage of ZDK5 + 775.094. This section includes one intermediate ventilation shaft and one shield shaft. The cover soil thickness ranges from 8.3 m to 24.9 m. The excavation diameter is 8850mm, the tunnel's outer diameter is 8500 mm, the inner diameter is 7700 mm, and the segment wedge measures 46 mm. The assembly method employs a staggered seam assembly.

2.2 Engineering Geological Conditions

The project site is situated in the Pearl River Delta Plain, characterized by a sea-land interaction alluvial plain. The geological conditions at the site are stable, with no unfavorable features such as faults identified. The topography of the site is flat, exhibiting a small relative elevation difference, and the ground elevation along the line generally ranges from 3.11 to 8.90 m.

The overlying strata primarily consist of Quaternary marine-terrestrial sedimentary layers and continental alluvial and alluvial facies strata, while the underlying bedrock comprises Sinian migmatite and mixed granite. The site presents a complex geological profile, featuring diverse strata such as miscellaneous fill, plain fill, cultivated soil, silt, muddy soil, fine sand, medium-coarse sand, gravel sand, gravel, plastic silty clay, hard plastic silty clay, residual soil layers, total weathered rock, strong weathered rock, medium weathered rock, micro weathered rock, among others. Alluvial-alluvial soil layers are often interspersed and layered. The majority of the shield tunnel traverses through the muddy soil stratum, characterized by high compressibility and low strength.

3 BIM Modeling Software

The characteristics of metro shield tunnel modeling encompass the following aspects: 1) Involves multiple specialties, requiring collaborative efforts across various disciplines; 2) Encompasses multiple software applications, necessitating seamless information transfer; 3) Involves a vast amount of data, requiring modeling efficiency; 4) Offers convenience in model modification; 5) Exhibits extensibility.

Table 1 [13] makes a comparative analysis of the advantages and disadvantages of the commonly used 3D geological BIM modeling software and its applicability. Table 2 makes a comparative analysis of the applicability of the commonly used BIM modeling software.

In summary, Revit emerges as the most suitable software for metro shield tunnel modeling, given its robust versatility, information integration capabilities, and extensibility. However, it exhibits a slight weakness in handling massive data, necessitating the use of complementary software to enhance efficiency.

Table 1. Comparison of 3D geological BIM modeling software.

Modeling Software	Advantages	Disadvantages	Scope of application
Revit	It can store and transfer the information of formation related parameters and materials, support the information interaction with other BIM software, and has abundant plug-ins for secondary development	Modeling efficiency is low, manual intervention is large, parameter and material assignment is complex, not intuitive and clear, data capacity is weak;	It is suitable for the situation of few stratum, simple terrain and flat stratum, and it is suitable for the situation that the application of model is not high and the result of model can be used for 3D display, stratum information storage and inquiry
Tekla	Support for information interaction with other BIM software; able to quickly create drill string model;	Modeling efficiency is low, parameters and materials can not be classified and assigned value storage and transfer, data capacity is weak;	It is suitable for the situation of few stratum, simple terrain and flat stratum, and it is suitable for the situation that the application of model is not high and the result of model can be used in 3D display
Civil3D	Data processing efficiency, simple operation, large data capacity, support Bourg operation and cutting processing, overall excavation (fill), earthwork balance real-time statistics;	The parameters and material information can not be stored and transmitted, and the vertical section can not display the situation of each stratum directly	For large-scale geological model creation, support the complex stratum terrain situation; The model results can be used for three-dimensional display, earthwork volume calculation, earthwork balance treatment, etc.
Bentley	Using geological data directly by reading GINT project documentation; Bentley has a range of BIM-specific software that can be accessed directly by Microstation, as well as each other	Bentley professional division is too detailed, leading to more difficult to learn;	Especially suitable for municipal engineering; support forward design; model results can be used for three-dimensional display

Table 2. Comparison of applicability of 3D geological BIM modeling software.

Modeling Software	Multi-specialty synergy	Data transitivity	Modeling efficiency	Modify convenience	Extensibility	Complete Series
Revit	Strong	Strong	General	General	Strong	Yes
Tekla	Weak	Strong	General	Strong	General	No
Civil3D	General	Strong	Strong	Strong	General	No
Bentley	Strong	General	Strong	General	Strong	Yes

4   Three-Dimensional Geological BIM Modeling Method Based on Shield Tunneling Simulation

4.1   Three-Dimensional Geological BIM Modeling Method

Dynamo is an auxiliary tool for parametric design based on Revit, assisting users in customizing algorithms to process data and generate geometric shapes, significantly enhancing modeling efficiency. The project adopts the Revit + Dynamo method to establish a 3D geological BIM model, as illustrated in Fig. 1 [14]. The main steps are as follows:

1) Organize survey data from each borehole and input the geological boundary point data into an Excel file, including 3D coordinates of the point, rock and soil properties, depth, thickness information of different soil layers, etc.
2) Utilize Dynamo to read the survey data from the Excel file and extract point data from each borehole.
3) Group point data in Dynamo based on geotechnical properties.
4) Generate triangulation surfaces for each soil layer using the point data.
5) Connect upper and lower triangular mesh surfaces to form mesh entities.
6) Import the mesh entity into Revit software to generate a 3D geological BIM model entity.

The resulting 3D geological BIM model is depicted in Fig. 2.

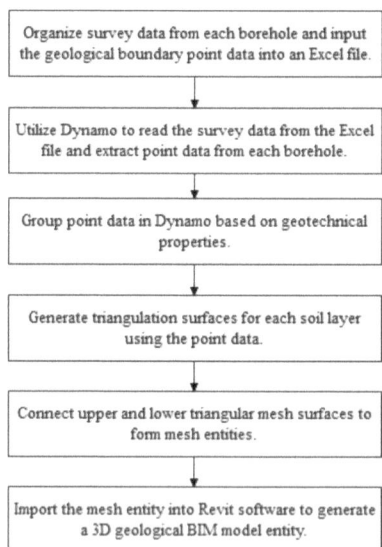

Fig. 1. Creation process on 3D geological BIM model.

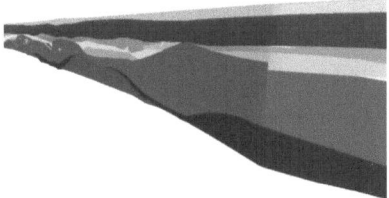

Fig. 2. 3D geological BIM model.

4.2 Information Attribute of 3D Geological BIM Model

In BIM model, the information attributes of shield tunnel engineering are added in the form of family parameters as the carrier of information communication.

Engineering Geological Information

Engineering geological information is divided into five categories: general information, test parameters, in-situ test, hydrogeology and bad geology [15], with a total of 27 sub-items of information, as shown in Table 3. The parameter interface of Revit family for information input is shown in Fig. 3.

Design Information

The design information of excavated soil includes 8 sub-items of information as shown in Table 4.

Construction Information

Construction information can be divided into three categories: time norm, construction measures and excavation parameters, with a total of 16 sub-items of information, as shown in Table 5.

Monitoring Information

According to different monitoring means, monitoring information includes 4 sub-items as shown in Table 6. The monitoring information is linked with the BIM model, which is a warning for shield construction.

After integrating the four types of information into the 3D geological BIM model, units such as investigation, design, construction, monitoring, and operation and maintenance can modify and utilize BIM model information within their respective authority. The communication flowchart of model information is illustrated in Fig. 4.

Table 3. Information of engineering geological.

Category	Information	Information attributes
General information	Stratigraphic number	Text type
	Geotechnical designation	Text type
	Times and causes	Text type
Test parameters	Natural moisture content	Numerical type
	Wet density	Numerical type
	Specific gravity of soil particle	Numerical type
	Saturation	Numerical type
	Natural porosity ratio	Numerical type
	Liquid index	Numerical type
	Plasticity index	Numerical type
	Compression modulus	Numerical type
	Cohesive force	Numerical type
	Angle of internal friction	Numerical type
	Compressive strength of rock	Numerical type
	sensitivity	Numerical type
In-situ test	Standard penetration	Numerical type
	Dynamic sounding	Numerical type
	Static sounding	Numerical type
	Acoustic test	Numerical type
Hydrogeology	Groundwater content	Text type
	Confined water pressure	Numerical type
	Permeability coefficient	Numerical type
	Water corrosion grade	Text type
Adverse geology	Liquefiable sand	Yes-no type
	Silty soil	Yes-no type
	Karst cave	Yes-no type
	Harmful gas	Yes-no type

4.3 Information Inquiry and Transfer Between BIM Model and Finite Element Model

The 3D geological BIM model and finite element model are created using different software with distinct formats, making direct connection of data information unfeasible. To address this issue, the IFC standard or secondary development technology is often employed for solving information transmission problems [16, 17]. However, these methods are limited to transferring only the geometric information of the model, leaving

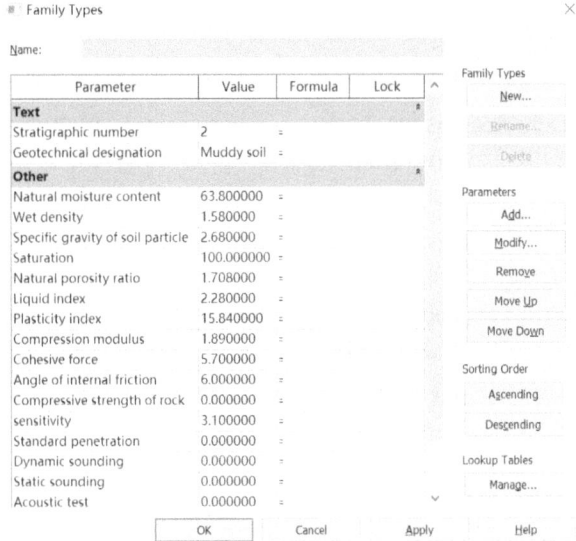

Fig. 3. Revit family parameter

Table 4. Design information of excavation soil.

Category	Information	Information attributes
Excavated soil design information	Elasticity Modulus	Numerical type
	Shear modulus	Numerical type
	Poisson's ratio	Numerical type
	Dynamic elastic modulus	Numerical type
	Dynamic shear modulus	Numerical type
	Dynamic Poisson's ratio	Numerical type
	Cohesive force	Numerical type
	Angle of internal friction	Numerical type

non-geometric information untouched. Extracting information from the 3D geological BIM model and achieving intelligent transmission of finite element model data represents a critical challenge in the integration of BIM models with shield construction simulation.

Building upon the foundation of establishing a comprehensive 3D geological BIM model, this paper presents a method for rapidly extracting BIM model information from any shield tunnel section, addressing information query and transfer challenges. The steps are illustrated in Fig. 5 [18]:

Table 5. Construction information.

Category	Information	Information attributes
Time norm	Driving speed	Numerical type
	Driving time per ring	Numerical type
	Assembly time per ring	Numerical type
	Process connection time	Numerical type
	Working time per ring	Numerical type
	Shift and maintenance time	Numerical type
	Number of rings drilled per day	Numerical type
Construction measures	Frequently questions	Text type
	Technical measure	Text type
	Organizational measure	Text type
	Muck improvement measures	Text type
Excavation parameters	Bin pressure	Numerical type
	Propulsion pressure	Numerical type
	Cutter speed	Numerical type
	Cutter torque	Numerical type
	Slag discharge	Numerical type

Table 6. Monitoring information.

Category	Information	Information attributes
Monitoring information	Abnormal of geological advance prediction	Yes-no type
	Abnormal of parameters of the shield machine	Yes-no type
	Abnormal of land subsidence	Yes-no type
	Abnormal of other	Text type

1) Collect and sort out engineering geological investigation data; 2) Use Revit + Dynamo to create 3D geological BIM model based on survey data, and integrate engineering geology, design, construction, monitoring and other information on the model; 3) Use Revit to establish BIM model of all shield segments, and mark the basic information such as segment ring number; 4) Merge the shield segment BIM model into the 3D geological BIM model according to the coordinates, and set the transparency of the 3D geological BIM model to 60%; 5) Find the segment with the specified ring number, identify the color quickly, and create a profile on the BIM model of the segment; 6) Directly query the geometric and non-geometric information of the 3D geological BIM

model on the BIM model displayed in the section, and transfer the information to the finite element model.

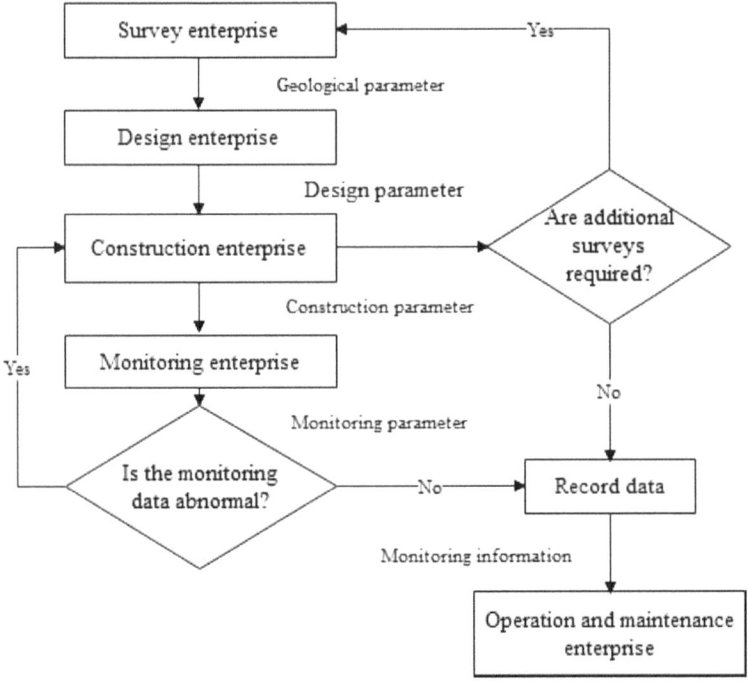

Fig. 4. Flow chart of model information communication.

The BIM model section corresponding to a shield ring number is shown in Fig. 6. The geological body where the shield segment is located can be directly identified on the section, and the spatial relationship between the shield segment and the geological body can be measured by measuring tools. Upon selecting the geological body, viewing family parameters allows for simultaneous querying of the comprehensive information integrated within that geological body, thus achieving information transfer within the BIM model.

The project utilized MIDAS/GTS finite element analysis software to model and partition the grid, and input boundary conditions, as depicted in Fig. 7. Subsidence deformation of the ground under different shield parameters is calculated to predict the distribution of ground subsidence in time and space, optimizing shield parameters. By establishing a 3D geological model for shield construction simulation, the project combines different geological models with finite element simulation to address safety simulation during shield construction and the selection of rational construction parameters under varying geological conditions. This integration enables the automation of information updates and the informatization of construction objects.

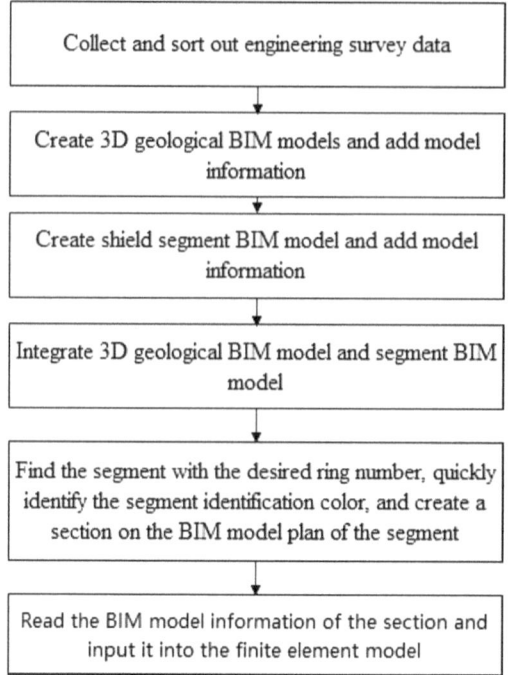

Fig. 5. Information transfer flow chart between BIM module and finite element model.

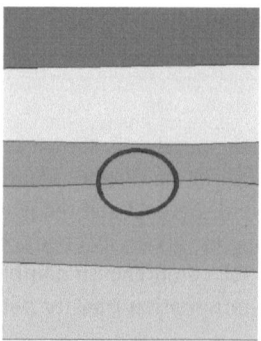

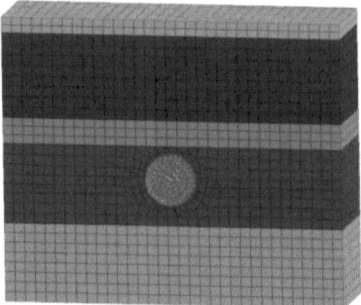

Fig. 6. Profile information of 3D geological BIM model.

Fig. 7. Finite element model.

5 Conclusion

1) This paper adopts the modeling approach of Revit + Dynamo, integrating survey, design, construction, and monitoring information through a three-dimensional geological BIM model. The generated information model meets the integrity and depth requirements of shield tunnel construction.

2) Building upon the 3D geological BIM model, a comparison with the segment BIM model is conducted. Using a sectional approach facilitates information exchange between BIM and finite element methods, resolving the issue of incompatible data interfaces across different software.

3) Simulation and analysis of shield parameters are performed under varying geological conditions, offering a method for selecting construction schemes. The predicted spatial and temporal distribution of deformations during construction provides theoretical support for monitoring plans.

4) The integration of the 3D geological modeling method with shield construction simulation addresses the automation of informationization and updates for shield construction projects, thereby enhancing the level of informationization in shield tunnel construction.

References

1. Koukoutas, S.P., Sofianos, A.I.: Settlements due to single and twin tube urban EPB shield tunnelling. Geotech. Geol. Eng. **33**(3), 487–510 (2015). https://doi.org/10.1007/s10706-014-9835-7
2. New Gubrist Tunnel: A Case for BIM. Tunnel **37**(5), 68–71 (2018)
3. Yang, H., Shi, H., Jiang, X.: Influence of construction process of double-line shield tunnel crossing frame structure on ground settlement. Geotech. Geol. Eng. **38**(2), 1531–1545 (2020). https://doi.org/10.1007/s10706-019-01109-3
4. Zongyang, C.: Interaction analysis of subway shield across construction site. Sichuan Build. Mater. **49**(06), 114–116 (2023)
5. Wanhong, L.: 3D geological modeling and application based on AUTO Civil 3D. People's Yangtze River **51**(8), 123–129 (2020). https://doi.org/10.16232/J.CNKI.1001-4179.2020.08.022
6. Zhu, Y.: Parametric modeling and application of tunnel based on BIM technology. Construction **54**(09), 25–30 (2022). https://doi.org/10.13402/J.Gcjs2022.09.111
7. Di, H.: Research on Integration of 3D Geological Modeling Based on BIM. Lanzhou University, Gansu (2019)
8. Bei, L.: Parametric modeling of tunnel based on BIM technology and its application research. Qingdao Technol. Univ. (2019). https://doi.org/10.27263/d.CNKI.GQUDC2019.000170
9. Xie, J.: Application of BIM modeling combined with finite element analysis in tunnel engineering. Shanxi Archit. **45**(3), 164–165 2019. https://doi.org/10.3969/J.ISSN.1009-6825.2019.03.086
10. Yujia, L.: FLAC3D complex geological body modeling based on BIM + Rhino. Pearl River Water Transp. **12**, 70–72 (2019)
11. Jung, Y.R., Nam, K.M., Kim, H.E., et al.: Analysis of correlation between shield TBM construction field data and settlement measurement data. J. Korean Tunn. Undergr. Space Assoc. **24**(1), 79–94 (2022). https://doi.org/10.9711/KTAJ.2022.24.1.079
12. Chaboo, C.S., Adam, S., Nishida, K., et al.: Architecture, construction, retention, and repair of faecal shields in three tribes of tortoise beetles (Coleoptera, Chrysomelidae, Cassidinae: Cassidini, Mesomphaliini, Spilophorini). ZooKeys **1177**, 87 (2023). https://doi.org/10.1016/j.tust.2003.11.008
13. Chu, S., Xia, M., Feng, M., et al.: Study on the method of creating 3D geological model of geotechnical engineering based on BIM technology. Tunnel Const. **39**(z1), 152–157 (2019). https://doi.org/10.3973/J.ISSN.2096-4498.2019.S1.022

14. Shanghai Jingdong Construction Development Co., Ltd.: A method of generating parameterized geological model based on BIM technology: CN201910487739.2, 25 October 2019
15. Xin, W.: Study on Engineering Geological Risk Identification of Shield Tunnel Based on BIM. Huazhong University of Science and Technology, Hubei (2018)
16. Sanmei, Y.: Research on the integration of 3D geological modeling and numerical simulation based on BIM. Hubei Univ. Technol. (2021). https://doi.org/10.27131/d.CNKI.UGC.2020.001052
17. Chen, G., Wu, J., Zhong, Y., et al.: Extension of 3D geological model based on IFC standard. Geotech. Mech. **41**(08), 2821–2828 (2020). https://doi.org/10.16285/J.RSM.November1785 2019
18. China Railway 19 Bureau Group Co., Ltd.: Guangzhou Institute of Science and Technology. A data transfer method and device for 3D geological BIM model and finite element model: CN202111306865.7, 18 March 2022

Research and Application of Dam Safety Monitoring Information Management System Based on GIS Platform

Yanan Li[1], Han Zhang[1(✉)], Zhiyong Yang[2], Yuxin Cao[1], and Yongqi Su[1]

[1] College of Water Resources and Hydropower, Sichuan University, Chengdu 610065, China
zhanghan@scu.edu.cn
[2] Chengdu Municipal Engineering Design and Research Institute, Chengdu 610065, China

Abstract. Hydropower stations, as a green energy source, have grown rapidly in China, and dam safety is critical to ensuring power generation efficiency. There is currently no safety monitoring and management platform that combines dam structure, geology, hydrology and other data. To this end, a dam safety monitoring information management system is built based on a GIS platform, with ArcGIS Engine serving as the GIS class library and the NET Framework serving as the programming model for program implementation. The system connects the GIS spatial database to the monitoring database, combines spatial analysis and data mining technology, and realizes functions such as 3D visual structures display, spatial analysis and evaluation of monitoring data, providing a more scientific and intelligent method for dam safety management. After verification in actual projects, the system can reasonably assess dam safety using measured data and projects, as well as make decision support recommendations.

Keyword: Dam · GIS · Safety monitoring · Information management system

1 Introduction

Whereas traditional dam safety information management systems primarily operate within a two-dimensional framework for managing monitoring instruments, Geographical Information System (GIS) is structured around data as the core, with three-dimensional (3D) visualization as a prominent feature. GIS systems enable positioning and querying of spatial data, with capabilities to collect, manage, analyze, and output various spatial data information [1]. This makes it easier to establish a safety monitoring and management system that integrates dam structure, geology, hydrology and other information, thereby resolving the problems of information islands and data dispersion that currently exist in most dam safety monitoring, and is critical for improving the level of dam safety management [2].

Pioneering research on dam safety monitoring information management systems began in the 1960s in developed countries like the United States, Japan, and Italy, with practical implementation starting in the 1970s. Among them, Italy has developed particularly quickly in this area and is at the international advanced level. Its monitoring

A. Bieliatynskyi et al. (Eds.): CSTTE 2023, LNCE 603, pp. 553–560, 2024.
https://doi.org/10.1007/978-981-97-5814-2_49

information management system enables real-time data collection, review, storage, and transmission, alongside strong online judgment and early warning functions [3].

Despite the fact that research into monitoring information management systems in China began late, it has advanced rapidly, owing to significant attention and support from both the government and businesses [4]. To meet the needs of rapid storage and real-time analysis of a large amount of monitoring data, the development of dam monitoring information management system based on GIS platform has attracted increasing attention. Xiao Zeyun et al. (2010) established the connection between monitoring instrument spatial information and monitoring data based on a GIS platform, and discussed issues like database design, spatial data expression, and the adaptability of monitoring data prediction models [5]. Wang Guowen et al. (2023) constructed a dam-strong earthquake monitoring and analysis system that combines information query, visualization display, and real-time monitoring using WebGIS technology [6].

The research on GIS platform-based dam safety monitoring information management systems is still in its early stages. Monitoring data is primarily analyzed using traditional methods, which underutilize the spatial analysis capabilities inherent in GIS. To address these limitations of GIS platform-based systems, this paper develops a GIS platform-based dam safety monitoring information management system based on previous research, which connects the GIS spatial database to the monitoring database to enable spatial analysis of monitoring data and other functions.

2 Safety Monitoring Information Management System

2.1 System Architecture Design

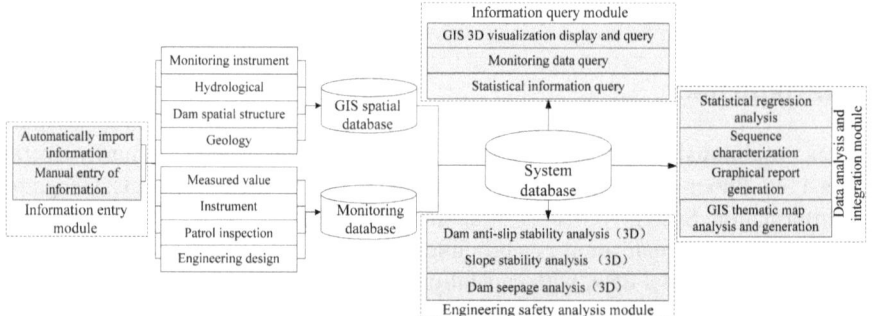

Fig. 1. Architecture of dam safety monitoring information management system based on GIS.

The primary goal of the dam safety monitoring information management system, built on the GIS platform, is to meet the needs of actual projects while also improving the efficiency of dam safety monitoring system management. Therefore, the design adheres strictly to the principles of practicality, dependability, security, maintainability, advancement, and standardization. Based on the design principles and the Client/Server (C/S) structure, this paper develops the overall architecture of the dam safety monitoring information management system using the GIS platform, as shown in Fig. 1. The system divides main functions of the client program into three modules: information query, data analysis and integration and engineering safety analysis. These modules carry out their functions by accessing data from the system database. The information entry module enters monitoring data either automatically or manually into the monitoring database and GIS database, which are then combined to form the system database.

2.2 System Functional Design

This paper delves deeply into actual projects, including Guandi (gravity dam), Changheba (earth-rock dam), Zilanba (gate dam), Lizhou (arch dam), the dam center of the Yalong river basin hydropower development company, and the dam management center of the Dadu river basin hydropower development company, to investigate the needs of dam safety monitoring managers for the use of monitoring information management systems. According to functional requirements articulated by users, the main functions of the dam safety monitoring information management system based on the GIS platform are divided into information entry, information review and transmission, information query, monitoring data analysis and compilation, engineering safety analysis and system setup and management, as shown in Fig. 2.

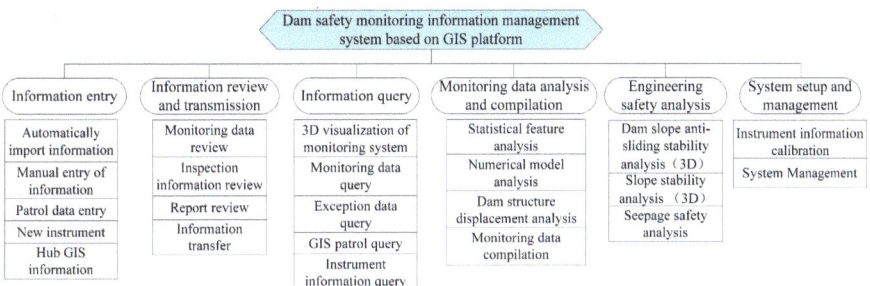

Fig. 2. Division of functional modules of system.

3 Key Technology Research

3.1 System Research and Development (R&D) Platform

GIS digitally represents the natural world, enabling the storage and processing of extensive geographic data in different periods, and has powerful comprehensive analysis capabilities of spatial information. The R&D benefits of GIS in dam safety monitoring information management systems are primarily reflected in the following three areas: (1) The GIS 3D simulation and query functions provide technical support for creating a 3D visualization of the hydropower station hub safety monitoring system. (2) GIS can efficiently process massive data, allowing for the rapid location of abnormal dam parts and timely feedback of abnormal data. (3) GIS's powerful spatial analysis function can perform visual domain analysis, underground seepage analysis, and engineering safety analysis in 3D data models.

The basic idea behind component GIS is to divide the major functional modules of GIS into multiple controls. Each control performs a different function, and controls that perform multiple functions are combined as needed to form an application system. This paper chooses ArcEngine, the most widely used component GIS development tool, as the GIS component for the dam safety monitoring information management system. It not only provides a powerful GIS class library with features such as layer display, 3D simulation, and spatial analysis, but it also increases system development autonomy.

The NET framework consists of the Common Language Runtime (CLR), a hierarchical collection of unified class libraries, and a componentized version of Active Server Pages. Compared to other software development technologies, NET allows for the use of a unified, component-oriented programming model that can well support each component's properties, events and methods, and developers do not need to understand the internal structure to realize collaboration between different applications and data access services at any time.

3.2 System Architecture

In order to fully utilize high-performance servers and the idle computing power of personal computers, this system uses the C/S architecture. The C/S architecture states that the client program sends the user's request to the server, which analyzes and processes it before returning the results to the client. The system database is installed on the server. When the client program runs, it performs related tasks by connecting to the server database. The C/S structure usually adopts a hierarchical design, with different levels operating independently of one another, making software upgrades and maintenance easier. The three-layer design structure of data access layer, business logic layer, and interface presentation layer is the most commonly used in current large and medium-sized application software, as shown in Fig. 3.

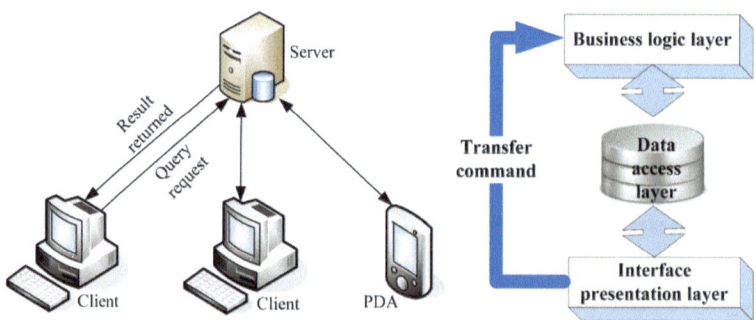

Fig. 3. Schematic diagram of C/S structure and program layering.

3.3 Links Between GIS Spatial Database and Monitoring Database

The system database consists of two databases: the GIS spatial database and the monitoring database. In order to manage and use these data scientifically and realize the rapid collection, query and analysis of monitoring information by the system, this paper combines the system functions to establish the logical relationship between the GIS spatial database and the monitoring database. That is, the connection between the attribute table in the GIS spatial database and the monitoring database table. The following example shows how to connect the two databases using the system query function.

When using the system, the first thing you see is the 3D visual monitoring system. The data information for the 3D model is entirely derived from the GIS spatial database. When you need to query monitoring data, simply click on the monitoring point to access the data stored in the monitoring database. In order to obtain accurate monitoring data for the corresponding monitoring points, the GIS spatial database table and the monitoring database table are linked using the instrument ID number. Each monitoring point in the system is unique, and it is labeled with an ID based on the type of instrument. For example, for dam appearance deformation monitoring, the displacement meters TP1, TP2, ..., TP10 can be assigned ID numbers 1001, 1002, ..., 1010, and the instrument number establishes the relationship between the GIS spatial database table and the monitoring database table, as shown in Fig. 4.

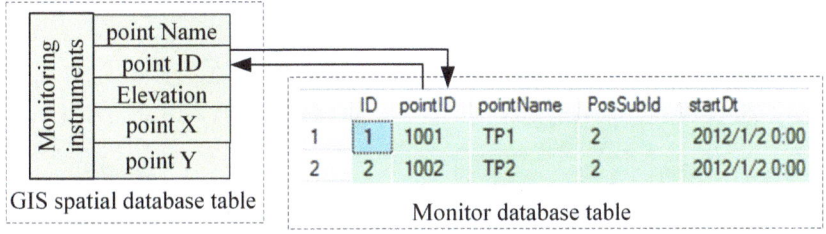

Fig. 4. Logical relationship between GIS spatial database table and monitoring database table.

4   Application of the System

4.1   Project Overview

A class II large (2) hydropower station project has an installed capacity of 195MW, and the barrage dam is designed as a first-class building. The primary hydraulic structures include gravel earth-corewall rockfill dams, flood discharge tunnels, drainage holes, water diversion systems, power plants, and so on. There are no comprehensive utilization requirements for flood control, shipping, or water supply.

The main monitoring items of this project include horizontal and vertical displacement of the dam surface, subsidence and horizontal displacement within the dam, foundation subsidence at the dam site, stress within the dam, stress and deformation of the anti-seepage wall, seepage of the dam body and foundation, slope stability on both sides and seepage around the dam, water level, water and air temperature, and earthquakes.

4.2 Application Effect

The dam safety monitoring information management system, built on the GIS platform, is used to realize 3D visualization of the monitoring system. The entire hub model can be arbitrarily scaled, rotated, and moved using the powerful topological relationships of GIS, as shown in Fig. 5, where the red point represents the project's dam appearance monitoring point.

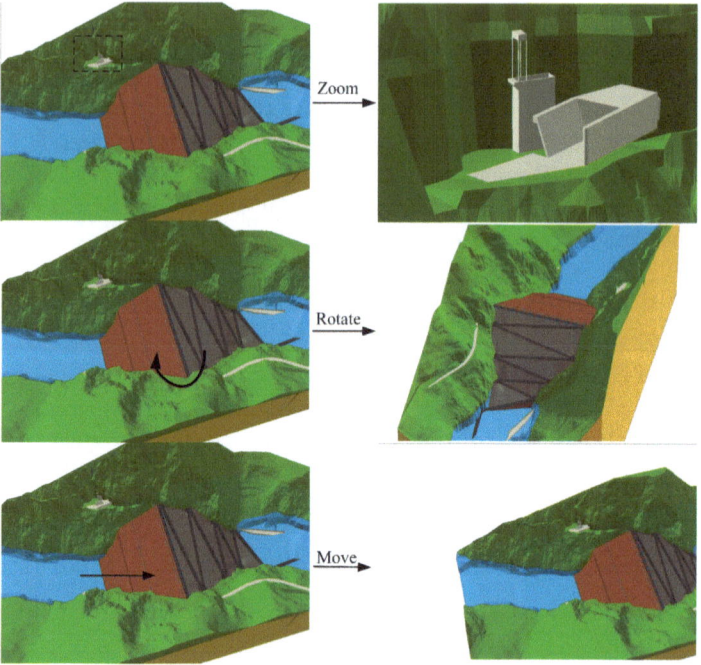

Fig. 5. System scaling, rotation, and movement functions.

We successfully applied the 3D stability analysis program for slopes developed using the rigid body limit equilibrium method to the project and obtained the most dangerous sliding surface by searching for given initial values. We then investigated the potential locations of the most dangerous sliding surfaces, accounting for changes in the elevation of the wetted surface as the water level in the reservoir fluctuated. By calling the measured values of seepage pressure buried in the dam body, a real-time wettability surface is generated, allowing for the analysis and calculation of the safety factor for dam slope stability while operating. In the downstream dam slope stability calculation diagram on November 28, 2009 (Fig. 6), the blue mesh surface is the most dangerous sliding surface obtained through the search, and the corresponding safety factor is 2.06. In comparison, the water storage safety appraisal report used the 2D simplified Bishop method to calculate dam slope stability under normal water storage conditions, yielding a result of 1.913. It is known from the water storage safety appraisal report that the 3D dam slope stability calculation result is slightly higher than the 2D calculation result.

This is mainly because the 3D calculation takes into account the lateral sliding force, and the sliding effect caused by the lateral sliding force is smaller than the sliding effect on the vertical section, so the safety factor will be slightly increased.

Fig. 6. The most dangerous sliding surface display.

5   Conclusion

This paper combines the actual situation of an earth-rock dam project, based on the GIS platform and uses C# as the development language to develop a dam safety monitoring information management system. The main conclusions are as follows:

1) The system divides the database into a GIS spatial database and a monitoring database, allowing for rapid collection and efficient management of spatial monitoring data. By applying GIS 3D visualization to monitoring data query, it successfully avoids the traditional management system that requires clicking on the tree menu to get monitoring data, but instead displays the whole monitoring system in 3D, and users can locate monitoring points through the functions of moving, zooming, rotating, roaming, which greatly improves the management efficiency of the dam monitoring system.

2) The GIS spatial analysis application successfully integrated 3D dam slope stability analysis as the system's safety analysis function. The stability analysis of the earth-rock dam slope is performed using engineering examples. The calculation results show that the 3D dam slope stability calculation calculations outperform the 2D calculations, which is consistent with the calculation rules. When combined with measured seepage pressure values in the dam body, real-time analysis of the dam slope's stability is realized, possible dangerous sliding surfaces are identified, and managers are prompted to take engineering measures to reinforce dangerous parts ahead of time to avoid disasters.

3) The 3D visualization of the dam safety monitoring information management system allows users to quickly grasp the operational status of various cascade hydropower stations in the whole basin. This remains the development path for the future dam safety monitoring information management system. The system described in this paper only applies arbitrary operations to points and lines. The ArcGIS Engine library has limitations when it comes to arbitrary surface generation and changes. If the technology can be broken, 3D deformation monitoring of dams, reservoir mountains, and strata can be easily implemented by combining monitoring data.

References

1. Abdelghani, L.: Modeling of dam-break flood wave propagation using HEC-RAS 2D and GIS: case study of Taksebt dam in Algeria. World J. Eng. (2023). https://doi.org/10.1108/WJE-10-2022-0405
2. Zhang, B., Shi, B., et al.: Summary of application and development trend of automation monitoring system for dam safety. Express Water Resour. Hydro. Inform. **43**, 68–73 (2022)
3. Oliveira, S., Alegre, A.: Seismic and structural health monitoring of Cabril dam. Software development for informed management. J. Civ. Struct. Health Monit. **10**, 913–925 (2020)
4. Zhou, S., Hu, W.: Research on construction of dam safety monitoring system under the background of digital twin water conservancy project. Pearl River **44**, 437–42+55 (2023)
5. Xiao, Z., Tian, B.: Research and application of dam safety monitoring system based on GIS platform. Adv. Sci. Technol. Water Resour. **30**, 48–52 (2010)
6. Wang, G., Zang, Y., et al.: Design and implementation of strong earthquake monitoring and analysis system for Tibet dam based on WebGIS. Xizang Sci. Technol. **1**, 66–69 (2023)

Computer Vision Technology in Cost Monitoring of Construction Projects

Xiaolin Ou[(✉)]

School of Architectural Engineering, Guangzhou Institute of Science and Technology,
Guangzhou 510080, Guangdong, China
876309002@qq.com

Abstract. With the continuous development of technology, the application of computer vision technology in cost monitoring of construction projects is becoming increasingly important. By utilizing computer vision technology, construction companies can monitor the progress of construction sites, material usage, and allocation of human resources in real-time, thereby better controlling project costs. The application of this technology can not only improve the efficiency of construction projects, but also reduce human errors and waste, which is of great significance for the successful completion of projects. Therefore, construction companies should actively adopt computer vision technology to improve the efficiency and accuracy of cost monitoring. In recent years, China's economy has continued to grow, continuously driving the development of infrastructure and manufacturing industries. However, at the same time, the production safety situation in China is becoming increasingly severe, and safety accidents continue to occur. One of the main causes of safety accidents is human misconduct, which includes workers wearing uniforms and safety helmets improperly during construction. Workers engage in some dangerous behaviors during construction, such as making phone calls, falling, and squatting for long periods of time. In response to the above issues, this article studies how the SSD (Single Shot MultiBox Detector) algorithm, CNN (Convolutional Neural Networks) algorithm, and YOLO (You only look once) algorithm can be applied to the behavior detection of construction workers, and compares them with existing object detection algorithms through experiments. The experimental data shows that the AP (Average precision), AP50, and AP75 values of the SSD-CNN algorithm in this article are 51.29%, 69.85%, and 54.81%, respectively. Among all the algorithms, the values of the three indicators rank in the top few.

Keyword: Cost Monitoring of Construction Projects · SSD Algorithm · Target Recognition and Detection · CNN Algorithm

1 Introduction

The environment at construction sites is usually very harsh, with various production materials arranged in a disorderly manner, and workers, equipment, and production materials moving together. So it is easy to cause building accidents at construction

A. Bieliatynskyi et al. (Eds.): CSTTE 2023, LNCE 603, pp. 561–571, 2024.
https://doi.org/10.1007/978-981-97-5814-2_50

sites. Many construction accidents are collision accidents, such as workers being hit by mobile equipment on construction sites. Even if workers wear high visibility clothing on site in accordance with existing safety regulations and standards, construction accidents may still occur. Therefore, adding additional safety measures to protect the safety of construction workers is of great research significance.

The purpose of this article is to explore and evaluate the feasibility and effect of applying computer vision technology in the cost monitoring of construction projects. Through the introduction of computer vision technology, functions such as automatic identification of workers, attendance records, working time tracking, work efficiency evaluation, and safety behavior monitoring can be realized. This computer vision-based cost monitoring method has many advantages, including reducing human error, improving data accuracy, real-time monitoring and evaluation of worker behavior, and improving safety.

In Chapter 3, this article introduces the CNN algorithm, the principle of object detection algorithm, and project schedule management. In Chapter 4, an experimental analysis of a construction worker object detection model based on the SSD-CNN algorithm is presented. Finally, a summary of the entire article is made.

2   Related Works

Experts have long conducted specialized research on cost monitoring for construction projects. Alizadehsalehi S has established a digital twin technology based approach to achieve real-time monitoring of construction progress [1]. Elghaish F's research has found that drones can be combined with 4D Building Information Modeling (BIM) to evaluate project progress and check compliance with geometric design models [2]. Keskin B aimed to systematically explore how building information models can change complex digital infrastructure environments (such as airports) by improving the connections and collaboration between key stakeholders and building technology solutions. It was found that the application of BIM can improve the utilization rate of the construction technology ecosystem and enhance the degree of process interconnection among various participating entities [3]. Dallasega P analyzed the advantages and disadvantages of production planning and management methods in construction enterprises through a systematic review of relevant literature [4]. Parsamehr M has established a decision-making system for construction industry enterprises based on BIM, and also pointed out the shortcomings in future research [5]. Rafsanjani H N revealed the important roles of virtual design and manufacturing and digital twins in the two technologies through comparative analysis, analyzed the development trends of the construction industry, and made cost predictions [6]. Nafe Assafi M adopted a 4D construction information model to reduce manual intervention, human error, and project progress. He used Auto desk Navisworks design software to design a 4D-BIM system. Finally, using this system, a simulation was conducted on a project under construction, which was delayed due to design errors and inefficient planning. Through simulation of actual engineering, good results have been achieved [7]. Crowther J studied the role of 4D-BIM in construction projects. There are 8 ways to build 4D-BIM to support project performance. It was found that among BIM coordinators, there was a lack of shared responsibility, a severe lack of understanding and training in 4D-BIM, and a complex process for effective execution

[8]. Chen X aims to provide an overview of the technologies used in the construction industry and the benefits they bring [9]. Han Y sorted out the current situation of the prefabricated building supply chain and predicted its future development trends by reviewing existing research [10]. Hou L has introduced the traditional open construction industry into innovative technological and collaborative models. Collaborative designers include architects, engineers, architectural experts, real estate managers, as well as providers of building materials, software, production equipment, assembly equipment, and more [11]. Oke A E aimed to improve its application level in the construction industry by evaluating the application of the Internet of Things in construction projects [12]. In order to identify and analyze the potential hazards that may pose a threat to human life and property in construction projects, as well as the associated safety risks, Namian M's research shows that the use of unmanned aerial vehicles can bring many hazards to construction projects that industry professionals are not aware of. The three most serious safety hazards are "collision with property", "collision with people", and "distraction" [13]. Statsenko L conducted systematic research on the construction industry, constructed a key technology system for Building 4.0, and conducted empirical research on its application in the construction industry. C4.0 is suitable for energy conservation, prefabricated buildings, sustainable development, safety and environmental protection, indoor comfort, and efficient asset utilization [14]. In order to investigate the application of cloud computing in Nigeria's construction industry for sustainable development, Oak A E used exploratory factor analysis method for empirical analysis. The survey results show that this method has significant advantages in terms of extensive application in information storage (location independence), high situational awareness ability, team collaboration ability, compatibility with advanced production equipment, and optimization of engineering plans [15]. Traditional construction project cost monitoring often relies on manual input and processing of data, lacking real-time cost information, making it difficult to detect and solve cost problems in a timely manner. Moreover, project cost monitoring may not be able to identify the reasons for cost fluctuations in a timely manner, resulting in ineffective measures to control costs. Construction projects involve a large number of workers and labor resources, and it is a complex task to manage and monitor the actual work conditions, attendance and working hours of these workers. These projects have certain safety risks, and workers need to comply with safety regulations and operating procedures to reduce the risk of accidents and personal injuries. It is essential to monitor the safety behavior and compliance of workers, as well as to detect and resolve potential safety hazards in a timely manner. In addition, construction projects involve a large amount of material and equipment resources, and improper use or abuse of these resources by workers may lead to waste and increased costs. Monitoring and controlling workers' use of resources and preventing and reducing waste of resources is an important challenge.

3 Methods

3.1 CNN Algorithm

The essence of CNN is a multi-layer perceptron, which can directly input images without the need for complex image preprocessing. It has the advantages of local connections and shared weights [16, 17]. Local connection refers to changing the connection method between the upper input unit and the hidden unit, and the input area connected by each hidden unit is the receptive field. Implicit units can only connect a small portion of adjacent regions of the upper input unit, which greatly reduces computational complexity [18]. Parameter sharing is based on the assumption that the importance of parameters to different points in the image is the same. Therefore, the same plane should share the same set of weights and biases, which has two benefits. On the one hand, repetition can be recognized without considering its position in the field of view. On the other hand, the number of parameters is greatly reduced, reducing computational complexity and time consumption. The weight sharing of CNN is similar to the image analysis method of biometrics, so convolutional neural networks have significant advantages in reducing the number of weights, especially when inputting multi-dimensional images, and are suitable for use in machine recognition images [19, 20].

3.2 Principle of Object Detection Algorithm

Object detection algorithm is a new method that introduces computer technology into the visual system. It can generate a candidate box in an image, and then classify and recognize the objects in the candidate box. This algorithm first inputs the image into the trained model, then operates on the input image to generate a candidate box, classifies the objects in the candidate box, and then uses the NMS (Non-Maximum Suppression) algorithm to remove excess candidate boxes. Figure 1 shows the identification process.

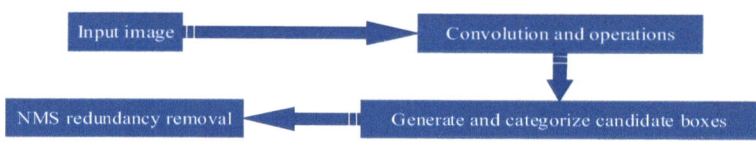

Fig. 1. Object Detection Process

The commonly used object detection algorithm models include region based R-series models and regression based models. The characteristics of the R series model are as follows: using an end-to-end convolutional network model for recognition and detection to ensure a certain degree of accuracy, but its timeliness and detection speed need to be improved. The characteristic of regression models is to transform detection tasks into regression problems, improving detection speed and meeting real-time requirements. This article uses the YOLO V3 algorithm under a regression model for object detection.

3.3 Project Progress Management

Project schedule management mainly manages the contract period goals of engineering projects. In order to ensure that the project can be completed as expected, it is necessary to prepare a reasonable and scientific schedule plan. Then, during the implementation process, it is possible to strictly follow the plan. If there are any discrepancies between the actual progress and the planned progress, it is necessary to identify the reasons and make adjustments or corrections to ensure the smooth progress of the project. Progress management is the most important part of project management, and it holds the same status as cost management and quality management. Good schedule management not only allows for early completion, but also saves a lot of manpower and material resources.

Currently, how to reduce the occurrence of safety accidents and strengthen the safety management of workers has become a complex and important task. Firstly, this article proposes a method to obtain the bounding rectangle Box of a character based on the human skeleton information extracted by a CNN network; Then the SSD algorithm can track the status of the character Box in the image in real time and set different IDs for each character. Based on the ID, it is possible to quickly lock characters with improper wearing and abnormal movements. Finally, different character images can be extracted based on the size of the character Box, and different character images as well as corresponding bone and ID information can be input to the server for recognition. The server can provide feedback on the problematic individuals to the client.

The SSD (Single Shot MultiBox Detector) network mainly includes the following core features:

(1) Using multi-scale feature maps for object detection

    The size of the six feature maps of SSD is not equal. Predicting on multiple feature maps with different resolutions can use convolution or pooling from shallow to deep to reduce the size of the feature maps. The receptive field of shallow small downsampling feature maps is relatively small and is generally used to predict small-scale objects; The receptive field of deep downsampling feature maps is larger and is generally used to predict large-scale objects.

(2) Using convolution for detection

    Compared to other single-stage network models, such as YOLO, which uses fully connected layers to predict the coordinates of bounding boxes and their corresponding classifications, SSD uses convolution to extract features from feature maps of different scales.

(3) Prior boxes with different scales and aspect ratios

The SSD algorithm sets different specifications of prior boxes on its feature maps of different scales, and the scale of the prior boxes increases linearly with the decrease of feature map size. The prior box scale of each layer is shown in Eq. (1).

$$S_k = S_{\min} + \frac{S_{\max} - S_{\min}}{m - 1}(k - 1), k \in [1, m] \tag{1}$$

Among them, S_k represents the proportion of the k-th prior box size of SSD to the image, while $S_{\min}$ and $S_{\max}$ represent the lowest and highest feature maps of 0.2 and 0.9, respectively, corresponding to the second and sixth feature maps. m represents the number of predicted feature layers, which should have been 6 here. However, since the first layer feature map is set separately, the calculated value here is 5. The first feature map is generally $S_{min}/2$, which is 0.1.

In order to reduce the computational cost of convolution operations, most mainstream lightweight neural network models currently use group convolution or depthwise separable convolution, but it also brings a lot of computational cost. By permeating the channels after group convolution, the information flow between groups can be better reflected, thereby enhancing the feature expression ability of the model. The essence of channel permutation is to achieve information exchange between grouped convolutional channels without increasing computational complexity. On this basis, the paper replaced traditional 3×3, 1×1 and other convolutional modules with group convolution and channel permutation, thereby achieving model compression.

The mean and variance during testing are based on the moving average distance during training, as follows:

(1) When testing or predicting, only a single data can be passed each time, and the model can use global statistics instead of batch statistics.
(2) When training each batch, a set (mean, variance) can be obtained.
(3) The global statistic is to calculate the mathematical expectation corresponding to these means and variances, and the specific formula is as follows:

$$E[x] \leftarrow E[\mu_i] \tag{2}$$

$$Var[x] \leftarrow \frac{m}{m-1}E[\sigma_i^2]Var[x] \leftarrow \frac{m}{m-1}E[\sigma_i^2] \tag{3}$$

Among them, μ_i and σ_i respectively represent the mean and standard deviation saved in the i-th batch processing, m is the batch size, and the coefficient $m/m-1$ is used to calculate unbiased square error estimation. At this point, BN (x) changes to:

$$BN(x_i) = \gamma \frac{x_i - E[x_i]}{\sqrt{Var[x_i] + \varepsilon}} + \beta \tag{4}$$

Among them, γ is the training parameter and β is the iteration step size.

4   Results and Discussion

The commonly used measurement indicators for detection include Recall, Precision, Average Accuracy (AP), and Intersection of Union (IoU). Precision is the proportion of TP (true positive) in the identified target image, and Recall is the proportion of TP in the identified target image to the target image in the test set. The threshold is set to 0.5, which means that when the intersection to union ratio is greater than or equal to 0.5, it is determined that the class it belongs to can be recognized, and then the most suitable box is selected and output through non maximum suppression. After training the model on a cluster for 500 epochs, good weight parameters were obtained. After training, the Precious, Recall, and F1 data for each category in the dataset of the model are shown in Fig. 2.

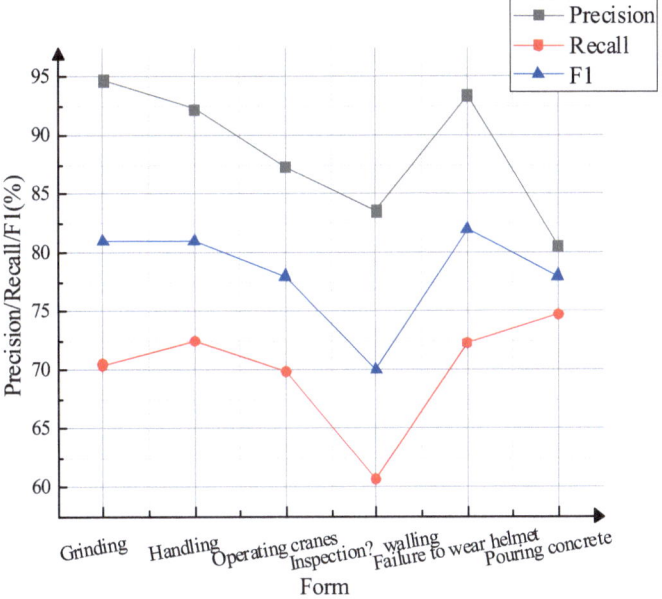

Fig. 2. Accuracy, recall, and F1 of trained categories

From Fig. 2, it can be seen that the accuracy, recall, and F1 of worker behavior object detection based on SSD algorithm and CNN algorithm are 80%–95%, 60%–75%, and 70%–82.5%, respectively. These indicators indicate that worker behavior object detection models based on SSD algorithm and CNN algorithm have certain performance in different aspects, and can help evaluate the performance of the model in worker behavior detection tasks. However, these indicators also indicate that the model's performance in terms of recall is relatively low and may require further improvement to improve the accuracy of identifying worker behavior.

The following conclusions can be drawn from the data in Table 1: AP50 refers to using a 50% recall rate as the benchmark when calculating AP in object detection tasks. This means that when calculating AP, only the accuracy when the recall rate reaches 50% is considered to evaluate the performance of the model when half of the targets are correctly detected. This indicator can help evaluate the performance of the model in high recall situations. AP50 is commonly used to evaluate the performance of object detection models on different datasets and can be used in conjunction with other AP values (such as AP75, AP90, etc.) to comprehensively evaluate the performance of the model. Among them, the values of AP, AP50, and AP75 for the SSD-CNN algorithm in this article are 51.29%, 69.85%, and 54.81%, respectively. Among all the algorithms, the values of the three indicators rank in the top, indicating that the algorithm has achieved excellent detection performance under different evaluation indicators.

Figure 3 shows the histogram values when abnormal behavior occurs in a video recording. The video consists of 700 frames, with the first 600 frames corresponding to normal numerical curves and 600–700 frames corresponding to abnormal situations.

Table 1. Comparison of Target Detection Performance in Construction Scenarios (%)

Algorithms	AP	AP50	AP75
SSD	35.2	59.1	37.62
YOLOv3	38.17	65.4	41.28
FCOS	46.72	68.99	49.14
R-CNN	47.26	69.39	51.01
NAS-FPN	46.84	67.72	49.18
SOLO	48.46	67.24	51.28
Retina Net	49.24	71.89	52.81
Faster R-CNN	50.19	74.35	55.34
TridentFast	50.15	72.84	54.2
Mask R-CNN	50.29	74.46	55.45
PointRend	50.57	74.66	55.4
YOLOX-S	47.85	67.74	52.52
SSD-CNN	51.29	69.85	54.81

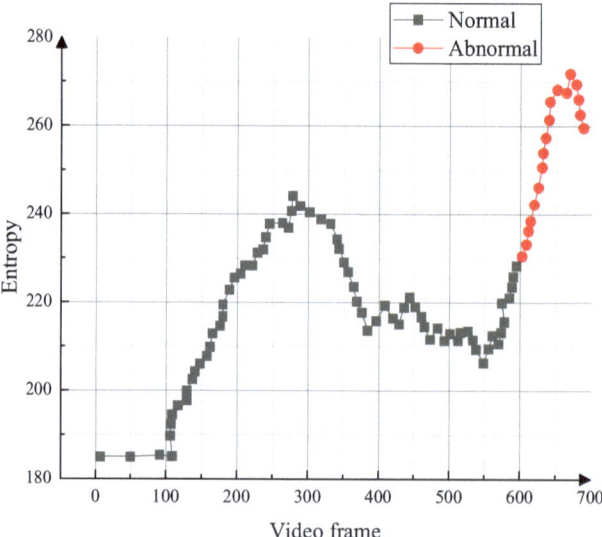

Fig. 3. Change curve of entropy

For normal behavior, the test is square, while for abnormal behavior, it is circular. It can be seen that when an ordinary action becomes an abnormal action, human actions become more irregular, and the initial value can also increase. After more than 700 frames, the human body has already left the camera. The histogram legitimate value detection algorithm based on SSD algorithm and CNN algorithm can accurately detect the time when abnormal behavior occurs.

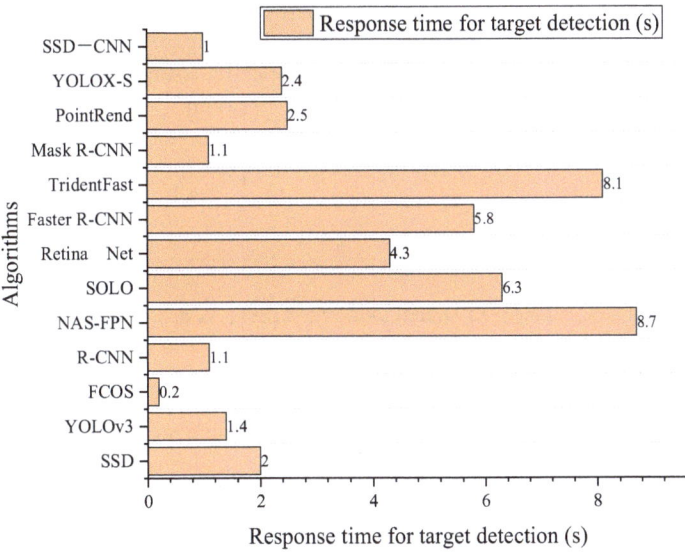

Fig. 4. Response time (s) for object detection using different algorithms

From Fig. 4, it can be seen that the response times of SSD CNN, Mask R-CNN, R-CNN, and FCOS are only 1s, 1.1s, 1.1s, and 0.2s, respectively, making them the fastest among all algorithms.

5   Conclusion

The traditional construction site inspection methods mainly rely on manual inspections and surveillance cameras, which have many problems such as low efficiency, high cost, and easy error. Therefore, detection technology based on artificial intelligence has become a new solution. The SSD algorithm, as an advanced object detection algorithm, has the advantages of fast detection speed, high accuracy, and strong adaptability, and is therefore widely used in construction site detection. The SSD algorithm discussed in this article can effectively detect various targets in construction sites, such as construction personnel, mechanical equipment, material stacking, etc., by performing multi-scale convolution operations on images. Compared with traditional detection methods, SSD algorithm can not only achieve comprehensive monitoring of construction sites, but also accurately identify and locate different targets, providing strong technical support for the management and monitoring of construction sites. Through the introduction of computer vision technology, this paper designs and implements a prototype of a cost monitoring system based on this technology. The research results show that computer vision technology can realize the functions of automatic identification of workers, tracking of working hours, work efficiency evaluation and safety behavior monitoring, and provide a novel, automated and accurate cost monitoring method. The research results of this paper emphasize the importance of technological innovation to improve construction project management. The introduction of computer vision technology has brought new ideas and methods to the cost monitoring of construction projects, and promoted innovation and development in the field of project management. In the cost monitoring of construction projects, there are still many specific issues that need to be studied in depth. For example, how to accurately identify workers' work status and behavior, how to effectively evaluate work efficiency and safety behavior, etc. Future research can conduct in-depth research on these problems and propose more specific solutions.

References

1. Alizadehsalehi, S., Yitmen, I.: Digital twin-based progress monitoring management model through reality capture to extended reality technologies (DRX)[J]. Smart and Sustainable Built Environment **12**(1), 200–236 (2021)
2. Elghaish, F., Matarneh, S., Talebi, S., et al.: Toward digitalization in the construction industry with immersive and drones technologies: a critical literature review. Smart and Sustainable Built Environment **10**(3), 345–363 (2021)
3. Keskin, B., Salman, B., Ozorhon, B.: Airport project delivery within BIM-centric construction technology ecosystems. Eng. Constr. Archit. Manag. **28**(2), 530–548 (2021)
4. Dallasega, P., Marengo, E., Revolti, A.: Strengths and shortcomings of methodologies for production planning and control of construction projects: a systematic literature review and future perspectives. Production Planning & Control **32**(4), 257–282 (2021)
5. Parsamehr, M., Perera, U.S., Dodanwala, T.C., et al.: A review of construction management challenges and BIM-based solutions: perspectives from the schedule, cost, quality, and safety management. Asian Journal of Civil Engineering **24**(1), 353–389 (2023)
6. Rafsanjani, H.N., Nabizadeh, A.H.: Towards digital architecture, engineering, and construction (AEC) industry through virtual design and construction (VDC) and digital twin. Energy and Built Environment **4**(2), 169–178 (2023)
7. Nafe Assafi, M., Hossain, M.M., Chileshe, N., et al.: Development and validation of a framework for preventing and mitigating construction delay using 4D BIM platform in Bangladeshi construction sector. Constr. Innov. **23**(5), 1255–1278 (2023)
8. Crowther, J., Ajayi, S.O.: Impacts of 4D BIM on construction project performance. Int. J. Constr. Manag. **21**(7), 724–737 (2021)
9. Chen, X., Chang-Richards, A.Y., Pelosi, A., et al.: Implementation of technologies in the construction industry: a systematic review. Eng. Constr. Archit. Manag. **29**(8), 3181–3209 (2022)
10. Han, Y., Yan, X., Piroozfar, P.: An overall review of research on prefabricated construction supply chain management. Eng. Constr. Archit. Manag. **30**(10), 5160–5195 (2023)
11. Hou, L., Tan, Y., Luo, W., et al.: Towards a more extensive application of off-site construction: a technological review. Int. J. Constr. Manag. **22**(11), 2154–2165 (2022)
12. Oke, A.E., Arowoiya, V.A.: Evaluation of internet of things (IoT) application areas for sustainable construction. Smart and Sustainable Built Environment **10**(3), 387–402 (2021)
13. Namian, M., Khalid, M., Wang, G., et al.: Revealing safety risks of unmanned aerial vehicles in construction. Transp. Res. Rec. **2675**(11), 334–347 (2021)
14. Statsenko, L., Samaraweera, A., Bakhshi, J., et al.: Construction 4.0 technologies and applications: a systematic literature review of trends and potential areas for development. Construction Innovation **23**(5), 961–993 (2023)

15. Oke, A.E., Kineber, A.F., Al-Bukhari, I., et al.: Exploring the benefits of cloud computing for sustainable construction in Nigeria. J. Eng. Desi. Technol. **21**(4), 973–990 (2023)
16. Deepalakshmi, P., Lavanya, K., Srinivasu, P.N.: Plant leaf disease detection using CNN algorithm. Int. J. Info. Sys. Model. Desi. (IJISMD) **12**(1), 1–21 (2021)
17. Wu, J.M.T., Li, Z., Herencsar, N., et al.: A graph-based CNN-LSTM stock price prediction algorithm with leading indicators. Multimedia Syst. **29**(3), 1751–1770 (2023)
18. Permana, S.D.H., Saputra, G., Arifitama, B., et al.: Classification of bird sounds as an early warning method of forest fires using convolutional Neural Network (CNN) algorithm. J. King Saud Uni.-Comp. Info. Sci. **34**(7), 4345–4357 (2022)
19. Chen, D.J.I.Z.: Automatic vehicle license plate detection using K-means clustering algorithm and CNN. J. Electr. Eng. Automat. **3**(1), 15–23 (2021)
20. Liu, J., Ban, W., Chen, Y., et al.: Multi-dimensional CNN fused algorithm for hyperspectral remote sensing image classification. Zhongguo Jiguang/Chinese Journal of Lasers **48**(16), 1–11 (2021)

Health Monitoring and Evaluation Method of Civil Engineering Structure Based on Machine Learning

Xinyao Wang(✉)

College of Harbour and Coastal Engineering, JiMei University, Xiamen, China
xinyaoyao.wang@foxmail.com

Abstract. The health monitoring and assessment of civil engineering structures are important for ensuring structural safety and can Sustainability is critical. Traditional monitoring methods often require a lot of time and resources, and have some limitations in real-time and accuracy. In order to overcome these problems, this paper proposes a method of health monitoring and evaluation of civil engineering structures based on machine learning. This method combines sensor data with machine learning algorithm to realize real-time and accurate monitoring and evaluation of structural health status. By training and learning the data of existing civil engineering structures, the correlation model between structural health and sensor data is established. Through field experiments and simulation verification, the results show that the proposed method can effectively detect structural abnormalities, early warning potential faults of the structure, and provide scientific decision-making basis for engineers, so as to improve the safety and reliability of civil engineering structures.

Keywords: Civil Engineering · Structural health monitoring · Machine learning · Sensor data · anomaly detection

1 Introduction

The safety and reliability of civil engineering structures have always been the focus in the field of engineering construction Important concerns Point. Traditional civil engineering structure monitoring methods, such as manual inspection and regular detection, have some limitations. First, manual inspection requires a lot of time and manpower investment, and it is difficult to achieve a comprehensive monitoring of the structure. Secondly, regular detection can only provide discrete data points, which can not reflect the health status of the structure in real time, so it may miss the early signals of structural abnormalities [1]. In addition, the traditional monitoring methods also have some difficulties in the detection and diagnosis of complex structural problems, such as micro deformation and hidden damage.

© The Author(s) 2024
A. Bieliatynskyi et al. (Eds.): CSTTE 2023, LNCE 603, pp. 572–581, 2024.
https://doi.org/10.1007/978-981-97-5814-2_51

Civil engineering structures carry various loads, such as natural environment (such as wind, earthquake, etc.) and human activities (such as vehicles, pedestrians, etc.), and are affected by climate change, material fatigue, structural aging and other factors for a long time [2]. These external and internal factors may cause damage, deformation or fatigue of the structure, and then adversely affect the safety and reliability of the structure. Among them, human activities such as vehicle traffic and weight placement, as a common source of load, the pressure on the structure can not be ignored. At the same time, with the passage of time, structural aging has also become a problem that can not be ignored. The combined action of these factors may lead to damage, deformation and even destruction of civil engineering structures, which will pose a serious threat to their safety and reliability.

In order to deeply explore the impact of these factors on civil engineering structures, this paper is committed to real-time and accurate detection of the state of the structure through the health monitoring and evaluation method based on machine learning. Machine learning algorithm can intelligently process and analyze a large number of sensor data, extract the characteristics of structural health state, and establish the model related to structural health. By learning the known structural health status and sensor data, the machine learning model can realize the real-time and accurate monitoring and evaluation of structural health status [3]. The practical significance of this research is that it can not only improve the safety of civil engineering structures, but also reduce the economic loss and social impact caused by structural damage, which has important theoretical value and practical significance.

2 Materials and Methods

2.1 Algorithm and Structural Damage Detection

Algorithm

First of all, as for the selection of machine learning algorithms, I plan to use two algorithms: support vector machine (SVM) and random forest. SVM algorithm has good performance in classification problem. It can separate different categories of samples by finding a hyperplane, and maximize the distance between different categories of samples. Random forest is an integrated learning method, which can improve the stability and accuracy of the model by constructing multiple decision trees and integrating their prediction results. The combination of these two algorithms can overcome the limitations of a single algorithm to a certain extent and improve the generalization ability of the model.

Structural Damage Detection

The most unfavorable combination means This part Impending Structural damage, When the structure is damaged, the structural parameters such as mass, stiffness and damping will change to varying degrees, which will lead to the changes of natural frequency, mode shape and impedance of the structure. Many existing damage detection methods are proposed based on these parameters. Zong Zhou Hong and others reviewed the research progress of civil engineering structural damage from the aspects of damage diagnosis, system identification, model modification and sensor layout; Rytter divided damage identification into four progressive levels in his doctoral dissertation: the first level determines whether there is damage in the structure; The second level determines the geometric location of the damage based on the first level; The third level quantifies the severity of damage on the basis of the second level; The fourth level predicts the remaining service life of the structure on the basis of the third level. Farrar and Worden divided damage identification into five processes, including SHM condition monitoring, NDT evaluation, statistical process control and damage prediction [4]. Giraldo divided damage detection and identification into three types in his doctoral dissertation: vibration based method, static based method and direct detection method of structure.

Structural damage detection can be transformed into a mathematical optimization problem in many cases. Genetic algorithm based on Darwin's natural selection and genetic theory is a typical global optimization method. Compared with traditional optimization methods, it has its own characteristics: it has the ability of self-organization, self-adaptive and self-learning, and can automatically discover environmental characteristics and laws according to environmental changes, so it can solve complex unstructured problems; It has essential parallelism and is very suitable for large-scale parallel operation; The search range is wide, which greatly reduces the possibility of falling into the local optimal solution; When solving various kinds of problems, only the objective function needs to be defined, without gradient and other traditional information [5]. Due to the above characteristics of simulating the natural evolution process, genetic algorithm has unique advantages in solving the multi peak complex objective function in damage identification.

Dynamic Control of Structural Damage

As shown in Fig. 1: the dynamic control chart of civil engineering structure health monitoring based on machine learning shows a comprehensive and fine maintenance process. Through GPS data collection by sensors, the system can obtain the state information of the structure in real time. These data are then compared with the state of the initial healthy structure, and the machine learning algorithm is used to analyze a large number of data, so as to accurately identify the abnormal changes or potential damage of the structure [6]. Once the damage affecting the structural safety is found, the system will quickly diagnose the damage, and further use the machine learning model to predict the development trend and impact degree of the damage.

Based on the results of damage diagnosis, the system will intelligently formulate maintenance plans and conduct cost assessment to ensure the economy and efficiency of maintenance work. Subsequently, the actual structural maintenance and repair work are carried out according to the plan to restore the structure to a healthy state. After the maintenance, the structure will enter the operation state again. At this time, the system will continue to conduct data collection and condition monitoring to ensure the long-term safety and stability of the structure [7].

This dynamic control process fully reflects the application value of machine learning in the health monitoring and evaluation of civil engineering structures, which not only improves the accuracy and efficiency of monitoring, but also provides scientific decision support for the maintenance and management of structure.

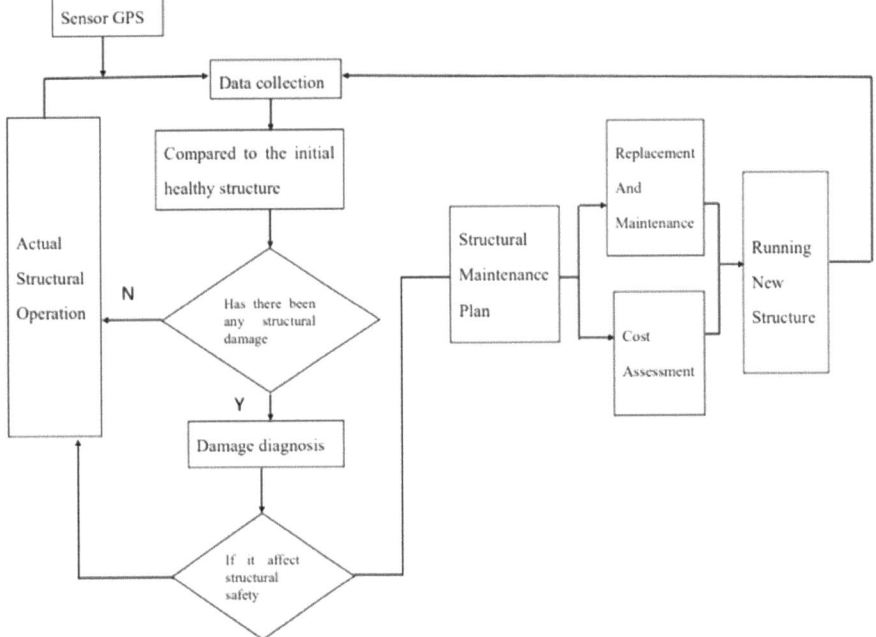

Fig. 1. Dynamic control process of structural health monitoring system

2.2 Structural Health Detection System

Figure 2 shows the full cycle maintenance and health monitoring system of building structure from design and construction to operation and use. In the early stage of the use of building structures, due to the short service life and high structural reliability, large-scale structural inspection and maintenance are not frequent. However, with the increase of service life, minor damages gradually accumulate. If these damages are not monitored and treated in time, they may pose a threat to the overall safety of the structure.

Therefore, it is necessary to introduce the structural health monitoring system. By installing the sensing system, we can monitor the state of the structure in real time, and the preventive maintenance management based on performance design can be implemented [8]. Such a system can not only reduce the cost increase caused by emergencies or improper maintenance, but also ensure the long-term safety and stability of the building structure in the use process.

Compared with the traditional maintenance management based on structural state, the performance-based structural health monitoring system pays more attention to the overall performance and long-term benefits of the structure. Through real-time data monitoring and analysis, we can more accurately predict the life of the structure and possible problems, so as to take corresponding maintenance measures in advance to ensure the safety and reliability of the building structure.

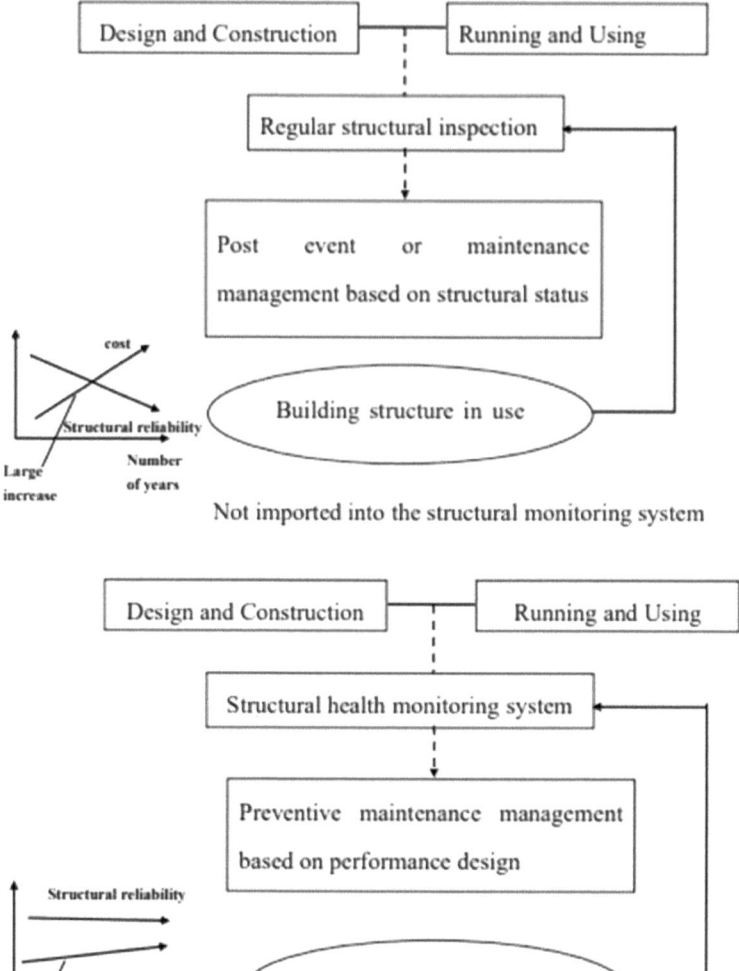

Fig. 2. Full cycle maintenance and health monitoring system

To sum up, the structural health monitoring system plays an increasingly important role in the maintenance and management of building structures [9]. Through the implementation of performance-based preventive maintenance management, we can realize the effective monitoring and maintenance of the building structure in the whole cycle, and ensure its safety and stability in the long-term use process.

3 Technology and Application

According to the real-time monitoring requirements of civil engineering structures, Real time monitoring on the machine, We will obtain the real-time data of the floor structure under different loads and environmental conditions. These data will cover the displacement, stress, strain, vibration and other aspects of the structure to comprehensively evaluate the performance and safety of the structure. First, we will monitor the displacement of the floor structure in real time through high-precision displacement sensors. The main beam calculation diagram shown in Fig. 3, These sensors will be installed in key positions, such as beams, plates and other stressed parts, to record the deformation of the structure under load in real time [10]. Through the analysis of displacement data, we can understand the overall stability and deformation trend of the structure, and timely find the potential safety hazards. Secondly, we will use the stress-strain sensor to monitor the stress and strain of the structure in real time. These sensors will be directly installed on

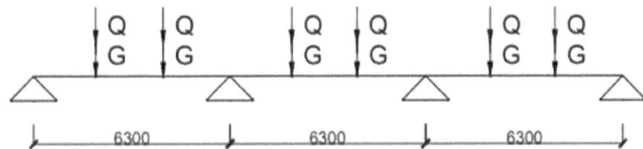

Fig. 3. Calculation diagram of main beam

Table 1. Bending moment design value of each section of main beam

Serial number	Load diagram	Mid span of edge span $\frac{K}{M1}$	Intermediate support $\frac{K}{MB(MC)}$	Mid span of Intermediate span $\frac{K}{M2}$
①		0.244 / 0.244 × 76.92 × 6.30 = 118.24	−0.267 / −0.267 × 76.92 × 6.30 = −129.39	0.067 / 0.067 × 76.92 × 6.30 = 32.47
②		0.289 / 0.289 × 151.2 × 6.30 = 275.29	−0.133 / −0.133 × 151.2 × 6.30 = −126.69	−0.133 / −0.133 × 151.2 × 6.30 = −126.69
③		$\approx \frac{1}{3}MB = -42.23$	−0.133 / −0.133 × 151.2 × 6.30 = −126.69	0.200 / 0.200 × 151.2 × 6.30 = 190.51
④		0.229 / 0.229 × 151.2 × 6.30 = 218.14	−0.311 (−0.089) / −0.311 × 151.2 × 6.30 = −296.25 (−0.089 × 151.2 × 6.30 = −84.78)	0.170 / 0.170 × 151.2 × 6.30 = 161.94
The most unfavorable internal force group	①+②	393.53	-256.08	-94.22
	①+③	76.01	-256.08	222.98
	①+④	336.38	-425.64 (-214.17)	194.41

materials such as steel and concrete to obtain the stress and strain distribution inside the structure. As shown in Tables 1 and 2, we will collect data on the bending moment and shear force of the main beam and other load-bearing parts that affect the structural stress and strain. Through the analysis of real-time data, we can understand the stress state of the structure under load, make a bending and shear envelope diagram as shown in Fig. 4, and identify the most unfavorable combination of internal forces, judge whether there are problems such as overload or fatigue, and take corresponding measures to repair or strengthen in time. In addition, vibration monitoring is also an important aspect of real-time monitoring of civil engineering structures. We will monitor the vibration of the structure in real time through acceleration sensors and other equipment to obtain the vibration frequency, amplitude and other parameters of the structure. These data can help us understand the response characteristics of the structure under dynamic load, evaluate the seismic performance of the structure, and provide a scientific basis for the optimal design and reinforcement of the structure. In the process of real-time monitoring, we will also consider the impact of environmental factors on structural performance. For example, environmental factors such as temperature and humidity may lead to changes

Table 2. Design value of shear force of each section of main beam

Serial number	Load diagram	Side span mid pan $\frac{K}{V_A^r}$	Intermediate support $\frac{K}{V_B^l(V_c^l)}$	Intermediate support $\frac{K}{V_B^r(V_c^r)}$
①	|G |G |G |G |G |G △ Ⱶ Ⱶ Ⱶ	0.733 / 56.38	−1.267 (−1.000) / −97.46 (−76.92)	1.000 (1.267) / 76.92 (97.46)
②	|Q |Q |Q |Q △ Ⱶ Ⱶ Ⱶ	0.866 / 130.94	−1.134 / 171.46	0
④	|Q |Q |Q |Q △ Ⱶ Ⱶ Ⱶ	0.689 / 104.18	−1.311 (−0.778) / −198.22 (−117.63)	1.222 (0.089) / 184.77 (13.46)
The most unfavorable combination of internal forces	①+②	187.32	-268.92	76.92
	①+④	160.56	-295.68 (-194.55)	261.69 (110.92)

in the performance of materials, and then affect the overall performance of the structure. Therefore, we will carry out real-time monitoring of environmental factors through temperature and humidity sensors and other equipment, and take environmental factors into account in data analysis, so as to more accurately evaluate the performance of the structure [11]. To sum up, by monitoring the displacement, stress, strain, vibration and other data of civil engineering structures in real time, and considering the environmental factors, we can comprehensively evaluate the performance and safety of the structure, and provide a scientific basis for the maintenance and management of the structure. This will help to find and solve potential safety hazards in time, and improve the reliability and durability of civil engineering structures. The following are obtained under the machine monitoring calculation Data and calculations, thus To judge structural health.

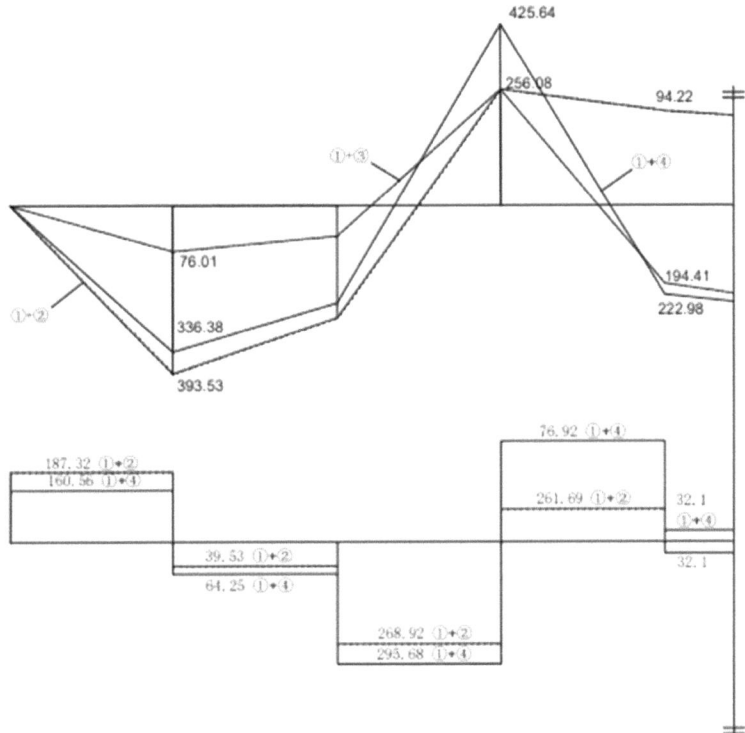

Fig. 4. Bending moment envelope diagram and shear force envelope diagram

4 Conclusions

Industrialized countries have invested a lot of money in the development of civil infrastructure. In order to maintain these investment values, proper maintenance must be paid attention to. SHM has emerged as a tool to support this work. Although many damage detection methods based on structural vibration response and system dynamic parameters have been developed, there are still many difficulties in the practical application of these methods due to the complexity of structural damage and the uncertainty of various influencing factors. SHM based on interdisciplinary will be a more advanced technology. In addition to profound structural knowledge, it also needs to understand the knowledge of other related disciplines [12]. Only the combination of structural vibration theory and signal processing, pattern recognition, artificial intelligence, control theory and material science can improve the accuracy of structural damage detection.

The practical significance of interdisciplinary methods in the field of structural health monitoring is very important. Firstly, by integrating the knowledge and technology of multiple disciplines, we can use the data fusion analysis of different types of sensors and comprehensively consider various factors, so as to improve the accuracy of structural damage identification of the monitoring system. Secondly, the application of intelligent algorithm can help to establish intelligent model, analyze and diagnose the monitoring data automatically, improve the monitoring efficiency and reduce the workload of manual intervention. In addition, there are complex interactions between the structure and the surrounding environment. Through the study of coupling effect, we can better understand the behavior of the structure under different environmental conditions, and improve the depth and accuracy of structural damage detection. In general, the application of interdisciplinary methods will promote the continuous innovation and progress of structural health monitoring technology, bring more opportunities and challenges to the field of building structures, and provide solid support for the sustainable development and safe operation of structures.

From the above analysis and summary of the latest literature on SHM and structural damage detection, we can see that there are still many problems to be solved, which need our continuous research and practice.

1) Multimodal data fusion and analysis: future research can focus on how to better integrate multimodal data obtained by different types of sensors, and develop more efficient data fusion and analysis methods. The importance of this field is that different types of sensors provide complementary information, and the comprehensive use of this information can improve the accuracy and reliability of structural damage detection. Solving this challenge will promote the SHM field to develop towards a more comprehensive and accurate monitoring direction.

2) Intelligent diagnosis and prediction model: future research can focus on developing more intelligent diagnosis and prediction models, combining machine learning, deep learning and other technologies to achieve more accurate and timely prediction of structural health. The importance of this field is that intelligent algorithms can improve the efficiency and automation level of the monitoring system, and provide more timely protection for structural safety. Solving this challenge will promote the SHM field to be intelligent and automated.

3) Research on structural behavior under coupling effect: future research can focus on the complex coupling effect between the structure and the surrounding environment, and further study the influence of these influencing factors on structural behavior. The importance of this field is that the coupling effect is a factor that can not be ignored in structural health monitoring. Only by fully understanding the coupling effect can

we more accurately evaluate the health status of structures. Solving this challenge will promote the SHM field to have a deeper understanding of the impact of the environment around the structure.

References

1. Wang, M., Li, J., Zhang, H.: Discussion on machine learning based health monitoring methods for civil engineering structures. Structural Engineer **34**(2), 45–56 (2018)
2. Zhang, H., et al.: Comparative analysis of traditional and machine learning based monitoring methods for civil engineering structures. Civil Engineering Technology **26**(4), 78–89 (2020)
3. Li, J., et al.: Application Research of machine learning in health monitoring of civil engineering structures. Structural Monitoring and Diagnosis **15**(3), 112–125 (2019)
4. Zhou, H., et al.: Review on damage detection of civil engineering structures based on structural parameters. Journal/Conference Name (2015)
5. Rytter, A.: Damage identification in structures: four levels of influence. Disruption/Conference Proceedings Name (2020)
6. Farrar, C.R., Worden, K.: An introduction to structural health monitoring. Phil. Trans. R. Soc. A **365**(1851), 303–315 (2015)
7. Giraldo, J.M.: Types of damage detection and identification methods in civil engineering structures. Disruption/Conference Proceedings Name (2017)
8. Gao, H., Li, H.N., Song, G.: A distributed computing strategy for structural health monitoring using deny sensor networks. Smart Structures and Systems **5**(5), 575–590 (2009)
9. Staszewski, W.J.: Structural Health Monitoring with Piezoelectric Wafer Active Sensors. Academic Press (2004)
10. Wang, S., Spencer, B.F., Jr.: Spatial wavelet based damage detection for structural health monitoring. J. Eng. Mech. **126**(7), 738–744 (2000)
11. Li, H.N.: Application of one-dimensional wavelet transform in civil engineering and potential in structural damage detection. Journal of Structural Engineering **128**(5), 622–629 (2002)
12. Sun, F., Wu, Z.S.: Damage identification of structural health monitoring based on wavelet packet transform and neural network. J. Vibrat. Eng. **19**(1), 45–51 (2006)

Author Index